The Chemistry of Bio-based Polymers: Second Edition

Scrivener Publishing
100 Cummings Center, Suite 541J
Beverly, MA 01915-6106

Publishers at Scrivener

Martin Scrivener (martin@scrivenerpublishing.com)
Phillip Carmical (pcarmical@scrivenerpublishing.com)

The Chemistry of Bio-based Polymers: Second Edition

Johannes Karl Fink
Montanuniversität Leoben, Austria

Scrivener

WILEY

Co-published by John Wiley & Sons, Inc., Hoboken, New Jersey, and Scrivener Publishing LLC, Salem, Massachusetts.
Published simultaneously in Canada

For general information on our other products and services or for technical support, please contact our Customer Care Department within the United States at (800) 762-2974, outside the United States at (317) 572-3993 or fax (317) 572-4002.

Wiley also publishes its books in a variety of electronic formats. Some content that appears in print may not be available in electronic formats. For more information about Wiley products, visit our web site at www.wiley.com.

For more information about Scrivener products please visit
www.scrivenerpublishing.com.

Cover design by Russell Richardson

Preface

Preface to Second Edition

The first edition of this text appeared in 2014. In the second edition, the most important literature which appeared after this date has been included. So the text is now updated up to 2019.

I want to thank the publisher, Martin Scrivener, for his interest in publishing the second edition. In addition, my thanks go to Jean Markovic, who made the final copyedit with utmost care.

J. K. F.
Leoben, September 9, 2019

Preface to First Edition

This book focuses on the chemistry of renewable polymers as well as low molecular compounds that can be synthesized from renewable polymers. As is well-documented, this issue has literally exploded in the literature because of growing awareness that conventional resources based on petroleum are limited.

After an introductory section to the general aspects of the field, the first part of the book deals with the chemistry of biodegradable polymeric types in five comprehensive chapters. The second part of the book deals with the applications (packaging and food, medical) of biodegradable polymers as well as the synthesis of low molecular compounds, including bio-based fuels.

The text focuses on the literature of the past decade. Beyond education, this book will serve the needs of industry engineers and specialists who have only a passing contact with the plastics and composites industries but need to know more.

How to Use This Book

Utmost care has been taken to present reliable data. Because of the vast variety of material presented here, however, the text cannot be complete in all aspects, and it is recommended that the reader study the original literature for more complete information.

The reader should be aware that mostly US patents have been cited where available, but not the corresponding equivalent patents in other countries. For this reason, the author cannot assume responsibility for the completeness, validity or consequences of the use of the material presented herein. Every attempt has been made to identify trademarks; however, there were some that the author was unable to locate.

Index

There are four indices: an index of trademarks, an index of acronyms, an index of chemicals, and a general index.

In the index of chemicals, compounds that occur extensively, e.g., "acetone," are not included at every occurrence, but rather when they appear in an important context.

Acknowledgements

I am indebted to our university librarians, Dr. Christian Hasenhüttl, Dr. Johann Delanoy, Franz Jurek, Margit Keshmiri, Dolores Knabl, Friedrich Scheer, Christian Slamenik, Renate Tschabuschnig, and Elisabeth Groß for support in literature acquisition. I also want to express my gratitude to all the scientists who have carefully published their results concerning the topics dealt with herein. This book could not have been otherwise compiled.

Last, but not least, I want to thank the publisher, Martin Scrivener, for his abiding interest and help in the preparation of the text.

Johannes Fink
Leoben, 10th October 2013

Contents

1

An Overview of Methods and Standards

Biocomposites are considered the next-generation materials as they can be made using natural/green ingredients to offer sustainability, eco-efficiency, and green chemistry (1–3). Nowadays, biocomposites are being utilized by numerous sectors, which include automobile, biomedical, energy, toys, sports, and others.

An effort has been made to provide a comprehensive assessment of the available green composites and their commonly used in order to make materials capable of meeting present and future demands. Various types of natural fibers have been investigated with polymer matrixes for the production of composite materials that are on par with the synthetic fiber composite. Also, the requirements for green composites in various applications from the viewpoint of variability of fibers available and their processing techniques have been detailed (4).

1.1 History of Biodegradable Plastics

In the late 1980s, biodegradable plastics came into use. However, these came to be misapplied in a number of situations. The misapplication of inappropriate or incompletely developed technology led to products which often did not meet performance claims and expectations. The so-called *first generation technologies* often lacked one or more of the following issues (5):

- Rate or extent of biodegradation, primarily due to limitations of starch incorporation,
- Necessary physical properties and related characteristics
- An economical means to effectively and efficiently manufacture starch-based blends,
- Intermediate product compatibility with conventional plastics product conversion processes, and
- Lower limits on film thickness caused by the use of non-gelatinized starch materials.

The synthesis, processing, and technology of renewable polymers has been reviewed (6–27). Furthermore, the state-of-the-art for food packaging applications has been reviewed (28–32). Using biomass for the production of new polymers can have both economic and environmental benefits (33).

Biomass-derived monomers can be classified into four major categories according to their natural resource origins (34):

1. Oxygen-rich monomers including carboxylic acids, e.g., lactic acid succinic acid, itaconic acid, and levulinic acid, but also ethers, such as furan,
2. Hydrocarbon-rich monomers including vegetable oils, fatty acids, terpenes, terpenoids and resin acids,
3. Hydrocarbon monomers, i.e., bio-olefins, and
4. Non-hydrocarbon monomers such as carbon dioxide.

Carbon dioxide is an interesting synthetic feedstock, which can be copolymerized with heterocycles such as epoxides, aziridines, and episulfides. In 1969, *Inoue* reported the zinc catalyzed sequential copolymerization of carbon dioxide and epoxides as a new route to poly(carbonate)s (9, 35). The reaction is shown in Figure 1.1.

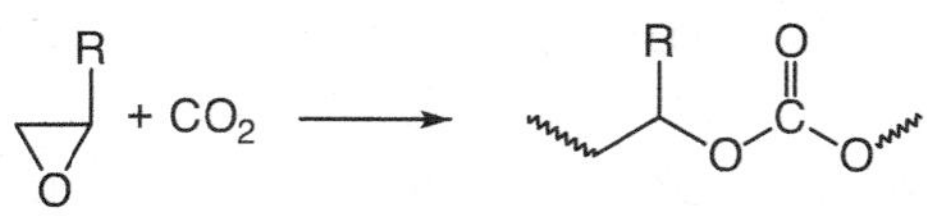

Figure 1.1 Reaction of carbon dioxide with epoxides (35).

Plants produce a wide range of biopolymers for purposes such as maintenance of structural integrity, carbon storage, and defense against pathogens as well as desiccation. Several of these natural polymers can be used by humans as food and materials, and increasingly as an energy carrier. Plant biopolymers can also be used as materials in certain bulk applications such as plastics and elastomers (36).

Lignin, suberin, vegetable oils, tannins, natural monomers like terpenes, and monomers derived from sugars are typically natural precursors for bio-based industrial polymers. Glycerol and ethanol also play a potential role as future precursors to monomers (37).

1.2 Green Chemistry

The principles and concepts of green chemistry are the subjects of several monographs (38–47). Recent progress in enzyme-driven green syntheses of industrially important molecules has been summarized (48). Studies in biotechnological production of pharmaceuticals, flavors, fragrance and cosmetics, fine chemicals, as well as polymeric materials (49) have been documented. Biocatalysis is a transformational technology uniquely suited to delivering green chemistry solutions for safer, efficient, and more cost-effective chemical synthesis.

The different catalytic processes for the conversion of terpenes, triglycerides and carbohydrates to valuable chemicals and polymers have been reviewed (50).

A basic task of green chemistry is to design chemical products and processes that use and produce less hazardous materials. The term *hazardous* covers several aspects, including toxicity, flammability, explosion potential and environmental persistence (51).

The synthesis of maleic anhydride illuminates a possibility of multiple pathways. Maleic anhydride can be synthesized both from benzene and from butene by oxidation. In the first route, a lot of carbon dioxide is formed as an undesirable byproduct. Thus, the first route is addressed as *atom uneconomic*. In Table 1.1, some uneconomic and economic reaction types in organic chemistry are opposed.

There were in total 12 basic principles in green chemistry (52–55). These principles are summarized in Table 1.2.

Table 1.1 Atom uneconomic and economic reaction types.

Economic	Uneconomic
Rearrangement reaction	Substitution reaction
Addition reaction	Elimination reaction
Diels-Alder reaction	Wittig reaction
Claisen reaction	Grignard reaction

Table 1.2 Basic principles of green chemistry (53).

Principle
Ensure that all material and energy inputs and outputs are as inherently nonhazardous as possible.
Better prevent waste than cleanup.
Minimize energy consumption and materials.
Maximize efficiency of mass, energy, space, and time.
Products, processes, and systems should be *output pulled* rather than *input pushed*.
Embedded entropy and complexity must be viewed as an investment when making design choices on recycle, reuse, or beneficial disposition.
The design goal should be targeted durability.
Unnecessary capacity or capability is not desirable.
Material diversity in multicomponent products should be minimized.
Development of products, processes, and systems must consider energy and materials flows.
The design should consider a commercial *afterlife*.
Material and energy inputs should be renewable.

Recently, the above-mentioned concept was extended (56). The special volume on green and sustainable chemistry and engineering has fourteen papers that were considered relevant to the present day issues and discussion, such as adequate use of raw materials and efficient energy, besides considering renewable sources for materials and energy; and changing economical canons towards circular economy. Businesses, governments and societies are facing a number of challenges along the pathway to sustainability for the well-being of future generations. Chemicals are ubiquitous in everyday activities. Their widespread presence provides benefits to societies' well-being, but can have some deleterious effects. To counteract such effects, green engineering and sustainable assessment in industrial processes have been gathering momentum in the last thirty years. Green chemistry, green engineering, eco-efficiency, and sustainability are becoming a necessity for assessing and managing products and processes in the chemical industry. Fourteen articles have been discussed, related to sustainable resource and energy use (five articles), circular economy (one article), cleaner production and sustainable process assessment (five articles), and innovation in chemical products (three articles) (56).

Catalytic processes from the viewpoint of green chemistry include catalytic reductions and oxidations methods, solid-acid and solid-base catalysis, as well as carbon-carbon bond formation reactions (57).

Novel concepts and techniques such as *bio-inspired* polymer design, *synthetically-inspired* material development are now considered to contribute to the development of natural monomers and polymers as a sustainable resource. These concepts and techniques that integrate materials synthesis, process and manufacturing options with eco-efficiency have been documented (58–62).

1.2.1 *Genetic Engineering*

The direct production of novel compounds in biomass crops in order to produce bioenergy as a coproduct seems to be a promising way to improve the economics of transgenic plants as biofactories (63).

Genetic engineering of plants may be used for the production of novel polymers and basic chemicals. These methods may help to alleviate the demands for limited resources and provide a platform to produce some desired compounds in bulk quantities.

Recent advances in enhancing the production of novel compounds in transgenic plants consist of a multigene transformation and the direction of the biosynthetic pathways to specific intracellular compartments.

Basically it appears feasible to produce interesting proteins, such as spider silk or collagen, novel carbohydrates, and biopolymers using transgenic plants. These compounds could replace petroleum-based plastics (63). However, there are pro and con arguments. For example, if transgenic plant factories should compete with conventional production processes, economic efficiency and sustainability are important. These factors depend on the future development of oil and energy prices.

On the other hand, societal factors such as the public acceptance of transgenic plants are also key factors (63). Chemicals that may be produced from biomass or in transgenic plants are listed in Table 1.3.

Cyanophycin is an attractive biopolymer with chemical and material properties that are suitable for industrial applications in the fields of food, medicine, cosmetics, nutrition, and agriculture (65). For the efficient production of cyanophycin, cyanophycin synthetases have been investigated and also fermentations and downstream processes have been elucidated.

The strategies for cyanophycin production in microbial strains, including *Escherichia coli, Pseudomonas putida, Ralstonia eutropha, Rhizopus oryzae,* and *Saccharomyces cerevisiae,* heterologously expressing different cyanophycin synthetase genes have been reviewed.

Also, the chemical and material properties of cyanophycin and its derivatives produced through biological or chemical modifications were addressed in the context of their industrial applications (65).

The biochemical pathways for producing 2(E)-heptenedioyl-CoA methyl ester have been described (66). Also, methods for enzymatically converting hept-2-enedioyl-CoA methyl ester to pimeloyl-CoA using a polypeptide have been shown.

Table 1.3 Chemicals from biomass or occurring in transgenic plants (64).

Compound	Remark
Succinic acid	Natural substances in plants
Fumaric acid	Natural substances in plants
Malic acid	Natural substances in plants
2,5-Furandicarboxylic acid	Oxidative dehydration of C6 sugars
3-Hydroxypropionic acid	Fermentation from sugar
Aspartic acid	Amination of fumaric acid
Glucaric acid	Oxidation of starch
Glutamic acid	Byproduct of sucrose production
Itaconic acid	Conversion of aconitic acid
Levulinic acid	Acid-catalyzed dehydration of cellulose
3-Hydroxybutyrolactone	Oxidation of starch
Glycerol	In plant oil
Sorbitol	Hydrogenation of sugars
Xylitol	Hydrogenation of sugars
Lysine	Fermentation
Proline	Fermentation
Arginine	A component of cyanophycin
Isomaltulose	From sucrose
Inulin	From chicory

1.3 Commercial Situation

The chemistry, important applications, and the market potential of intrinsically biodegradable polymers have been reviewed (67). One method for evaluating the potential demand for biodegradable polymers is to review the applications and necessary pricing to penetrate various end uses. Each application end use has a price hurdle associated with it.

However, the true market potential for biodegradable plastics will depend on:

- The selling price of the material,
- Environmental pressure,
- Legislation,
- Establishment of standards for degradability,
- The development of composting infrastructure, and
- The ability to overcome the problem of potentially contaminating the pool of recyclable materials.

These factors are difficult to predict since there are external forces that may not be universally applied in the same manner. The interest in biodegradable plastics has continuously grown as the conventional resources based on petroleum are beginning to decrease. The last two decades of the twentieth century saw a paradigm shift from biostable to biodegradable materials.

For example, in the next couple of years, many of the permanent prosthetic devices used for temporary therapeutic applications will be replaced by biodegradable devices that could help the body to repair and regenerate the damaged tissues (68).

Finding applications for renewable polymers that lead to mass production and price reduction poses a major contemporary challenge. This can be attained by improving the end performance of the biodegradable polymers (69).

The complexities of renewable supply chains have been elucidated (70). In particular, polymers manufactured from renewable feedstocks will augment various industrial markets such as plant material used as a renewable ingredient in paint manufacture, partially substituting for crude oil derivative ingredients. Polymer industrial supply chains have been identified and the market opportunity for renewable polymers has been estimated.

The developments in the field of renewable polymers illustrate how business models can link producers and customers through the development of new technologies and products (71). Initially, the companies assumed that reducing the costs and increasing the production will guarantee success of biopolymers in the market. However, some unconventional hurdles emerged. Companies have build markets for biopolymers and to assure customers that biopolymers are in fact produced sustainably.

Several companies have identified new market opportunities for biopolymers, designed distinctive types of business models to seize these opportunities, and developed ways to create an increased value by communicating performance advantages and the reduction of the environmental impact to downstream entities.

However, because they did not include societal factors in their efforts to define the term *sustainable*, a significant risk emerges that their sustainable value propositions may not endure without further refinements (71).

1.4 Environmental Situation

Polymer waste management options are shown in Figure 1.2. The utilization of waste polymers by mechanical recycling and incineration has ecological limitations.

The impact of biodegradable polymers on the environment and on the society has been detailed (72).

Landfills contain a tremendous amount of plastic waste. As the plastics degrade, pollutants leach into the soil and gases escape into the air. In response to this issue, the concept of recycling has been introduced into the consumption cycle. Recycling generally involves processing of the used materials into new products. However, the processing of waste can be economically ineffective, as it entails various mechanisms, such as:

- Collecting the waste,
- Sorting the waste according to provided specifications, and
- Processing the waste into materials that can be used in new products in the final stage.

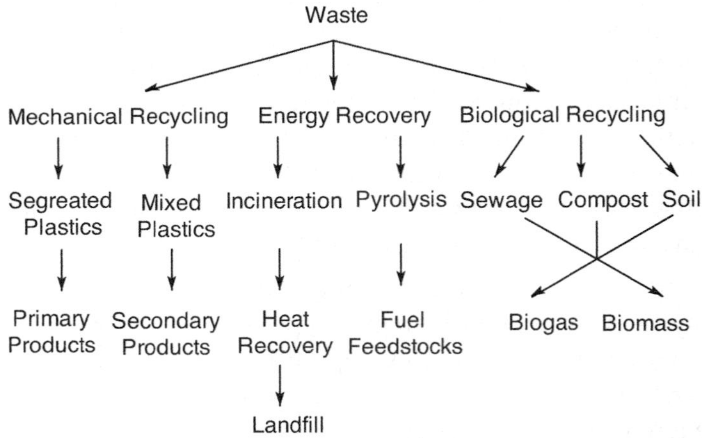

Figure 1.2 Polymer waste management options (73, 74).

Aside from the ineffective processes, recycling is not widely available in all communities and if available, often is not mandatory. Thus, many individuals either do not have a convenient venue for recycling or simply choose not to recycle (75).

Recycling has its disadvantages as well. The sorting and shipping of the plastic waste to the appropriate recycling facility is costly, both monetarily and environmentally. Different types of plastics must be recycled separately because they do not cooperate to form a stable reusable plastic. In addition, many plastics have a limited recyclable life. For example, recycling plastic water bottles can result in a lower grade plastic that cannot be converted into a new plastic water bottle.

Other methods of preventing this pollution include the partial use of biodegradable materials in plastic products. Certain auxiliary elements, made of biodegradable material, are then incorporated within the container. However, the remainder of the container is substantially plastic based on petroleum. If these mixed products are included with other plastics for recycling, they can contaminate the product and render it unusable (75).

Materials such as paper, paperboard, plastics, and even metals are presently used in enormous quantities in the manufacture of articles such as containers, separators, dividers, lids, tops, cans, and other packaging materials.

Most of the currently used food packaging materials are non-degradable and are creating serious environmental problems. New technologies are being explored and developed to study the complex interaction between the food packaging materials and food (76). For example, a nanocomposite of cellulose constitutes environmentally friendly packaging, which is easily recycled by combustion and requires low power consumption in production.

There are several such biodegradable materials which are available at a low price, have good mechanical properties and allow disposal in the soil. This is advantageous because biological degradation produces only carbon dioxide, water, and inorganic compounds. Also, it has been discovered that biodegradable plastics made of such materials can be disposed of together with organic waste. The widespread use of biopolymers in the place of standard plastics would help to reduce the weight of waste. Therefore, biodegradable materials take part in the natural cycle *from nature to nature* and play an important role for environmental sustainability (76).

Modern processing and packaging technology allows a wide range of liquid and solid goods to be stored, packaged, and shipped in packaging materials while being protected from harmful elements such as gases, moisture, light, microorganisms, vermin, physical shock, crushing forces, vibration, leaking, or spilling. Many of these materials are characterized as being disposable, but actually have little, if any, functional biodegradability. For many of these products, the time for degradation in the environment can span decades or even centuries (77).

Each year, over 100 billion aluminum cans, billions of glass bottles, and thousands of tons of paper and plastic are used in storing and dispensing soft drinks, juices, processed foods, grains, beer and other products. In the United States approximately 5.5 million tons of paper are consumed each year in packaging materials, which represents about 15% of the total annual domestic paper production.

Packaging materials are all, to varying extents, damaging to the environment. For example, the manufacture of poly(styrene) (PS) products involves the use of a variety of hazardous chemicals and starting materials, such as benzene, a known mutagen and a probable carcinogen. Chlorofluorocarbons have also been used in the manufacture of blown or expanded PS products. Chlorofluorocarbons have been linked to the destruction of the ozone layer.

Due to widespread environmental concerns, there has been significant pressure on companies to discontinue the use of PS products in favor of more environmentally safe materials. Some groups have favored the use of products such as paper or other products made from wood pulp. However, there remain drawbacks to the sole use of paper due to the tremendous amount of energy that is required to produce it. A strong need to find new, easily degradable materials that meet necessary performance standards remains (77).

The concept of sustainable bio-based products is as follows (78): A bio-based product derived from renewable resources should have a recycling capability and a triggered biodegradability. This means that it is stable in the course of service time. However, it should biodegrade after disposal under the specific conditions of composting. This composting procedure should also be commercially viable.

The general situation of compostable polymer materials has been described in detail in a monograph (79).

From a waste management perspective, high-barrier, multilayer, biodegradable food packaging could be a useful replacement for current multilayered packaging that is non-recyclable and non-degradable (80).

It has been envisioned that a biodegradable thermoplastic starch and poly(β-hydroxyalkanoate) (PHA) layered material could be a promising target. In a study an attempt was made to quantify the greenhouse gas trade-offs associated with using the proposed biodegradable packaging and identify the important design considerations (80).

The study also considered the impacts of landfill methane capture efficiency, which is an important aspect as biodegradable packaging may release methane when disposed of in a landfill whereas non-biodegradable packaging is inert (80). However, a key result is that when food waste is included in the system boundaries, it contributes over 50% of the greenhouse gas emissions associated with the system, regardless of whether the package is biodegradable or not.

This shows that even for biodegradable packaging, reducing food waste is a key design consideration. In fact, the negative environmental impacts associated with disposal of a PHA-biodegradable thermoplastic starch packaging in landfill with low gas capture rates can actually be offset if the package reduces food wastage (beef) by approximately 6%. The overarching result is that a PHA-biodegradable thermoplastic starch food packaging only delivers positive greenhouse gas outcomes if it reduces food wastage or increases the viability of biological food waste processing (80).

1.4.1 *Problems with Bio-based Composites*

Bio-based composites often exhibit unsatisfactory properties, such as, or resulting from (81):

- Inadequate processing conditions, resulting in filler agglomeration and poor filler dispersion within the matrix,
- Variations in natural fiber properties, often due to geographical and seasonal variability,
- Anisotropy of the natural fibers themselves,
- High linear coefficient of thermal expansion for natural fibers, and
- Incompatibility between typically hydrophilic natural fibers and hydrophobic polymer matrices, resulting in poor interfacial adhesion between the phases.

The chemical modification of a natural fiber is often performed in order to enhance the properties of the interface between fiber and matrix. A more recent method of modification involves the deposition of a coating of nanosized cellulose onto the natural fibers or dispersing a nanosized cellulose in natural fiber reinforced composites. This method has been shown to improve the fiber-matrix interface and the overall mechanical performances. Such composites have been addressed as hierarchical, multiscale, nanoengineered, or nanostructured composites. The state-of-the-art in this field has been reviewed (81).

Natural polymer blends and nanocomposites as well as natural fiber reinforced composites and other aspects have been reviewed in a monograph (82).

1.4.2 Biodegradation

Biodegradable polymers are commonly those that undergo a microbially induced degradation, i.e., a chain scission. This will result in mineralization, photodegradation, oxidation, and hydrolysis. In this way, the polymer is altered in the course of the degradation process (83).

Biodegradation reactions are typically enzyme catalyzed and occur in aqueous media. Natural macromolecules containing hydrolyzable linkages, such as protein, cellulose and starch, are generally susceptible to biodegradation by the hydrolytic enzymes of microorganisms.

A few man-made polymers, however, are also biodegradable. The hydrophilic character of the polymers greatly affects their biodegradability. A general rule is that polar polymers are more readily biodegradable. Other important polymer characteristics that affect the biodegradability include crystallinity and chain flexibility.

Besides being able to biodegrade, it is often important for a polymer to exhibit certain physical properties such as stiffness, flexibility, water resistance, strength, elongation, temperature stability, or gas permeability (84).

The intended application often dictates the necessary properties. For example, in the case of sheets and films for usage as packaging materials, the desired criteria of performance include elongation, printability, imperviousness to liquids, temperature stability, etc.

Since the number of biodegradable polymers is limited, it is often difficult, or impossible, to select a single polymer or copolymer that meets all the desired performance criteria.

Polymers that have a high glass transition temperature (T_g) are difficult to be blown into films or are too brittle for use as a packaging material. In contrast, polymers with a low glass transition temperature show low softening and melting temperatures. This makes the fabrication of sheets and films difficult without self- adhesion. In addition, such sheets may lack adequate water permeation, which is necessary for packaging applications (84).

1.4.2.1 *Details of the Mechanism of Degradation*

In general, biodegradable plastics are believed to undergo enzymatic biodegradation through the following processes (85):

- A polymer degrading enzyme is first adsorbed onto the surface of the polymer material. This enzyme is a substance secreted extracellularly by a certain kind of microorganism.
- The enzyme then breaks down chemical bonds in polymer chains, such as ester, glycosidic, and peptide bonds, by a hydrolysis reaction.
- As a result, the polymer material is reduced in molecular weight and even decomposes to a low molecular weight compound unit by the degrading enzyme.
- Finally, decomposed products are further metabolized and utilized by various microorganisms and converted into carbon dioxide, water, and bacterial cell components.

1.4.2.2 *Controlling the Rate of Biodegradation*

The rate of biodegradation can be controlled. The essential ingredients for such a composition are a biodegradable polymer, a carbodiimide compound, an ultraviolet ray absorbent, and an antioxidant (85).

Examples of carbodiimide compounds are summarized in Table 1.4.

N,N′-Dicyclohexylcarbodiimide and *N,N′*-diisopropylcarbodiimide are particularly preferable because of their industrial availability. Polycarbodiimides can be synthesized from isocyanates by conventional methods (86). As carbodiimidization catalyst, 3-methyl-1-phenyl-2-phospholene-1-oxide is used.

Antioxidants are collected in Table 1.5 and examples of ultraviolet absorbers are summarized in Table 1.6. Some ultraviolet absorbers are also shown in Figure 1.3.

These compounds can be used as additives to tailor the properties of the polymers if desired.

Table 1.4 Carbodiimides (85).

Monocarbodiimides
N,*N*´-dicyclohexylcarbodiimide
N,*N*´-Diisopropylcarbodiimide
Dimethylcarbodiimide
Diisobutylcarbodiimide
dioctylcarbodiimide
tert-Butylisopropylcarbodiimide
Diphenylcarbodiimide
Di-*tert*-butylcarbodiimide
Di-β-naphthylcarbodiimide
Isocycanate Monomers for Polycarbodiimides
4,4´-Dicyclohexylmethane diisocyanate
Isophorone diisocyanate
Tetramethylxylylene diisocyanate

Table 1.5 Antioxidants (85).

Hindered Phenol-based Antioxidants
4,4´-Methylene-bis-(2,6-di-*tert*-butylphenol)
Octadecyl-3-(3,5-di-*tert*-butyl-4-hydroxyphenyl)propionate
Phosphite-based Antioxidants
Tris-(2,4-di-*tert*-butylphenyl)phosphite
Bis-(2,4,di-*tert*-butylphenyl)pentaerythritol-diphosphite
Bis-(2,6-di-*tert*-butyl-4-methylphenyl)pentaerythritol-diphosphite

Table 1.6 Ultraviolet absorbers (85).

Benzotriazole-based Compounds
2-(2´-Hydroxy-5´-methylphenyl)benzotriazole
2-(2´-Hydroxy-5´-*tert*-butylphenyl)benzotriazole
2-(2´-Hydroxy-3´-*tert*-butyl-5´-methylphenyl)-5-chlorobenzotriazole
2-(3,5-Di-*tert*-butyl-2-hydroxyphenyl)-5-chlorobenzotriazole
2-[2´-Hydroxy-3´,5´-bis(α,α-dimethylbenzyl)phenyl]benzotriazole
Triazine-based Compounds
2-(4,6-Diphenyl-1,3,5-triazine-2-yl)-5-[(hexyl)oxy]phenol
2-[4,6-Bis(2,4-dimethylphenyl)-1,3,5-triazine-2-yl]-5-[(octyl)oxy]phenol
2,4,6-Triamino-1,3,5-triazine
2,4-Diamino-6-phenyl-1,3,5-triazine
2,4-Diamino-6-methyl-1,3,5-triazine
2,4-Diamino-6-(2-(dodecasylamino)ethyl)-1,3,5-triazine
2,4-Diamino-6-(o-methoxyphenyl)-1,3,5-triazine
4,6-Diamino-1,2-dihydro-2,2-dimethyl-1-(2,6-xylyl)-1,3,5-triazine
2,4-Diamino-6-(2-methoxyethyl)-1,3,5-triazine
2-Amino-4-ethyl-1,3,5-triazine
2-Amino-4-phenyl-1,3,5-triazine
2-Amino-4-ethyl-6-methyl-1,3,5-triazine

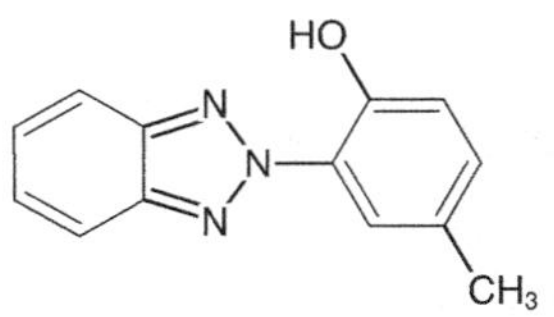

2-(2´-Hydroxy-5´-methyl phenyl)benzotriazole

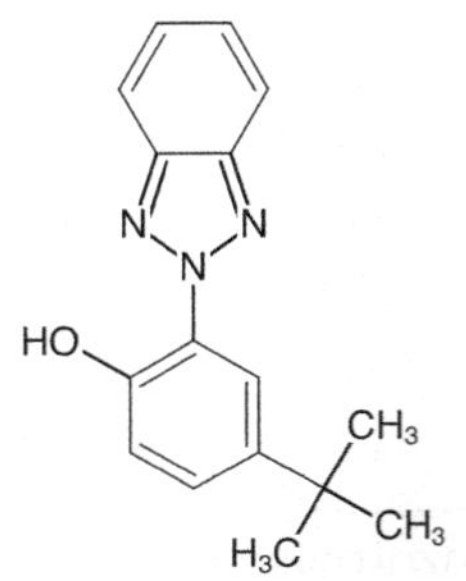

2-(2´-Hydroxy-5´-*tert*-butyl phenyl)benzotriazole

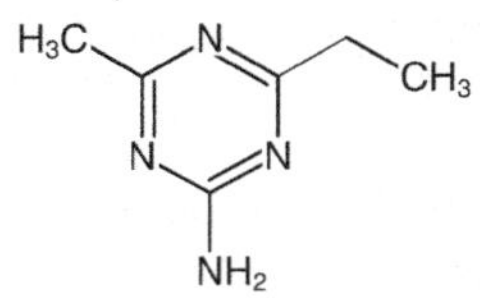

2-Amino-4-ethyl-6-methyl-1,3,5-triazine

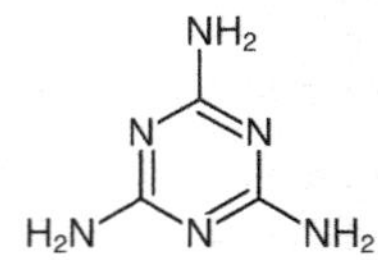

2,4,6-Triamino-1,3,5-triazine

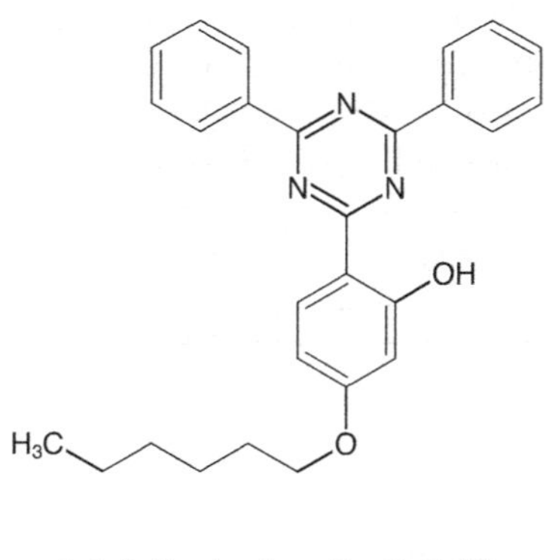

2-(4,6-Diphenyl-1,3,5-triazine-2-yl)-5-[(hexyl)oxy]phenol

Figure 1.3 Ultraviolet absorbers.

1.5 Properties of Biodegradable Polymers

Biopolymers are polymers that degrade through the action of living organisms. Such polymers include a range of synthetic polymers such as poly(ester)s, poly(ester amide)s, poly(carbonate)s, and others (84). In Table 1.7 the mechanical properties of E-glass and some natural fibers are compared.

Table 1.7 Mechanical properties (78).

Material	Density /[$g\,cm^{-3}$]	E-Modulus /[GPa]
E-glass	2.55	73
Hemp	1.48	70
Flax	1.4	60–80
Jute	1.46	10–30
Sisal	1.33	38
Coir	1.25	6
Cotton	1.51	12

As can be seen from Table 1.7, in the case of hemp and flax, the elastic modulus is comparable to E-glass and the density is even smaller than that of E-glass.

1.6 Special Methods of Synthesis

The methods of synthesis of biodegradable polymers can be subdivided into conventional methods, click chemistry, enzymatic polymerization and polycondensation, chemoenzymatic polymerization, vine-twining polymerization, and bacterial synthesis. These methods will be detailed subsequently.

1.6.1 Conventional Methods

Biodegradable polymers can be synthesized using renewable resources by conventional methods (33). The recent advances in the synthesis of polymers from renewable resources have been shown (87).

For example, yttrium complexes are suitable compounds for the synthesis of poly(lactide)s (33).

Also, a chiral Schiff base/aluminium alkoxide can be used as initiator for the stereoelective polymerization of *rac*-(*D*,*L*)-lactide (88). The compound is shown in Figure 1.4.

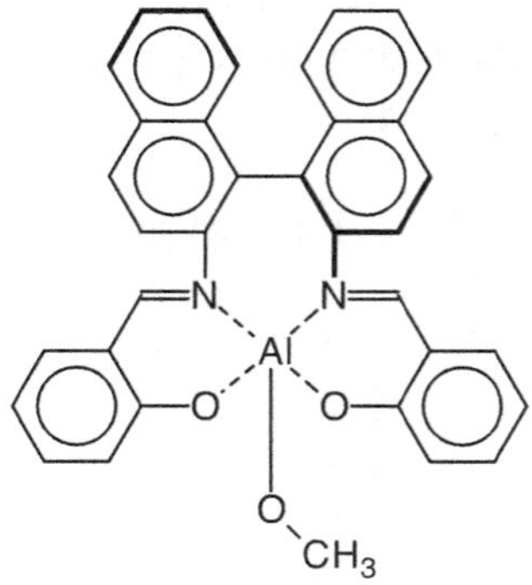

Figure 1.4 Chiral Schiff base (88).

A rather high stereoelectivity is observed, as a polymer with an 88% enantiomeric enrichment in the *D* units can be isolated at 19% conversion. At higher conversions a stereo-complex from *D*- and *L*-enriched stereocopolymers is formed. A narrow molecular weight distribution of M_w/M_n in the range of 1.05–1.30 is observed even at very high conversions. Thus, the polymerization reaction shows living type features (88).

1.6.2 *Click Chemistry*

Click chemistry is a powerful strategy that relies mainly on the construction of carbon-heteroatom bonds. It has been described in the present meaning in 2001 (89).

Applications are found in a wide variety of methods in modern chemistry, including biocompatible synthesis (90). For example, adhesive polymers can be formed when polyvalent azides and alkynes are assembled into crosslinked polymer networks by copper-catalyzed 1,3-dipolar cycloaddition (91). The polycondensation is efficiently promoted by copper ions.

As another example, poly(glycolide)s have been described that include a polymeric alkynyl-substituted glycolide (92). The alkynyl groups provide reactive sites for further functionalization of the polymer, for example, by reaction with azide derivatives. The alkynyl and azide groups react via the click chemistry mechanism to form functional groups covalently bonded to the polymer via a triazole link. The polymers are biodegradable and can be used to deliver drugs or other therapeutic substances at controlled release rates.

The thiol-ene click reactions are considered a versatile and broadly applicable tool in polymer science (93). This reaction can be used as a click reaction for the synthesis of monomers as well as of dendrimers. Also, polysaccharides can be derivatized via a click reaction (94).

1.6.3 Enzymatic Polymerization

The field of enzymatic polymerization has been reviewed in detail (95–97). Enzymes have been industrially established in large-scale synthesis and degradation, such as the use of:

- Glucose isomerase in food industry,
- Cellulase for textile finishing,
- Lipase, protease, and cellulase in detergents, and
- Catalysts in chemical and pharmaceutical industries.

Enzymatic polymerization is an important issue for green polymer chemistry to save energy in production processes and to reduce the formation of undesired byproducts since the reaction is mostly selective. In general, an enzyme catalyzed reaction proceeds much faster than a conventional reaction, by lowering the activation energy.

Typical polymers that can be synthesized by an *in-vitro* enzymatic catalysis are summarized in Table 1.8. The basic concept of the

Table 1.8 Polymers by *in-vitro* enzymatic catalysis.

Enzyme types	Polymers
Oxidoreductases	Polyphenols, polyanilines, vinyl polymers
Transferases	Polysaccharides, cyclic oligosaccharides, polyesters
Hydrolases	Polysaccharides, polyesters, poly(carbonate)s, polyamides, polyphosphates, polythioesters

in-vitro enzymatic polymerization can be explained as follows: In the course of an *in-vitro* enzymatic polymerization, a monomer is treated by the catalyst enzyme as an unnatural or artificial substrate. But the substrate must be recognized and form a complex with the enzyme in order to allow a reaction.

The polymerization of catechol under high-pressure homogenization was catalyzed by laccase from *Myceliophthora thermophila* (98). This polymer was used for the green coloration of textile substrates.

The oxidation reactions were conducted using different forms of laccase, i.e., native laccase, PEGylated laccase and PEGylated laccase immobilized onto an epoxy resin. These three enzyme forms were deposited inside a polyester fabric bag during the experiments. The amount of polymer obtained was similar when using the three enzyme forms and its dispersion in a water/dimethyl sulfoxide mixture led to powder particles of about 30–60 *nm*.

It could be shown that the oxidation of catechol conducted under high-pressure homogenization can be an efficient methodology for the *in-situ* coloration of textiles. The polymers produced by this methodology strongly stained the textile container, revealing this experimental setup as a promising greener coloration/coating methodology involving milder conditions than that normally used in textile processes (98).

2,5-Furandicarboxylic acid-based semi-aromatic poly(amide)s were synthesized using enzymatic polymerization (99). These polymers are bio-based alternatives to poly(phthalamide)s, which are petrol-based semi-aromatic polyamides. From a commercial perspective, they have interesting properties as high-performance materials and engineering thermoplastics. It is even more appealing to explore novel 2,5-furandicarboxylic acid-based polyamides with added functionality for the development of sustainable functional materials.

Here, a set of 2,5-furandicarboxylic acid-based heteroatom polyamides have been successfully produced via a Novozyme 435 (N435)-catalyzed polymerization of bio-based dimethyl 2,5-furandicarboxylate with potentially heteroatom diamines, i.e., 4,9-dioxa-1,12-dodecanediamine, diethylenetriamine, and 3,3-ethylenediiminopropylamine (99). The enzymatic polymerization reactions were performed both in solution and in bulk. The latter approach is more sustainable and results in higher molecular weight products. Among the tested heteroatom diamines, N435 showed the highest catalytic activity toward 4,9-dioxa-1,12-dodecanediamine. Furthermore, it was found that all obtained 2,5-furandicarboxylic acid-based heteroatom polyamides are amorphous materials with a relatively high thermal stability. These heteroatom polyamides show glass transition temperatures ranging from 41°C to 107°C (99).

Enzymatic ring-opening polymerization is an appealing method for the preparation of poly(lactide)s and poly(lactone)s (100). These reactions are typically carried out at relatively high temperatures of 60–130°C.

However, there is a deficiency of enzyme-compatible solvents for such thermally demanding biocatalytic processes. A series of short-chained glycol-grafted ionic liquids were prepared based on a phosphonium, imidazolium, pyridinium, ammonium, or piperidinium cationic headgroup.

Most of these glycol-grafted ionic liquids exhibit relatively low dynamic viscosities (33–123 *mPa s* at 30°C), coupled with excellent short-term thermal stabilities with decomposition temperatures in the 318°C–403°C range (100). Significantly, the long-term thermal stability under conditions matching those for enzymatic ring-opening polymerization synthesis (130°C for 7 *d*) is excellent for several of these task-specific ionic liquids.

Using a Novozym 435-catalyzed ring-opening polymerization, these ionic liquids were demonstrated to be viable solvents for the enzymatic production of reasonable yields (30–48%) of high molecular mass ($M_w \sim 20$ *kDa*) poly(*L*-lactide) and poly(ε-caprolactone) compared to solventless conditions (12–14 *kDa*) (100).

1.6.3.1 *Polycondensation*

The recent developments in lipase-catalyzed synthesis of polyesters have been reviewed (101). A series of diacids, such as succinic acid, glutaric acid, adipic acid, and sebacic acid, and diols, such as 1,4-butanediol, 1,6-hexanediol, and 1,8-octanediol have been polymerized in solution and in bulk using lipase as a catalyst (102, 103).

Polymerization reactions with longer chain length monomers show a higher reactivity than reactions of shorter chain length monomers (103). The lipase-catalyzed reaction of an alcohol with a vinyl ester proceeds much faster than with an alkyl ester (104).

The dependence of the origin of the particular lipase on the conversion in the ring-opening polymerization of 15-pentadecanolactone is shown in Table 1.9.

Table 1.9 Origin of lipase and conversion of 15-pentadecanolactone (101).

Lipase Type	Conversion/[%]
None	0
Rhizopus japanicus (lipase RJ)	<5
Hog liver (HLE)	<5
Penicillium roqueforti (lipase PR)	12
Aspergillus niger (lipase A)	16
Candida rugosa (lipase CR)	21
PPL	27
Lipase CC	54
Pseudomonas cepacia (lipase PC)	90
Lipase PF	97

1.6.4 Chemoenzymatic Polymerization

The mechanism of catalysis of green biocatalysts and metal catalysts is quite different, but if both are mutually compatible, then it allows the application concurrently in the same reaction system. The combination of these different types of catalysts is also known as chemoenzymatic method. Utilizing the advantages of enzymes, the chemoenzymatic method has been developed for the synthesis of various block copolymers, which are otherwise difficult to prepare.

The combination of the lipase-catalyzed ring-opening polymerization of lactones and the atom transfer radical polymerization allows a versatile synthesis of block copolymers consisting of a polyester chain and a vinyl polymer chain (105). Also, branched polymers have been produced by the chemoenzymatic technique (106).

1.6.4.1 Cyclodextrins

Cyclodextrins can polymerize cyclic esters such as lactones and lactides (107). They can initiate the polymerization of cyclic esters in bulk without any solvents to give products in high yields.

Cyclodextrins are cyclic oligosaccharides. α-Cyclodextrin is a 6-membered ring, β-cyclodextrin is a 7-membered ring, and γ-cyclodextrin is an 8-membered ring. The structure of α-cyclodextrin is shown in Figure 1.5.

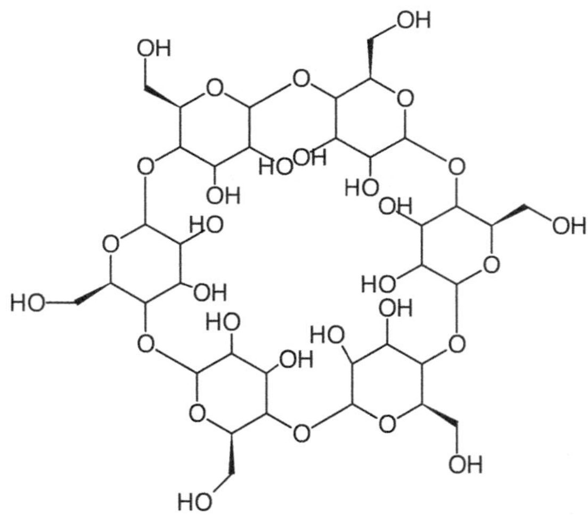

Figure 1.5 α-Cyclodextrin.

It has been found that cyclodextrins are active not only as initiators and catalysts but also support certain architectures of the final product similar to chaperone proteins.

Chaperones are proteins that assist the folding or unfolding process and thus the assembly or disassembly of other macromolecular structures. However, they do not appear in those macromolecular structures (108).

The cyclodextrin may encircle a linear polymer chain so that the chain assumes the proper conformation and avoids coagulation. So, a cyclodextrin mimics the strategy that a living system uses to form polymers. It is believed that such a system could provide an environmentally friendly route to produce biodegradable functional polymers (107).

Multi-responsive cyclodextrin vesicles have been prepared and shown to be self-assembled by supramolecular bola-amphiphiles, consisting of (*N,N′*-bis(ferrocenylmethylene)-diaminohexane a guest, and as host, i.e., γ-hydroxybutyric-β-cyclodextrin (109). The vesicles may serve as redox-responsive systems.

Bola-amphiphiles are amphiphilic molecules that have hydrophilic groups at both ends of a sufficiently long hydrophobic hydrocarbon chain. Basically, a bola-amphiphile looks like the schematic shown in Figure 1.6.

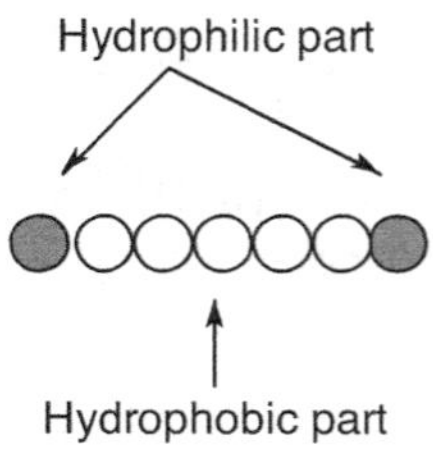

Figure 1.6 Schematic illustration of a bola-amphiphile.

1.6.5 Vine-Twining Polymerization

The Vine-twining polymerization is a method for the preparation of well-defined supramolecules, which are amylose polymer inclusion complexes (110). The method consists of the enzymatic polymerization of α-*D*-glucose-1-phosphate catalyzed by phosphorylase in the presence of various synthetic polymers such as poly(ether)s, poly(ester)s, poly(ester ether)s, and amphiphilic block copolymers. The 1H-NMR spectra of the polymers indicated structures composed from amylose and guest polymers.

The preparation of inclusion complexes composed of amylose and hydrophobic poly(carbonate)s have been achieved by vine-twining polymerization (111, 112). The structure of such complexes is shown in Figure 1.7.

This is a phosphorylase-catalyzed enzymatic polymerization of an α-*D*-glucose-1-phosphate from a maltoheptose in the presence of poly(carbonate)s. Poly(carbonate)s with a shorter methylene chain

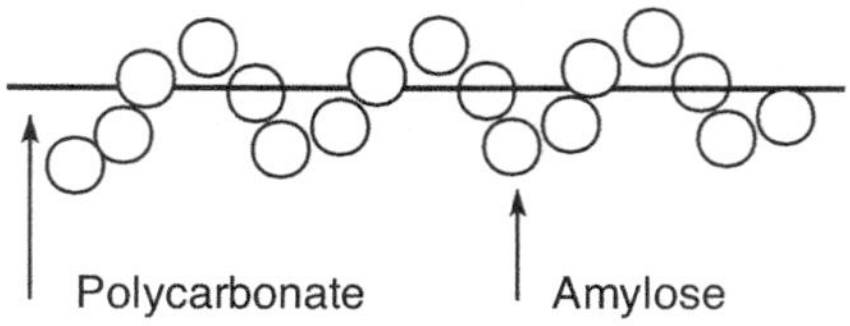

Figure 1.7 Inclusion complexes (112).

length, such as poly(tetramethylene carbonate), is more favorable as the guest polymer to form the inclusion complex with amylose.

It has been shown that amylose selectively includes poly(tetrahydrofuran) or poly(δ-valerolactone) from a mixture of two resemblant poly(ether)s, i.e., poly(oxetane) and poly(tetrahydrofuran), or also a mixture of two resemblant polyesters in vine-twining polymerization (113). The highest priority on the inclusion of amylose showed poly(tetrahydrofuran). Also, a selective inclusion according to the molecular weight of a poly(tetrahydrofuran) was demonstrated (114).

The enzymatic polymerization was investigated at 80°C using a primer-grafted poly(γ-glutamic acid) in the presence or absence of poly(*L*-lactic acid) as a guest polymer for the inclusion by amylose (115). The produced amylose-grafted poly(γ-glutamic acid)s formed microparticles by cooling the mixtures at room temperature after the enzymatic polymerization in either the presence or absence of poly(*L*-lactic acid). The particle sizes, which were evaluated by scanning electron microscopy (SEM), were dependent on the feed ratios of poly(*L*-lactic acid) (115).

It has been found that amylosic inclusion complexes, which were obtained by vine-twining polymerization using a designed guest polymer, i.e., an amphiphilic triblock copolymer poly(2-methyl-2-oxazoline-block-tetrahydrofuran-block-2-methyl-2-oxazoline), exhibited gel and film formation properties (116).

The characterization of the products suggested that enzymatically elongated amylose chains were complexed with the poly(tetrahydrofuran) block in the triblock copolymer (116).

Here, the outer poly(2-methyl-2-oxazoline) blocks constructed hydrophilic spaces among the inclusion complex segments. Furthermore, the presence of such outer blocks affected the lower regularity of crystalline alignment among the inclusion complex segments in the products. Probably, such higher-order structures induced the formation of supramolecular soft materials such as gels and films (116).

1.6.6 *Bacterial Synthesis*

Bacteria can synthesize a wide range of biopolymers. The key aspects of the production of bacterial biopolymers have been reviewed (117, 118). It is expected that a better understanding of polymer biosynthesis and material properties can lead to an increased use of bacterial biopolymers.

PHAs are biopolymers that can be synthesized by microorganisms such as the bacterium *Burkholderia xenovorans* LB400 (119). In particular, LB400 cells are capable of synthesizing poly(3-hydroxybutyrate) (PHB) from glucose.

An attempt has made to produce and characterize electrospun fibers obtained from bacterial PHB (119). Here, the bacterial strain LB400 was grown in M9 minimal medium using xylose and mannitol (10 $g\,l^{-1}$) as the sole carbon sources and NH_4Cl (1 $g\,l^{-1}$) as the sole nitrogen source.

The so-obtained biopolymer-based films were used to produce fibers by electrospinning. The diameter and the morphology of the microfibers were analyzed by SEM and also their thermogravimetric properties were investigated. Bead-free fibers using both PHBs were obtained with diameters of less than 3 μm. The surface morphology of the microfibers based on PHBs obtained from both carbon sources was different, even though their thermogravimetric properties were found to be similar. These results indicate that the carbon source may determine the fiber structure and properties (119).

1.7 Biodegradability Standards

There are several standards available in order to determine the biodegradability of plastic materials. Technical reports have been given that analyze a set of standards, documents and other reports, related to bio-based products (120, 121). The report is limited to matters concerning bio-based products, and hence excludes traditional products, energy applications and food.

In January 2018, the European Union released its strategy for a more sustainable plastics industry to be achieved by the year 2030 (122). Besides promoting plastic recycling, the strategy discusses the opportunities and the risks of biodegradable plastics.

These standards are summarized in Table 1.10. In addition, other certification schemes that are used in various other countries have been collected in the literature (123).

Table 1.10 Biodegradability Standards.

Number	Title
ISO 10210:2012	Plastics – Methods for the preparation of samples for biodegradation testing of plastic materials (124)
ISO 13975:2019	Plastics – Determination of the ultimate anaerobic biodegradation of plastic materials in controlled slurry digestion systems – Method by measurement of biogas production (125)
ISO 14851:2019	Determination of the ultimate aerobic biodegradability of plastic materials in an aqueous medium – Method by measuring the oxygen demand in a closed respirometer (126)
ISO 14852:2018	Determination of the ultimate aerobic biodegradability of plastic materials in an aqueous medium – Method by analysis of evolved carbon dioxide (127)
ISO 14853:2005	Plastics – Determination of the ultimate anaerobic biodegradation of plastic materials in an aqueous system – Method by measurement of biogas production
ISO 16929:2018-04	Plastics – Determination of the degree of disintegration of plastic materials under defined composting conditions in a pilot-scale test (128)

1.7.1 *Guidelines for the Development of Standards*

There are guidelines for the recovery and recycling of plastics waste (145). The standard ISO 15270:2008 provides a guideline for the

Table 1.10 (contd.) Biodegradability Standards.

Number	Title
ISO 15270:2008	Plastics – Guidelines for the recovery and recycling of plastics waste
ISO 14855-1-2012	Determination of the ultimate aerobic biodegradability of plastic materials under controlled composting conditions – Method by analysis of evolved carbon dioxide – Part 1: General method (129)
ISO 14855-2-2018	Determination of the ultimate aerobic biodegradability of plastic materials under controlled composting conditions – Method by analysis of evolved carbon dioxide – Part 2: Gravimetric measurement of carbon dioxide evolved in a laboratory-scale test (130)
ISO 17556:2019	Plastics – Determination of the ultimate aerobic biodegradability of plastic materials in soil by measuring the oxygen demand in a respirometer or the amount of carbon dioxide evolved (131)
ISO 17088:2012	Specifications for compostable plastics (132)
ISO 23517	Plastics – Biodegradable mulch films for use in agriculture and horticulture – Requirements and test methods (133)
ISO 15985:2014	Plastics – Determination of the ultimate anaerobic biodegradation and disintegration under high-solids anaerobic digestion conditions – Method by analysis of released biogas (134)
ISO 18830:2016	Plastics – Determination of aerobic biodegradation of non-floating plastic materials in a seawater/sandy sediment interface – Method by measuring the oxygen demand in closed respirometer (135)
ISO 14852:2018-09	Determination of the ultimate aerobic biodegradability of plastic materials in an aqueous medium - Method by analysis of evolved carbon dioxide (136)

Table 1.10 (contd.) Biodegradability standards.

Number	Title
ASTM D5338-15	Standard Test Method for Determining Aerobic Biodegradation of Plastic Materials Under Controlled Composting Conditions (137)
ASTM D5511-18	Standard Test Method for Determining Anaerobic Biodegradation of Plastic Materials Under High-Solids Anaerobic-Digestion Conditions (138)
ASTM D5526-18	Standard Test Method for Determining Anaerobic Biodegradation of Plastic Materials Under Accelerated Landfill Conditions (139)
ASTM D5929-18	Standard Test Method for Determining Biodegradability of Materials Exposed to Source-Separated Organic Municipal Solid Waste Mesophilic Composting Conditions by Respirometry (140)
ASTM D5988-18	Standard Test Method for Determining Aerobic Biodegradation of Plastic Materials in Soil (141)
ASTM D6400-04	Standard Specification for Compostable Plastics (142)
ASTM D6691-17	Standard Test Method for Determining Aerobic Biodegradation of Plastic Materials in the Marine Environment by a Defined Microbial Consortium or Natural Sea Water Inoculum (143)
ASTM D7475-11	Standard Test Method for Determining the Aerobic Degradation and Anaerobic Biodegradation of Plastic Materials under Accelerated Bioreactor Landfill Conditions (144)

Table 1.10 (contd.) Biodegradability Standards.

Number	Title
DIN EN 13432	Packaging – Requirements for packaging recoverable through composting and biodegradation – Test scheme and evaluation criteria for the final acceptance of packaging
DIN EN 14995	Plastics – Evaluation of compostability – Test scheme and specifications
OECD 306B	Biodegradability in seawater

development of standards and specifications covering plastics waste recovery, including recycling.

This standard establishes the different options for the recovery of plastics waste arising from pre-consumer and post-consumer sources. It also establishes the quality requirements that should be considered in all steps of the recovery process, and provides general recommendations for inclusion in material standards, test standards and product specifications. Consequently, the process stages, requirements, recommendations and terminology presented in the standard are intended to be of general applicability.

Proper methods for the preparation of test samples that can be used for the determination of the ultimate aerobic and anaerobic biodegradability of plastic materials in an aqueous medium, soil, controlled compost or anaerobic digesting sludge have been standardized. These should result in an improved reproducibility of the test results during the assessment of the biodegradability (124).

1.7.2 Specifications for Compostable Plastics

The ASTM D6400-04 specification covers plastics and products made from plastics that are designed to be composted in municipal and industrial aerobic composting facilities (142). Properties that are required to predict whether plastics products will compost satisfactorily, including biodegrading, are specified. In this way, products can be assessed that will compost satisfactorily in commercial and municipal composting facilities.

1.7.3 *Ultimate Anaerobic Biodegradability*

A method of the assessment of the ultimate anaerobic biodegradability of plastic materials in a controlled anaerobic slurry digestion system with a solids concentration of less than 15% has been described. Such a solid concentration is often present after the treatment of sewage sludge, livestock feces, or garbage. The rate of conversion of the organic carbon into carbon dioxide and methane, i.e., biogas, can be obtained (146).

On the other hand, the evaluation of the ultimate anaerobic biodegradability of plastics under high-solid content of more than 20% of total solids and static non-mixed conditions is available (147). This is based on the measurement of the evolved biogas and the degree of disintegration at the end of the test procedure. In particular, the method is designed to simulate typical anaerobic digestion conditions for the organic fraction of mixed municipal solid waste.

1.7.4 *Aerobic Biodegradability*

Biodegradation of a plastic within a composting unit is an important phenomenon because it may affect the decomposition of other materials enclosed by the plastic and the resulting quality and appearance of the composted material. Biodegradation of plastics will also allow the safe disposal of these plastics through large, professionally managed composting plants and well-run residential units, where thermophilic temperatures are achieved. A procedure, ASTM D5338-15, has been developed to permit the determination of the rate and degree of aerobic biodegradability of plastic products when placed in a controlled composting process (137).

1.7.4.1 *Oxygen Control*

The ultimate aerobic biodegradability of plastic materials in an aqueous medium can be controlled by measuring the oxygen demand in a closed respirometer (148).

ISO 18830:2016 specifies a test method to determine the degree and rate of aerobic biodegradation of plastic materials when settled on marine sandy sediment at the interface between seawater and the seafloor, by measuring the oxygen demand in a closed respirometer (135).

This test method is a simulation under laboratory conditions of the habitat found in different seawater/sediment-areas in the sea, e.g., in a benthic zone, where sunlight reaches the ocean floor (photic zone) that, in marine science, is called the sublittoral zone.

The determination of biodegradation of plastic materials buried in marine sediment is outside the scope of this International Standard. The conditions described in this International Standard may not always correspond to the optimum conditions for the maximum degree of biodegradation to occur.

The measurement of the aerobic biodegradation can also be obtained by monitoring the carbon dioxide evolution. This is not in the scope of this International Standard but of ISO 19679 (149). This test method is a simulation under laboratory conditions of the habitat found in different seawater/sediment-areas in the sea, e.g., in a benthic zone where sunlight reaches the ocean floor (photic zone) that, in marine science, is called the sublittoral zone.

Furthermore, the ultimate aerobic biodegradability of plastic materials in soil can be performed by measuring the oxygen demand in a closed respirometer or by the amount of carbon dioxide evolved. The method is designed to yield an optimum degree of biodegradation by adjusting the humidity of the test soil (150, 151). If a non-adapted soil is used as an inoculum, the test simulates the biodegradation processes which take place in a natural environment. If a pre-exposed soil is used, the method can be used to investigate the potential biodegradability of a test material.

The standard ASTM D6954-18 is used for the exposure of plastics that can degrade by a combination of oxidation and biodegradation (152). This standard should permit the comparison and ranking of the overall rate of the environmental degradation of plastics in the course of thermal or photo oxidation. Each degradation stage is independently evaluated to allow a combined evaluation of a polymer's environmental performance.

1.7.4.2 Carbon Dioxide Control

The ISO standards 14852 and 14855 (153, 154) have been recently revised in the form of ISO 14855-1:2012 (129). This standard describes a method for the determination of the ultimate aerobic biodegradability of organic plastics materials under controlled composting conditions by the measurement of carbon dioxide evolved and the degree of disintegration of the plastic at the end of the test.

This method is designed to simulate the typical aerobic composting conditions for the organic fraction of solid mixed municipal waste. The test material is exposed to an inoculum which is derived from compost. The composting takes place in an environment wherein temperature, aeration and humidity are closely monitored and controlled. The test method is designed to yield the percentage conversion of the carbon in the test material to evolved carbon dioxide as well as the rate of conversion.

A variant of this method uses a mineral bed, i.e., vermiculite inoculated with thermophilic microorganisms obtained from compost with a specific activation phase, instead of a mature compost. This variant is designed to yield the percentage of carbon in the test substance converted to carbon dioxide and the rate of conversion.

Part 2 of this standard specifies a method for determining the ultimate aerobic biodegradability of plastic materials under controlled composting conditions by gravimetric measurement of the amount of carbon dioxide evolved. The method is designed to yield an optimum rate of biodegradation by adjusting the humidity, aeration and temperature of the composting vessel. This method applies to the following materials (155):

1. Natural or synthetic polymers and copolymers, and mixtures of these,
2. Plastic materials that contain additives such as plasticizers or colorants,
3. Water-soluble polymers, and
4. Materials that, under the test conditions, do not inhibit the activity of microorganisms present in the inoculum.

1.7.4.3 Measurement of Biogas

The ISO 14853:2005 Standard specifies a method for the determination of the ultimate anaerobic biodegradability of plastics by anaerobic microorganisms (156). The conditions described in the standard do not necessarily correspond to the optimum conditions for the maximum degree of biodegradation to occur. Instead, the test calls for exposure of the test material to sludge for a period of up to

60 *d*, which is longer than the normal sludge retention time (25–30 *d*) in anaerobic digesters, though digesters at industrial sites can have much longer retention times. The method applies to the same materials as in Standard ISO 14855-2-2018 (130).

1.7.5 Biodegradability of Plastics in Seawater

Many industrial wastewaters with a variety of chemicals are reaching the sea either by direct discharge or via estuaries and rivers in which the residence times are low compared with the period necessary for complete biodegradation. Because of the growing awareness of the need to protect the marine environment against increasing loads of chemicals and the need to estimate the probable concentration of chemicals in the sea, test methods for biodegradability in seawater have been developed.

In addition, the use of plastics aboard ships is on the rise and the use of the sea as a trash dumping site is no longer a possibility (143). Also, for this reason, the disposal of plastic materials while at sea has emerged as a major issue. Biodegradable plastics could help with regard to the safe disposal of plastic materials at sea.

The OECD 306B test deals with the biodegradability of plastics in seawater (157). This test is also known as the Zahn-Wellens/EVPA Test and is included in a series of standards. Here, the shake flask method is used.

The test substance is used in the test medium at a concentration of 5–40 $mg\, l^{-1}$ of total organic carbon dissolved. No inoculum is added in addition to the microorganisms already present in the seawater. Stock solutions for mineral nutrients are shown in Table 1.11.

A test medium is prepared by adding 1 *ml* of each of the stock solutions in Table 1.11 to 1 *l* of pretreated seawater. The solution of the test substance in the test medium is incubated under agitation in the dark or in diffuse light under aerobic conditions at 15–20°C. Five flasks should be used:

- Two for the test suspension,
- Two for the blank seawater, and
- One for procedure control.

Optionally, one additional flask is used containing a reference compound for procedure control and another containing the test

Table 1.11 Aqueous stock solutions for mineral nutrients.

Mineral	Amount/[$g\,l^{-1}$]
Solution A	
Potassium dihydrogen orthophosphate, KH_2PO_4	8.50
Dipotassium hydrogen orthophosphate, K_2HPO_4	21.75
Disodium hydrogen orthophosphate dihydrate, $Na_2HPO_4 \times 2H_2O$	33.30
Ammonium chloride, NH_4Cl	0.50
Solution B	
Calcium chloride, $CaCl_2$	27.50
Solution C	
Magnesium sulfate heptahydrate, $MgSO_4 \times 7H_2O$	22.50
Solution D	
Iron(III) chloride hexahydrate, $FeCl_3 \times 6H_2O$	0.25

substance and a sterilizing agent, e.g., mercury chloride, for abiotic sterile control. As reference compounds, sodium benzoate, sodium acetate or aniline may be used.

The recommended maximum test duration is around 60 *d*. The degradation is monitored by measurements of the total organic carbon content. A typical reaction curve is shown in Figure 1.8.

In addition, the closed bottle method has been described. This method uses similar conditions as the previously described method, but the experiments are performed in the dark. In addition, the degradation is followed by the change of oxygen by a chemical or electrochemical method. Here, many more test samples are used.

Another test method has been developed to assess the rate and degree of aerobic biodegradation of plastics exposed to marine microorganisms (143). Aerobic biodegradation is determined by measuring the amount of biogas produced during such an exposure. It has been stated that there is no similar or equivalent ISO standard.

Non-floating products made from biodegradable plastics, including packaging and coatings, can be tested under the conditions of the marine environment (158). In particular, environments are shallow and deep saltwater and brackish water. The products should exhibit satisfactory performances in terms of disintegration during

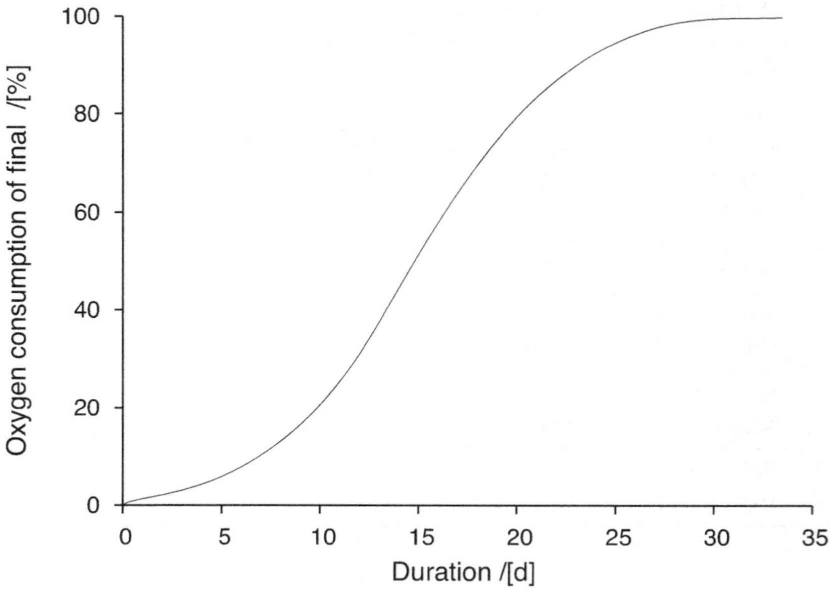

Figure 1.8 Typical reaction curve.

marine degradation, inherent biodegradation, and environmental toxicity.

1.8 Test of the Biological Origin

Biologically derived compounds may be distinguished from similar compounds produced from a petrochemical source or from fossil fuel carbon by dual carbon-isotopic fingerprinting. This method can distinguish chemically identical materials (159). In addition, such methods have gained some importance in discriminating between natural and industrial pollution (160).

The isotopes ^{14}C and ^{13}C give complementary information to this problem. The radiocarbon dating isotope ^{14}C with its nuclear half life of 5730 *y*, clearly allows differentiating between fossil and biospheric feedstocks (161).

The basic assumption in radiocarbon dating is that the constancy of the ^{14}C concentration in the atmosphere leads to the constancy of the concentration of ^{14}C in living organisms. When dealing with an isolated sample, the age of a sample can be approximately deduced by the relationship (159)

$$t = -\frac{5730}{0.693}\ln(\frac{A}{A_0}) \quad . \tag{1.1}$$

Here, t is the age, 5730 years is the half lifetime of radiocarbon, and $\frac{A}{A_0}$ is the specific ^{14}C activity of the sample and of a modern standard (162).

However, because of atmospheric nuclear testing since 1950 and the burning of fossil fuel since 1850, ^{14}C has acquired a second, geochemical time characteristic. Its concentration in atmospheric CO_2, and thus in the living biosphere, approximately doubled at the peak of nuclear testing in the mid-1960s. Afterwards, it has gradually returned to the steady-state baseline isotope rate ^{14}C of ca. 1.2×10^{-12} (159).

The stable carbon isotope ratio $^{13}C/^{12}C$ is another route of checking the origin. This ratio in a biosourced material results from the $^{13}C/^{12}C$ ratio in atmospheric carbon dioxide at the time the carbon dioxide is fixed and also reflects the precise metabolic pathway. Regional variations may occur.

Terrestrial plants and marine carbonates show differences in the isotope ratio. Sometimes large variations due to isotopic fractionation effects are observed.

The major cause of differences in the carbon isotope ratio in plants is closely associated with differences in the pathway of photosynthetic carbon metabolism in the plants, particularly the reaction occurring during the primary carboxylation.

The most significant reaction type is the photosynthetic mechanism. There are two classes of the photosynthetic cycle, the Calvin-Benson photosynthetic cycle and the Hatch-Slack photosynthetic cycle.

The Calvin-Benson photosynthetic cycle is dominant in hardwoods and conifers. The primary CO_2 fixation or carboxylation reaction involves the enzyme ribulose-1,5-diphosphate carboxylase and the first stable product is a 3-carbon compound (163).

On the other hand, the Hatch-Slack photosynthetic cycle is important for tropical grasses, corn and sugarcane. Phosphenol-pyruvate carboxylase is responsible for the primary carboxylation reaction.

The first stable carbon compound is a C_4 acid, which is subsequently decarboxylated (164, 165). The fundamentals and systematics of the non-statistical distributions of isotopes in natural compounds have been reviewed (166).

The ^{13}C measurement scale was originally defined by a zero set by belemnite limestone. The values are given in parts per thousand deviations from this material. These values are calculated as follows:

$$\delta^{13}C = \frac{\left(^{13}C/^{12}C\right)_{sample} - \left(^{13}C/^{12}C\right)_{standard}}{\left(^{13}C/^{12}C\right)_{standard}} \times 1000 \tag{1.2}$$

For example, biologically derived 1,3-propanediol, and compositions comprising biologically derived 1,3-propanediol, may be completely distinguished from their petrochemically derived counterparts on the basis of ^{13}C and dual carbon-isotopic fingerprinting (159).

Also, the average carbon isotope contents of a series of glycerol samples have been determined. The results indicated that it is possible to distinguish the glycerol obtained from the glycerides produced in plants following the C_3 and C_4 carbon fixation pathways (167). Natural ethyl butyrate can be differentiated from industrially manufactured ethyl butyrate by the content of ^{14}C (168).

Trade names appearing in the references are shown in Table 1.12.

Table 1.12 Trade names in references.

Trade name Description	Supplier
Acrawax®	Lonza Group AG, Basel, Switzerland
Amide wax (159)	
Acronal® 4F	BASF AG
Poly(*n*-butylacrylate) (159)	
Acronal®	BASF
Acrylic resins (159)	
Adcote® 313	Morton International
Polyethyleneimine (159)	
Adcote® 50T4983	Morton International
Ethylene acrylic acid dispersion (159)	
Adcote® 50T4990	Morton International
Ethylene acrylic acid dispersion (159)	
Airflex® (Series)	Air Products and Chemicals, Inc.
Vinyl acetate/ethylene copolymer emulsions (159)	
Aquathane® 97949	Reichhold
Nonsulfonated urethane dispersion (159)	
Aquathane® 97959	Reichhold
Nonsulfonated urethane dispersion (159)	
Biolice®	Limagrain
Biodegradable polymer (75)	
Biomax®	DuPont
Sulfonated aliphatic-aromatic copolyesters (159)	
Bionolle®	Showa Highpolymer Co.
Poly(butylene succinate) (159)	
Biopol®	Zeneca
Biodegradable hot melt adhesive (PHV/B) (159)	
Bynel® (Series)	DuPont
Anhydride modified ethylene vinyl acetate resin, adhesion promoter (159)	
Carboset® CR-760	B.F. Goodrich Co.
Anionic acrylate-styrene dispersion (159)	
Citroflex® A-4	Morflex, Inc.
Acetyltri-*n*-butyl citrate (159)	
Cozeen™ 303N	Freeman Industries LLC
Natural coating (75)	

Table 1.12 (cont) Trade names in references.

Trade name Description	Supplier
Cymel® (Series) Amino resins (159)	Cytec
Devcon® Acrylic adhesive (91)	ITW Performance Polymers
Dispercoll® (Series) Sulfonated polyester urethane dispersions (159)	Bayer Corp.
Eastar® Bio Compostable copolyester (159)	Eastman Chemical Company
EcoFoam® Biodegradable materials for use in packaging (159)	National Starch and Chemical Co.
EcoPLA® Poly(lactide) (159)	Cargill Dow LLC
EnPol® Poly(1,4-butylene succinate) (159)	IRe Chemical Ltd.
EnviroFil® Modified starch (159)	ENPAC LLC.
Escorez® (Series) Tackifying resins (EVA) (159)	ExxonMobil Chemical Co.
Flexthane® (Series) Nonsulfonated urethane dispersions (159)	Air Products and Chemicals, Inc.
Hycar® (Series) Amine-terminated butadiene-acrylonitrile (159)	Lubrizol Advanced Materials Inc.; B.F. Goodrich Co.
Jagotex® KEA Acrylic dispersions (159)	Jager Co.
Kynar® Poly(vinylidene fluoride) (159)	Arkema Inc.
Luphen® D 200 A Nonsulfonated urethane dispersions (159)	BASF
MicroMid® (Series) Poly(amide) dispersions (159)	Union Camp Corp.
Myvacet® (Series) Acetylated monoglycerides of modified fats (159)	Kerry Group Services Ltd.
Myvaplex® 600 2,3-Dihydroxypropyl octadecanoate (glycerol monostearate) (159)	Kerry Group Services Ltd.
Myverol® Propylene glycol monostearate (159)	Kerry Group Services Ltd.

Table 1.12 (cont) Trade names in references.

Trade name Description	Supplier
NatureFlex™ NE 30 Heat-sealable compostable film (75)	Innovia Films Ltd.
NeoRez® (Series) Nonsulfonated urethane dispersions (159)	Zeneca Resins
Papermatch® Masterbatch to impart paper-like properties to polymer films (159)	A Shulman, Inc.
Piccotex® LC-55WK Styrene resin emulsion (159)	Hercules Inc.
Polectron® 430 Vinylpyrrolidone/styrene copolymer emulsion fluid (159)	ISP Corp.
Quilastic® (Series) Nonsulfonated urethane dispersions (159)	Merquinsa
Resyn® (Series) Vinyl acetate homopolymer dispersions (159)	ICI
Rhoplex® (Series) Acrylic latex (159)	Evonik, Rohm & Haas
Rucoflex® Polyester polyol (159)	Bayer
Sancure® Nonsulfonated urethane dispersions (159)	Lubrizol Advanced Materials Inc.
Synthemul® (Series) Carboxylated acrylic copolymer (159)	Reichhold
Tacolyn® 5001 Styrene resin dispersion (159)	Hercules Inc.
Tecoflex® (Series) Urethane elastomers (159)	Lubrizol Advanced Materials Inc.
Texigel® (Series) Silanated anionic acrylate-styrene polymer dispersions (159)	Scott Bader Inc.
Tronox® 470 Titanium dioxide pigment (159)	Tranox Ltd.
Tween® 20 Sorbitan monolaurate (159)	Uniqema
Tween® 40 Sorbitan monopalmitate (159)	Uniqema
Tween® 60 Sorbitan monostearate (159)	Uniqema

Table 1.12 (cont) Trade names in references.

Trade name Description	Supplier
Tween® 85	Uniqema
Sorbitan monooleate (159)	
Vancryl® (Series)	Air Products & Chemicals Inc.
Anionic acrylate-styrene dispersions (159)	
Vycar®	Lubrizol
Poly(vinylidene chloride) emulsion (159)	
Weld-It®	Devcon Inc.
All purpose adhesive (91)	

References

1. V. Thakur, *Green Biorenewable Biocomposites: From Knowledge to Industrial Applications*, Apple Academic Press, Toronto, 2015.
2. P.M. Visakh and S. Lüftl, eds., *Polyethylene-Based Biocomposites and Bionanocomposites*, John Wiley & Sons, Hoboken, New Jersey; Scrivener Publishing, Beverly, Massachusetts, 2016.
3. P.M. Visakh and M. Poletto, *Polypropylene-Based Biocomposites and Bionanocomposites*, John Wiley & Sons, Hoboken, New Jersey; Scrivener Publishing, Beverly, Massachusetts, 2018.
4. G.S. Mann, L.P. Singh, P. Kumar, and S. Singh, *Journal of Thermoplastic Composite Materials*, Vol. in press, p. 0892705718816354, 2018.
5. D.C. Schiltz, Biodegradable starch-based polymer compositions, US Patent 5 449 708, September 12, 1995.
6. H. Chum, *Polymers from Biobased Materials*, Noyes Data Corp., Park Ridge, NJ, 1991.
7. U. Albertsson, ed., *Degradable Aliphatic Polyesters*, Vol. 157 of *Advances in Polymer Science*, Springer, Berlin, Heidelberg, 2001.
8. M.S. Lindblad, Y. Liu, A.-C. Albertsson, E. Ranucci, and S. Karlsson, Polymers from renewable resources in *Degradable Aliphatic Polyesters*, Vol. 157 of *Advances in Polymer Science*, pp. 139–161. Springer, Berlin, Heidelberg, 2001.
9. C. Williams and M. Hillmyer, *Polymer Reviews*, Vol. 48, p. 1, January 2008.
10. S. Pilla, ed., *Handbook of Bioplastics and Biocomposites Engineering Applications*, Wiley Scrivener, Hoboken, NJ, Salem, Mass, 2011.
11. V. Mittal, ed., *Renewable Polymers Synthesis, Processing, and Technology*, Scrivener Publishing LLC, Beverly, MA, 2012.
12. A. Morelli, D. Puppi, and F. Chiellini, *Journal of Renewable Materials*, Vol. 1, p. 83, April 2013.
13. S. Ebnesajjad, ed., *Handbook of Biopolymers and Biodegradable Plastics: Properties, Processing and Applications*, William Andrew, Oxford, 2013.
14. F. Flores, J.A. Dobado, J. Isac-García, and F.J. Martín-Martínez, *Lignin and Lignans as Renewable Raw Materials: Chemistry, Technology and Applications*, Vol. 1 of *Wiley Series in Renewable Resource*, Wiley, Hoboken, 2015.
15. A. Pandey, R. Höfer, M. Taherzadeh, M. Nampoothiri, and C. Larroche, eds., *Industrial Biorefineries and White Biotechnology*, Elsevier Ltd., Amsterdam, Netherlands, 2015.
16. S. Madbouly, C. Zhang, and M.R. Kessler, eds., *Bio-based Plant Oil Polymers and Composites*, William Andrew, an imprint of Elsevier, Kidlington, Oxford, UK, 2015.
17. M. Khalaf, ed., *Green Polymers and Environmental Pollution Control*, Apple Academic Press, Toronto, 2016.

18. K.M. Zia, A. Noreen, M. Zuber, S. Tabasum, and M. Mujahid, *International Journal of Biological Macromolecules*, Vol. 82, p. 1028, 2016.
19. F. Cavani, S. Albonetti, F. Basile, and A. Gandini, eds., *Chemicals and Fuels from Bio-based Building Blocks*, Wiley-VCH Verlag GmbH & Co, Weinheim, 2016.
20. S.E. Hosseini and M.A. Wahid, *Renewable and Sustainable Energy Reviews*, Vol. 57, p. 850, 2016.
21. Z. Fang and J. Smith, Richard L., eds., *Production of Biofuels and Chemicals from Lignin*, Biofuels and Biorefineries, Springer, Berlin, 2016.
22. S.A. Ashter, *Introduction to Bioplastics Engineering*, William Andrew, an imprint of Elsevier, Oxford, UK, 2016.
23. R. Łukasik, ed., *High Pressure Technologies in Biomass Conversion*, Royal Society of Chemistry, United Kingdom, 2017.
24. V.K. Thakur, M.K. Thakur, and M.R. Kessler, eds., *Structure and Chemistry*, Vol. 1 of *Handbook of Composites from Renewable Materials. Advanced Applications*, John Wiley & Sons,, Hoboken, NJ; Scrivener Publishing LLC, Beverly, MA, 2017, 2017.
25. V.K. Thakur, M.K. Thakur, and M.R. Kessler, eds., *Nanocomposites: Advanced Applciations*, Vol. 8 of *Handbook of Composites from Renewable Materials. Advanced Applications*, John Wiley & Sons, Inc., Hoboken, NJ; Scrivener Publishing LLC, Beverly, MA, 2017.
26. N. Shimpi, ed., *Biodegradable and Biocompatible Polymer Composites*, Elsevier Science, 2017.
27. S. Kurtz, ed., *PEEK Biomaterials Handbook*, William Andrew, Oxford, 2nd edition, 2019.
28. R. Tharanathan, *Trends in Food Science & Technology*, Vol. 14, p. 71, March 2003.
29. V. Siracusa, P. Rocculi, S. Romani, and M.D. Rosa, *Trends in Food Science & Technology*, Vol. 19, p. 634, December 2008.
30. B.M. Cherian, G.M. de Olyveira, L.M.M. Costa, A.L. Leão, M.R. de Morais Chaves, S.F. de Souza, and S. Narine, Bionanocomposites for natural food packing in P.M. Visakh, S. Thomas, L.B. Iturriaga, and P.D. Ribotta, eds., *Advances in Food Science and Technology*, chapter 9, pp. 265–299. Wiley-Scrivener, 2013.
31. R. Busquets, ed., *Emerging Nanotechnologies in Food Science*, Elsevier, Amsterdam, Netherlands, 2017.
32. D.R. Heldman, D.B. Lund, and C. Sabliov, eds., *Handbook of Food Engineering*, CRC press, Boca Raton, FL Cambridge, England, 3rd edition, 2019.
33. M.J.-L. Tschan, E. Brulé, P. Haquette, and C.M. Thomas, *Polymer Chemistry*, Vol. 3, p. 836, 2012.
34. K. Yao and C. Tang, *Macromolecules*, Vol. 46, p. 1689, March 2013.

35. S. Inoue, H. Koinuma, and T. Tsuruta, *Journal of Polymer Science Part B: Polymer Letters*, Vol. 7, p. 287, April 1969.
36. J.B. van Beilen and Y. Poirier, *The Plant Journal*, Vol. 54, p. 684, May 2008.
37. A. Gandini, *Macromolecules*, Vol. 41, p. 9491, December 2008.
38. P. Anastas, *Green Chemistry: Theory and Practice*, Oxford University Press, Oxford, 1998.
39. J.H. Clark and D.J. Macquarrie, *Handbook of Green Chemistry and Technology*, Blackwell Science, Oxford, England and Malden, MA, 2002.
40. M. Lancaster, *Green Chemistry: An Introductory Text*, Royal Society of Chemistry, Cambridge, 2010.
41. R. Luque, *Green Chemistry*, Nova Science Publishers, Hauppauge, NY, 2012.
42. J. Pohlman, *Green Chemistry*, Tap the power, knowledge is power, Wisconsin Legislative Reference Bureau, Madison, Wis., 2012.
43. J. Clark and F. Deswarte, eds., *Introduction to Chemicals from Biomass*, Wiley, Chichester, West Sussex, 2015.
44. S. Sharma, ed., *Green Chemistry for Dyes Removal from Wastewater: Research Trends and Applications*, Scrivener Publishing/Wiley, Hoboken, New Jersey, Salem, Massachusetts, 2015.
45. M. Lancaster, *Green Chemistry : An Introductory Text*, Royal Society of Chemistry, Cambridge, UK, 3rd edition, 2016.
46. G. Rothenberg, *Catalysis: Concepts and Green Applications*, Wiley-VCH Verlag GmbH & Co. KGaA, Weinheim, 2nd edition, 2017.
47. S. Muniyasamy, K. Mohanrasu, A. Gada, T.C. Mokhena, A. Mtibe, and T. Boobalan, *Integrating Green Chemistry and Sustainable Engineering*, John Wiley & Sons, Ltd., 2019.
48. J. Tao and R.J. Kazlauskas, *Biocatalysis for Green Chemistry and Chemical Process Development*, John Wiley & Sons, Hoboken, NJ, 2011.
49. R. Mathers and M.A.R. Meier, eds., *Green Polymerization Methods Renewable Starting Materials, Catalysis and Waste Reduction*, Wiley-VCH, Weinheim, Chichester, 2011.
50. P. Gallezot, *Catalysis Today*, Vol. 121, p. 76, March 2007.
51. M. Lancaster, Principles and concepts of green chemistry in *Green Chemistry: An Introductory Text*, chapter 1, pp. 1–20. Royal Society of Chemistry, Cambridge, 2010.
52. M. Poliakoff, J.M. Fitzpatrick, T.R. Farren, and P.T. Anastas, *Science*, Vol. 297, p. 807, August 2002.
53. P.T. Anastas and J.B. Zimmerman, *Environmental Science & Technology*, Vol. 37, p. 94A, March 2003.
54. W. McDonough, M. Braungart, P.T. Anastas, and J.B. Zimmerman, *Environmental Science & Technology*, Vol. 37, p. 434A, December 2003.

55. P.T. Anastas and N. Eghbali, *Chemical Society Reviews*, Vol. 39, p. 301, 2009.
56. F.J. Lozano, R. Lozano, P. Freire, C. Jiménez-Gonzalez, T. Sakao, M.G. Ortiz, A. Trianni, A. Carpenter, and T. Viveros, *Journal of Cleaner Production*, Vol. 172, p. 227, 2018.
57. R. Sheldon, *Green Chemistry and Catalysis*, Wiley-VCH, Weinheim, 2007.
58. C.K.S. Pillai, *Designed Monomers and Polymers*, Vol. 13, p. 87, 2010.
59. J.A. Rosewitz, H.A. Choshali, and N. Rahbar, *Cement and Concrete Composites*, Vol. 96, p. 252, 2019.
60. D. Ji and J. Kim, *ACS Nano*, Vol. 13, p. 2773, 2019.
61. D. Zhao, J.F. Kim, G. Ignacz, P. Pogany, Y.M. Lee, and G. Szekely, *ACS Nano*, Vol. 13, p. 125, 2019.
62. W. Xu, J. Liu, T. Chen, X. Jiang, X. Qian, Y. Zhang, Z. Jiang, and Y. Zhang, *Small*, Vol. in press, p. 1901582, 2019.
63. F. Börnke and I. Broer, *Current Opinion in Plant Biology*, Vol. 13, p. 353, June 2010.
64. J.B. van Beilen, *Biofuels, Bioproducts and Biorefining*, Vol. 2, p. 215, May 2008.
65. J. Du, L. Li, and S. Zhou, *Biotechnology Advances*, 2019.
66. A.V.E. Conradie and A.L. Botes, Methods, reagents and cells for biosynthesizing compounds, US Patent 9 896 702, assigned to Invista North America S.A.R.L. (Wilmington, DE), February 20, 2018.
67. Anonymous, Commercial outlook for biodegradable polymers, PERP Program PERP-94/95-S13, Nexant, Process Evaluation/Research Planning (PERP), White Plains, NY, USA, 1997.
68. L.S. Nair and C.T. Laurencin, *Progress in Polymer Science*, Vol. 32, p. 762, August 2007.
69. S. Joseph, M.J. John, L.A. Pothen, and S. Thomas, Raw and renewable polymers in P. Eyerer, M. Weller, and C. Hübner, eds., *Polymers - Opportunities and Risks*, Vol. 2 of *The Handbook of Environmental Chemistry*, pp. 55–80. Springer, Berlin Heidelberg, 2009.
70. T.L. Sporleder, P.D. Goldsmith, J. Cordier, and P. Godin, *International Food and Agribusiness Management Review*, Vol. 14, p. 35, 2011.
71. A. Iles and A.N. Martin, *Journal of Cleaner Production*, Vol. 45, p. 38, April 2013.
72. T.P. Haider, C. Völker, J. Kramm, K. Landfester, and F.R. Wurm, *Angewandte Chemie International Edition*, Vol. 58, p. 50, 2019.
73. G. Scott, *Polymers and the Environment*, Royal Society of Chemistry, Cambridge, 1999.
74. G. Scott, *Polymer Degradation and Stability*, Vol. 68, p. 1, April 2000.

75. C. Tolibas-Spurlock, B. Allred-Forsman, and D. Agnew-Von Moos, Compostable container for storing fluids, US Patent 8 245 848, assigned to Sacred Green, Inc. (Chehalis, WA), August 21, 2012.
76. J. Wróblewska-Krepsztul, T. Rydzkowski, G. Borowski, M. Szczypiński, T. Klepka, and V.K. Thakur, *International Journal of Polymer Analysis and Characterization*, Vol. 23, p. 383, 2018.
77. J.A. Bowden and C.C. Johnston, Biodegradable or compostable containers, US Patent 7 967 904, assigned to New Ice Limited (Isle of Man), June 28, 2011.
78. A.K. Mohanty, M. Misra, and L.T. Drzal, *Journal of Polymers and the Environment*, Vol. 10, p. 19, 2002.
79. E. Rudnik, *Compostable Polymer Materials*, Elsevier, Amsterdam London, 2008.
80. L.S. Dilkes-Hoffman, J.L. Lane, T. Grant, S. Pratt, P.A. Lant, and B. Laycock, *Journal of Cleaner Production*, Vol. 180, p. 325, 2018.
81. J.J. Blaker, K.-Y. Lee, and A. Bismarck, *Journal of Biobased Materials and Bioenergy*, Vol. 5, p. 1, March 2011.
82. Y. Long, *Biodegradable Polymer Blends and Composites from Renewable Resources*, Wiley, Hoboken, NJ, 2009.
83. A.K. Mohanty, M. Misra, and G. Hinrichsen, *Macromolecular Materials and Engineering*, Vol. 276-277, p. 1, 2000.
84. K. Khemani, P.J. Andersen, S.K. Hodson, and H. Schmidt, Biodegradable polymer films and sheets suitable for use as laminate coatings as well as wraps and other packaging materials, US Patent 6 573 340, assigned to Biotec Biologische Naturverpackungen GmbH & Co. KG (DE), June 3, 2003.
85. I. Takahashi and H. Iida, Biodegradable plastics composition, molded article of the composition, and method of controlling biodegradation rate, US Patent 7 816 424, assigned to Nisshinbo Industries, Inc. (Tokyo, JP), October 19, 2010.
86. Y. Imashiro and I. Takahashi, Process for producing polycarbodiimide resin, US Patent 5 373 080, assigned to Nisshinbo Industries, Inc. (Tokyo, JP), December 13, 1994.
87. A. Gandini and T.M. Lacerda, *Progress in Polymer Science*, Vol. 48, p. 1, 2015. Topical Volume on Polymer Chemistry.
88. N. Spassky, M. Wisniewski, C. Pluta, and A. Le Borgne, *Macromolecular Chemistry and Physics*, Vol. 197, p. 2627, September 1996.
89. H.C. Kolb, M.G. Finn, and K.B. Sharpless, *Angewandte Chemie International Edition*, Vol. 40, p. 2004, 2001.
90. J.E. Moses and A.D. Moorhouse, *Chemical Society Reviews*, Vol. 36, p. 1249, 2007.

91. V. Fokin, M.G. Finn, and K.B. Sharpless, Polymeric materials via click chemistry, US Patent 8 101 238, assigned to The Scripps Research Institute (La Jolla, CA), January 24, 2012.
92. G.L. Baker, M.R. Smith, III, X. Jiang, and E.B. Vogel, Functionalization of polyglycolides by "click" chemistry, US Patent Application 20 090 054 619, assigned to Board of Trustees of Michigan State University, East Lansing (MI), February 26, 2009.
93. O. Türünç, *Efficient Routes to Degradable and non-Degradable Renewable Polymers from Fatty Acids.* PhD thesis, Fakultät für Chemie und Biowissenschaften Karlsruher Institut für Technologie, University of Karlsruhe, Germany, 2012.
94. G.-L. Zhao, J. Hafrén, L. Deiana, and A. Córdova, *Macromolecular Rapid Communications*, Vol. 31, p. 740, April 2010.
95. S. Kobayashi and A. Makino, *Chemical Reviews*, Vol. 109, p. 5288, November 2009.
96. S. Kobayashi and H. Uyama, Enzymatic polymerization in *Encyclopedia of Polymer Science and Technology*. Wiley Blackwell (John Wiley & Sons), 2011.
97. A. Douka, S. Vouyiouka, L.-M. Papaspyridi, and C.D. Papaspyrides, *Progress in Polymer Science*, Vol. 79, p. 1, 2018.
98. J. Su, J. Noro, J. Fu, Q. Wang, C. Silva, and A. Cavaco-Paulo, *Journal of Cleaner Production*, Vol. 202, p. 792, 2018.
99. D. Maniar, K.F. Hohmann, Y. Jiang, A.J.J. Woortman, J. van Dijken, and K. Loos, *ACS Omega*, Vol. 3, p. 7077, 2018.
100. H. Zhao, L.O. Afriyie, N.E. Larm, and G.A. Baker, *RSC Advances*, Vol. 8, p. 36025, 2018.
101. S. Kobayashi, *Macromolecular Rapid Communications*, Vol. 30, p. 237, February 2009.
102. S. Okumura, M. Iwai, and Y. Tominaga, *Agricultural and biological chemistry*, Vol. 48, p. 2805, 1984.
103. A. Mahapatro, B. Kalra, A. Kumar, and R.A. Gross, *Biomacromolecules*, Vol. 4, p. 544, May 2003.
104. H. Uyama and S. Kobayashi, *Chemistry Letters, Japan*, Vol. 23, p. 1687, 1994.
105. U. Meyer, A.R.A. Palmans, T. Loontjens, and A. Heise, *Macromolecules*, Vol. 35, p. 2873, April 2002.
106. J.W. Peeters, A.R.A. Palmans, E.W. Meijer, C.E. Koning, and A. Heise, *Macromolecular Rapid Communications*, Vol. 26, p. 684, May 2005.
107. A. Harada, M. Osaki, Y. Takashima, and H. Yamaguchi, *Accounts of Chemical Research*, Vol. 41, p. 1143, September 2008.
108. R.J. Ellis, *Trends in Biochemical Sciences*, Vol. 31, p. 395, July 2006.
109. H. Zhang, J. Shen, Z. Liu, A. Hao, Y. Bai, and W. An, *Supramolecular Chemistry*, Vol. 22, p. 297, May 2010.
110. Y. Kaneko and J.-I. Kadokawa, *The Chemical Record*, Vol. 5, p. 36, 2005.

111. Y. Kaneko, K. Beppu, and J.-I. Kadokawa, *Biomacromolecules*, Vol. 8, p. 2983, October 2007.
112. Y. Kaneko, K. Beppu, and J.-I. Kadokawa, *Macromolecular Chemistry and Physics*, Vol. 209, p. 1037, May 2008.
113. Y. Kaneko, K. Beppu, T. Kyutoku, and J.-I. Kadokawa, *Polymer Journal*, Vol. 41, p. 279, 2009.
114. Y. Kaneko, K. Beppu, and J.-I. Kadokawa, *Polymer Journal*, Vol. 41, p. 792, 2009.
115. J.-I. Kadokawa, S. Orio, and K. Yamamoto, *RSC Advances*, Vol. 9, p. 16176, 2019.
116. J.-I. Kadokawa, K. Yano, S. Orio, and K. Yamamoto, *ACS Omega*, Vol. 4, p. 6331, 2019.
117. M.P. Coughlan and L.G. Ljungdahl, Comparative biochemistry of fungal and bacterial cellulolytic enzyme systems, in J. Millet, ed., *Biochemistry and Genetics of Cellulose Degradation*, number 43 in FEMS Symposia, pp. 11–30, Delft, the Netherlands, 1988. Federation of European Microbiological Societies, Academic Press, Inc.
118. G.-Q. Chen, *Plastics from Bacteria: Natural Functions and Applications*, Springer, Heidelberg; New York, 2010.
119. F. Acevedo, P. Villegas, V. Urtuvia, J. Hermosilla, R. Navia, and M. Seeger, *International Journal of Biological Macromolecules*, Vol. 106, p. 692 , 2018.
120. CEN, Biobased products - overview of standards, Review of Standards CEN/TR 16208, European Committee for Standardization (CEN), Brussels, Belgium, 2011.
121. P. Kamath, Biobased products, Review of Standards EN 16575, European Committee for Standardization (CEN), Brussels, Belgium, 2014.
122. European Commission, A European Strategy for Plastics in a Circular Economy, Technical report, European Union, 2018.
123. A. Krzan, S. Hemjinda, S. Miertus, A. Corti, and E. Chiellini, *Polymer Degradation and Stability*, Vol. 91, p. 2819, December 2006.
124. Plastics – Methods for the preparation of samples for biodegradation testing of plastic materials, ISO Standard 10210:2012, International Organization for Standardization, Geneva, Switzerland, 2012.
125. Plastics – Determination of the ultimate anaerobic biodegradation of plastic materials in controlled slurry digestion systems – Method by measurement of biogas production, ISO Standard 13975, International Organization for Standardization, Geneva, Switzerland, 2019.
126. Determination of the ultimate aerobic biodegradability of plastic materials in an aqueous medium – Method by measuring the oxygen demand in a closed respirometer, ISO Standard 14851, International Organization for Standardization, Geneva, Switzerland, 2019.

127. Determination of the ultimate aerobic biodegradability of plastic materials in an aqueous medium – Method by analysis of evolved carbon dioxide, ISO Standard 14852, International Organization for Standardization, Geneva, Switzerland, 2018.
128. Plastics – Determination of the degree of disintegration of plastic materials under defined composting conditions in a pilot-scale test, ISO Standard 16929:2018-04, International Organization for Standardization, Geneva, Switzerland, 2018.
129. Determination of the ultimate aerobic biodegradability of plastic materials under controlled composting conditions – Method by analysis of evolved carbon dioxide – Part 1: General method, ISO Standard 14855-1-2012, International Organization for Standardization, Geneva, Switzerland, 2005.
130. Determination of the ultimate aerobic biodegradability of plastic materials under controlled composting conditions – Method by analysis of evolved carbon dioxide – Part 2: Gravimetric measurement of carbon dioxide evolved in a laboratory-scale test, ISO Standard 14855-2, International Organization for Standardization, Geneva, Switzerland, 2018.
131. Plastics – Determination of the ultimate aerobic biodegradability of plastic materials in soil by measuring the oxygen demand in a respirometer or the amount of carbon dioxide evolved, ISO Standard 17556, International Organization for Standardization, Geneva, Switzerland, 2019.
132. Specifications for compostable plastics, ISO Standard 17088:2012-06, International Organization for Standardization, Geneva, Switzerland, 2012.
133. Plastics – Biodegradable mulch films for use in agriculture and horticulture – Requirements and test methods, ISO/AWI Standard 23517, International Organization for Standardization, Geneva, Switzerland, 2019.
134. Plastics – Determination of the ultimate anaerobic biodegradation under high-solids anaerobic-digestion conditions – Method by analysis of released biogas, ISO Standard 15985, International Organization for Standardization, Geneva, Switzerland, 2014.
135. Plastics – Determination of aerobic biodegradation of non-floating plastic materials in a seawater/sandy sediment interface – Method by measuring the oxygen demand in closed respirometer, ISO Standard 18830, International Organization for Standardization, Geneva, Switzerland, 2016.
136. Determination of the ultimate aerobic biodegradability of plastic materials in an aqueous medium - Method by analysis of evolved carbon dioxide, ISO Standard 2018, International Organization for Standardization, Geneva, Switzerland, 2018.

137. Subcommittee: D20.96, Standard test method for determining aerobic biodegradation of plastic materials under controlled composting conditions, incorporating thermophilic temperatures, ASTM Standard D5338-15, ASTM International, West Conshohocken, PA, 2015. http://www.astm.org/Standards/D5338.htm.
138. Standard test method for determining anaerobic biodegradation of plastic materials under high-solids anaerobic-digestion conditions, ASTM Standard D5511, ASTM International, West Conshohocken, PA, 2018.
139. Standard test method for determining anaerobic biodegradation of plastic materials under accelerated landfill conditions, ASTM Standard D5526, ASTM International, West Conshohocken, PA, 2018.
140. Standard test method for determining biodegradability of materials exposed to source-separated organic municipal solid waste mesophilic composting conditions by respirometry, ASTM Standard D5929-18, ASTM International, West Conshohocken, PA, 2018. https://www.astm.org/Standards/D5929.htm.
141. Standard test method for determining aerobic biodegradation of plastic materials in soil, ASTM Standard D5988-18, ASTM International, West Conshohocken, PA, 2018.
142. Standard specification for compostable plastics, ASTM Standard D6400-04, ASTM International, West Conshohocken, PA, 2004.
143. Standard test method for determining aerobic biodegradation of plastic materials in the marine environment by a defined microbial consortium or natural sea water inoculum, ASTM Standard D6691-17, ASTM International, West Conshohocken, PA, 2009.
144. Standard test method for determining the aerobic degradation and anaerobic biodegradation of plastic materials under accelerated bioreactor landfill conditions, ASTM Standard D7475, ASTM International, West Conshohocken, PA, 2011.
145. Plastics – Guidelines for the recovery and recycling of plastics waste, ISO Standard 15270:2008, International Organization for Standardization, Geneva, Switzerland, 2008.
146. Plastics – Determination of the ultimate anaerobic biodegradation of plastic materials in controlled slurry digestion systems – Method by measurement of biogas production, ISO Standard 13975:2012, International Organization for Standardization, Geneva, Switzerland, 2012.
147. Plastics – Determination of the ultimate anaerobic biodegradation and disintegration under high-solids anaerobic-digestion conditions – Method by analysis of released biogas, ISO Standard 15985:2004, International Organization for Standardization, Geneva, Switzerland, 2004.

148. Determination of the ultimate aerobic biodegradability of plastic materials in an aqueous medium – Method by measuring the oxygen demand in a closed respirometer, ISO Standard 14851:1999, International Organization for Standardization, Geneva, Switzerland, 1999.
149. Plastics – Determination of aerobic biodegradation of non-floating plastic materials in a seawater/sediment interface – Method by analysis of evolved carbon dioxide, ISO Standard 19679, International Organization for Standardization, Geneva, Switzerland, 2016.
150. Subcommittee: D20.96, Standard test method for determining aerobic biodegradation of plastic materials in soil, ASTM Standard D5988-12, ASTM International, West Conshohocken, PA, 2012.
151. Plastics – Determination of the ultimate aerobic biodegradability of plastic materials in soil by measuring the oxygen demand in a respirometer or the amount of carbon dioxide evolved, ISO Standard 17556:2012, International Organization for Standardization, Geneva, Switzerland, 2012.
152. Standard guide for exposing and testing plastics that degrade in the environment by a combination of oxidation and biodegradation, ASTM Standard D6954-18, ASTM International, West Conshohocken, PA, 2018. http://www.astm.org/Standards/D6954.htm.
153. Determination of the ultimate aerobic biodegradability of plastic materials in an aqueous medium – Method by analysis of evolved carbon dioxide, ISO Standard 14852:1999, International Organization for Standardization, Geneva, Switzerland, 1999.
154. Determination of the ultimate aerobic biodegradability of plastic materials under controlled composting conditions – Method by analysis of evolved carbon dioxide, ISO Standard 14855, International Organization for Standardization, Geneva, Switzerland, 2005.
155. Determination of the ultimate aerobic biodegradability of plastic materials under controlled composting conditions – Method by analysis of evolved carbon dioxide – Part 2: Gravimetric measurement of carbon dioxide evolved in a laboratory-scale test, ISO Standard 14855-2-2018, International Organization for Standardization, Geneva, Switzerland, 2018.
156. Plastics – Determination of the ultimate anaerobic biodegradation of plastic materials in an aqueous system – Method by measurement of biogas production, ISO Standard 14853:2005, International Organization for Standardization, Geneva, Switzerland, 2005.
157. OECD, Biodegradability in seawater, OECD Guidelines for the Testing of Chemicals 306, OECD Publishing, Paris, France, 1992.

158. Standard specification for non-floating biodegradable plastics in the marine environment, ASTM Standard D7081-05, ASTM International, West Conshohocken, PA, 2005.
159. R.M. Gohil, N.M. Hasty, R.A. Hayes, J.V. Kurian, Y. Liang, E.J. Stancik, M. Strukelj, and S.Y.-T. Tseng, Aliphatic-aromatic polyesters, and articles made therefrom, US Patent 7 888 405, assigned to E. I. du Pont de Nemours and Company (Wilmington, DE), February 15, 2011.
160. A.A. Rosen and M. Rubin, *Water Pollution Control Federation*, Vol. 37, p. 1302, September 1965.
161. L.A. Currie, Source apportionment of atmospheric particles in J. Buffle and H.P. van Leeuwen, eds., *Environmental Particles*, Vol. 1 of *IUPAC Environmental Analytical Chemistry Series*, pp. 3–74. Lewis Publishers, Inc., Chelsea, MI, 1992.
162. Y.-P. Hsieh, *Soil Science Society of America Journal*, Vol. 56, p. 460, March 1992.
163. J.A. Bassham, A.A. Benson, and M. Calvin, *J. Biol. Chem.*, Vol. 185, p. 781, August 1950.
164. C.R. Slack and M.D. Hatch, *The Biochemical Journal*, Vol. 103, p. 660, March 1967.
165. M.D. Hatch, *Photosynthesis Research*, Vol. 73, p. 251, 2002.
166. H.-L. Schmidt, *Naturwissenschaften*, Vol. 90, p. 537, December 2003.
167. G. Fronza, C. Fuganti, P. Grasselli, F. Reniero, C. Guillou, O. Breas, E. Sada, A. Rossmann, and A. Hermann, *Journal of Agricultural and Food Chemistry*, Vol. 46, p. 477, February 1998.
168. B. Byrne, K.J. Wengenroth, and D.A. Kruger, *Journal of Agricultural and Food Chemistry*, Vol. 34, p. 736, July 1986.

PART I

BIO-BASED POLYMERS DEGRADATION AND CHEMISTRY

2

Vinyl-Based Polymers

2.1 Polyolefins

Cleaner methods for cationic olefin polymerization have been reviewed (1). The sustainable synthesis of monomers for cationic polymerization has also been dealt with. Tris(pentafluorophenyl)gallium and aluminum, respectively, are active coinitiators for the production of medium-high molecular weight polymers of styrene and isobutene under aqueous reaction conditions (2).

Poly(ethylene) (PE) and biodegradable mulches for agricultural applications have been reviewed (3–5).

A more recent concept in biodegradable polymers is to make non-biodegradable synthetic polymers, such as poly(ethylene), biodegradable (6).

PE compositions can be made biodegradable under certain circumstances. A composition is summarized in Table 2.1.

Table 2.1 Biodegradable poly(ethylene) composition (7).

Ingredient	Amount /[*mg*]	Amount /[*ppm*]
Cellulose	0.4375	18.04
Ammonium nitrate	0.0625	2.58
Agar-agar	0.125	5.15
Yeast	0.125	5.15
Poly(ethylene) powder	24250	999969.07

The components of Table 2.1 are homogeneously mixed with 5 *ml* of boiling water, maintained at 100°C to form a slurry, and the resulting slurry is kept undisturbed for 12 *h* to get a biodegradable additive polymer composition.

Other biodegradable compositions are summarized in Table 2.2.

Table 2.2 Biodegradable PE-based compositions.

Type	Reference
Poly(ethylene furanoate-*co*-adipate)	(8)
Poly(ethylene succinate) copolymers	(9)
Poly(butylene succinate)	(10)

2.1.1 *Degradability*

In order to test the biodegradability according to ASTM D 5988-03 (11), some 24 *g* of the polymer composition was mixed with 500 *g* of soil. The biodegradation of the polymer composition was measured based on the amount of carbon dioxide evolved. It was observed that 659 *mg* of CO_2 is evolved during a period of 45 *d*, confirming that the polymer composition has undergone a biodegradation reaction (7).

The biodegradable polymer composition can be mixed directly with virgin polymers for making end products such as carrier bags, garbage disposal bags, hospital disposables, packaging film, and thermoformed plastics. The degradation products are obviously nontoxic, hence the composition is safe for the environment, living beings and food. Furthermore, the products do not alter the pH value of the soil.

Besides PE, other industrial polymers can be used such as poly(propylene) (PP), poly(styrene), and poly(vinyl chloride).

2.1.2 *Degradation Mechanism*

It is believed that the biodegradability of the compositions is based on a nucleophilic substitution reaction. Carbonyl groups are particularly susceptible to nucleophilic attachment to carboxyl carbon due to the tendency of oxygen to acquire electrons even at the expense of gaining a negative charge, in addition to the relatively unhindered transition state leading from the trigonal reactant to the tetrahedral intermediates. The carboxyl group provides the site for nucleophilic attack in the polymer chain and also increases the number of hydrogens attached to the α-carbon. Nucleophilic substitution takes place with the nascent nitrogen, nascent oxygen and nascent hydrogen. Ions are provided by the reactions that take place between an amide and water.

The nucleophilic substitution reaction may take place when a durable synthetic polymer is mixed with agar or yeast, cellulose, amides and water. The hydroxyl groups present in the cellulose become attached to the hydroxyl groups of agar and yeast in a link resembling a glycosidic linkage.

Consequently, the chain of the durable polymer contains a number of weak C–C bonds or C–H bonds. This results in a weak polymer chain containing monomeric units, making it susceptible to biodegradation (7). So, when the weakened polymer comes into contact with the soil, the monomeric units of the polymer act as nutrients for the bacteria present in the soil. Consequently, a rapid biodegradation of the polymer takes place in the soil.

It should be emphasized that these biodegradable compositions are not a mere admixture of the ingredients that result in an aggregation of their properties but the compositions have synergistically enhanced properties with regards to biodegradation (7).

A review of the degradation behavior of biodegradable polymers and related composites has been presented, with particular concern for multilayer films (12). Here, the processing of biodegradable polymeric films and the manufacturing and properties of multilayer films based on biodegradable polymers are discussed.

Also, the recent developments in thermogel research with a focus on synthesis and self-assembly mechanisms, gel biodegradability, and applications for drug delivery, cell encapsulation and tissue engineering are discussed in a monograph (13).

2.1.3 *Prodegradants*

Polyolefins are usually formulated to be resistant to oxidation and biodegradation, as antioxidants and stabilizers are included (14). However, they can be made oxo-biodegradable by the use of prodegradant additives (15, 16). Most active prodegradants are based on metal ions that allow redox reactions.

Organozirconates and organotitanates have been described as active prodegradants (17, 18). The prodegradant behavior of organotitanate and organozirconate adducts with organopyrophosphate compounds exhibits a remarkable synergism with regards to the compostability of olefin polymers.

The environmental fate of linear low density poly(ethylene) films designed for mulching purposes and loaded with different prodegradant additives has been studied (19).

The conditions were chosen to mimic a general crop season in the Mediterranean region. The samples showed a relatively low extent of degradation based on monitoring the carbonyl index, molecular weight variation, extractability by solvent, changes in the onset of the decomposition temperature and the crystallinity.

During a soil burial biodegradation test lasting 27 months, samples were withdrawn at time intervals and characterized by means of structural and thermal analysis. This method allows watching any progress of oxidative degradation as a direct effect of the incubation in an active microbial environment.

The data indicate an initial abiotic oxidation via a free radical chain reaction. This reaction is promoted by a prodegradant additive that acts on hydroperoxides and peroxide moieties that are present initially in the polymer bulk. Afterwards, free radical cascade reactions result in a degradation under relatively mild conditions. A synergistic effect of a microbial metabolization coupled with biotically mediated oxidation of the original abiotically fragmented samples is suspected (19).

The prodegradant effect of talc nanoparticles on the degradation process of PP films was studied (20). Two talc minerals with different intrinsic characteristics and iron content were used for the study. Films with 0%, 1% and 5% of talc were exposed to three different weathering scenarios: Indoor, outdoor and accelerated aging.

The extent of degradation in films was evaluated by analyzing the chemical and surface structure as well as optical, thermal and mechanical properties. The presence of talc induced a greater degradation in the nanocomposite films. Also, this phenomenon was increased with the concentration of the nanoparticles. Samples having talc with a higher iron content were more sensitive to weathering since iron impurities act as promoters of the degradation process.

So, an iron impure talc is a good environmentally friendly alternative as a prodegradant additive for the degradation of PP and its sustainability since it does not contain heavy metals (20).

2.2 Poly(styrene) Elastomers

Elastomeric styrene-containing compounds, such as styrene-butadiene-styrene block copolymers or styrene-isoprene-styrene block copolymers, including its hydrogenated homologues, all of them containing vegetal or even mineral oils in their formulations, can be made biodegradable (21). This is achieved by the admixture of brewer's yeast or mother yeast in amounts of 1–4%.

Advantageously, modified yeasts can be used to reduce the typical odor of yeast, such as partially hydrolyzed yeasts (22).

The yeast can be added to the raw materials before mixing and melting to obtain the composition. Alternatively, the yeast can be added before compounding process.

For example, the styrenic block copolymer, natural oil and brewer's yeast are mixed to obtain a semifluid product that is extruded at a temperature of about 120–180°C in such a way to obtain granules or pellets. The biodegradable compound can be used in an additional extrusion or molding process to get the final biodegradable product. Thus, a final polymeric material-based product is formed that biodegrades in a reasonable period of time in suitable composting sites (21).

2.3 Poly(vinyl alcohol)

The biodegradability of vinyl ester-based polymers with a special emphasis on poly(vinyl acetate) and poly(vinyl alcohol) (PVA) has been reviewed (23, 24). Also, the general physical and chemical properties of these polymers and how the relationships of these properties can influence the biodegradability were discussed.

2.3.1 *Plasticized Compositions*

For certain uses, it is desirable to prepare biodegradable PVA plasticized with biodegradable plasticizers that specifically assist in lowering the polymer melt temperature and the extrusion temperature and at the same time maximize water solubility by minimizing crystallinity. A PVA composition has been described that is transesterified with a polyol as plasticizer (25).

The melt extrusion process for making PVA films must be carefully managed, as PVA exhibits a tendency to undergo decomposition in an extruder due to screw shear and residence time under heat. The decomposition is believed to begin at less than or about 200°C.

Plasticizers are used with PVA in melt extruding to lower the extrusion temperature. Commonly, polyols such as glycerol or glycerol triacetate can be used for this purpose. However, glycerol triacetate is a poor plasticizer for PVA of higher degree of hydrolysis, because it becomes less compatible.

Examples of polyol ester plasticizers are summarized in Table 2.3. The acetate esters of glycerol are sometimes also known as monoacetin, diacetin, and triacetin.

Table 2.3 Polyol ester plasticizers (25).

Compound	Compound
Ethylene glycol monoacetate	Ethylene glycol monobutyrate
Glycerol acetate	Glycerol diacetate
Glycerol monopropionate	Glycerol dipropionate
Glycerol monolactate	Glycerol dilactate
Trimethylolpropane acetate	Trimethylolpropane diacetate
Trimethylolpropane lactate	Trimethylolpropane dilactate
Sorbitol acetate	Sorbitol diacetate
Sorbitol triacetate	Sorbitol tetra acetate
Sorbitol monolactate	Sorbitol dilactate

Chemical bonding of the plasticizer onto the main chain hydroxyl groups of the PVA would be helpful to overcome the problems of compatibility. Thus, acetate esters of glycerol are preferred, namely glycerol mono- and di-acetate and mixtures thereof. Both glycerol mono- and di-acetate are subject to transesterification both within the glycerol ester and with PVA (25). Commercially available glycerol mono- and di-acetate are mixtures with different levels of acetylation.

Blends consisting of PVA, glycerol and xanthan or gellan were prepared by laboratory extrusion (26). The mechanical properties were compared with the properties of neat PVA foil and their biodegradability in soil, compost and both activated and anaerobic sludge were assessed.

The elongation at break of all produced blends is relatively close to that of neat PVA. The biodegradability levels correspond to the content of natural components. Furthermore, the mineralization of the samples with the highest carbohydrate proportion of 42% was found to be 50–78%, depending on the type of environment. A complete biodegradation of the samples was found in activated sludge (26).

Biodegradable plastics were fabricated from a blend of jackfruit seed powder starch and PVA (27). Results concerning the mechanical properties of the fabricated biodegradable plastics show that by increasing the composition of jackfruit seed powder starch, the tensile strength of the biodegradable plastics was significantly decreased.

The biodegradability properties of the samples were determined by monitoring the structure of fabricated biodegradable plastics by using Fourier transform infrared spectroscopy (FTIR) (27). The results indicated that the structural properties were changed after eigth weeks of soil burial test by comparing them with control samples. The biodegradable plastics with a high content of jackfruit seed powder starch have better biodegradability characteristics.

Super-tough biodegradable films could be prepared using a PVA/poly(*N*-vinyl-2-pyrrolidone) (PVP) (1:1) blend with plasticizers of glycerol, sorbitol, and a one to one mixture of these compounds (28).

The effect of plasticization on the structural, thermal, and mechanical properties of the PVA/PVP blend films was studied (28). Fourier transform infrared spectra indicated good miscibility of the two components due to the hydrogen bonding between the PVA and PVP molecules. The addition of plasticizers reduced the interaction between the polymers, evidenced by an increase in the intensity of PVA diffraction peaks observed by X-ray diffraction (XRD).

The thermal degradation of the blends increased as a function of the plasticizer used. Glycerol affected the thermal degradation behavior more than sorbitol and the plasticizer mixtures. The incorporation of the plasticizers promoted the growth of PVA crystals, as evidenced by XRD patterns and the enthalpy of fusion obtained by differential scanning calorimetry (DSC) measurements. The introduction of sorbitol to the binary blend increased the toughness by seven times and imparted simultaneous and pronounced improvements to maximum tensile stress and elongation at break (28).

In-situ nanofibrillar reinforced blends with thermoplastic PVA as matrix and poly(lactic acid) (PLA) as dispersed fibrils were successfully prepared by a simple extrusion-stretching process (29). The microstructure and mechanical properties of the melt processed thermoplastic PVA/PLA composites were characterized by FTIR, thermogravimetry scanning electron microscopy (SEM) and tensile tests. The results indicated that there is hydrogen bonding between thermoplastic PVA and PLA which enhances PLA dispersion as well as the interfacial adhesion with thermoplastic PVA matrix.

The PLA content was varied from 10% to 30%, resulting in reinforced thermoplastic PVA matrices with fibrils of 78 to 150 *nm*. Compared with thermoplastic PVA, thermoplastic PVA/PLA composites exhibit remarkable improvement in mechanical properties and thermal stability (29).

Glycerol and PVA are commonly used as plasticizers of starch (30). Their relatively low molecular weight, compared with starch, contributes to a good processability. Rosin is a renewable product whose incorporation in the starch/PVA matrix induces processing aid and reinforcing effects. Its relatively high molecular weight might prevent its migration to the surface of the final product.

Biodegradable potato starch/PVA samples containing different concentrations of rosin were prepared by melt mixing in order to study the enhancement of the properties of native starch films (30).

Water content, solubility in water, mechanical properties, microstructure and dynamic mechanical analysis of the samples were studied. The addition of 8% rosin to starch/PVA blends led to tensile strength values higher than 10 *MPa* and elongation at break values close to 2000%, values comparable to those offered by conventional polymers used in food packaging, e.g., low density poly(ethylene). Furthermore, starch compounds have a low cost and a high biodegradability (30).

Biodegradable nanocomposites were prepared from plasticized starch-polyvinyl alcohol composite incorporated with cellulose nanocrystals (31). The nanocomposite films were characterized by mechanical, thermal, and barrier analysis methods, as well as atomic force microscopy (AFM). The geometrical characteristics of the cellulose nanocrystals confirmed that the thickness of cellulose nanocrystals was less than 100 *nm*.

The generated nanocomposites exhibited an improvement in the mechanical and barrier properties. The films showed decrease in solubility, water absorption, water vapor permeability, and strain at break. However, contact angle, ultimate tensile strength, glass transition temperature, and melting point increased after the addition of the cellulose nanocrystals. The results suggest that cellulose nanocrystals can be a good replacement for mineral reinforcement such as silicates (31).

2.3.2 *Hydrogels*

Poly(vinyl alcohol) (PVA) hydrogels offer many advantages over poly(ethylene glycol) (PEG)-based hydrogels. Here, pendant hydroxyl groups along the backbone imparts more versatility with respect to various modifications of the macromer. This may be the attachment of degradable segments, active agents, or hydrophobic groups.

Biodegradable and biocompatible hydrogels based on PVA and methods of preparation have been described (32). The hydrogels can be used for a number of biomedical applications, including implants, embolic agents, wound-healing dressings, adhesion prevention, sealants, bulking agents, coatings for biomaterials, and delivery of biologically active compounds.

The hydrogels can be formed either *in vivo* or *ex vivo*. One method of preparation involves the use of prepolymers. The prepolymers have a PVA backbone and pendant moieties bearing a polymerizable group.

The hydrogels can be designed to degrade as fast as a few hours to more than one year. The use of a hydroxyethyl methacrylate lactate as the biodegradable crosslinkable groups will provide a faster degradation than the use of a methacrylate ester such as 2-(methacryoyloxy)ethyl succinate (32).

Other factors that will affect the degradation rate are the density of the pendant chain bearing the degradable region, the length of the degradable region, the hydrophobicity of the network, and the crosslinking density (32).

Hydrogels are a promising candidate for applications in biomedicine and bioengineering, but their mechanical properties often restrict their applications. To improve the mechanical performance of PVA hydrogels, sodium carboxymethyl cellulose (CMC) and graphene oxide were introduced into them (33).

A series of composite hydrogels were prepared, composed of PVA, CMC, and graphene oxide with epichlorohydrin as a chemical crosslinking agent. FTIR and XRD were used to characterize the chemical structures of graphene oxide and the hydrogel. Dynamic mechanical analysis results showed the synergistic enhancement effects of CMC and graphene oxide on the PVA hydrogel.

The swelling process of the hydrogels also fit well with the second-order kinetic equation. SEM results suggested that the neat mesh structure facilitated superior mechanical properties in the hydrogels (33).

PVA/chitosan) hydrogels containing 1% and 3% of lignin nanoparticles were prepared with a freeze-thaw procedure (34). Results from microstructural, thermal and mechanical characterization of lignin-nanoparticles-based PVA/chitosan demonstrated that the lowest amount of lignin nanoparticles of 1% was beneficial, whereas the presence of agglomerates at a higher lignin nanoparticle content limited the effect.

A different swelling behavior was observed for hydrogels containing lignin nanoparticles with respect to PVA/chitosan, due to the formation of a porous honeycomblike structure (34). A synergic effect of chitosan and lignin nanoparticles was revealed in terms of antioxidative response by 1,1-diphenyl-2-picryl-hydrazyl, c.f., Figure 2.1, activity of migrated substances, whereas results from antimicrobial tests confirmed lignin nanoparticles as effective against gram-negative bacteria (*E. coli*) when compared to gram-positive (*S. aureus* and *S. epidermidis*) strains. The obtained results suggested the possible use of produced PVA/chitosan hydrogels incorporating lignin nanoparticles in many different sectors such as drug delivery, food packaging and wound dressing (34).

2.3.3 *Antibacterial Film*

The synthesis of an antibacterial biodegradable nanocomposite film of PVA-biogenic silver has been reported (35). The biosynthesized

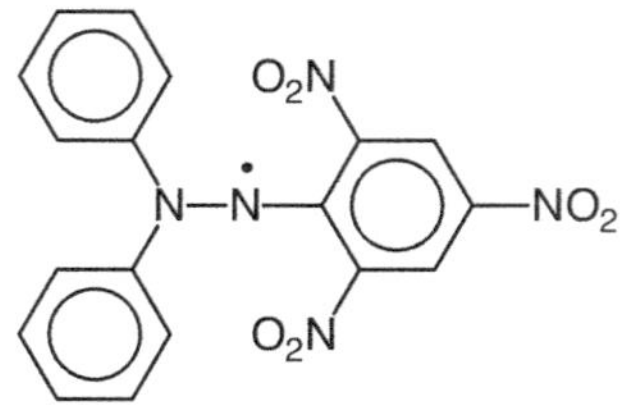

Figure 2.1 1,1-Diphenyl-2-picryl-hydrazyl.

silver nanoparticles were used for development of nanocomposite film. A *Ficus benghalensis* leaf extract was used to synthesize silver nanoparticles.

Fresh leaves of *F. benghalensis* were collected and thoroughly washed by deionized water and dried at room temperature. Then, 25 *g* of leaves were cut and boiled in deionized water to prepare the leaf extract. A 1 *mmol* silver nitrate solution was prepared in 45 *ml* of deionized water. Next, 5 *ml* leaf extract was added drop by drop into the silver nitrate solution with constant stirring at 50°C–60°C for the preparation of the silver nanoparticles.

Nanocomposite film was prepared by taking 8% of PVA dissolved in deionized water at 90°C–100°C at constant stirring. After some time a transparent solution was prepared. Subsequently, the biosynthesized silver nanoparticles colloidal solution was added into the PVA solution at a concentration of 0 μl, 200 μl, and 500 μl. Then, the solution was transferred onto a Teflon coated glass plate and dried at 80°C. The so-prepared films with different concentrations of silver nanoparticles were washed with water and further dried at room temperature.

The above-described method is cost-effective, environmental friendly and rapid. The PVA-biogenic silver nanocomposite film was characterized by AFM to determine the morphology of the film surface and to confirm the incorporation of the silver nanoparticles.

The antibacterial activity of the nanocomposite film was investigated against *Salmonella typhimurium* by means of disk diffusion method. The PVA-biogenic silver nanocomposite film shows excellent antibacterial activity against *S. typhimurium*. This nanocomposite film could be a prominent material for food packaging (35).

2.4 Poly(vinyl butyral)

Technically, poly(vinyl butyral) (PVB) is produced from PVA by an acetalization process with butanal. The reaction is shown in Figure 2.2.

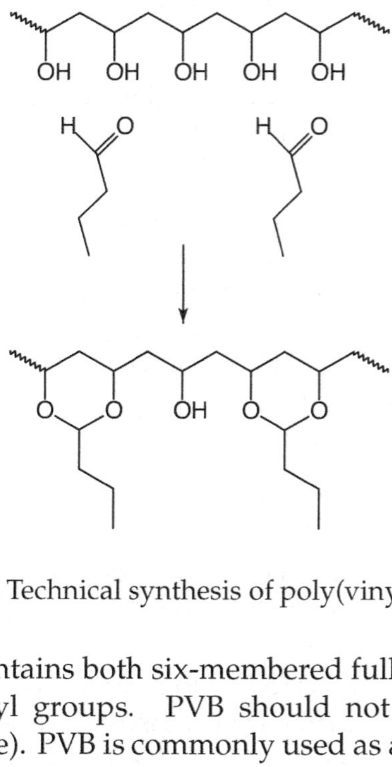

Figure 2.2 Technical synthesis of poly(vinyl butyral).

The polymer contains both six-membered full acetal groups separated by hydroxyl groups. PVB should not be confused with poly(vinyl butyrate). PVB is commonly used as a hot melt adhesive for safety glasses.

PVB is a natural polymer produced in *Burkholderia sacchari* bacteria. Biofilms can be produced by casting processes. The films have been characterized by a variety of methods. The properties are highly influenced by the presence of a plasticizer, i.e., PEG. The thermal properties are comparable to domestic and industrial packing materials (36).

The direct electrospinning of PVB onto human dermal fibroblasts using a portable device has been reported (37). Electrospinning was carried out by using a handheld electrospinning device. The device contained a linear actuator to form a syringe pump and a converter model to provide high voltage. A pack of 10 rechargeable AA batteries powered the device. A pen extension could be attached onto the positive electrode through a Teflon tube. A metal plate was used as the ground electrode (37, 38).

For applications in ultrafiltration membranes, bioimaging, and diagnostics, a special method of synthesis of PVB has been suggested (39). There, swollen PVA particles are added to a butyraldehyde solution in ethanol. This method gives polymers with high contents of butyral moieties up to 84%.

A maximum of butyral groups are achieved using swollen PVA particles at 95 *vol*% and pH 1. The reverse dissolution method has several advantages, such as (39):

- Mild reaction condition of 70°C,
- No predissolution of PVA at high temperatures,
- No vacuum removal of water during reaction, and
- Low water consumption in the post-treatment stage.

Acetal compounds based on PVA are widely used. The biodegradation behavior of PVA modified by formaldehyde, *n*-butyraldehyde, glyoxaldehyde and glutaraldehyde was investigated (40). It has been established that the biodegradation levels of all the acetal modified samples decreased.

The level of biodegradation seems to be systematically dependent on the degree of the acetal of PVA. For example, the biodegradation of poly(vinyl formal) is better than poly(vinyl butyral) with the same degree of acetals. Furthermore, the biodegradation levels of poly(vinyl glyoxal) are lower than poly(vinyl glutaral), which has the same degree of crosslinking (40).

2.4.1 *Blends with Poly(3-hydroxybutyrate)*

Miscibility and the morphology of blends of poly(3-hydroxybutyrate) (PHB) and PVB were studied by DSC and SEM (41).

PHB is a semicrystalline polymer while PVB can be interpreted as a random copolymer that contains both vinyl butyral and vinyl alcohol moieties.

Two glass transition temperatures can be seen in the amorphous phase for all PVB and PHB blends. This suggests an immiscibility of the polymers. However, partial miscibility was observed in 1:1 blends of PVBs with a content of 25–36% vinyl alcohol. This was concluded by an inward shift of the glass transition temperatures of the two phases relative to the glass transition temperatures of the neat components.

2.4.2 *Blends with Poly(lactic acid)*

The miscibility of solution cast blends of PLA and PVB has been reported (42). PLA was blended with PVB through a solution casting method using chloroform as solvent.

The films obtained were characterized for miscibility by DSC, tensile testing and FTIR spectroscopy. DSC measurements showed that the glass transition temperature associated with PLA and PVB are not dependent on the composition of the blend. Two glass transition temperatures in the blends indicate an immiscibility of the base materials. Mechanical analysis showed that the tensile strength and elongation decreased due to blending (42).

Blends of poly(caprolactone) (PCL) with PVB were studied for biodegradation in soil and by bacterial strains of *Bacillus subtilis* and *Escherichia coli*. Films of PCL/PVB blends of different compositions were buried in soil in the laboratory and in the outside environment (43, 44).

Blends with a low polyester concentration of less than 30% PCL were clear and transparent and no spherulite formation was observed. Above 30% PCL spherulites could be observed. The size of the spherulites increased with increasing PCL concentration.

The films were examined for changes in the physical appearance and mass loss every week. Weight loss was observed in all the blends. Blends rich in PCL showed more degradation, which was faster in the natural environment than in the laboratory.

The physical appearance and microscopic examination revealed that the films deteriorated in soil. Blends in the *B. subtilis* strain showed more degradation as compared to the *E. coli* strain (44).

Blends of PCL and cottonseed protein (45) plasticized with cottonseed oil were made and analyzed for their mechanical, adhesive, and thermal properties (46).

The addition of water-washed cottonseed meal to PCL increased the Young's modulus but decreased the tensile strength and elongation at break of PCL. Furthermore, the addition of cottonseed oil to the PCL/water-washed cottonseed meal blend kept the tensile strength about the same but enhanced the elongation. The PCL blends with water-washed cottonseed meal and cottonseed protein isolate gave about the same mechanical properties, both somewhat better than the PCL/soy protein isolate blend.

As plasticizers, cottonseed oil performed slightly better than coconut oil, both better than PEG. The addition of water-washed cottonseed meal and cottonseed oil, up to a PCL:water-washed cottonseed meal:plasticizer ratio of 60:40:20, did not change the adhesive performance of PCL on fiberboard. Thus, the combination of PCL/cottonseed protein/cottonseed oil seems to be a viable bioplastic, and one possible application for this material may be in the hot melt adhesive area (46).

2.4.3 *Paper Coatings*

Biopolymeric coatings were applied to paperboard in order to improve its water barrier ability (47). Two coating solution compositions were prepared: Whey protein isolate/cellulose-based films and PVB/zein films.

For the first type, whey protein isolate and glycerol were dissolved in water with cellulose xanthate. In addition, glutaraldehyde was added as crosslinking agent. The solution was cast, dried, and insolubilized by entrapment in regenerated cellulose. The films were combined with beeswax into a bilayer coating system and then applied to paperboard by heating compression.

The second coating solution was prepared by dissolving PVA and zein in 70% ethanol with glutaraldehyde and butyraldehyde. The PVB/zein solution could be sprayed on the paperboard.

Both treatments improve the water barrier properties of paperboard by decreasing the water vapor transmission rate by 77–78%. The results suggest that the surface coating by biodegradable polymers may be utilized for the manufacture of paperboard containers even in industrial applications (47).

The water vapor transmission rates of poly(3-hydroxybutyrate-*co*-4-hydroxybutyrate) (PHBV) and PHBV/nanocomposite-coated papers were measured at various levels of relative humidity and temperature (48). The coating of PHBV on base paper was performed with two different methods, and the more effective one for lowering water vapor transmission rate values was utilized for coating the PHBV/nanocomposites on the paper.

Nanocomposites of PHBV were prepared with a commercial montmorillonite Cloisite 30B, and a novel modified clay was obtained via a solution blending process. To prepare a series of modified clays, β-butyrolactone was intercalated in the Cloisite 30B clay by a ring-opening polymerization (48). The morphology of the clay and nanocomposites was revealed by XRD. Additionally, the dispersion and phase behavior of the clay in the PHBV matrix was observed with transmission electron microscopy.

It was found that the coating method and clay exfoliation were the most important factors affecting water vapor permeability. The water vapor barrier of the coated papers was improved significantly if the surface of the base substrate was pre-spray-coated with a PHBV suspension prior to the laminate coating of the PHBV film with a hot press. The papers coated with exfoliated PHBV/nanocomposites exhibited even lower water vapor transmission rate values. Overall, the coating treatments lowered the water vapor transmission rate values (48).

2.4.4 *Fibers*

Electrospinning of a suspension of hydroxyapatite and PVB leads to the formation of fibers with an average diameter of 937–1440 *nm* (49).

The diameter of the fiber was found to be influenced by the applied voltage and the spinning distance. In contrast, the injection flow rate did not significantly affect the diameter. The electrospinning method can successfully reduce the commercial hydroxyapatite particle size in the range of 400–1100 *nm* to smaller than 100 *nm*.

These fibers were converted into hydroxyapatite nanoparticles with a size smaller than 100 *nm* by a calcination treatment at 600°C. The dispersion of the finally calcined hydroxyapatite nanoparticles was improved significantly when an anionic surfactant, sodium dodecyl sulfate, was used (49).

Charged PVB hollow fiber ultrafiltration membranes were fabricated with perfluorosulfonic acid as additive by a wet-spinning method (50). Perfluorosulfonic acid with *N*-methyl-2-pyrrolidone promotes an instantaneous demixing mechanism and the precipitation rate increases with the addition of perfluorosulfonic acid. Fibers fabricated with 50% ethanol in aqueous solution as a bore liquid for oil well applications show an asymmetrical double fingerlike structure. When the fibers are fabricated with a 90% *N*-methyl-2-pyrrolidone aqueous solution as bore liquid a single fingerlike structure is formed. The break strain and Young´s modulus values of the fibers increase when the perfluorosulfonic acid content increases regardless of the bore fluid (50).

Hollow fiber membranes from PVA were fabricated (51). There, the streaming potential was studied using different electrolyte solutions, including NaCl, KCl, $CaCl_2$, and $MgCl_2$.

The effects of ionic strength, ion valence and pH value on the streaming potential of the fiber membrane were investigated. It was demonstrated that the PVB membrane accumulates a weak negative charge due to the specific adsorption of ions.

The streaming potential, the ζ-potential and the surface charge density of the membrane depend strongly on the salt concentration and the type and valence of ions. An isoelectric point of around 3.0 in the monovalent salt solutions and 3.5 in divalent electrolytes was found (51).

Electrospun nanofibers of several polymers, including PVB, were used in a surface acoustic wave resonator (52). The surface acoustic wave resonator is used as an ultrafast response humidity sensor.

The experiments indicated that PVB is suitable as a matrix to form nanofibers together with poly(aniline) (PANI) by electrospinning. Electrospun PANI/PVB nanofibers exhibited a very high sensitivity of 75 kHz/% relative humidity from 20–90% relative humidity. Furthermore, an ultrafast response of 1–2 s for humidification and desiccation was reported. In addition, the sensor could detect humidity levels as low as 0.5% relative humidity (52).

2.4.5 *Membranes*

Microporous PVB membranes could be prepared by supercritical CO_2-induced phase separation (53). Characterization be SEM indicated that the process parameters, such as the concentration of PVB in the casting solution, the CO_2 pressure, and also temperature and depressurization time, had significant effects on the structure of the membrane.

An increase of the PVB concentration decreases the pore size. The top surface of the membrane becomes dense when temperature and CO_2 pressure is increased. An increase of the depressurization time causes a porous top surface (53).

PVB and cationic PVB was blended with poly(vinylidene fluoride) (PVDF). These blends were used to fabricate hydrophilic ultrafiltration membranes (54). A visual inspection method and a glass transition temperature method were applied to study the miscibility of the blends. PVDF is immiscible with PVB, but with cationic PVB it is partially miscible.

Dynamic contact angle experiments showed that the hydrophilicity of the blend membranes was significantly improved with the addition of PVB and cationic PVB. The permeation of water increases with the content of PVB and cationic PVB (54).

A thermally induced phase separation method was adopted to manufacture a PVDF blend hollow fiber microfiltration membrane with a tenon-connection structure, which is used as a fastener in Chinese traditional building by the stronger friction between rabbet and mortise, for water treatment by blending PVB (55).

In this microstructure formation process, both the distribution of PVB within the matrix, and its regulatory role were investigated. The comprehensive effect of rabbeting and PVB adhesion on strength was elucidated. The maximum breaking strength reached 17.08 MPa.

The results showed that a suitable PVB will markedly reduce the thickness of skin layer close to the outside surface even for the solid-liquid phase separation while the excessive PVB results in a thick skin layer (55). Besides the strength, PVB markedly improved the hydrophilicity and permeability of the membrane.

The breaking strength is more than 8 MPa, the maximum pure water flux is as high as 929.23 $l m^{-1} h^{-1}$ and the rejection rate of carbon particles was found to be nearly 100%; even the polymer content was only 22%. In addition, the membrane showed a β-phase and an improved antifouling capacity.

In summary, a simple way was presented to regulate microstructure, enhance mechanical strength, flux and antifouling capacity of the membrane via the thermally induced phase separation process for water treatment (55).

2.4.6 *Solar Cells*

The use of solar cells is rapidly expanding because they provide a sustainable energy resource (56). Solar cells can typically be categorized into two types based on the light-absorbing material used, i.e., bulk or wafer-based solar cells and thin film solar cells.

Wafer-based solar cells often comprise a series of about 180 and about 240 μm thick self-supporting wafers that are soldered together.

Such a panel of solar cells is encapsulated by polymeric encapsulants to form a solar cell assembly. This assembly may be further sandwiched between two protective outer layers to form a weather resistant module. The protective outer layers are formed from sufficiently transparent material to allow photons to reach the solar cells. In modules, where PVB is used as the encapsulant material, it has been found that the PVB tends to discolor over time when in contact with an oxidizable metal component.

2.4.6.1 Chelating Agents and UV Absorbers

Formulations that contain a chelating agent and an UV absorber have been developed to overcome these difficulties (56). Suitable chelating agents and ultraviolet absorbers are listed in Table 2.4. Some of these compounds are shown in Figure 2.3.

Table 2.4 Chelating agents and UV absorbers (56).

Chelating agent	UV absorber
Ethylenediamine tetraacetic acid	2,2,6,6-Tetramethylpiperidine
Ethylene diamine monoacetic acid	2,2,6,6-Tetramethylpiperidinol
Ethylene diamine diacetic acid	2-(Dimethylamino) pyridine
Ethylene diamine triacetic acid	4-(Dimethylamino) pyridine
Ethylene diamine	*N*-Butyl piperidine
Tris(2-aminoethyl)amine	*N,N*-Diethyl cyclohexylamine
Diethylenetriaminepentacetic acid	

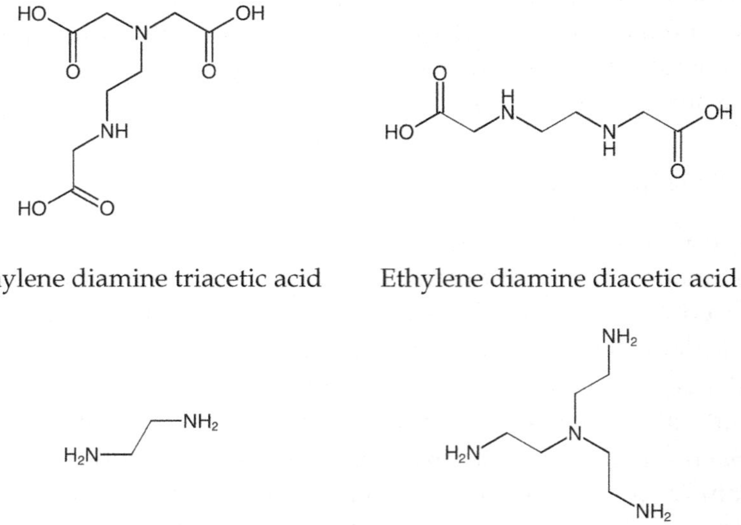

Ethylene diamine triacetic acid | Ethylene diamine diacetic acid

Ethylene diamine | Tris(2-aminoethyl)amine

Figure 2.3 Chelating agents.

Ethylenediamine tetraacetic acid (EDTA) is preferably used as chelating agent. EDTA effects a greatly reduced yellowness index.

2.4.6.2 *Dye-Sensitized Solar Cells*

Due to liquid electrolyte limitation in dye-sensitized solar cells, gel polymer electrolytes were prepared and optimized (57). Poly(vinyl butyral-*co*-vinyl alcohol-*co*-vinyl acetate) acted as host polymer and tetrapropylammonium iodide as dopant salt. The prepared gel polymer electrolytes with different tetrapropylammonium iodide salt content were characterized using electrochemical impedance spectroscopy, FTIR and X-ray diffractogram.

An optimum ionic conductivity of 1.93 $mS\,cm^{-1}$ was obtained with 40% of tetrapropylammonium iodide salt added along with the lowest activation energy of 0.064 eV. The temperature-dependent ionic conductivity behavior of the gel polymer electrolytes was studied and it was found that the systems obeyed Arrhenius behavior in the temperature range of 303–373 K. Structural studies of gel polymer electrolytes were performed using XRD and FTIR and confirmed the existence of the complexation of the copolymers and the tetrapropylammonium iodide salt (57).

Gel polymer electrolytes were used to assemble the dye-sensitized solar cells. A power conversion efficiency of 4.615% was achieved with a maximum short circuit current density of 13.585 $mA\,cm^{-2}$, an open circuit voltage (Voc) of 0.678 V and a fill factor of 50.1% at 40% tetrapropylammonium iodide salt under AM 1.5 (100 $mW\,cm^{-2}$) illumination (57).

2.4.7 *Adhesive for Safety Glass*

The increase in fatal road accidents, natural disasters, and even terrorist attacks around the world have contributed to the improvement of public security. Windows can be particularly hazardous because of cutting fragments expelled during breakage or an explosion, which may induce injury (58).

PVB is commonly used in the manufacture of polymer sheets that can be used as interlayers in light-transmitting laminates, such as safety glass or polymeric laminates (59). Safety glass is often used to provide a transparent barrier in architectural and automotive openings. Its main function is to absorb energy, such as that caused by a blow from an object, without allowing penetration through the opening or the dispersion of shards of glass, thus minimizing damage or injury to the objects or persons within an enclosed area.

Plasticized PVB inherently tends to stick to itself during storage or transportation. The strength of the self-adhesion of plasticized PVB can reach such levels that it is impossible to separate if the sheets or pellets are not refrigerated.

This blocking creates great difficulties in the handling of the sheets during manufacture, transportation, and lamination. Also, it becomes difficult to feed PVB pellets continuously into an extruder.

Actually, these problems can be minimized by formulations to which an antiblocking agent is added. Useful antiblocking agents may be bifunctional surface-modifying agents that comprise an antiblocking segment and a compatibility segment.

Fatty acid amides are an important class of polymer additives used as slip agents or lubricants to prevent unwanted adhesion. Actually, the addition of such amides for the purpose of antiblock additives for polymer sheets, has often been found to adversely affect the optical characteristics of the polymer sheet in glass laminates as well as adhesion of the polymer sheet to glass.

However, a recent study showed that it is possible to get amide-based formulations that do not suffer from these drawbacks. The amides are used together with a plasticizer. Suitable amides are shown in Table 2.5.

Table 2.5 Amides as slipping agents (59).

Amide
Erucamide
Behenamide Oleyl palmitamide
Stearyl erucamide
Erucyl stearamide
Hydroxystearamide
Oleic acid diethanolamide
Stearic acid diethanolamide
Poly(ethylene glycol) oleic amide
Octadecanamide

Secondary monoamides particularly are preferred. A particularly preferred secondary monoamide is *N*-oleyl palmitamide, an amide with a double bond geometry. Preferably the amide is incorporated into the plasticized PVB at a concentration of 0.1–0.3%. The glass adhesion of a PVB sheet produced from the formulated pellets is substantially unaffected by the presence of the amide antiblocking agent (59).

Nanocellulose was used as additive for security glazing. It has a low ecological footprint, improved safety glazing properties and was based on PVB (58). Following the processing of many different recipes for layers based on both PVB and nanocellulose polymers, intercalary films were assembled with glass using hot pressing. The results of the three-point bending experiments were promising. Breaking loads were approximately 8000 *N* for the two nanocellulose samples, which were close to the results of the sample with PVB only. Furthermore, the obtained composites possessed a transparency close to that of PVB alone. Finally, nanocellulose overtop PVB had a surface mass as low as one eighth of that of the PVB (58).

Trade names appearing in the references are shown in Table 2.6.

Table 2.6 Trade names in references.

Trade name Description	Supplier
Bio-PDO™	DuPont
1,3-Propanediol from biodegradable polymer (7)	
Butvar™	Solutia Inc.
Poly(vinyl butyral) (59)	
Ecocite®	DuPont
Poly(vinyl butyral) copolymer (25)	
Elvaloy® (Series)	DuPont
n-Butyl acrylate copolymers (25)	
Elvanol® (Series)	DuPont
Poly(vinyl alcohol) (25, 32)	
Gelvatol®	Monsanto
Poly(vinyl alcohol) (32)	
Luviskol® VA 73 W	BASF AG
50% Solution of a copolymer of vinylpyrrolidone and vinylacetate (70:30) in water (32)	
Mowiol®	Kuraray Europe GmbH
Poly(vinyl alcohol) (32)	
Polyviol®	Wacker
Poly(vinyl alcohol) (32)	
Vinol® 107	Air Products
Hydrolyzed poly(vinyl alcohol) (32)	

References

1. S.P. Lewis and R.T. Mathers, Advances in acid mediated polymerizations in V. Mittal, ed., *Renewable Polymers: Synthesis, Processing, and Technology*, chapter 3, pp. 69–173. Wiley Blackwell (John Wiley & Sons), October 2011.
2. R.T. Mathers and S.P. Lewis, *Journal of Polymer Science Part A: Polymer Chemistry*, Vol. 50, p. 1325, April 2012.
3. S. Kasirajan and M. Ngouajio, *Agronomy for Sustainable Development*, Vol. 32, p. 501, April 2012.
4. S. Bandopadhyay, L. Martin-Closas, A.M. Pelacho, and J.M. DeBruyn, *Frontiers in Microbiology*, Vol. 9, p. 819, April 2018.
5. A. Hidangmayum, A. Singh, V. Kumar, and P. Dwivedi, *EurAsian Journal of BioSciences*, Vol. 12, p. 487, 2018.
6. G.E. Luckachan and C.K.S. Pillai, *Journal of Polymers and the Environment*, Vol. 19, p. 637, 2011.
7. S. Sumanam, Biodegradable polymer composition, US Patent 8 026 301, assigned to BNT Force Biodegradable Polymers Pvt Ltd. (Chennai, IN), September 27, 2011.
8. L. Papadopoulos, A. Magaziotis, M. Nerantzaki, Z. Terzopoulou, G.Z. Papageorgiou, and D.N. Bikiaris, *Polymer Degradation and Stability*, Vol. 156, p. 32, 2018.
9. S. Qiu, K. Zhang, Z. Su, and Z. Qiu, *Polymer Testing*, Vol. 66, p. 64, 2018.
10. O. Valerio, M. Misra, and A.K. Mohanty, *Journal of Polymers and the Environment*, Vol. 26, p. 1078, 2018.
11. Subcommittee: D20.96, Standard test method for determining aerobic biodegradation of plastic materials in soil, ASTM Standard D5988-12, ASTM International, West Conshohocken, PA, 2012.
12. R. Scaffaro, A. Maio, F. Sutera, E.F. Gulino, and M. Morreale, *Polymers*, Vol. 11, p. 651, April 2019.
13. P.L. Chee, D.J. Young, and X.J. Loh, Degradation behaviour of biodegradable thermogels in X.J. Loh and D.J. Young, eds., *Biodegradable Thermogels*, Vol. 2 of *Biomaterials Science Series*, chapter 7, pp. 113–132. Royal Society of Chemistry, 2019.
14. I. Jakubowicz, N. Yarahmadi, and H. Petersen, *Polymer Degradation and Stability*, Vol. 91, p. 1556, July 2006.
15. A.-C. Albertsson, C. Barenstedt, and S. Karlsson, *Polymer Degradation and Stability*, Vol. 37, p. 163, January 1992.
16. A.J. Sipinen and D.R. Rutherford, *Journal of Environmental Polymer Degradation*, Vol. 1, p. 193, July 1993.
17. R.F. Grossman, Compostable olefin polymer compositions, composites and landfill biodegradation, US Patent Application 20 090 253 324, assigned to Biotech Products, LLC, Randolph (NJ), October 8, 2009.

18. R.F. Grossman, Compostable olefin polymer compositions, WO Patent Application 2 009 126 178, assigned to Biotech Products LLC, October 15, 2009.
19. A. Corti, M. Sudhakar, and E. Chiellini, *Journal of Polymers and the Environment*, Vol. 20, p. 1007, December 2012.
20. P.B. Linares, L.A. Castillo, and S.E. Barbosa, *Journal of Polymers and the Environment*, Vol. 27, p. 1, 2019.
21. B. Cardinali, Biodegradable polymers and compounds, US Patent 8 349 914, assigned to Technofilm S.p.A. (Sant'Elpidio a Mare, IT), January 8, 2013.
22. B. Cardinali, Biodegradable product obtained from compounds of thermoplastic polymers, US Patent 8 293 814, assigned to Tecnofilm S.p.A. (Sant Elpidio a Mare, IT), October 23, 2012.
23. M. Amann and O. Minge, Biodegradability of poly(vinyl acetate) and related polymers in B. Rieger, A. Künkel, G.W. Coates, R. Reichardt, E. Dinjus, and T.A. Zevaco, eds., *Synthetic Biodegradable Polymers*, Vol. 245 of *Advances in Polymer Science*, pp. 137–172. Springer-Verlag, November 2012.
24. M. Teodorescu, M. Bercea, and S. Morariu, *Polymer Reviews*, Vol. 58, p. 247, 2018.
25. D.C. Urian, P.A. Morken, and D.L. Visioli, Poly(vinyl alcohol) composition comprising a polyol, US Patent 7 781 506, assigned to E.I. du Pont de Nemours and Company (Wilmington, DE), August 24, 2010.
26. M. Kopčilová, J. Hubáčková, J. Růžička, M. Dvořáčková, M. Julinová, M. Koutný, M. Tomalová, P. Alexy, P. Bugaj, and J. Filip, *Journal of Polymers and the Environment*, Vol. 21, p. 88, March 2013.
27. F.A. Lothfy, M.F. Haron, and H.A. Rafaie, *Journal of Polymer Science and Technology*, Vol. 3, p. 1, 2019.
28. P. Boonsuk, K. Kaewtatip, S. Chantarak, A. Kelarakis, and C. Chaibundit, *Journal of Applied Polymer Science*, Vol. 135, p. 46406, 2018.
29. Z. Guo, Y. Ba, D. Liu, and W. Pan, *International Journal of Nanomanufacturing*, Vol. 14, p. 291, 2018.
30. D. Domene-López, M. Guillén, I. Martin-Gullon, J. García-Quesada, and M. Montalbán, *Carbohydrate Polymers*, Vol. 202, p. 299, 2018.
31. N. Noshirvani, W. Hong, B. Ghanbarzadeh, H. Fasihi, and R. Montazami, *International Journal of Biological Macromolecules*, Vol. 107, p. 2065 , 2018.
32. T. Hirt, T. Holland, V. Francis, and H. Chaouk, Degradable poly(vinyl alcohol) hydrogels, US Patent 6 710 126, assigned to BioCure Inc. (Norcross, GA), March 23, 2004.
33. F. Xue, X. He, S. Cai, J. Nie, Z. Shi, and X. Wang, *Journal of Applied Polymer Science*, Vol. 136, p. 47644, 2019.

34. W. Yang, E. Fortunati, F. Bertoglio, J. Owczarek, G. Bruni, M. Kozanecki, J. Kenny, L. Torre, L. Visai, and D. Puglia, *Carbohydrate Polymers*, Vol. 181, p. 275, 2018.
35. R.M. Tripathi, R.N. Pudake, B.R. Shrivastav, and A. Shrivastav, *Advances in Natural Sciences: Nanoscience and Nanotechnology*, Vol. 9, p. 025020, June 2018.
36. D.F. Parra, J.A. Fusaro da R. Rodrigues, P. Ponce, and A.B. Lugão, *Pakistan Journal of Biological Sciences*, Vol. 8, p. 1041, 2005.
37. C.-Y. Chui, P.-A. Mouthuy, and H. Ye, *Biotechnology letters*, Vol. 40, p. 737, 2018.
38. P.-A. Mouthuy, L. Groszkowski, and H. Ye, *Biotechnology Letters*, Vol. 37, p. 1107, 2015.
39. B. Yang, R. Liu, J. Huang, and H. Sun, *Industrial & Engineering Chemistry Research*, Vol. 52, p. 7425, June 2013.
40. H.Z. Zhang, B.L. Liu, R. Luo, Y. Wu, and D. Lei, *Polymer Degradation and Stability*, Vol. 91, p. 1740, 2006.
41. W. Chen, D.J. David, W.J. MacKnight, and F.E. Karasz, *Polymer*, Vol. 42, p. 8407, September 2001.
42. J.R. Khurma, D.R. Rohindra, and R. Devi, *The South Pacific Journal of Natural and Applied Sciences*, Vol. 23, p. 22, 2005.
43. D.R. Rohindra and J.R. Khurma, *The South Pacific Journal of Natural and Applied Sciences*, Vol. 21, p. 47, 2003.
44. D.R. Rohindra, P. Sharma, and J.R. Khurma, *Macromolecular Symposia*, Vol. 224, p. 323, April 2005.
45. Wikipedia contributors, Cottonseed — Wikipedia, the free encyclopedia, https://en.wikipedia.org/w/index.php?title=Cottonseed&oldid=883059157, 2019. [Online; accessed 15-June-2019].
46. H.N. Cheng, C.V. Ford, and Z. He, *International Journal of Polymer Analysis and Characterization*, Vol. 24, p. 389, 2019.
47. J. Han, S. Salmieri, C. Le Tien, and M. Lacroix, *Journal of Agricultural and Food Chemistry*, Vol. 58, p. 3125, March 2010.
48. M. Farmahini-Farahani, H. Xiao, A. Khan, Y. Pan, and Y. Yang, *Industrial & Engineering Chemistry Research*, Vol. 54, p. 11277, 2015.
49. S.M. Zakaria, S.H. Sharif Zein, M.R. Othman, and J.A. Jansen, *Journal of Biomedical Materials Research Part A*, Vol. 101A, p. 1977, July 2013.
50. W.-Z. Lang, J.-P. Shen, Y.-X. Zhang, Y.-H. Yu, Y.-J. Guo, and C.-X. Liu, *Journal of Membrane Science*, Vol. 430, p. 1, March 2013.
51. J. Qi, Y.-q. Wang, and Y.-r. Qiu, *Journal of Central South University*, Vol. 20, p. 1490, June 2013.
52. Q. Lin, Y. Li, and M. Yang, *Analytica Chimica Acta*, Vol. 748, p. 73, October 2012.
53. Z. Li, H. Tang, X. Liu, Y. Xia, and J. Jiang, *Journal of Membrane Science*, Vol. 312, p. 115, April 2008.

54. B.Y. Wan, J. Liao, Z. Chen, G.Q. Wang, L. Wang, and J.L. Wang, *Journal of Macromolecular Science, Part B*, Vol. 51, p. 1851, 2012.
55. Z. Cui, W. Li, H. Zeng, X. Tang, J. Zhang, S. Qin, N. Han, and J. Li, *European Polymer Journal*, Vol. 111, p. 49, 2019.
56. R.L. Smith, K.M. Stika, and J.S. Wall, Poly(vinyl butyral) encapsulant comprising chelating agents for solar cell modules, US Patent 8 338 699, assigned to E I du Pont de Nemours and Company (Wilmington, DE), December 25, 2012.
57. N.K. Farhana, S. Ramesh, and K. Ramesh, *Materials Science in Semiconductor Processing*, Vol. 91, p. 414 , 2019.
58. C. Maury, F. Crispino, and Éric Loranger, *BioResources*, Vol. 14, p. 4157, 2019.
59. W. Chen, A.N. Smith, and A. Karagiannis, Poly(vinyl butyral) pellets, US Patent 7 491 761, assigned to Solutia Incorporated (St. Louis, MO), February 17, 2009.

3

Acid and Lactone Polymers

3.1 Poly(lactic acid)

The synthesis, structure and properties of poly(lactic acid) (PLA) have been summarized in a monograph (1). The opportunities and challenges of the production of lactic acid-based polymers and related oligomers have been discussed that can act as precursors for catalytic synthesis of PLA (2).

PLAs have been produced for many years. PLAs resemble clear poly(styrene) (PS) and have good gloss and clarity for aesthetic appeal, along with physical properties well suited for use as fibers, films, and thermoformed packaging. PLA is biocompatible and have been used extensively in medical and surgical applications, i.e., sutures and drug delivery devices (3).

PLA can exist as two optically active forms: *L*-PLA and *D*-PLA (4). PLA can have any suitable *D*/*L* ratio. The ratio of *D*/*L* plays an important role for the thermal and physical properties of the PLA. PLA having a high *L*-enantiomer content typically is more crystalline and has a longer degradation time, whereas PLA having a high *D*-enantiomer content is more amorphous and has a shorter degradation time (4).

3.1.1 Production Processes

3.1.1.1 Lactic Acid

Lactic acid is a consolidated bioproduct in the world market. It has many applications, such as (5): The production of biodegradable polymers, substitution of plastics from oil, and uses in medicine. Besides this, new applications are being discovered every year, especially in chemical industries as a building-block molecule. The lactic acid market is constantly growing and the fact that the final products are able to comply with environmental laws as green, renewable and biodegradable products contributes to the tendency of continuous growth in the next few years (5).

Lactic acid bacteria have extensive industrial applications as producers of lactic acid, as probiotics, as biocontrol agents and as biopreservatives (6). Lactic acid can be obtained under acidic conditions by lactic acid bacteria strains (7). Several attempts have been made to improve the acid tolerance of lactic acid bacteria by using both genome-shuffling approaches and rational design based on known mechanisms of lactic acid tolerance and gene deletion in the yeast strains. The production of *d*-lactic acid by *Lactobacillus delbrueckii* ssp. *delbrueckii* from orange peel waste has been described (8).

Fermentation of food waste. A thermal pretreatment was involved in co-fermentation of food waste and waste activated sludge to enhance the production of lactic acid (9). First, sole food waste was observed as the most suitable substrate employing thermal pretreatment for the generation of lactic acid. The fermentation time for reaching the maximal plateau could be significantly shortened at a corresponding thermal pretreatment temperature. The enhancement of lactic acid yield was in accordance with the acceleration of solubilization and hydrolysis. Furthermore, the physicochemical characteristics of fermentative substrate and surface morphology of the fermentation mixture varied with the pretreatment temperatures. Further investigations revealed that the proportions of key microorganisms such as *Bacillus* and *Lactobacillus* were changed by the thermal pretreatment (9).

Production from Whey. Whey is an important byproduct of the dairy industry containing mainly lactose and proteins. Only a few studies have been conducted on the co-utilization of lactose and proteins in whey for chemical production (10).

To establish an efficient bioconversion method for the comprehensive utilization of cheese whey, a low-cost and labor-saving method for *D*-lactic acid production from cheese whey powder (CWP) was investigated and optimized in this study. CWP could be directly utilized by *Lactobacillus bulgaricus* CGMCC 1.6970 without the addition of extra nutrients.

To fully utilize proteins in CWP, different proteases were selected to hydrolyze CWP, and the highest *D*-lactic acid concentration was obtained using neutral protease pretreated CWP as substrate. The *D*-Lactic acid concentration could be further improved by adding a small amount of yeast extract. Based on the optimization, 70.70 $g\,L^{-1}$ *D*-lactic acid with an average productivity of 1.47 $g\,L^{-1}h^{-1}$ was obtained from batch fermentation. In fed-batch fermentation, a high *D*-lactic acid concentration of 113.18 $g\,L^{-1}$ was obtained with an average productivity of 2.36 $g\,L^{-1}h^{-1}$. All the fermentations were carried out under an open condition without sterilization. So, a simple and economic strategy for CWP valorization and *D*-lactic acid production was presented (10).

3.1.1.2 Poly(lactic acid)

Atactic PLA has been prepared by polymerization of a lactide racemate. It has an amorphous structure. Isotactic PLA has been prepared by the polymerization of *D*-lactide or *L*-lactide and is crystalline. Syndiotactic PLA can be obtained from cyclic dimer of lactic acid, e.g., meso-lactide (11). Production processes of PLA resins include (12):

1. A process of subjecting lactic acid or a mixture of lactic acid and an aliphatic hydroxy carboxylic acid as a material directly to dehydration polycondensation (13),
2. A process of ring-opening polymerization in which a cyclic dimer of lactic acid (lactide) is melted and polymerized (14, 15),
3. A process of ring-opening polymerization in which cyclic dimers of lactic acid and an aliphatic hydroxy carboxylic acid, such as lactide or glycolide and ε-caprolactone, are subjected to melt polymerization in the presence of a catalyst (16),

4. A process of subjecting a mixture of lactic acid, an aliphatic dihydric alcohol and an aliphatic dibasic acid directly to dehydration polycondensation (17),
5. A process of subjecting a PLA and a polymer derived from an aliphatic dihydric alcohol and an aliphatic dibasic acid to condensation in the presence of an organic solvent (18), and
6. A process of producing a polyester by subjecting lactic acid to dehydration polycondensation in the presence of a catalyst, in which solid-phase polymerization is carried out at least in part of the steps.

The resins may be produced by copolymerization in the coexistence of a small amount of an aliphatic polyhydric alcohol such as trimethylolpropane or glycerol, an aliphatic polybasic acid such as butanetetracarboxylic acid, or a polyhydric alcohol such as a polysaccharide (12).

The production of syndiotactic PLA starts with a cyclic dimer of lactic acid, meso-lactide, i.e., the cyclic diester of lactic acid, which has two centers of asymmetry with the opposite configurations *R* and *S*. Meso-lactide is shown in Figure 3.1.

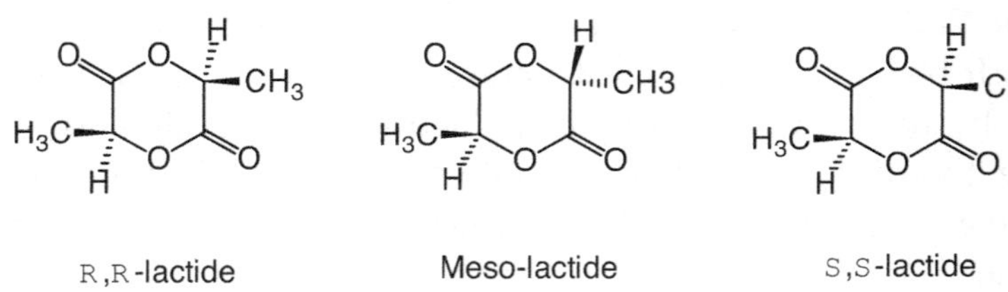

Figure 3.1 Lactides.

Meso-lactide can be obtained from racemic PLA by depolymerization at 180–200°C under reduced pressure and distilling off the meso-lactide (19). Also, lactic acid can be used directly in the presence of catalyst such as tin octanoate and lithium carbonate (20).

The boiling points of pure meso-lactide and *D*,*L*-lactide differ by about 7°C over a wide range of pressures (19).

For the polymerization reaction, the same oxygen acyl bond in a series of meso-lactide molecules are cleaved by a ring-opening reaction to form a polymer containing a series of units with stereocenters of the opposite stereochemistry (11). The ring-opening is catalyzed by chiral complexes, which function to exhibit kinetic preference for reaction at one oxygen acyl bond site per molecule to the extent of fostering formation and joining of a series of the same units with respect to relative location of enantiomorphoric site and carbonyl and oxygen moieties.

The chiral complex can be a chiral metal alkoxide. A suitable catalyst is prepared by reacting a chiral ligand, c.f. Figure 3.2, or the opposite enantiomer and aluminum isopropyloxide.

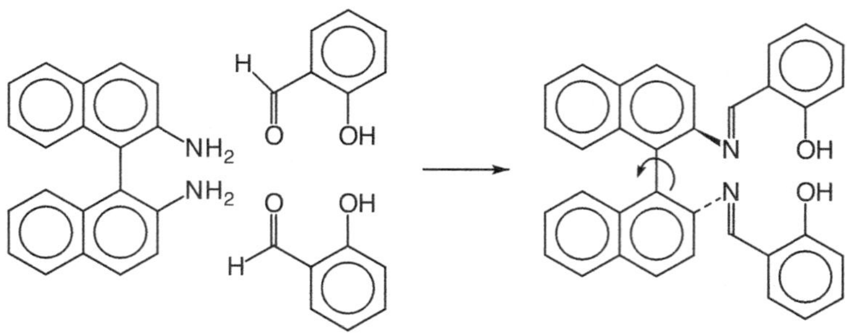

Figure 3.2 Preparation of a chiral ligand as catalyst precursor (11, 21).

These types of chiral Schiff bases can be obtained by the condensation of 2,2′-diamino-1,1′-binaphthalene with salicyl aldehydes. 1,2-Diaminocyclohexane can also be used as base material (21). In addition, methods for polymerization and copolymerization of lactic acid compounds have been reported by Hiltunen in a series of papers (22–24).

Besides an *L*-lactic acid polymer, hydroxyl- and carboxyl-terminated telechelic prepolymers can be prepared by the addition of small amounts of 1,4-butanediol and adipic acid, respectively. The polymerization is carried out in a melt with tin octoate as the catalyst (25).

Since lactic acid is a hydroxy carboxylic acid, copolymers from other hydroxy carboxylic acids can be formed, which are listed in Table 3.1. Selected hydroxy acids are shown in Figure 3.3.

The lactic acid/hydroxy carboxylic acid copolymer is composed of 85–100 *mol%* *L*- or *D*-lactic acid unit and 0–15 *mol%* hydroxy carboxylic acid unit (26). Preferably, a PLA resin can be obtained

Table 3.1 Hydroxy carboxylic acid monomers or lactones (26).

Monomer	Molar mass /[$g\ mol^{-1}$]
Lactic acid	90.08
Glycolic acid	76.05
Hydroxybutyric acid	104.1
Hydroxyvaleric acid	118.13
6-Hydroxycaproic acid	132.16
γ-Butyrolactone	86.09
γ-Heptalactone	128.17
δ-Valerolactone	100.12

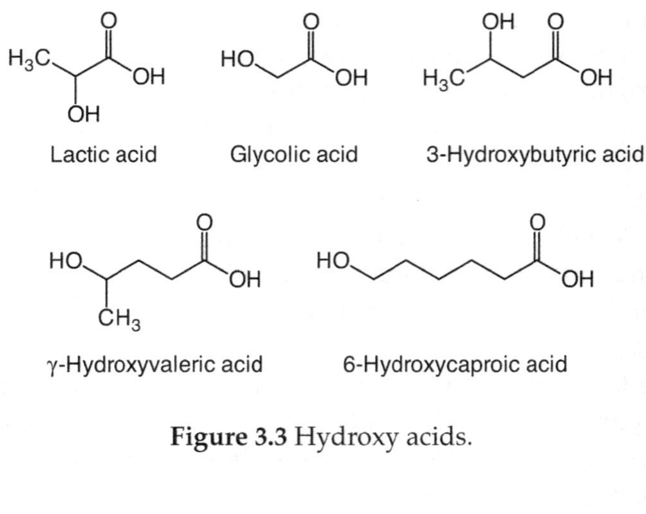

Figure 3.3 Hydroxy acids.

by selecting a starting material having a necessary structure from a lactide that is a cyclic lactic acid dimer, a glycolide that is a cyclic glycolic acid dimer, caprolactone, etc., and subsequent ring-opening polymerization of the starting material (26). The lactide can occur as a lactide that is a cyclic lactic acid dimer and various mixtures of their stereo isomers.

Also, lactic acid polymers can be synthesized from oligomers by the addition of 1,6-hexamethylene diisocyanate and 2,2′-bis(2-oxazoline) as chain extenders (27). In addition, the molecular weights of the resins may be increased by using a binding agent (polymer chain extender) such as a diisocyanate compound (12).

Unfortunately, PLAs exhibit major weaknesses such as brittleness as well as low thermal resistance of 58°C and moisture-related degradation, which is a drawback for a lot of commercial applications (3).

However, PLA can be polymerized with diisocyanates to form poly(ester-urethane)s or with caprolactones to form poly(*L*-lactic acid-*co*-ε-caprolactone-urethane)s. Such copolymers can be tailored as hard plastics or flexible elastomers and can have unique glass transition temperatures from −70°C to +60°C, tensile modulus from 2 MPa to 2000 MPa, and % strain from 1% to about 1000% (4).

Materials for PLA resins can be synthetically prepared from saccharides derived from grains which are recycling capable resources, such as corn and potatoes, or lactic acid as fermented products of such saccharides. In addition, PLA resins are easily hydrolyzed in natural environments and decomposed by the action of microorganisms and ultimately become water and carbon dioxide gas (12).

PLAs can be softened by the addition of plasticizers, by blending with soft polymers, or using comonomers. However, when blending soft polymers, the suitable soft polymers are limited to biodegradable resins such as poly(butylene succinate). Moreover, such biodegradable resins have to be added in large quantities to impart sufficient flexibility. The addition of large quantities may impair the desired properties of PLA. Also, comonomers change the physical properties such as melting point and heat resistance, owing to the decrease in crystallinity and glass transition temperature (28).

PLA-based compositions with a plasticizer and a crystal nucleating agent have been reported (26). Plasticizers are collected in Table 3.2.

Table 3.2 Plasticizers for poly(lactic acid) (26, 28).

Compound
Methyl diglycol butyl diglycol adipate
Benzyl methyl diglycol adipate
Benzyl butyl diglycol adipate
Ethoxycarbonyl methyl dibutyl citrate
Glycerin ethylene oxide ether, end-capped with acetate groups
Ester of succinic acid and triethylene glycol monomethyl ether
Ester of adipic acid and diethylene glycol monomethyl ether
Ester of 1,3,6-hexatricarboxylic acid and butyl diglycol

When plasticizers are added to PLA, a good flexibility is obtained. However, the resin is remarkably changed in flexibility at a temperature close to the glass transition point because the resin is in an amorphous state. In order to solve this problem, the PLA is crystallized by adding a crystal nucleus agent such as talc. However, this method poses the problem that the rate of crystallization performed by heat treatment after the resin is molded is insufficient, and when a crystal nucleus agent such as talc is added in a large amount, the transparency of a sheet or film after thermal molding is deteriorated and a plasticizer bleeds in the case of allowing a molded article to stand under a high-temperature and high-humidity environment.

As crystal nucleating agent, ethylene bis-12-hydroxystearic acid amide or hexamethylene bis-12-hydroxystearic acid amide can be used (26). Ethylene bis-12-hydroxystearic acid amide is more preferable from the viewpoint of the moldability, heat resistance, impact resistance and anti-blooming property of the biodegradable resin composition (29).

Also, the biodegradation of lactic acid-based poly(ester-urethane)s has been investigated in detail (30). Since PLAs are biodegradable, they suffer from a stability against hydrolysis (31). Carbodiimides are widely used as a stabilizer against hydrolysis for an ester-based resin. However, compounds with a low molecular weight, such as a monocarbodiimide, have problems in causing an environmental pollution. Poly(carbodiimide)s have been used to solve these problems, but the use of such a compound may become difficult for an application where hue may be a problem due to yellowing during processing.

An antioxidant is used in the synthesis of a carbodiimide compound. The antioxidant is added during the synthesis. A suitable carbodiimide compound can be synthesized as follows (31):

Preparation 3–1: Into a flask equipped with a stirring motor, a nitrogen gas bubbling tube and a condenser, 100 parts by weight of 4,4′-dicyclohexylmethane diisocyanate and 0.5 part by weight of 3-methyl-1-phenyl-2-phospholene-1-oxide are charged. The mixture was subjected to the carbodiimidization reaction at 185°C for 24 *h*, with nitrogen gas bubbling. A carbodiimide with an NCO content of 2.4% was obtained.

Natural rubber from *Hevea brasiliensis* has been regarded as a good toughening agent for PLA. Natural rubber can enhance the hydrophobility of the surface of PLA; however, the water absorption in the bulk of PLA is not decreased (32). The amount of water absorption increases with increasing content of natural rubber. So, the hydrolysis occurs in the entire bulk of the samples and is not confined in the surface.

The systematic development and analysis of the modeling framework has been summarized that allows a direct mapping between the PLA production process conditions and rheological properties of the polymer melt. To achieve this, the framework builds upon three distinct elements that approach the production process from different scales (33):

1. A macroscopic deterministic model of *l,l*-lactide ring-opening polymerization taken from the literature,
2. Microscopic stochastic simulation of the polymerization process based on a hybrid Monte Carlo approach, and
3. Mesoscopic public domain model of polymer chain reptation dynamics.

Based on the input reaction conditions, the macro-scale model predicts *l,l*-lactide conversion and averaged molar mass of PLA, while the micro-scale and meso-scale simulations allow prediction of full molar mass distribution and melt viscosity of the product. The developed predictive tool is validated by the literature data, i.e., experimentally measured rheological characteristics of three commercial PLA samples with different molecular architectures (33).

Moreover, a comprehensive global sensitivity analysis has been carried out to support exploration of the process conditions space in relation to target polymer melt properties. Computational efficiency of the developed model achieved so far foreshadows its potential use as soft sensor for molar mass distribution and melt viscosity in the optimization and control of PLA production (33).

The synthesis of PLA was studied using the direct melt condensation polymerization method in an experimental microwave system where external cooling was applied to keep both temperature and microwave power continuous and constant during the polymerization (34). These factors and reaction time, catalyst type, and catalyst amount were considered to determine the appropriate process conditions. In addition, some experiments were carried out under similar process conditions using microwave-assisted ring-opening polymerization and conventional heating methods.

These methods were compared in terms of process performance and various properties of the oligomeric PLA. According to results, polymerization rate, polymer yield, average-molecular weight, dispersity, and thermal properties of PLA products were similar in both microwave methods but higher than conventional heating counterparts. Furthermore, low crystallinity of PLA products synthesized by microwave methods showed the effect of microwave on the optical purity of products. Finally, energy saving was accomplished by 73% and an increase of polymer production rate by 37.7% in microwave direct melt condensation polymerization method (34).

3.1.2 *Fibers*

3.1.2.1 *Forcespinning Technique*

A systematic study has been done for the process development and optimization of poly(*d*,*l*-lactic acid) (PDLLA) submicrometer fibers utilizing the centrifugal spinning technique known as forcespinning (35). The study analyzes the effect of polymer concentration (8%, 10%, and 12%) and angular speed on the fiber morphology, diameter distribution, and fiber yield.

The increase in polymer concentration and angular speed favored the formation of continuous and homogeneous submicrometer fibers with an absence of bead formation and higher output. The optimal conditions were established considering the morphological characteristics that exhibit a greater surface area (homogeneous and submicrometer fibers); and they were achieved at a polymer concentration of 10% at an angular speed ranging from 8,000 *rpm* to 10,000 *rpm*.

The optimization of PDLLA submicrometer fiber fabrication lays the groundwork for scaling up the process and serves as a platform to further develop promising applications of PDLLA nonwoven mats, particularly in the biomedical area such as in scaffolds for tissue engineering (35).

3.1.2.2 Triboelectric Nanogenerators

A fully biodegradable triboelectric nanogenerator was developed based on gelatin film and electrospun PLA nanofiber membrane (36). By optimizing the material properties of the gelatin and PLA polymer films, an output voltage up to 500 *V*, a short circuit current density of 10.6 $mA\ m^{-2}$ and a maximum power density over 5 $W\ m^{-2}$ were achieved with a device dimension of 4×4 cm^2.

The performance of the biodegradable triboelectric nanogenerators with different material combinations under various test conditions were investigated and analyzed. The biodegradable triboelectric nanogenerators show excellent mechanical stability and reliability upon cyclical contact for up to 15,000 times. Biodegradation experiments showed that all the materials of the triboelectric nanogenerator could be degraded completely into water in about 40 *d*. The biodegradable triboelectric nanogenerators provide a promising green micro-power source for environment monitoring, biomedical implants through harvesting energy from wind, etc. Moreover, they can dissolve with no adverse effect to the environment (36).

3.1.2.3 Surface Modification of Fibers

The major aspects of kenaf fibers haves been reviewed (37). A kenaf fiber is shown in Figure 3.4.

Kenaf originates mostly from Iraq, Tanzania, Jamaica, South Africa, Cuba and Togo. Kenaf fiber is extracted from bark of the kenaf tree. The fibers are extracted both from the bast and the core of the bark (37). Some properties of kenaf fibers are shown in Table

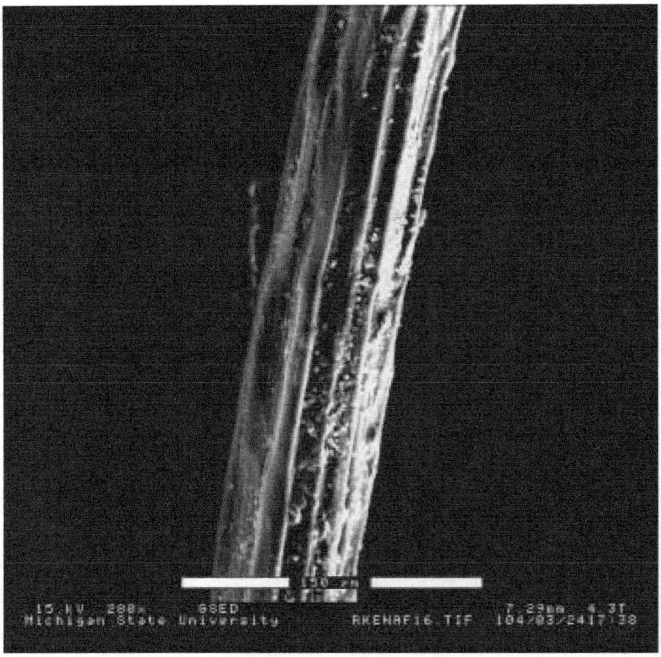

Figure 3.4 Kenaf fiber. Reprinted from (38) with permission from Elsevier.

3.3. Kenaf fibers can be used in both thermoplastic and thermoset resins for reinforcement.

Table 3.3 Properties of kenaf fibers (37).

Fiber from	Length/[mm]	Diameter/[μm]
Bast	2.3	22
Core	0.7	22

3.1.3 Influence of Fabrication Methods and Kenaf Fiber Length

Biocomposites from kenaf fiber and soy-based polymers have been fabricated by extrusion, followed by injection or compression molding (38). It turned out that compression molded specimens have a similar modulus in comparison to injection molded specimens at room temperature. However, the samples exhibit a higher heat deflection temperature and a higher notched Izod impact strength.

These improvements are explained by an increase in the modulus at high temperature and also by fiber bridging effects.

Furthermore, the modulus, impact strength, and the heat deflection temperature of the fiber reinforced biocomposites increase with the fiber length, the fiber content and the orientation of the fibers.

The fractured fiber length on an impact fracture surface increases with increasing fiber length and fiber content. This finding suggests that the role of fiber bridging effects is predominant on the impact strength of the biocomposites (38).

3.1.4 *Kenaf Fibers for Reinforcement of PP*

Kenaf fibers can also be used for the reinforcement of poly(propylene) (PP) (39). In a study, the optimal fabrication method for such materials was investigated. It turned out that a compression molding process is most promising. A layered sifting of a microfine PP powder and chopped kenaf fibers is used. An adequate reinforcement is achieved with a kenaf fiber content of 30–40%.

A coupling agent based on maleic anhydride, such as Epolene (maleated PP), results in an improved adhesion of fiber to matrix. The kenaf-maleated PP composites are comparatively cheap and exhibit a higher modulus than sisal, coir, and E-glass (39).

3.1.5 *Films*

Poly(lactic acid) (PLA) was melt mixed in a laboratory extruder with poly(butylene adipate-*co*-terephthalate) and poly(butylene succinate) in the presence of polypropylene glycol diglycidyl ether that acted as both plasticizer and compatibilizer (40).

Then, the process was scaled up in a semi-industrial extruder by preparing pellets having different content of a nucleating agent (LAK301, an aromatic sulfonate derivative, c.f. Figure 3.5). All of the formulations could be processed by blowing extrusion and the obtained films showed mechanical properties dependent on the content of the nucleating agent.

Scanning electron microscopy (SEM) micrographs of cryogenic fractures films with different amounts of nucleating agent are shown in Figure 3.6.

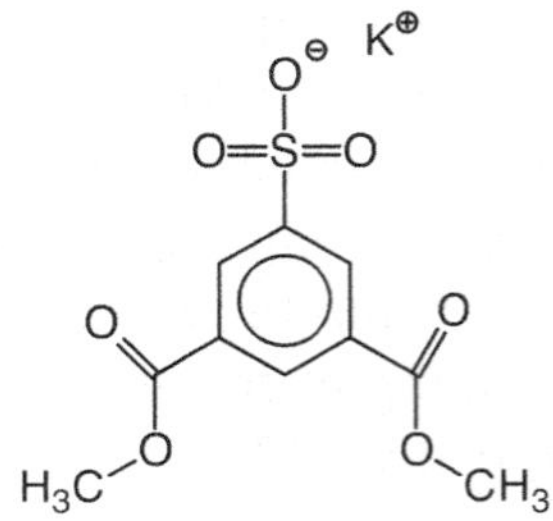

Figure 3.5 LAK301, 5-sulfoisophthalic acid sodium salt.

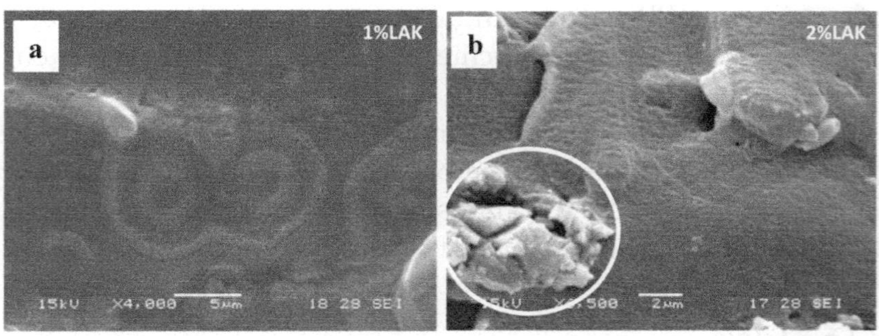

Figure 3.6 SEM micrographs of cryogenic fractures with different amounts of nucleating agent (40): (a) 1% LAK; (b) 2% LAK; (c) 3% LAK; (d) 4% LAK.

In particular, the tearing strength showed a maximum like trend in the investigated composition range. The films prepared with both kinds of blends showed a tensile strength in the range of 12–24 *MPa*, an elongation at break in the range of 150–260% and a significant crystallinity (40).

An approach was investigated for incorporating cellulose nanocrystals into a PLA matrix by examining two different cellulose nanocrystal addition approaches (41). The first approach consisted of melt blended PLA and cellulose nanocrystals in a three-piece internal mixer. The second method involved the direct dry-mixing of PLA and cellulose nanocrystals in a high intensity mixer. The compounded materials were then blown into films and compared in terms of particle dispersion, optical, thermal, molecular weight and barrier properties. A good distribution was achieved by both the direct dry blending and melt blending methods. However, some agglomerations were still present that resulted in a reduced transparency of the composite films.

Furthermore, the PLA/cellulose nanocrystal films thermally degraded during the blending processes, as indicated by a reduction in their molecular weights due to chain scission. More chain scission occurred in melt blended films than the dry blended counterparts (41).

The addition of only 1% cellulose nanocrystals into the PLA matrix by the direct dry blending and melt blending approaches improved the water barrier properties of PLA films by 30% and 24%, and oxygen barrier properties by 60% and 39%, respectively. The inferior performance of the melt blending films compared to the direct dry blended counterparts could be attributed to more chain scissions in these films. Thus, the direct dry blending process appeared to be the better approach for incorporating cellulose nanocrystals into the PLA matrix (41).

3.1.6 *Fertilizer Solutions*

A superabsorbent composite material to be used in agricultural and horticultural applications has been described (42). The composite is designed to retain and release fertilizer solutions into the soil in a controlled manner, permitting resource optimization. Because of its ability to absorb and release large amounts of saline water, a natural superabsorbent hydrogel derived from cellulose was chosen.

Potassium nitrate was chosen to model the fertilizer. PLA was added to the final composition in order to delay the solution release. The composite material was obtained using easily available and low-cost starting materials and using a simple manufacturing process, using a standard mixer. After being analyzed for morphological (SEM), physical (X-ray diffraction), chemical (energy dispersive X-ray spectroscopy), and thermal properties (thermogravimetry and differential scanning calorimetry (DSC)), the material was tested using two different Mediterranean cultivations (Pomodoro di Morciano di Leuca and Cicoria Otrantina) and two different kinds of soil (red and white soils).

The analysis revealed different water release characteristics for different soils. These findings have been confirmed by measuring plant growth for both species, as well as fruit yield of the tomatoes (42).

3.1.7 Reinforced Composites

Composites from PLA and cotton linter or maple wood fiber have been prepared by compounding and injection molding (43). A reduction of the crystallinity in the PLA matrix was observed due to thermal processing.

The composites showed an excellent improvement with regard to stiffness and toughness. On the other hand, the tensile strength and the flexural strength remained unchanged and the strain at break values were found to be reduced.

Characterization by thermal mechanical analysis (DMA) indicated that the incorporation of the fibers cause a decrease of the mechanical loss factor. This results in better damping capabilities. SEM studies of the impact fractured surfaces of the composites with cotton linter showed a debonding cavitation at the matrix-fiber interface.

Composting studies of the composites up to 90 *d* revealed that the degradation onset time was increased with an increase of the fiber content. In contrast, the maximum degree of degradation and the rates of degradation were decreased in comparison to neat PLA.

The composites with cotton linter showed a quicker onset of biodegradation and a higher degree of biodegradation after 90 *d* than the maple wood fiber-based composites. This is possibly due to a quicker thermal hydrolysis during processing and composting.

3.1.8 *Nanocomposites*

3.1.8.1 *Nanofillers*

High-lignin-containing cellulose nanocrystals were used as nanofiller for a PLA matrix (44). These cellulose nanocrystals could be successfully isolated from hydrothermally treated aspen fibers and freeze-dried and compounded with PLA by extrusion and injection molding. As a comparison, PLA composites containing commercial lignin-coated cellulose nanocrystals (BioPlus-L™) were also produced.

The chemical composition of lignin-coated and lignin-containing cellulose nanocrystals is shown in Table 3.4.

Table 3.4 Chemical composition of lignin-coated and lignin-containing cellulose nanocrystals (44).

			Amount/[%]		
Sample	Glucan	Xylan	Mannan	Klason lignin	Ash
HLCNCs	51.4	0.2	–	48.6	0.2
BLCNCs	70.4	18.7	0.3	10.2	0.3

HLCNCs: High-lignin-containing cellulose nanocrystals
BLCNCs: Lignin-coated cellulose nanocrystals (BioPlus-L™)

The high-lignin-containing cellulose nanocrystals showed higher crystallinity, larger surface area, lower degree of agglomeration, and more hydrophobic surfaces compared to commercial lignin-coated cellulose nanocrystals, as characterized by electron microscopy, surface area measurements, thermal analysis, spectroscopy and water contact angle measurements.

The effect of lignin and cellulose nanocrystal morphology on the mechanical, thermal and viscoelastic properties and nanocrystal/polymer interfacial adhesion of nanocomposites was investigated with tensile test, DSC and DMA.

In comparison to neat PLA, the Young's modulus, elongation at break, and toughness of PLA with 2% high-lignin-containing cellulose nanocrystals were improved by 14%, 77%, and 30%, respectively. High-lignin-containing cellulose nanocrystals and commercial lignin-coated cellulose nanocrystals act as nucleating fillers, increasing the degree of crystallinity of PLA in nanocomposites. The presence of lignin nanoparticles in the compositions increased the compatibility/adhesion between cellulose nanocrystals and polymer matrix, which increased the storage modulus (44).

3.1.8.2 3D Printing

Thermoplastic materials that are printable into complex structures are widely employed for 3D printing. PLA is among the most promising polymers used for 3D printing, owing to its low cost, biodegradability, and nontoxicity (45). However, PLA is electrically insulating and mechanically weak; this limits its use in a variety of 3D printing applications. This study demonstrates a straightforward and environment-friendly method to fabricate conductive and mechanically reinforced PLA composites by incorporating graphene nanoplatelets.

To fully utilize the superior electrical and mechanical properties of graphene, liquid-exfoliated graphene nanoplatelets are dispersed in isopropyl alcohol without the addition of any surfactant and combined with PLA dissolved in chloroform. The graphene nanoplatelet-PLA composites exhibit improved mechanical properties, e.g, an improvement in the tensile strength by 44% and maximum strain by 57% even at a low graphene nanoplatelet threshold concentration of 2%. The graphene nanoplatelet-PLA composites also exhibit an electrical conductivity of over 1 $mS\,cm^{-1}$ at >1.2%.

The graphene nanoplatelet-PLA composites can be 3D-printed into various features with electrical conductivity and mechanical flexibility (45).

Carbon-based nanostructures-multiwalled carbon nanotubes, graphene nanoplatelets, and a combination of both fillers were incorporated into PLA to overcome certain limitations of conventional 3D manufacturing based on insulating materials (46). Solution blending and melt mixing were the two fabrication methods adopted for the preparation of the tested samples. A comparison of the morphological, rheological, and electrical properties of the resulting nanocomposites was carried out. In addition, for the same weight concentrations, the influence of physical and geometrical features (i.e., functionalization and aspect ratio) of the embedded fillers was also investigated. Rheological methods were applied to control the quality of fillers dispersion in the PLA matrix. The rheological percolation threshold was considered as reference in order to evaluate the internal structure of nanodispersions. Transmission electron microscopy visualization, combined with rheological characterizations, was used for efficient control of the nanofiller dispersion.

DC electrical conductivity experiments revealed that lower electrical percolation thresholds and higher values of electrical conductivity were achieved using fillers with a larger aspect ratio and melt mixing, respectively (46). Moreover, given the possibility of obtaining complex and appropriate shapes for electromagnetic compatibility applications, electromagnetic response of the nanocomposites at the highest filler concentration was investigated in the *GHz* and *THz* regions.

It was found that the electromagnetic shielding efficiency of nanocomposites strongly depended on the aspect ratio of the nanofillers, whereas the type of processing technique did not have a significant effect. Therefore, a careful choice of methods and materials must be made to address the final application for which these materials are to be used (46).

3.1.9 Membranes

Membrane science has an emerging role as a very well-recognized eco-friendly technology. In order to meet the complete concept of sustainability, however, greener strategies still need to be put in place regarding the preparation of the membranes (47).

A lotus-leaf-inspired superhydrophobic PLA porous membrane was developed for oil-water separation (48). The fabrication was based on fused deposition modeling three-dimensional printing and subsequent chemical etching and the decoration of PS nanospheres.

A superhydrophobic PLA fractal surface with a water contact angle of 151.7° and a low water adhesion force of 21.8 μN could be achieved (48). The membrane pore size could be easily adjusted from 40 μm to 600 μm using a computer-aided design program to optimize the separation performance.

The maximal oil-water separation efficiency of 99.4% was achieved with a pore size of 250 μm, which also exhibited a high flux of 60 $kl\,m^{-2}h^{-2}$ (48).

PLA green flat sheet membranes were prepared using ethyl lactate as a green solvent and water as a nonsolvent (47). The morphology, thickness, contact angle, mechanical properties, Fourier transform infrared spectroscopy and degree of swelling were determined for studying the properties of the produced membranes. A systematic study was performed testing PLA membranes in pervaporation for the separation of methanol/methyl *tert*-butyl ether azeotropic mixture and evaluating their performance by varying feed temperature and vacuum degree.

These findings revealed that the membrane morphology changed from fingerlike to spongy-like and finally to dense-like structure by acting on the evaporation time during the preparation, with an improvement in the overall mechanical properties. The PLA dense membrane produced with an evaporation time of 7 *min* was successfully tested in pervaporation, exhibiting a preferred permeation towards methanol with a highest selectivity value of more than 75. Also, an Arrhenius-type dependence between flux and temperature was found (47).

3.1.10 Antibacterial Composites

Poly(lactic acid) (PLA) is a biodegradable and biocompatible polyester. However, this biomaterial suffers from some shortcomings related to the absence of both bioactivity and antibacterial capacity (49).

Composites of PLA with either graphene oxide or thermally reduced graphene oxide were prepared by melt mixing to overcome these limitations. PLA composites with both graphene oxide and thermally reduced graphene oxide inhibited the attachment and proliferation of *Escherichia coli* and *Staphylococcus aureus* bacteria depending on the kind and amount of filler. It was also shown that by applying an electrical stimulus to the percolated PLA/thermally reduced graphene oxide, the antibacterial behavior can be dramatically increased.

MTT analysis showed that while all the PLA/graphene oxide composites were more cytocompatible with osteoblast-like cells (SaOS-2) than pure PLA, only low content of thermally reduced graphene oxide was able to increase this property (49).

An MTT assay is a colorimetric assay for assessing cell metabolic activity (50). Cellular oxidoreductase enzymes may, under defined conditions, reflect the number of viable cells present. These enzymes are capable of reducing the tetrazolium dye MTT 3-(4,5-dimethylthiazol-2-yl)-2,5-diphenyltetrazolium bromide to its insoluble formazan, which has a purple color. The reaction is shown in Figure 3.7.

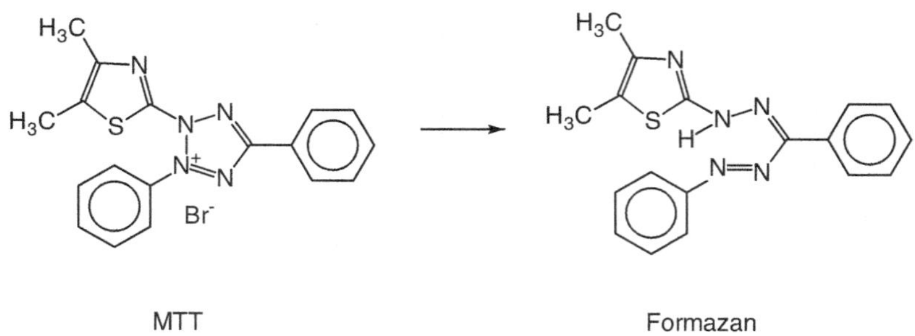

Figure 3.7 MTT reduction (50).

These tendencies could be related to the surface properties of the resulting polymer composites such as polarity and roughness. In this way, the addition of graphene oxide and thermally reduced graphene oxide into a PLA matrix allows the development of multifunctional composites for potential applications in biomedicine (49).

3.1.11 Laminated Composites from Kenaf Fiber

Laminated composites from kenaf fiber reinforced PLA were prepared by compression molding (51). The mechanical and thermal properties of these composites were assessed depending on the modification of the kenaf fiber by alkaline and silane treatments, such as 3-aminopropyltriethoxysilane, which acts as a coupling agent.

3-Aminopropyltriethoxysilane bears ethoxy groups that can hydrolyze in water to form a silanol. The silanol reacts with the hydroxyl group of the kenaf fiber to form stable covalent bonds with the surface of the fibers. In addition, the amine groups in the silane may react with the terminal carboxylic groups of a hydrolyzed PLA polymer (52). The treatment with silane is shown in Figure 3.8.

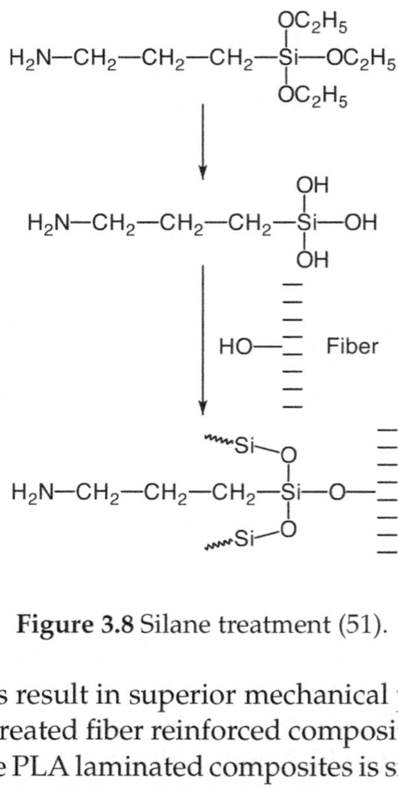

Figure 3.8 Silane treatment (51).

Both treatments result in superior mechanical properties in comparison to the untreated fiber reinforced composites. The deflection temperature of the PLA laminated composites is significantly higher than that of the resin. So, standard PLA resins are very useful in the manufacture of kenaf fiber reinforced laminated biocomposites (51).

3.1.12 *Copolyesters*

Fully bio-based poly(*L*-lactide)-*b*-poly(ricinoleic acid)-*b*-poly(*L*-lactide) triblock copolyesters have been described (53). Methyl ricinoleate, produced from castor oil, is condensed in the presence of a small amount of 1,3-propanediol. This results in α,ω-hydroxy-terminated poly(ricinoleic acid) with a molar mass of 11 $kg\,mol^{-1}$. The polymerization of *L*-lactide initiated from the terminal hydroxyl moieties of the α,ω-hydroxy-terminated poly(ricinoleic acid) results eventually in triblock copolymers.

A transesterified PLA product has been described that has been transesterified with a diol and at least one natural oil. The transesterified PLA product has surprisingly improved properties compared to unmodified PLA including (54):

1. A liquid state at room temperature,
2. A reduced melting temperature, and
3. An increased solubility in a variety of solvents.

Also disclosed was a method for producing the transesterified PLA product. The transesterified PLA polymers have desirable improved properties compared to PLA such as a liquid or waxy solid state at room temperature and an improved solubility in a variety of solvents. Soybean oil and castor oil were used for the reaction.

The transesterified PLA product is economical to produce and includes a high content of renewable resources. In addition, the product is biodegradable (54).

3.1.13 *Transparent Crystalline Poly(lactic acid)*

In molding techniques, the crystallization rate and crystallinity of PLA are relatively low due to its slow nucleation (55, 56). This causes long molding cycles for the production of products, c.f. Table 3.5.

The crystallization behavior of amorphous PLA was investigated by exposing it to acetone vapor (57). Acetone vapor effectively induced the crystallization of the amorphous PLA. The crystallization of a $10 \times 20 \times 0.55$ *mm* specimen at 25°C is completed within 24 *min*.

The crystallization rate is increased with temperature. A high crystallinity of about 40% can be obtained. The specimens crystallized by the acetone vapor exhibit a higher transparency than those prepared by annealing or immersion methods.

The crystallization is induced by the diffusion of the acetone into the amorphous phase of PLA. The diffusion obeys a *Fick* type diffusion, as suggested by polarized optical photomicrograph measurements. The crystallization behavior is shown in Table 3.5.

Table 3.5 Crystallization data of PLA (57).

Method	T/[°C]	Time/[*min*]	Crystallization/[%]
Untreated	–	–	1.6
Acetone vapor	25	60	40.9
Acetone vapor	40	20	42.7
Acetone vapor	50	20	44.0
Acetone immersion	25	10	43.4
Annealing	90	270	43.5

3.1.14 Laminated Biocomposites

Laminated biocomposites were prepared from biodegradable PLA and pineapple leaf fibers by compression molding by the film stacking method (58). The pineapple leaf fibers were either untreated or surface-treated. The mechanical properties of the laminated composites were found to be significantly improved by a chemical treatment.

Both silane- and alkali-treated fiber reinforced composites show superior mechanical properties in comparison to untreated fibers. The incorporation of the pineapple leaf fibers effect a considerable increase of the storage modulus, i.e., stiffness. Also, the heat defection temperature of the composites was significantly higher.

DSC measurements suggest that the surface treatment affects the crystallization properties of the PLA matrix.

SEM of the fiber surface and the fracture surfaces of the composites indicate a significant fiber-matrix interface adhesion (58).

The treatment by NaOH makes the surface rougher. This occurs due to the partial removal of hemicellulose, lignin, etc. In contrast, the silane treatment rather smoothes the fiber surface.

3.1.14.1 Fiber Composites

Chopped glass fibers and recycled newspaper cellulose fibers can be used to reinforce PLA base materials for their use as composites (59). The composites can be fabricated in a twin screw extruder and an injection molding device.

The tensile and flexural moduli of the reinforced composites are found to be significantly higher in comparison to the virgin PLA resin. The storage and loss moduli of the composites are found to be increased, but the mechanical loss factor decreases.

3.2 Poly(glycolic acid)s

3.2.1 Glycolic Acid

Typical methods for producing glycolic acid are (60):

1. Reacting carbon monoxide, formaldehyde and water in the presence of a strongly acidic catalyst under high-temperature and high-pressure conditions,
2. Reaction of chloroacetic acid with sodium hydroxide,
3. Cannizzaro reaction between glyoxal available by oxidation of ethylene glycol and a strong alkali to form a glycolate salt, and then adding an acid to liberate glycolic acid from the resulting glycolate salt,
4. Reaction between glyoxal available by oxidation of ethylene glycol and water in the presence of an inorganic catalyst,
5. Catalytic oxidation of ethylene glycol in the presence of a noble metal catalyst and oxygen, and
6. Oxidative esterification of ethylene glycol with methanol and oxygen to obtain methyl glycolate and then hydrolyzing into glycolic acid.

More recent methods are described subsequently: Glycolic acid can be synthesized by contacting carbon monoxide and formaldehyde with a catalyst containing an acidic polyoxometalate compound encapsulated within the pores of a zeolite. The zeolite has cages larger than the acidic polyoxometalate compound, and has pores with a diameter smaller than the diameter of the acidic polyoxometalate compound (61).

Polyoxometalate compounds comprise polyoxometalate anions, the structures of which are based on a plurality of condensed metal-oxide species. With protons as counter-ions, they exhibit Brønsted acidity. There are a number of known polyoxometalate structures, for example, the Wells-Dawson, Anderson and Keggin forms (62).

Common examples of acidic polyoxometalate compounds are heteropolyacids. Examples of heteropolyacids include silicotungstic acid, $H_4SiW_{12}O_{40}$, and phosphotungstic acid, $H_3PW_{12}O_{40}$, which adopt the Keggin structure. Another example is iodomolybdic acid, $H_5Mo_6IO_{24}$, which adopts the Anderson structure. A further example is a different polymorph of phosphotungstic acid which adopts the Wells-Dawson structure, $H_6P_2W_{18}O_{62}$.

Another process producing glycolic acid starts from formaldehyde and hydrogen cyanide (63). These compounds are reacted to produce glycolonitrile. The glycolonitrile is subsequently converted to an aqueous solution of ammonium glycolate using an enzyme catalyst having nitrilase activity derived from *Acidovorax*, *Rhodococcus*, *Nocardia*, or *Alcaligenes* (64). Eventually, glycolic acid is recovered in the form of the acid or salt from the aqueous ammonium glycolate.

On the other hand, the glycolonitrile can be transformed into glycolic acid directly or via a glycolate salt. This occurs in somewhat alkaline medium (60).

3.2.2 *Polymers, Copolymers, and Blends*

Poly(glycolic acid) has been synthesized by a sequential melt-solid ring-opening polymerization of glycolide. High molecular weight poly(glycolic acid) can be produced and the method is expected to be applicable for industrial mass production (65). Blends of PLA and poly(glycolic acid) have been synthesized with the aid of microwave radiation.

A simple production method has been described that avoids challenging lactonization steps and converts a low molecular weight aliphatic polyester, consisting of hydroxy acids and a comonomer, whose molecular weight has been increased by step-growth polymerization reactions (66).

3.2.3 *Condensation Polymer of Glycerol*

A biodegradable condensation polymer of glycerol and a diacid, sebacic acid, has been developed. Sebacic acid was chosen based on toxicological aspects. Sebacic acid is the natural metabolic intermediate in ω-oxidation of medium- to long-chain fatty acids. Both glycerol and sebacic acid are approved for medical applications. Poly(glycerol sebacate) partly mimics the structural elements of collagen and elastin (67).

Preparation 3–2: Poly(glycerol sebacate) was synthesized by the polycondensation of equimolar amounts of glycerol and sebacic acid at 120°C. The preferred molar ratios of glycerol to diacid range from 1:1 to 1:1.5. Biocompatible catalysts may be used to reduce reaction temperature, shorten reaction time, and increase individual chain length. An example of an FDA-approved catalyst is stannous octoate.

The polycondensation yields a transparent, nearly colorless elastomer that does not swell or dissolve in water (67). The chemical analysis reveals that the polymerization reaction has a yield of close to 100% (67).

The hydroxyl groups may be modified by *tert*-butyl, benzyl, or other hydrophobic groups in order to reduce the rate of degradation. On the other hand, the modification with hydrophilic groups, e.g., sugars or amino acids, will increase the rate of degradation (67).

During the interaction of the polymer with cells, biomolecules may be controlled by coupling bioactive agents to the hydroxyl groups (68, 69).

The dynamics of poly(lactic-*co*-glycolic acid) (PLGA) and its glass transition behavior was investigated by performing variable temperature terahertz time-domain spectroscopy experiments (70). It could be demonstrated that the behavior of PLGA is in line with the potential energy surface interpretation by Goldstein as well as the free volume theory of Fox and Flory, and that both concepts together can be used to understand amorphous polymers.

The isothermal crystallization of poly(glycolic acid) has been studied using terahertz and infrared spectroscopy and X-ray measurements (71). The changes in the intermolecular interactions in poly(glycolic acid) during the isothermal crystallization were monitored using terahertz spectroscopy, which is an efficient technique for analyzing the higher-order structure of polymers. In the terahertz spectra, the temporal difference in the intensity observed in the isothermal crystallization is due to the difference in the vibrational origins of two bands at 192 cm^{-1} and 65 cm^{-1}. The band at 192 cm^{-1} primarily originates from the intramolecular vibrational mode, i.e., twisting of the local structure of the poly(glycolic acid) molecular chain.

The band at 65 cm^{-1} results from the intermolecular vibration mode of the C=O and H–C bonds between the polymer chains (71).

3.3 Butyrolactone-Based Vinyl Monomers

Some renewable cyclic analogs of methyl methacrylate (MMA) belong to the group of vinyl butyrolactones and are considered to be very interesting monomers (72).

3.3.1 *Tulipalin A*

α-Methylene-γ-butyrolactone is also known as Tulipalin A, which is a natural product that can be isolated from tulips (73). It is present in relatively high concentrations in white tulips.

Tulipalin A belongs to the class of sesquiterpene lactone family and is considered to be the cyclic analog of the most common vinyl monomer, MMA. Due to its exo-methylene double bond, it can be used as a naturally originated vinyl monomer for the synthesis of biopolymers. Its reactivity in free-radical polymerization is comparable or even higher than that of MMA. The increase in free radical reactivity presumably comes from the ring strain of the molecule (74). Specialty polymers from this naturally occurring tulipalin A monomer have been reviewed (75).

The synthesis of the homopolymer of tulipalin A and its properties have been described (76). It is a brittle glassy material with a glass transition temperature, T_g, of 195°C.

The reaction of tulipalin A by atom transfer radical polymerization has been described (77). Copolymers derived from α-methylene-γ-butyrolactone, styrene, methyl methacrylate, and acrylonitrile have also been described (74). The copolymers can be used as protective layers in multilayer articles that include UV-sensitive substrate materials.

Methylene lactone monomers have been considered for use in dental resins (78). The effect of the more reactive methylene lactone monomer on mechanical properties and the degree of conversion of the polymers was examined. The addition of a small amount of α-methylene-γ-butyrolactone to bis-glycidyl methacrylate significantly increases the conversion. However, additional α-methylene-γ-butyrolactone of 10–30% does not further increase the conversion. Therefore, an incompatibility between the two monomers is assumed (78).

3.3.1.1 *Superabsorbents*

Superabsorbent hydrogels were prepared and examined with respect to their swelling properties and stimuli-responsive character (79). The hydrogels were synthesized by copolymerization of either sodium or potassium 4-hydroxy-2-methylenebutanoate products resulted from saponification of tulipalin A with acrylamide in the presence of *N,N′*-methylenebis(acrylamide) as crosslinking agent.

The effect of temperature, pH and ionic strength of solutions on the swelling behavior of the hydrogels was examined (79). Reversible swelling character of hydrogels in buffer and water was proved as well. The experiments demonstrated the reversible ability of hydrogels to swell upon immersing in ultrapure water and deswell in a buffer solution (79).

The phytotoxicological aspects of superabsorbent hydrogels were investigated (80). Poly(acrylamide-*co*-sodium 4-hydroxy-2-methylenebutanoate) was prepared from tulipalin A. This was compared with borate-crosslinked poly(vinyl alcohol) (PVA), poly(acrylamide), and poly(acrylamide-co-sodium acrylate) conventional hydrogels. Oat (*Avena sativa*) and radish (*Raphanus sativus* L. subvar. *radicula* Pers.) were selected as the representatives of monocotyledonous and dicotyledonous plants, respectively.

Tulipalin A-based superabsorbent hydrogels revealed superior properties. The results confirmed an excellent suitability of tulipalin A-based hydrogels for application as reservoirs of water during plant stress conditions. Values of fresh matter (yield) and shoot height of the examined plants growing in soil amended with these hydrogels were ca 10% higher than those of plants growing in soil without hydrogels.

Borate-crosslinked PVA hydrogels (containing increasing amount of borax crosslinker) revealed a harmful effect on plants. The negative effect was observed on most of the investigated properties, increasing with content of the hydrogel in soil and concentration of the borax in it (80).

3.3.2 α-Methylene-γ-valerolactone

Acrylic monomers can be derived from biomass (81). An early synthesis of α-methylene-γ-valerolactone involves two steps (82, 83). The first step is the carboxylation of γ-butyrolactone with methyl methoxymagnesium carbonate (*Stiles'* reagent) to produce the acid. Next, the acid is briefly treated with a mixture of aqueous formaldehyde and diethylamine, followed by a separate treatment of the crude product with sodium acetate in acetic acid. The first step requires 6–7 *h* and affords almost quantitative yields, whereas the second step can be accomplished in less than 30 *min* but with yields of only 50%.

In particular, α-methylene-γ-valerolactone can be synthesized in a two-step process from levulinic acid. Levulinic acid can be obtained by the degradation of cellulose. In the first step, a hydrogenation of levulinic acid to γ-valerolactone occurs. Group 1 and 2 elements supported on SiO_2 are excellent catalysts (81, 83). Catalytically active compounds for the synthesis of α-methylene-γ-butyrolactone are summarized in Table 3.6.

Table 3.6 Catalytically active compounds (83).

Oxide compound	Hydroxide compound	Carbonate compound
Magnesium oxide		
Barium oxide	Barium hydroxide	
Lanthanum oxide		
	Potassium hydroxide	
Cadmium oxide		
Rubidium oxide		
	Lithium hydroxide	
	Strontium hydroxide	
	Sodium hydroxide	
	Calcium hydroxide	
		Barium carbonate

The activity of the catalysts in Table 3.6 using a wide variety of conditions of reactions have been extensively tabulated (83). The best overall catalyst turned out to be 8% $Ba(OH)_2/SiO_2$, which gave an 11% yield of α-methylene-γ-butyrolactone at 37% conversion of the γ-butyrolactone at 340°C.

Also, the gas phase synthesis of methylene lactones at a temperature in the range of 300–340°C and a pressure of 0.1–0.5 *M Pa* using catalysts derived from hydrotalcite precursors has been documented (84). Another catalyst for the hydrogenation of levulinic acid is based on ruthenium. The catalytic performance has been studied at 4 *M Pa* H_2 and 200°C (85).

The second step is a heterogeneous gas phase catalytic condensation of formaldehyde with γ-valerolactone. The reaction is depicted in Figure 3.9.

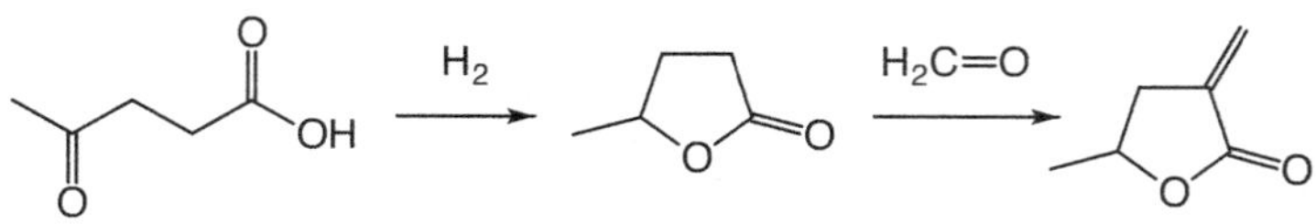

Figure 3.9 Synthesis of α-methylene-γ-valerolactone from levulinic acid (81).

The reaction is very selective to lactone derivatives; however, the process suffers from a rapid deactivation of the catalyst. But the catalyst can be regenerated under relatively mild conditions. Thus, the regeneration of the catalyst can be readily accomplished by a nitrogen or air purge at the reaction temperature of 200°C.

α-Methylene-γ-valerolactone is an attractive monomer for adding temperature stability to polymers. Previous economic limitations have been overcome due to the method of synthesis described above (81).

3.3.2.1 Emulsion Polymerization

The kinetics of the free radical emulsion polymerization of α-methylene-γ-valerolactone has been investigated (86). Stable polymer latices could be prepared. A homogeneous nucleation is the dominant path for particle formation. Also, the miniemulsion copolymerization with styrene as comonomer has been investigated. Both the reversible addition-fragmentation chain transfer (RAFT) miniemulsion polymerization and the RAFT bulk polymerization are well controlled and copolymers with a narrow polydispersity are formed.

In addition, the RAFT *ab initio* emulsion copolymerization with styrene as comonomer was investigated (87). Since the lactone is comparatively water-soluble a dramatic influence on the polymerization kinetics and the copolymer properties was found.

α-Methylene-γ-valerolactone and α-methylene-γ-butyrolactone have been successfully polymerized by a living polymerization method (88). As amphiphilic catalyst, a nucleophilic silyl ketene acetal and an electrophilic silylium moiety is used. The compound is shown in Figure 3.10. As a counter ion, trityl tetrakis(pentafluorophenyl)borate is used.

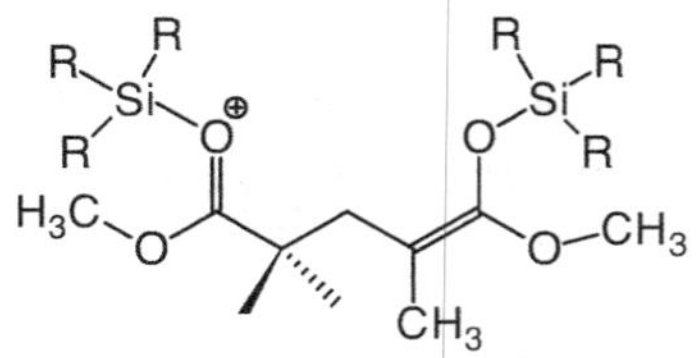

Figure 3.10 Amphiphilic catalyst for living polymerization of methylene lactones (88).

Also, the anionic polymerization of α-methylene-γ-valerolactone and α-methylene-γ-butyrolactone has been reported (89).

γ-Valerolactone as such can be used as a solvent for biomass processing, which results in significant improvements in the yields of product and, moreover, a simplified process for producing biomass-derived chemicals such as lactic acid, furfural, and hydroxymethylfurfural (90).

Lactic acid itself is a base product for the production of various chemicals by catalytic routes. These routes are shown in Figure 3.11.

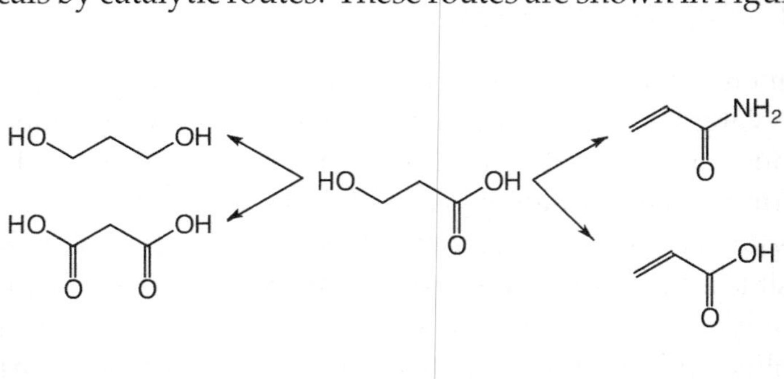

Figure 3.11 Chemicals from lactic acid (91).

3.4 Poly(caprolactone)

Poly(caprolactone) (PCL) is a biodegradable aliphatic polyester with a relatively low melting point and a very low glass transition temperature. It is so named because it is formed by polymerizing ε-caprolactone. The glass transition temperature of PCL is −60°C and the melting point is only 60°C. Because of this, PCL and other similar aliphatic polyesters with low melting points are difficult to process by conventional techniques such as film blowing and blow molding. Films made from PCL are tacky when extruded and have low melt strength over 130°C. Also, the slow crystallization of this polymer causes the properties to change over time. Blending PCL with other polymers improves the processability of PCL. One common PCL is TONE, manufactured by Union Carbide. Other manufacturers of PCL include Daicel Chemical Industries Ltd. and Solvay.

ε-Caprolactone, c.f. Figure 3.12, is a seven-member ring compound that is characterized by its reactivity. The ring-opening takes place at the carbonyl group. ε-Caprolactone is typically made from

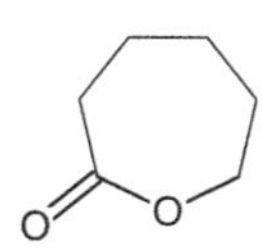

Figure 3.12 ε-Caprolactone.

cyclohexanone by a peroxidation process. PCL is a polyester made by polymerizing ε-caprolactone. Higher molecular weight PCL may be prepared using a wide variety of catalysts such as aluminum alkyls, Grignard reagents, metal amides, etc. PCL is typically prepared by initiation with an aliphatic diol, which forms a terminal end group (92).

Recently, the biocatalytic synthesis of caprolactone by oxidative lactonization has been reported (93). The enzymatic catalysis to access such compounds represents a green alternative to organometallic catalysis and, although Baeyer-Villiger (94) monooxygenases have been used for this purpose, limitations exist with regard to co-factor regeneration, enzyme stability, and issues with product/substrate inhibition. Alcohol dehydrogenases catalyze the interconversion of primary and secondary alcohols into aldehydes and ketones, respectively.

Cellophane multilayer films have been prepared by coating with PCL and chitosan (95). The effects on barrier and mechanical properties were investigated. The multilayer films exhibit a much better water vapor barrier than cellophane and have high oxygen barriers. The mechanical properties of the multilayer films are slightly decreased.

A series of 6-membered ring aluminum complexes bearing a Schiff base, benzotriazole phenolate, aminophenolate, carboxamide phenolate, and carboxylate phenolate were synthesized (96). Their applications in the polymerization of ε-caprolactone were studied.

The decreasing catalytic rates for ε-caprolactone polymerization were ranked as follows: Aluminum complexes bearing aminophenolate, benzotriazole phenolate, carboxylate phenolate, Schiff bases, and finally carboxamide phenolate. The Al complex with aminophenolate revealed an appreciably higher polymerization rate of 10 to 100 fold, and a shorter induction period than the other 6-membered ring Al complexes in the polymerization of ε-caprolactone (96).

Further investigation of the steric effect in the phenolate ring of aminophenolate ligands demonstrated that Al complexes with *tert*-butyl groups on the 2,4 positions of the phenolate ring had the highest catalytic activity, followed by that of the cumyl groups. Also, it was found that the hydrogen substituents were responsible for the largest decrease in the catalytic activity of the Al complexes. The bulky substituents on the amino group increased the catalytic activity of associated Al complexes as well as the aminophenolate ligands with the third chelating atoms on the amino group. However, the aminophenolate ligands with the fourth chelating atoms on the amino group slightly decreased the catalytic activity (96).

Seven bis-ligated magnesium complexes resulting from the reaction of ketoimine pro-ligands with dibutyl magnesium could be isolated in yields of 59% to 70% (97). The complexes consisted of bis-ligated homoleptic ketoiminates coordinated in distorted octahedral geometry around the magnesium centers. The basic pro-ligand compounds are summarized in Table 3.7.

The complexes were investigated for their ability to initiate the

Table 3.7 Ketoimines.

Compound
N-Quinolyl-2-imino-3-methyl-pent-4-one
N-Quinolyl-2-imino-5,5-dimethyl-hex-4-one
N-Quinolyl-3-iminohept-5-one

ring-opening polymerization of *l*-lactide to PLA and ε-caprolactone to PCL in the presence of 4-fluorophenol as co-catalyst.

A series of trimethylene terephthalate ε-caprolactone block copolymers with different contents of trimethylene terephthalate as rigid, and PCL as flexible segments have been synthesized from dimethyl terephthalate, 1,3-bio-propanediol, and PCL in a two-step process involving transesterification and polycondensation in the melt (98).

In copolymers with 35% and 45% of flexible segments, the crystalline phase is formed during annealing above the glass transition temperature of copolymer. During cooling at the standard rate, these copolymers do not crystallize. It was also found that incorporation of flexible segments, due to the macrophase separated structure, cause a decrease of the melting point and glass transition temperature, along with the tensile modulus and tensile strength. At the same time, an increase in the value of the elongation at break is observed (98).

Trade names appearing in the references are shown in Table 3.8.

Table 3.8 Trade names in references.

Trade name Description	Supplier
Elvacite® 2041 Acrylic resin (74)	Lucite International Inc.
Flexan® (Series) Poly(styrene sulfonate) (74)	National Starch
Tinuvin® 1577 2-(4,6-diphenyl-1,3,5-triazin-2-yl)-5-hexyloxy-phenol (74)	Ciba Inc.

References

1. M.L.D. Lorenzo and R. Androsch, eds., *Synthesis, Structure and Properties of Poly(lactic acid)*, Vol. 279 of *Advances in Polymer Science* book series, Springer Nature, Cham, Switzerland, 2018.
2. J.M. Nduko and S. Taguchi, Microbial production and properties of LA-based polymers and oligomers from renewable feedstock in Z. Fang, R.L. Smith, Jr, and X.-F. Tian, eds., *Production of Materials from Sustainable Biomass Resources*, pp. 361–390. Springer Singapore, Singapore, 2019.
3. B.D. Favis, P. Sarazin, J. Li, and Z. Yuan, Microporous articles comprising biodegradable medical polymers, method of preparation thereof and method of use thereof, US Patent 8 007 823, assigned to Corporation de l'Ecole Polytechnique de Montreal (Montreal, CA), August 30, 2011.
4. A. Prasad, Polishing pad comprising biodegradable polymer, US Patent 7 264 641, assigned to Cabot Microelectronics Corporation (Aurora, IL), September 4, 2007.
5. R.A. de Oliveira, A. Komesu, C.E.V. Rossell, and R.M. Filho, *Biochemical Engineering Journal*, Vol. 133, p. 219, 2018.
6. M. Lübeck and P.S. Lübeck, *FEMS Microbiology Letters*, Vol. 366, 01 2019.
7. M. Singhvi, T. Zendo, and K. Sonomoto, *Applied Microbiology and Biotechnology*, Vol. 102, p. 5911, 2018.
8. I. De la Torre, M. Ladero, and V.E. Santos, *Applied Microbiology and Biotechnology*, Vol. 102, p. 10511, 2018.
9. J. Li, W. Zhang, X. Li, T. Ye, Y. Gan, A. Zhang, H. Chen, G. Xue, and Y. Liu, *Bioresource Technology*, Vol. 247, p. 890, 2018.
10. P. Liu, Z. Zheng, Q. Xu, Z. Qian, J. Liu, and J. Ouyang, *Process Biochemistry*, Vol. 71, p. 18 , 2018.
11. G.W. Coates, Syndiotactic poly(lactic acid), US Patent 6 608 170, assigned to Cornell Research Foundation, Inc. (Ithaca, NY), August 19, 2003.
12. M. Goino, Y. Kakida, and S. Obuchi, Biodegradable resin composition, US Patent 7 714 048, assigned to Toho Chemical Industry Co., Ltd. (Tokyo, JP), May 11, 2010.
13. K. Enomoto, M. Ajioka, and A. Yamaguchi, Polyhydroxycarboxylic acid and preparation process thereof, US Patent 5 310 865, assigned to Mitsui Toatsu Chemicals, Incorporated (Tokyo, JP), May 10, 1994.
14. P. Salzberg, Optically active homopolymers containing but one antipodal species of an α-monohydroxy monocarboxylic acid, US Patent 2 758 987, assigned to DuPont (Wilmington DE), August 14, 1956.

15. M. Fujiwara, J.L. Hedrick, D.P. Sanders, and M. Yasumoto, Cyclic carbonyl compounds with pendant carbonate groups, preparations thereof, and polymers therefrom, US Patent 8 207 351, assigned to International Business Machines Corporation (Armonk, NY) Central Glass Co., Ltd. (Ube-shi, Yamaguchi, JP), June 26, 2012.
16. R.G. Sinclair, Copolymers of *L*-lactide and epsilon caprolactone, US Patent 4 057 537, assigned to Gulf Oil Corporation (Pittsburgh, PA), November 8, 1977.
17. T. Kashima, T. Kameoka, C. Higuchi, M. Ajioka, and A. Yamaguchi, Aliphatic polyester and preparation process thereof, US Patent 5 428 126, assigned to Mitsui Toatsu Chemicals, Inc. (Tokyo, JP), June 27, 1995.
18. I. Kitada, C. Higuchi, M. Ajioka, and A. Yamaguchi, Preparation process of degradable block copolyesters, EP Patent 0 712 880, assigned to Mitsui Toatsu Chemicals, May 22, 1996.
19. M. Müller, J. Hess, W.-G. Schnell, D. Bendix, and G. Entenmann, Meso-lactide, US Patent 5 214 159, assigned to Boehringer Ingelheim GmbH (Ingelheim am Rhein, DE), May 25, 1993.
20. K. Kubota and Y. Murakami, Process for producing lactides and process for purifying crude lactides, US Patent 5 463 086, assigned to Dainippon Ink and Chemicals, Inc. (Tokyo, JP), October 31, 1995.
21. K. Bernardo, S. Leppard, A. Robert, G. Commenges, F. Dahan, and B. Meunier, *Inorg. Chem.*, Vol. 35, p. 387, January 1996.
22. K. Hiltunen, J.V. Seppälä, and M. Härkönen, *Macromolecules*, Vol. 30, p. 373, February 1997.
23. K. Hiltunen, J.V. Seppälä, and M. Härkönen, *Journal of Applied Polymer Science*, Vol. 63, p. 1091, 1997.
24. K. Hiltunen, J.V. Seppälä, and M. Härkönen, *Journal of Applied Polymer Science*, Vol. 64, p. 865, 1997.
25. K. Hiltunen, M. Härkönen, J.V. Seppälä, and T. Väänänen, *Macromolecules*, Vol. 29, p. 8677, 1996.
26. A. Takenaka and S. Nomoto, Biodegradable resin composition, US Patent 7 652 085, assigned to Kao Corporation (Tokyo, JP), January 26, 2010.
27. J. Tuominen, J. Kylmä, and J. Seppälä, *Polymer*, Vol. 43, p. 3, 2002.
28. Y. Fujita and O. Sawa, Ester compound, plasticizer for biodegradable aliphatic polyester resin, and biodegradable resin composition, US Patent 7 166 654, assigned to Daihachi Chemical Industry Co., Ltd. (Osaka, JP), January 23, 2007.
29. A. Takenaka and H. Kishimoto, Biodegradable resin composition, US Patent 8 252 852, assigned to Kao Corporation (Tokyo, JP), August 28, 2012.
30. K. Hiltunen, J.V. Seppälä, M. Itävaara, and M. Härkönen, *Journal of Polymers and the Environment*, Vol. 5, p. 167, 1997.

31. I. Takahashi, H. Iida, and N. Nakamura, Carbodiimide composition with suppressed yellowing, a stabilizer against hydrolysis and a thermoplastic resin composition, US Patent 7368493, assigned to Nisshinbo Industries, Inc. (Tokyo, JP), May 6, 2008.
32. Y. Huang, C. Zhang, Y. Pan, Y. Zhou, L. Jiang, and Y. Dan, *Polymer Degradation and Stability*, Vol. 98, p. 943, May 2013.
33. A. Zubov and G. Sin, *Chemical Engineering Journal*, Vol. 336, p. 361, 2018.
34. B.T. Ergan and M. Bayramoğlu, *Chemical Engineering Communications*, Vol. 205, p. 1665, 2018.
35. V. Padilla-Gainza, G. Morales, H. Rodríguez-Tobías, and K. Lozano, *Journal of Applied Polymer Science*, Vol. 136, p. 47643, 2019.
36. R. Pan, W. Xuan, J. Chen, S. Dong, H. Jin, X. Wang, H. Li, and J. Luo, *Nano Energy*, Vol. 45, p. 193, 2018.
37. M. Thiruchitrambalam, A. Alavudeen, and N. Venkateshwaran, *Rev. Adv. Mater. Sci.*, Vol. 32, p. 106, 2012.
38. W. Liu, L.T. Drzal, A.K. Mohanty, and M. Misra, *Composites Part B: Engineering*, Vol. 38, p. 352, April 2007.
39. M. Zampaloni, F. Pourboghrat, S. Yankovich, B. Rodgers, J. Moore, L. Drzal, A. Mohanty, and M. Misra, *Composites Part A: Applied Science and Manufacturing*, Vol. 38, p. 1569, June 2007.
40. N. Mallegni, T.V. Phuong, M.-B. Coltelli, P. Cinelli, and A. Lazzeri, *Materials*, Vol. 11, 2018.
41. S.S. Karkhanis, N.M. Stark, R.C. Sabo, and L.M. Matuana, *Polymer Engineering & Science*, Vol. 58, p. 1965, 2018.
42. P. Calcagnile, T. Sibillano, C. Giannini, A. Sannino, and C. Demitri, *Journal of Applied Polymer Science*, Vol. 136, p. 47546, 2019.
43. C. Way, D.Y. Wu, D. Cram, K. Dean, and E. Palombo, *Journal of Polymers and the Environment*, Vol. 21, p. 54, March 2013.
44. L. Wei, U.P. Agarwal, L. Matuana, R.C. Sabo, and N.M. Stark, *Polymer*, Vol. 135, p. 305, 2018.
45. M. Kim, J.H. Jeong, J.-Y. Lee, A. Capasso, F. Bonaccorso, S.-H. Kang, Y.-K. Lee, and G.-H. Lee, *ACS Applied Materials & Interfaces*, Vol. 11, p. 11841, 2019.
46. G. Spinelli, P. Lamberti, V. Tucci, R. Kotsilkova, S. Tabakova, R. Ivanova, P. Angelova, V. Angelov, E. Ivanov, R. Di Maio, C. Silvestre, D. Meisak, A. Paddubskaya, and P. Kuzhir, *Materials*, Vol. 11, p. 2256, November 2018.
47. F. Galiano, A.H. Ghanim, K.T. Rashid, T. Marino, S. Simone, Q.F. Alsalhy, and A. Figoli, *Clean Technologies and Environmental Policy*, Vol. 21, p. 109, 2019.
48. R. Xing, R. Huang, W. Qi, R. Su, and Z. He, *AIChE Journal*, Vol. 64, p. 3700, 2018.

49. P. Arriagada, H. Palza, P. Palma, M. Flores, and P. Caviedes, *Journal of Biomedical Materials Research Part A*, Vol. 106, p. 1051, 2018.
50. Wikipedia contributors, MTT assay — Wikipedia, the free encyclopedia, https://en.wikipedia.org/w/index.php?title=MTT_assay&oldid=889392768, 2019. [Online; accessed 5-June-2019].
51. M.S. Huda, L.T. Drzal, A.K. Mohanty, and M. Misra, *Composites Science and Technology*, Vol. 68, p. 424, February 2008.
52. A.M.P. Dupraz, J.R. de Wijn, S.A.T. v.d. Meer, and K. de Groot, *Journal of Biomedical Materials Research*, Vol. 30, p. 231, 1996.
53. T. Lebarbé, E. Ibarboure, B. Gadenne, C. Alfos, and H. Cramail, *Polymer Chemistry*, Vol. 4, p. 3357, 2013.
54. G.B. Kenion, A.P. Mgaya, B. Ramalingam, and Z. Chang, Transesterification of polylactic acid with natural oils, US Patent 10 294 328, assigned to Henkel IP & Holding GmbH (Düsseldorf, DE), May 21, 2019.
55. T. Miyata and T. Masuko, *Polymer*, Vol. 39, p. 5515, October 1998.
56. M. Yasuniwa, S. Tsubakihara, K. Iura, Y. Ono, Y. Dan, and K. Takahashi, *Polymer*, Vol. 47, p. 7554, October 2006.
57. N. Naga, Y. Yoshida, K. Noguchi, and S. Murase, *Open Journal of Polymer Chemistry*, Vol. 3, p. 29, 2013.
58. M.S. Huda, L.T. Drzal, A.K. Mohanty, and M. Misra, *Composite Interfaces*, Vol. 15, p. 169, January 2008.
59. M. Huda, L. Drzal, A. Mohanty, and M. Misra, *Composites Science and Technology*, Vol. 66, p. 1813, September 2006.
60. H. Hinago, H. Nagahara, and T. Aoki, Method for producing glycolic acid, US Patent 8 106 238, assigned to Asahi Kasei Chemicals Corporation (Tokyo, JP), January 31, 2012.
61. Y. Sun, H. Wang, and Z. Liu, Process for the production of glycolic acid, US Patent 8 299 297, assigned to Dalian Institute of Chemical Physics, Chinese Academy of Sciences (Liaoning, CN) BP P.L.C. (London, GB), October 30, 2012.
62. I. Kozhevnikov, Catalysis by polyoxometalates in *Catalysis for Fine Chemical Synthesis*, Vol. 2. John Wiley & Sons, Ltd, New York, 2003.
63. R. DiCosimo, A. Panova, J.S. Thompson, R.D. Fallon, F.G. Gallagher, T. Foo, X. Li, G.C. Fox, J.J. Zaher, M.S. Payne, and D.P. O'Keefe, Process for producing glycolic acid from formaldehyde and hydrogen cyanide, US Patent 8 110 382, assigned to E. I. du Pont de Nemours and Company (Wilmington, DE), February 7, 2012.
64. R. DiCosimo, A. Panova, S.D. Arthur, and H.K. Chenault, Process for enzymatically converting glycolonitrile to glycolic acid, US Patent 7 927 847, assigned to E. I. du Pont de Nemours and Company (Wilmington, DE), April 19, 2011.
65. H. Sato, F. Kobayii, Y. Ichikawa, and Y. Oishi, *Kobunshi Ronbunshu*, Vol. 69, p. 60, 2012.

66. L. Nurmi and T. Gadda, Method of producing glycolic acid polymers, US Patent 10 227 446, assigned to Valtion Teknillinen Tutkimuskeskus, March 12, 2019.
67. Y. Wang, G. Ameer, and R. Langer, Biodegradable polymer, US Patent 7 722 894, assigned to Massachusetts Institute of Technology (Cambridge, MA), May 25, 2010.
68. D.A. Barrera, E. Zylstra, P. Lansbury, Jr., and R. Langer, *Journal of the American Chemical Society*, Vol. 115, p. 11010, 1993.
69. J.L. West and J.A. Hubbell, *Macromolecules*, Vol. 32, p. 241, 1999.
70. T.A. Shmool and J.A. Zeitler, *Polym. Chem.*, Vol. 10, p. 351, 2019.
71. F. Nishimura, H. Hoshina, Y. Ozaki, and H. Sato, *Polymer Journal*, Vol. 51, p. 237, 2019.
72. M.J.-L. Tschan, E. Brulé, P. Haquette, and C.M. Thomas, *Polymer Chemistry*, Vol. 3, p. 836, 2012.
73. M.W. van Rossum, M. Alberda, and L.H. van der Plas, *Phytochemistry*, Vol. 49, p. 723, October 1998.
74. J. Pickett, Q. Ye, and D. Steiger, Tulipalin copolymers, US Patent 7 932 336, assigned to SABIC Innovative Plastics IP, B.V. (Bergen op Zoom, NL), April 26, 2011.
75. S. Agarwal, Q. Jin, and S. Maji, Biobased polymers from plant-derived tulipalin A, in P.B. Smith and R.A. Gross, eds., *Biobased Monomers, Polymers, and Materials*, Vol. 1105 of *ACS Symposium Series*, pp. 197–212. American Chemical Society, January 2012.
76. M.K. Akkapeddi, *Macromolecules*, Vol. 12, p. 546, July 1979.
77. J. Mosnáček and K. Matyjaszewski, *Macromolecules*, Vol. 41, p. 5509, 2008.
78. J. Stansbury and J. Antonucci, *Dental Materials*, Vol. 8, p. 270, July 1992.
79. J. Kollár, M. Mrlík, D. Moravčíková, B. Iván, and J. Mosnáček, *European Polymer Journal*, Vol. 115, p. 99 , 2019.
80. P. Rychter, D. Rogacz, K. Lewicka, J. Kollár, M. Kawalec, and J. Mosnáček, *Advances in Polymer Technology*, Vol. 2019, 2019.
81. L.E. Manzer, *Applied Catalysis A: General*, Vol. 272, p. 249, September 2004.
82. J. Martin, P.C. Watts, and F. Johnson, *Journal of the Chemical Society D: Chemical Communications*, p. 27, 1970.
83. D.R. Coulson, L.E. Manzer, and N. Herron, Process for the preparation of alpha-methylene lactones, US Patent 6 313 318, assigned to E.I. du Pont de Nemours and Company (Wilmington, DE), November 6, 2001.
84. L.E. Manzer and K. Kourtakis, Gas phase synthesis of methylene lactones using catalysts derived from hydrotalcite precursors, US Patent 7 348 442, assigned to E. I. du Pont de Nemours and Company (Wilmington, DE), March 25, 2008.
85. W. Luo, U. Deka, A.M. Beale, E.R. van Eck, P.C. Bruijnincx, and B.M. Weckhuysen, *Journal of Catalysis*, Vol. 301, p. 175, May 2013.

86. G. Qi, M. Nolan, F.J. Schork, and C.W. Jones, *Journal of Polymer Science Part A: Polymer Chemistry*, Vol. 46, p. 5929, September 2008.
87. S. Xu, J. Huang, S. Xu, and Y. Luo, *Polymer*, Vol. 54, p. 1779, March 2013.
88. G.M. Miyake, Y. Zhang, and E.Y.-X. Chen, *Macromolecules*, Vol. 43, p. 4902, June 2010.
89. Y. Hu, L.O. Gustafson, H. Zhu, and E.Y.-X. Chen, *Journal of Polymer Science Part A: Polymer Chemistry*, Vol. 49, p. 2008, May 2011.
90. D.M. Alonso, S.G. Wettstein, and J.A. Dumesic, *Green Chemistry*, Vol. 15, p. 584, 2013.
91. P. Gallezot, *Catalysis Today*, Vol. 121, p. 76, March 2007.
92. K. Khemani, P.J. Andersen, S.K. Hodson, and H. Schmidt, Biodegradable polymer films and sheets suitable for use as laminate coatings as well as wraps and other packaging materials, US Patent 6 573 340, assigned to Biotec Biologische Naturverpackungen GmbH & Co. KG (DE), June 3, 2003.
93. V. Snieckus and P. Richardson, *Synfacts*, Vol. 15, p. 0348, 2019.
94. K. Balke, M. Kadow, H. Mallin, S. Saß, and U.T. Bornscheuer, *Organic & Biomolecular Chemistry*, Vol. 10, p. 6249, 2012.
95. C. Shi, S. Zhang, M. Li, W. Sun, G. Fan, Y. Jin, J. Yang, and T. Dong, *Journal of Applied Polymer Science*, Vol. 130, p. 1805, 2013.
96. W.-Y. Lu, Y.-C. Chang, C.-J. Lian, K.-H. Wu, M.Y. Chiang, H.-Y. Chen, C.-C. Lin, and B.-T. Ko, *European Polymer Journal*, Vol. 114, p. 151, 2019.
97. R.M. Slattery, A.E. Stahl, K.R. Brereton, A.L. Rheingold, D.B. Green, and J.M. Fritsch, *Journal of Polymer Science Part A: Polymer Chemistry*, Vol. 57, p. 48, 2019.
98. S. Paszkiewicz, A. Szymczyk, I. Irska, D. Pawlikowska, and E. Piesowicz, *Journal of Applied Polymer Science*, Vol. 136, p. 47341, 2019.

4

Ester and Amide Polymers

4.1 Poly(ester)s

4.1.1 Poly(hydroxyalkonate)s

Most plastics, materials, fuels and other organic chemicals are derived from fossil fuel feedstocks at the moment (1). Due to the finite nature and foreseeable depletion potential of these raw materials, concerted efforts are being explored to find sustainable alternatives to the fossil fuel feedstock-derived products. Among these, bioplastics and oligomers derived from fermentation of the renewable plant biomass are promising candidates to replace fossil-fuel-derived plastics. Bioplastics are a class of storage polymers synthesized by microorganisms. Natural plastics can also be produced via a biochemical process that combines fermentative production of monomers or oligomers, followed by a chemical synthesis process to produce a variety of polymers. These polymers, in particular poly(hydroxyalkanoate)s, are futuristic biomaterials owing to their biodegradability and biocompatibility.

Poly(hydroxyalkanoate)s have physicochemical properties that are similar to petrochemical-based plastics (1).

Poly(hydroxyalkonate)s are natural polymers and they are getting attention for drug delivery due to its biodegradability and biocompatible properties (2). Poly(hydroxyalkonate) with different monomer compositions has different physical and chemical properties which further affect drug loading and release.

Poly(hydroxyalkonate) polymers of different monomer composition can be synthesized using various approaches such as metabolic engineering, precursor feedings, and others.

Poly(hydroxyalkonate) polymers in their intrinsic form or block copolymers forms can be used to prepare different structures such as microsphere, naonparticles, nanofiber, and scaffold, etc. Drug and pharmaceutically active molecule loaded poly(hydroxyalkonate) material carrier can be delivered through oral, intravenous or subcutaneous route. Selection of drug delivery route depends on the nature and required dose of drug.

Drug release from the polymeric material occurs through osmosis, diffusion or biodegradation mechanism. Composition of polymeric material affects its drug release behavior. The various trends in poly(hydroxyalkonate) polymer production and their application in drug delivery have been discussed (2).

4.1.2 *Methyl-10-undecenoate*

Methyl-10-undecenoate, c.f. Figure 4.1, has been described as a raw material for the synthesis of renewable semicrystalline polyesters and poly(ester amide)s (3). Diester, ester-amide, monoamide and diamide linkages have been synthesized from methyl-10-undecenoate.

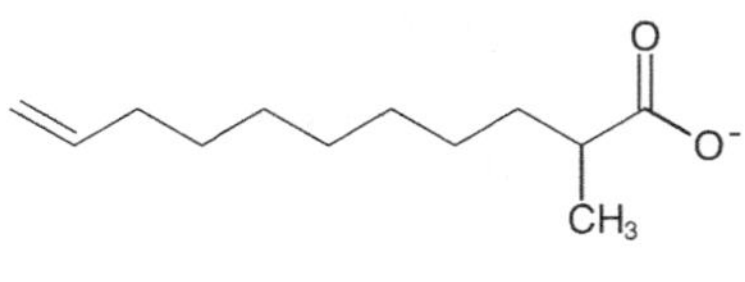

Figure 4.1 Methyl-10-undecenoate.

The synthesis runs via transesterification, amidation and thiol-ene reactions to get aliphatic diols containing ester; these diols are then reacted with a bio-based methyl diester. Most of these polyesters display a good thermal stability with temperature at 5% weight loss in the range of 330–350°C.

The incorporation of amide functions in the polyester backbones results in semicrystalline materials with melting points of 22–127°C and a complex melting behavior due to polymorphism and melting crystallization processes (3).

The issues of functional biopolymers have been collected in a monograph (4). An efficient synthesis strategy for the preparation of two renewable polyesters and one renewable polyamide via catalytic oxyfunctionalization of methyl-10-undecenoate, a castor oil derived platform chemical, has been described (5). The keto-fatty acid methyl ester can be synthesized applying a cocatalyst-free Wacker oxidation process using a high-pressure reactor system. For this purpose, catalytic amounts of palladium chloride are used in the presence of a dimethylacetamide/water mixture and molecular oxygen as sole reoxidant.

The thus derived monomers (hydroxy-esters, amine-esters) are synthesized from the obtained keto-fatty acid methyl ester through a Baeyer-Villiger oxidation and subsequent transesterification, reduction, or reductive amination, respectively.

The resulting step-growth monomers were then studied in homopolymerizations using 1,5,7-triazabicyclo[4.4.0]dec-5-ene, c.f. Figure 4.2, and titanium(IV) isopropoxide as transesterification catalyst, yielding polymers with molecular weights (M_n) up to 15 *kDa* (5).

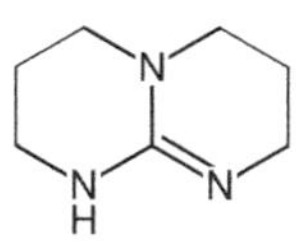

Figure 4.2 1,5,7-Triazabicyclo[4.4.0]dec-5-ene.

Also, the copolymerization of unsaturated bio-based esters with ethylene has been detailed (6, 7). The process of copolymerization of methyl-10-undecenoate with ethylene runs as follows (7):

Preparation 4–1: In a high-pressure autoclave a catalyst and silver trifluoromethanesulfonate (AgOTf) are dissolved in CH_2Cl_2. After the addition of the methyl-10-undecenoate the autoclave is purged with ethylene, closed, and stirred under a constant overpressure of ethylene (5 *bar*) at room temperature for 18 *h*. After cooling to room temperature, the pressure is released, MeOH/HCl is added and the mixture is stirred for 2 *h* to precipitate the polymeric product. The supernatant is removed and the polymer is washed with MeOH. The polymer is dissolved in CH_2Cl_2, precipitated by the addition of MeOH, and the supernatant is removed.

The solvents are removed under reduced pressure and the copolymer is obtained as a colorless sticky solid.

Renewable monomers derived from eugenol, c.f. Figure 4.3, and 10-undecenoic acid were synthesized via thiol-ene click and nucleophilic substitution reactions (8).

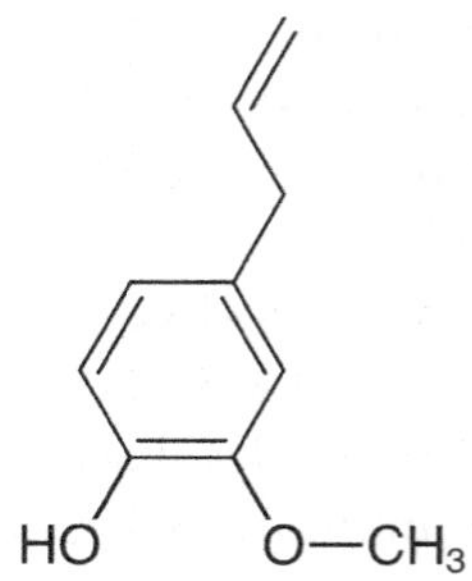

Figure 4.3 Eugenol.

With these monomers, a series of thermoplastic polyesters with tunable thermomechanical properties were prepared via a two-step melt polycondensation. The prepared polyesters exhibit weight-average molecular weights in the range of 21 *kDA* to 48 *kDA*, with polydispersity values between 1.8 and 2.1.

The thermal and mechanical properties of the polymers were studied. All prepared polyesters proved to be thermally stable at temperatures up to 300 °C. Furthermore, the aromatic and semi-aromatic polyesters exhibit fully amorphous behaviors with glass transition temperatures ranging between −34.13°C and 6.97°C.

The incorporation of aromatic eugenol into the polyester chains and the density of the benzene rings were also found to result in a significant improvement in the mechanical properties, including Young's modulus of *MPa* 11.6 to 44.2 *MPa* and an elongation at break in the range of 33.6–106.7%. The relatively high tan δ and the low storage modulus indicate that eugenol-based polyesters have superior viscosity properties (8).

10-Undecenoic acid, based on castor oil, was used as raw material for the synthesis of alternating copolymers (9). ω-Unsaturated fatty esters as alkyl 10-undecenoate were prepared by the esterification of 10-undecenoic acid with alkyl alcohol. A series of comblike copolymers were synthesized by the free radical polymerization from maleic anhydride and alkyl 10-undecenoates copolymers with different length of alkyl side chains in toluene solution.

The copolymers were obtained with a low molecular weight in a range of 3370–12,240 *Da*. Their structural characterization indicated the formation of alternating copolymers. Differential scanning calorimetry (DSC) characterization revealed that these comblike copolymers were amorphous to semicrystalline behavior by increasing the length of side chains (9).

The hydrolytic and enzymatic degradation of poly(10,11-epoxyundecanoic acid) with hydroxyl groups along the main chain and poly(11-hydroxyundecanoate)diol, were studied and compared to that of poly(caprolactone) (PCL) diol (10). The changes after polyester degradation in sample weight, molecular weight, chemical constitution, thermal properties, crystallinity and morphology were monitored by size exclusion chromatography (SEC), nuclear magnetic resonance spectroscopy (NMR), DSC, thermogravimetry (TG), environmental scanning electron microscopy (ESEM), and atomic force microscopy.

A significant decrease of the molecuar weight of poly(10,11-epoxyundecanoic acid) in both hydrolytic and enzymatic degradation was detected versus poly(11-hydroxyundecanoate)diol and PCL. This was related to its hydrophilic character, due to the presence of hydroxyl pendant groups, and its superior amorphous character. An accelerated degradation in acidic media, monitored by SEC and 1H NMR spectroscopy, showed complete degradation only in the case of poly(10,11-epoxyundecanoic acid) and PCL (10).

α-Diimine nickel catalysts show appreciable activities in the polymerizations of propylene, 1-hexene, 1-octene and 1-decene. Also, the catalysts were demonstrated to mediate efficient propylene copolymerizations with methyl-10-undecenoate and 10-undecen-1-ol, affording polar functionalized polyolefins (11). In addition, imine/phosphine-oxide ligands allow an efficient copolymerization of ethylene with methyl-10-undecenoate and 6-chloro-1-hexene (12). These monomers are shown in Figure 4.4.

A thioglycerol route to bio-based bis-cyclic carbonates has been

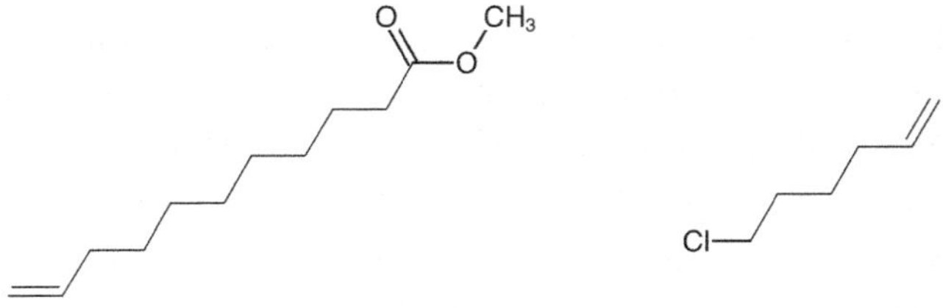

Methyl-10-undecenoate 6-Chloro-1-hexene

Figure 4.4 Monomers.

presented. This is based on the preparation poly(hydroxyurethane) and post-functionalization (13).

Sulfur-substituted cyclic carbonates have been synthesized from thioglycerol, fatty acids and sugar derivatives (13). In this methodology, a sulfur atom is inserted in the β-position of 5-membered ring cyclic carbonates via a two-step synthesis including the thiol-ene coupling of thioglycerol on fatty acid derivatives, followed by a transcarbonation.

A similar strategy was adopted to prepare glycolipid-based cyclic carbonates in order to bring biodegradability into the final poly(hydroxyurethane)s (13). The so-formed monomers were characterized by NMR spectroscopies, high performance liquid chromatography and DSC. The enhanced reactivity of sulfur-substituted cyclic carbonates was demonstrated through a ^{1}H-NMR spectroscopy kinetic study of a model reaction with hexylamine.

Fatty acid- and glycolipid-based sulfur-substituted bis-cyclic carbonates were then polymerized with diamines in a solvent using a catalyst-free process. Fourier transform infrared spectroscopy (FTIR), NMR, SEC, DSC and TG were performed to investigate the chemical structure, molar masses and thermal properties of the poly(hydroxyurethane). Finally, the so-formed poly(hydroxyurethane)s were post-functionalized by sulfonation with *m*-chloroperoxybenzoic acid (m-CPBA), taking advantage of thioether functions (13).

4.1.3 Poly(butylene adipate) Copolyesters

Monomers containing the 1,4-cyclohexylene unit are interesting because they are potentially bio-based and rigid enough to improve the glass transition temperature as well as the melting temperature (14). The preparation of poly(1,4-cyclohexylenedimethylene adipate) has been described.

The material has been compared to similar polymers, e.g., poly(butylene adipate), poly(butylene-1,4-cyclohexanedicarboxylate), and poly(1,4-cyclohexylenedimethylene-1,4-cyclohexanedicarboxylate).

The effect of the rigidity and stereochemistry of the 1,4-cyclohexylene unit on the thermal properties has been assessed. This effect reaches a maximum level when the cyclic moieties are the diacid or the diester and when trans isomers predominate (14).

4.1.4 Poly(hydroxyalkanoate)s

Poly(hydroxyalkanoate)s have been the focus of attention as a biodegradable and biocompatible substitute for the corresponding conventional nondegradable plastics (15). However, the costs of the large-scale production of these polymers have barred their widespread use. Now, new fermentation strategies have been developed for the efficient production of poly(hydroxyalkanoate)s.

Poly(hydroxyalkanoate)s are semicrystalline, thermoplastic polyester compounds that can either be produced by synthetic methods or by a variety of microorganisms such as bacteria and algae. Traditionally known bacterial poly(hydroxyalkanoate)s include poly(3-hydroxybutyrate), poly(hydroxybutyric acid), and poly(3-hydroxybutyrate-*co*-valerate) (16).

The ability to biodegrade readily in the presence of microorganisms has been demonstrated in numerous instances (16). Unfortunately, they are somewhat fragile polymers, which tend to exhibit brittle fracture. In addition, they are not tough, ductile or flexible polymers. The high melting points require processing temperatures that may contribute to an extensive thermal degradation in the melt.

The biosynthesis of medium-chain-length poly(hydroxyalkanoate)s requires oxygen; however, little is known about how medium-chain-length poly(hydroxyalkanoate) production is impacted in low dissolved oxygen environments (17). The growth and the medium chain length of poly(hydroxyalkanoate) under microaerophilic bioreactor conditions were observed in *Pseudomonas putida* LS46 when grown on carbon substrates to varying extent for medium-chain-length poly(hydroxyalkanoate) synthesis.

Poly(hydroxyalkanoate)s can be modified with long side chains. These types are virtually amorphous owing to the recurring higher alkyl side chains that are regularly spaced along the backbone.

However, the crystalline fraction has a very low melting point as well as an extremely slow crystallization rate, which seriously limits their potential of use.

Using poly(hydroxyalkanoate) blends can help to tailor the properties somewhat. Immiscible blends show an improved apparent biodegradability when compared to miscible blends. The biodegradability is mostly controlled by the morphology of the blend. Blends of poly(hydroxyalkanoate)s with other biodegradable polymers also usually show improved biodegradability in comparison to pure poly(hydroxyalkanoate)s (18).

Blending, however, often does not readily resolve the issue of the mechanical fragility and the lack of flexibility of high crystalline poly(hydroxyalkanoate)s while maintaining the biodegradable nature of these materials (16). A significant improvement in toughness has been observed for blends containing poly(butylene succinate-*co*-adipate) (Bionolle™). The changes in the mechanical properties are shown in Table 4.1.

Table 4.1 Mechanical properties of blends (16).

PHBHX	Bionolle™	Tensile modulus /[*MPa*]	Fracture toughness /[$kJ\,m^{-2}$]
0	100	217	518
30	70	208	465
50	50	248	401
70	30	275	179

PHBHX Poly(3-hydroxybutyrate)-*co*-3-hydroxyhexanoate (10.8%)
Bionolle™ Poly(butylene succinate-*co*-adipate)

4.1.5 *Poly(hydroxybutyrate)*

Poly(β-hydroxy butyric acid) is a thermoplastic polyester that is useful as a biodegradable plastics material. It is accumulated by many bacteria as an energy reserve material in the form of granules within the bacterial cells (19, 20).

It finds applications in many fields; for example, as a medical material, such as a surgical suture and a fixing material for treatment of bone fracture, and as a material for use in slow-releasing systems for slowly releasing pharmaceuticals and agricultural chemicals.

Particularly, as synthetic plastics have recently raised a serious social problem from the standpoint of environmental pollution and recycling of resources, poly(3-hydroxybutyrate) has attracted attention as a biopolymer not dependent on petroleum (20).

Poly(hydroxybutyrate) was first discovered in 1926 as a constituent of the bacterium *Bacillus megaterium* (21). Since then, poly(hydroxybutyrate) has been found in a wide variety of different genera of gram-negative and gram-positive bacteria (22).

The microorganisms produce poly(hydroxyalkanoate)s using R-β-hydroxyacyl-Coenzym-A enzymes as the direct metabolic substrate for a poly(hydroxyalkanoate) synthase.

Poly(β-hydroxy butyric acid) can be extracted from a suspension of bacterial cells, i.e., *Alcaligens eutrophus*, by causing the cells to flocculate, by pH modification, and then extracting the polymer from the flocculated cells. Flocculation of the cells renders subsequent separation of the poly(β-hydroxy butyric acid) solution from the cell debris more facile. Preferably, lipids are extracted from the flocculated cells before the contact with the extraction solvent is performed.

After agitation of the aqueous slurry with the extraction solvent, the two liquid phases may be separated: The cell debris remains in the aqueous phase while the poly(β-hydroxy butyric acid) is dissolved in the solvent phase. Suitable nonaqueous solvents can be chloroform, 1,2-dichloroethane, and methylene chloride (19, 23).

In addition, *Aeromonas caviae* can produce a random copolymer of 3-hydroxybutyric acid and 3-hydroxyhexanoic acid under aerobic conditions, when fed sodium salts of alkanoic acids of even carbon numbers and olive oil (24).

Conventionally, poly(3-hydroxybutyric acid) can be prepared by the ring-opening polymerization of β-butyrolactone (25). The ring-opening polymerization is carried out in the presence of a distannoxane derivative, e.g., 1-chloro-3-ethoxy-1,1,3,3-tetrabutyldistannoxane, a tin carboxylate compound, e.g., tin dioctanoate, and a alkyltin oxide, e.g., dibutyltin oxide. The catalysts help to get an increased molecular weight of the polymer. In this way, poly(3-hydroxybutyric acid) can be produced in high yield at low cost through simple production steps.

A practical route to a highly syndiotactic poly(3-hydroxybutyrate) (PHB) using a one-pot reaction has been presented (26). The stereoselective and controlled polymerization of racemic β-butyrolactone can be achieved by an initiator that is prepared *in-situ* from yttrium(III) isopropoxide and a bisphenoxide ligand.

Also, a bacterial pathway for the synthesis of PHB by *Ralstonia eutropha* H16 into the diatom *Phaeodactylum tricornutum* has been described. The PHB production is feasible in a microalgal system. There, the bioplastic accumulates in granule-like structures in the cytosol of the cells, as shown by light and electron microscopy (27).

The successful biochemical production of poly(3-hydroxybutyrate-*co*-3-hydroxyvalerate) in plants or bacteria requires the generation of substrates that can be utilized by biosynthetic enzymes. For the 3-hydroxybutyrate component, the substrate naturally exists in plants in sufficient concentration in the form of acetyl-CoA (28). For the 3-hydroxyvalerate, the starting substrate is propionyl-CoA.

A variety of different methods have been presented for optimizing levels of substrates employed in the biosynthesis of copolymers of 3-hydroxybutyrate and 3-hydroxyvalerate in plants and bacteria via manipulation of normal metabolic pathways using recombinant DNA techniques (29).

About 24 *h* incubation of *Azomonas* (A.) macrocytogenes isolate KC685000 in a 14 *l* fermenter produced 22% PHB per cell dry weight biopolymer using 1 *vvm* aeration, 10% inoculum size, and an initial pH of 7.2 (30). To control the fermentation process, logistic and Luedeking-Piret models were used to describe the cell growth and PHB production, respectively.

The best method for recovery of PHB was a chemical digestion method using sodium hypochlorite alone (30).

The production of PHB by purple non-sulfur bacteria using genetic engineering was first reported in 2019 (31). A PHB depolymerase gene-disrupted *Rhodobacter sphaeroides* HJ strain was constructed. This strain showed about 2.9-fold higher volumetric PHB production than that of the parent HJ (pLP-1.2) strain after 5 days of culture. Eight different genes involved in PHB biosynthesis in the genome of *R. sphaeroides* 2.4.1 could be identified.

4.1.6 *Poly(hydroxyvalerate)*

Poly(3-hydroxyvalerate) is commonly produced as a granular component in bacterial cells of various species (32). Based on 16S rDNA gene sequence analysis, strain PJC48 was identified as a *Bacillus* species.

A study was done to screen for a high-yield strain that can produce poly(3-hydroxyvalerate) efficiently and to increase the yield of poly(3-hydroxyvalerate) by optimizing the fermentative process (32).

A high-producing strain could be identified based on Nile red staining. Characterization by NMR spectroscopy of the poly(3-hydroxyvalerate) produced by PJC48 revealed that it consisted of (R)-3-hydroxyvalerate monomers. The suggested model was validated by response surface methodology. Optimization of the poly(3-hydroxyvalerate) yield resulted in an increase of 32.75% compared to a control, with a maximum production of 1.64 $g\,l^{-1}$ after 48 h (32).

4.1.6.1 *Copolymer of Hydroxybutyrate and Hydroxyvalerate*

In addition to the homopolymer of hydroxyvalerate, more studies concerning copolymers are available.

Bacterial Synthesis. A soft aliphatic polyester that may be used in manufacturing the polymer blends of a poly(3-hydroxybutyrate-*co*-4-hydroxybutyrate) (PHBV) copolymer, which is manufactured using a microbial-induced fermentation. One such PHBV copolyester is manufactured by Monsanto and has a glass transition temperature of about 0°C and a melting point of about 170°C.

In the fermentation process of manufacturing PHBV, a single bacterium species converts corn and potato feedstocks into a copolymer of poly(hydroxybutyrate) and hydroxyvalerate constituents. By manipulating the feedstocks, the proportions of the two polymer segments can be varied to make different grades of material. All grades are moisture resistant while still being fully biodegradable. The world producers of PHBV are Monsanto, with its Biopol® product, and Metabolix®, with its various grades of poly(hydroxyalkanoate)s (33). In January 2017 Metabolix Inc. changed its name to Yield10 Bioscience Inc.

A marine bacterium, *Halomonas* sp. SF2003, was grown on agro-industrial effluents as the sole carbon sources and it was able to produce PHB with a productivity of 1.3 $g\,l^{-1}$in 40 h of culture and a number average molar weight of 342 kDa (34). With the addition of valeric acid in the substrates, poly(3-hydroxybutyrate-*co*-3-hydroxyvalerate) with controlled proportion of hydroxyvalerate monomers were obtained. Their thermal and mechanical characteristics were investigated as a function of the amount of hydroxyvalerate and showed a decrease of the glass transition and melting temperatures and in the Young's modulus when the hydroxyvalerate content increased (34).

Reactive Extrusion. Reactive extrusion for grafting poly(β-hydroxybutyrate-*co*-valerate) has been used. Other polymers suitable for grafting are poly(butylene succinate) (PBSU) or PCL (35). The grafting of 2-hydroxyethyl methacrylate with a peroxide is shown in Figure 4.5. 1,1,4,4-Tetramethyltetramethylene bis(*tert*-butyl peroxide) (Lupersol® 101) is a preferred peroxide (35).

The polymer compositions are water-responsive and thus useful as components in flushable and degradable articles. They are biodegradable compositions. Articles made from the compositions can be degraded in aeration tanks by aerobic degradation and anaerobic digestors by anaerobic degradation in wastewater treatment plants. Therefore, articles made from these compositions will not significantly increase the volume of sludge accumulated at wastewater treatment plants (35).

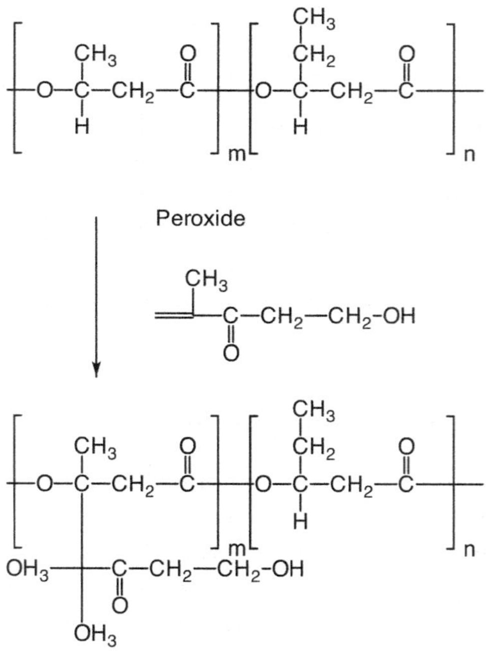

Figure 4.5 Schematic of grafting reaction (35).

Functionalization. The lipase-mediated functionalization of poly(3-hydroxybutyrate-*co*-3-hydroxyvalerate) with ascorbic acid into an antioxidant active biomaterial has been described (36). Lipase is an enzyme that catalyzes the hydrolysis of fats, i.e., lipids (37, 38). Lipases are a subclass of the esterases.

The copolymer was produced using engineered *Escherichia coli* YJ101, and further functionalized with ascorbic acid using *Candida antarctica* lipase B mediated esterification (36).

The ascorbic acid functionalized copolymer showed a lower degree of crystallinity of 9.96%, a higher thermal degradation temperature of 294.97°C and a hydrophilicity of 68°, as compared to the copolymer not functionalized. Furthermore, the functionalized biomaterial showed a 14% scavenging effect on 1,1-diphenyl-2-picryl-hydrazyl, and 1.6-fold increase in biodegradability as compared to the copolymer not functionalized.

Such an improvement of the polymer properties by adding functional groups could be a good approach to increase their biodegradability, economic value and important applications in the medical field (36).

4.1.7 *Poly(3-hydroxyhexanoic acid)*

The commercial production of poly(3-hydroxybutyrate-*co*-3-hydroxyhexanoate) *Aeromonas hydrophila* was examined in a 20 m^3 fermentor. The cells were first grown by using glucose as a carbon source, and the biosynthesis was triggered by the addition of lauric acid under conditions of limited nitrogen or phosphorus. The first grown cells in a medium containing 50 $g\,l^{-1}$ of glucose were further cultivated after the addition of 50 $g\,l^{-1}$ of lauric acid with limited phosphorus content; a final cell concentration of 50 $g\,l^{-1}$ and a concentration of 3-hydroxyhexanoate of 25 $g\,l^{-1}$ was achieve. This corresponds to a productivity of 0.54 $g\,l^{-1}h^{-1}$. The copolymer is a random copolymer (39).

Bacteria that can produce 3-hydroxyhexanoate-based polymers are summarized in Table 4.2.

The *in-vivo* tissue reactions and biodegradations of poly(3-hydroxybutyrate-*co*-3-hydroxyhexanoate) and other polymers have been evaluated by subcutaneous implantation in rabbits. The results revealed that the degradation rate increased in the order

Table 4.2 3-Hydroxyhexanoate polymers producing bacteria.

Bacterium	Reference
Alcaligens eutrophus	(19)
Aeromonas caviae	(24)
Aeromonas hydrophila	(39, 40)

of poly(3-hydroxybutyrate), poly(3-hydroxybutyrate-*co*-3-hydroxyhexanoate), poly(lactic acid) (PLA). During the implantation period, the crystallinity of poly(3-hydroxybutyrate-*co*-3-hydroxyhexanoate) increased from 19% to 22% and then dropped to 14%. The results suggest that a rapid degradation occurs in the amorphous region rather than in crystalline region (41).

The degree of crystallinity of solvent cast copolymer films decrease from 60 to 18% when the fraction of 3-hydroxyhexanoic acid is increased from 0 to 25 *mol*%. This finding suggests that the 3-hydroxyhexanoic acid units are excluded from the crystalline phase of 3-hydroxybutyric acid. The rates of enzymatic erosion increase markedly with the increase of 3-hydroxyhexanoic acid content and reach a maximum value at 15 *mol*% 3-hydroxyhexanoic acid (24).

Many poultry eggs are discarded worldwide because of infection, i.e., avian flu, or the presence of high levels of pesticides. The possibility of adopting egg yolk as a source material to produce poly(hydroxyalkanoate) biopolymer was examined (42).

Cupriavidus necator Re2133/pCB81 was used for the production of poly(3-hydroxybutyrate-*co*-3-hydroxyhexanoate), a polymer that would normally require long-chain fatty acids as carbon feedstocks for the incorporation of the 3-hydroxyhexanoate monomers (42). The optimal medium contained 5% egg yolk oil and ammonium nitrate as a nitrogen source, with a carbon-to-nitrogen ratio of 20.

Time course monitoring using the optimized medium was conducted for 5 *d*. Biomass production was 13.1 $g\,l^{-1}$, with 43.7% copolymer content. A comparison with other studies using plant oils and the current study using egg yolk oil revealed similar polymer yields. Thus, discarded egg yolks could be a potential source of poly(β-hydroxyalkanoate) (42).

A bionanocomposite of poly(3-hydroxybutyrate-*co*-3-hydroxyhexanoate) and cellulose nanocrystals was prepared by the cast-film method (43). The influence of the cellulose nanocrystal content on the morphology, crystallization, as well as thermal, mechanical, and barrier properties (oxygen and water vapor) of the bionanocomposite was investigated.

The results showed that the crystallization behavior of poly(3-hydroxybutyrate-*co*-3-hydroxyhexanoate) was changed with the addition of cellulose nanocrystals, resulting in an improvement in the thermal stability and mechanical properties of the bionanocomposite. In comparison to poly(3-hydroxybutyrate-*co*-3-hydroxyhexanoate), the water-vapor permeability coefficient of the bionanocomposite increased by 207.82% and the oxygen permeability coefficient decreased by 18.46%. The bionanocomposite showed optimal mechanical properties when the cellulose nanocrystal content was 2.3% (43).

4.1.7.1 Soil Burial Biodegradation

The soil burial biodegradation behavior of poly(3-hydroxybutyrate-*co*-3-hydroxyhexanoate) was studied based on its copolymer composition determined by thermally assisted hydrolysis and methylation-gas chromatography (44).

The change in copolymer composition of the film samples during soil burial degradation test was monitored in detail by methylation-gas chromatography. As a result, the copolymer compositions of the 3-hydroxyhexanoate units gradually increased, while those of 3-hydroxybutyrate units decreased with soil burial time, almost correlating with recovery weight. Based on these results, the biodegradation mechanism of poly(3-hydroxybutyrate-*co*-3-hydroxyhexanoate) was elucidated with considering the influence of the crystallinity on the biodegradability. Furthermore, local differences in biodegradability of a given poly(3-hydroxybutyrate-*co*-3-hydroxyhexanoate) film after soil burial test were also successfully evaluated from the copolymer compositions obtained by methylation-gas chromatography (44).

4.1.7.2 *Particle Fillers to Improve the Thermomechanical and Fire Retardant Properties*

The poor thermomechanical and flame-retardant properties of biopolymers are limiting their application and potential exploitation as sustainable polymers (45). The use of agricultural byproducts as a functional filler for biopolymers is presented here to address the production of environmentally friendly and economically sustainable biocomposites.

Hemp hurd and alfalfa particles were melt blended with a poly(3-hydroxybutyrate-*co*-3-hydroxyhexanoate) copolymer. The addition of 30% of filler achieves an improvement of 150% in stiffness. The same composites show an increase in the heat deflection temperature of more than 100°C. Also, the flame-retardant properties were evaluated and showed strong reductions in flame spread rates of 40% and the combustion kinetics of 30% (45).

The composites presented show an opportunity for the production of functional and sustainable materials through the valorization of agricultural byproducts (45).

4.1.8 *Poly(β-hydroxyoctanoate)*

A latex of poly(β-hydroxyoctanoate) can be obtained from the bacterium *Pseudomonas oleovorans* grown on sodium octanoate at a high cell density (46). In the course of purification, sodium hypochlorite was used. It was observed that latex stabilization occurred spontaneously due to the persistence around the polymer granules of the murein sacculus, which envelopes the bacterial cell. It was shown that the optimal conditions for the bacteria digestion correspond to a sodium hypochlorite concentration of 21–26 $mmol\ g^{-1}$.

The resulting films display the typical properties of a thermoplastic elastomer. Depending on the conditions of purification, fully amorphous poly(β-hydroxyoctanoate) films can also be obtained. Nanocomposite materials can be prepared when using this latex as a matrix and additionally using a colloidal suspension of hydrolyzed starch or cellulose whiskers as a natural and biodegradable filler. The properties are strongly dependent on the aspect ratio of the whiskers (47).

Poly(3-hydroxyoctanoate) with pendant carboxylic groups has been prepared by the chemical modification of unsaturated bacterial polyesters. The oxidation of the pendant alkene groups is complete and thereby a loss in molecular weight of polymer was observed. The introduction of pendant carboxylic groups enhances the hydrophilic character of the polymers (48).

Blends of bacterial poly(β-hydroxyoctanoate) and poly(hydroxybutyrate) can be prepared by dissolving the two polyesters in chloroform and then by casting the mixture. The casted blends are not miscible. This results in two-phase systems in which the nature of the continuous phase is dependent on the composition. The mechanical properties of these materials could be predicted using a model based on the concept of percolation (49).

Medium-chain-length poly(3-hydroxyalkanoate)s, such as poly(3-hydroxyoctanoate), exhibit a low crystallinity and are elastomeric in character. The synthesis of block copolymers containing poly(ethylene glycol), poly(3-hydroxy butyrate), and poly(3-hydroxy octanoate) has been reported (50).

The synthesis uses transamidation reactions with primary amine yield equimolar amounts and poly(3-hydroxy butyrate) with amine ends (50). Poly(3-hydroxy octanoate) reacts with the modified poly(3-hydroxy butyrate) containing the amine end to give the block copolymers.

The biodegradability of a series of plastic blends containing at least some biodegradable material was investigated (51). The materials were exposed to activated sludge for 150 *d*, and changes in mass, molecular weight average, and tensile properties were measured. Poly(β-hydroxybutyrate-*co*-hydroxyvalerate) showed a considerable loss of mass and a significant loss of tensile strength; however, a thermoplastic elastomer formed from poly(β-hydroxyoctanoate) did not degrade (51).

Another study has reported that poly(3-hydroxybutyrate-*co*-3-hydroxyoctanoate) was extensively mineralized in digester sludge, septage sediments, and landfill reactors, with half-lives less than 30 *d* (52).

On the other hand, PCL was not significantly mineralized in digester sludge over 122 *d*. In the landfill reactors, PCL mineralization was slow and was preceded by a long time lag of >200 *d*.

This suggests that the mineralization of PCL is limited by its rate of hydrolysis (52).

4.1.8.1 Controlled Release of Buprofezin

The extensive application of pesticides has caused a lot of environmental pollution and health problems, prompting the development of highly efficient and lowly-toxicity pesticide formulations (53).

Buprofezin-loaded poly(3-hydroxybutyrate-*co*-3-hydroxyhexanoate) microspheres were prepared by an oil-in-water emulsion solvent evaporation method. Buprofezin is an insecticide used for control of insect pests (54). It is a growth regulator, acting as an inhibitor of chitin synthesis. Buprofezin is shown in Figure 4.6.

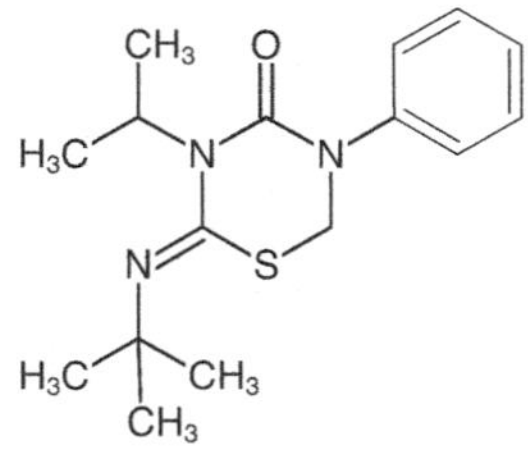

Figure 4.6 Buprofezin.

Under the optimal conditions (poly(3-hydroxybutyrate-*co*-3-hydroxyhexanoate) 7.07% (w/v) and poly(vinyl alcohol) (PVA) 1.84% (w/v)), the spherical and monodispersed microspheres were obtained. The average particle size, pesticide loading, and encapsulation efficiency of the optimized microspheres were 1.2 μm, 15.68%, and 78%, respectively. Release of 80% buprofezin from the microspheres in pH 5 (192 h) was faster than that in pH 7 (228 h) and 8 (204 h).

Moreover, a kinetic analysis indicated that the buprofezin release was done in a non-Fickian diffusion manner ($0.43 < n < 0.85$) and the release mechanism was the combined effects of pesticide diffusion and hydrolysis of polymer. The bioassay and toxicity results showed that encapsulation of buprofezin could exhibit high efficacy in the target organism and low-toxicity in the non-target organism. Therefore, these results demonstrated that buprofezin-loaded poly(3-hydroxybutyrate-*co*-3-hydroxyhexanoate) microspheres with good stability could be successfully prepared (53).

4.1.9 Poly(γ-glutamic acid)

Poly(γ-glutamic acid) is an edible polypeptide excreted by certain bacteria that is presently accessible on a semi-industrial scale (55). Chemically, it is a nylon-4 derivative bearing a carboxylic side group attached to the fourth carbon of the repeating unit.

Although this biopolymer is being exploited in the food, agriculture and cosmetic sectors, its use in massive applications as packaging or coating is still unknown mainly because of its instability in wet environments and unsuitability to be processed by common techniques (55).

Proton exchange membranes have been fabricated from poly(γ-glutamic acid) nanofibers with an excellent performance (56). They have great significance for the commercial application of proton exchange membrane fuel cells.

The nanofibers were first fabricated by solution blowing with the help of PLA and designed to form amino acid arrays as efficient proton channels for proton exchange membranes (56). Nanofibers with 50% poly(γ-glutamic acid) exhibit a high proton conductivity of 0.572 $S\,cm^{-1}$ at 80 °C and 50% relative humidity, and 1.28 $S\,cm^{-1}$ at 40°C and 90% relative humidity.

Then, the poly(γ-glutamic acid)/poly(lactic acid) nanofibers were incorporated into sulfonated poly(ether sulfone) to prepare proton exchange membranes, which showed a remarkable performance in comparison to a Nafion membrane. The composite membrane with 30% nanofibers exhibited the highest proton conductivity of 0.261 $S\,cm^{-1}$ at 80 °C and 100% relative humidity. The direct methanol fuel cells with this membrane showed a maximum power density of 202.3 $mW\,cm^{-2}$ among all of the proton exchange membranes, thus showing a great application potential in the field of proton exchange membranes (56).

Gallic acid is one of the main phenolic components occurring naturally in plants and has been a subject of increasing interest owing to its antioxidant, antimutagenic and anticarcinogenic properties. Gallic acid is shown in Figure 4.7.

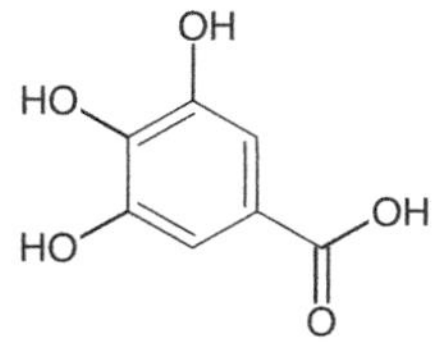

Figure 4.7 Gallic acid.

A rapid and cost-effective analytical procedure has been described for the determination of gallic acid (57). A poly(glutamic acid) graphene modified electrode was fabricated by the electro-polymerization of glutamic acid on reduced graphene oxide modified paraffin-impregnated graphite electrode. The electrochemical behavior of gallic acid at the modified sensor was studied by voltammetric and amperometric techniques under optimized conditions in an acetate buffer at pH 5 (57).

The electrode showed a good linear response for the determination of gallic acid over the range of 0.03–480 μM with 0.01 μM as the detection limit for voltammetric technique, and the amperometric technique showed a linear range of 1–17 μM with 0.33 μM as the detection limit. The electrode also showed good stability and reproducibility with a sensitivity of 0.97 $\mu M \, \mu A^{-1}$ (57).

Chemical modification of poly(γ-glutamic acid) involving esterification or amidation of the carboxylic group is an approach to obtain materials with potential interest as fibers, films or hydrogels (55). Electrostatic coupling of poly(γ-glutamic acid) with hydrophobic cationic species may also be beneficial for generating non-water-soluble stable derivatives.

Antibacterial films have been made from ionic complexes of poly(γ-glutamic acid) and ethyl lauroyl arginate chloride (58). The antibacterial activity of films made of these complexes against gram-positive (*L. monocytogenes* and *S. aureus*) and gram-negative (*E. coli* and *S. enterica*) bacteria was preliminary evaluated and was found to be very high against the former and only moderate against the latter. The bactericide activity displayed by the complexes was directly related to the amount of ethyl lauroyl arginate chloride that was released from the film into the environment. Ethyl lauroyl arginate is shown in Figure 4.8.

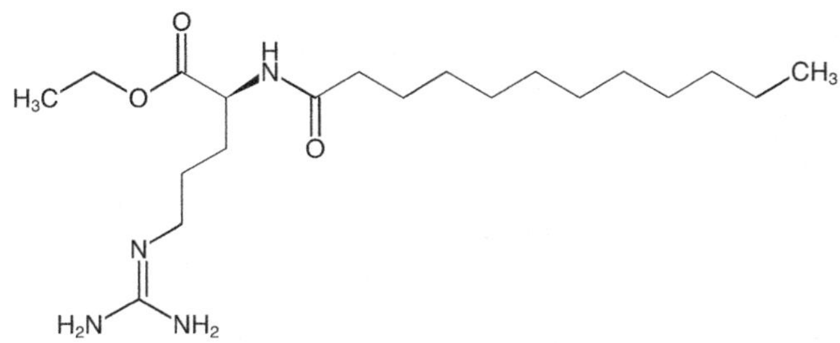

Figure 4.8 Ethyl lauroyl arginate.

The preparation of derivatives from poly(γ-glutamic acid) shows some difficulties, including its scarce solubility in organic solvents. A derivatization procedure for poly(γ-glutamic acid) has been reported that is based on the use of its tetrabutylammonium salt (59). The modified solubility of this polymer is provided by counter ion exchange and it led to the synthesis of poly(α-ethyl-γ-glutamate), poly(α-benzyl-γ-glutamate), and poly(α-n-butyl-γ-glutamate) under smoother conditions and an almost complete functionalization degree.

4.1.10 Poly(butylene succinate)

The synthesis of PBSU can be achieved by using microwaves. 1,3-Dichloro-1,1,3,3-tetrabutyldistannoxane is used as a catalyst (60). Optimally, a polymer with a weight average molecular weight of 23 kD was obtained within 20 *min*.

A wide variety of butylene succinate containing copolymers have been described. Comonomer units are butylene adipate (61, 62) or ethylene terephthalate (63). These compounds are shown in Figures 4.9 and 4.10.

Poly(butylene succinate-*co*-butylene adipate) can be obtained from 1,4-butanediol and dimethyl esters of succinic and adipic acids through a two-step process of transesterification and polycondensation (62). A catalyst for transesterification is tetra-n-butyl-titanate used at a temperature of 200°C (61). The change of the molar mass with reaction time is shown in Figure 4.11. As can be seen from

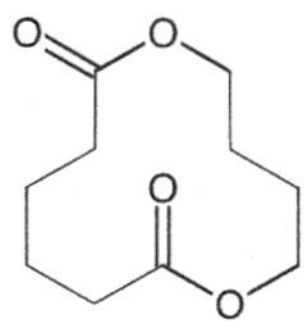

Figure 4.9 Butylene adipate.

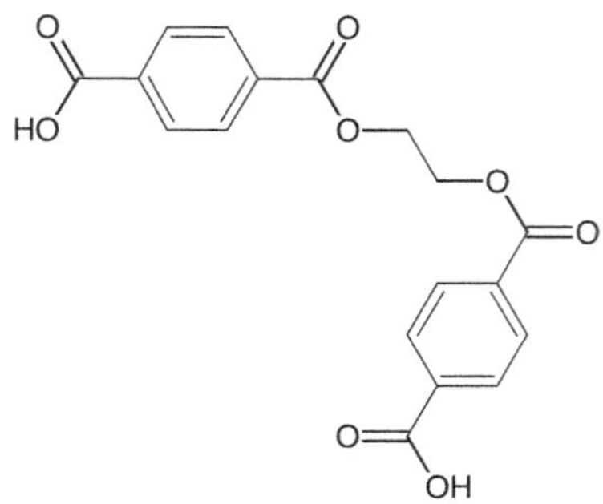

Figure 4.10 Ethylene terephthalate.

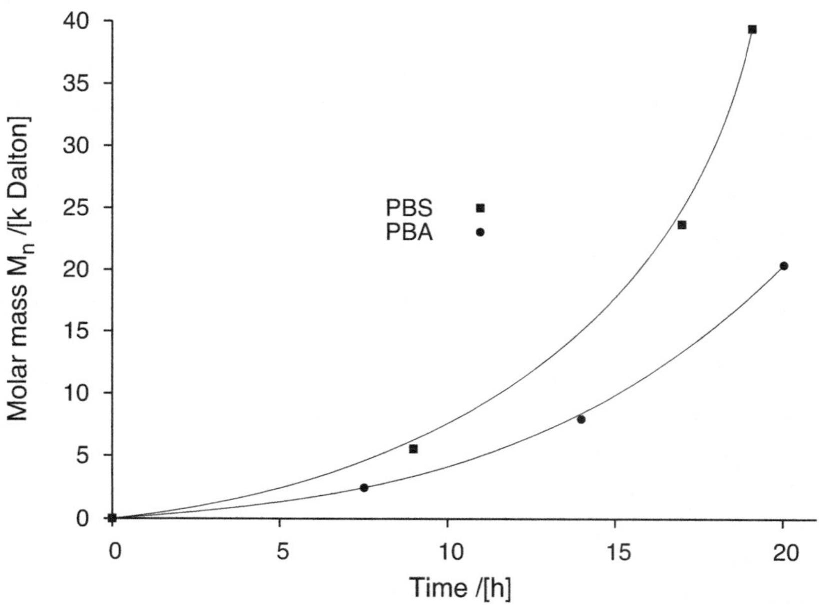

Figure 4.11 Change of the molar mass with reaction time (61).

Figure 4.11, the reaction of PBSU is faster than poly(butylene adipate). Obviously, dimethyl succinate is more reactive.

The biodegradability of copolyesters from butanediol and varying ratios of succinic acid and adipic acid expressed in terms of weight loss are shown in Figure 4.12.

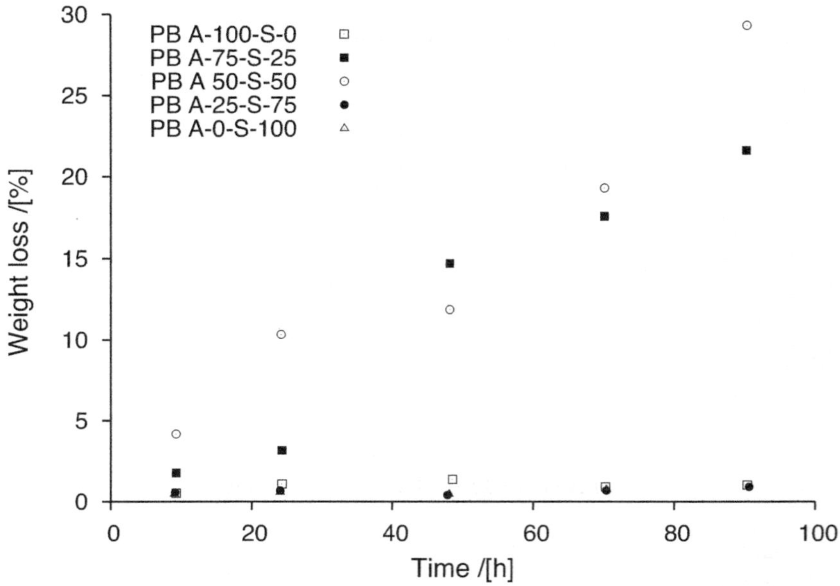

Figure 4.12 Biodegradability of copolyesters from Butanediol (B), Succinic acid (S), and Adipic acid (A) (61). Numbers refer to the monomer content in *mol%*.

Candida cylindracea lipase was used for biodegradation at a temperature of 30°C. The homopolymers do not show any significant weight loss. This probably occurs due to the high degree of crystallinity of the homopolymers. However, the copolymers with 50 *mol%* succinate degrade.

When hexamethylene diisocyanate is used as a chain extender, high molecular weight polyesters can be obtained. The chain extension reaction results in a significant increase in the molecular weight of polyester and also in an increased tensile strength. The rate of biodegradation is rather affected by the crystallinity, rather than by the molar mass.

Thermoplastic starch can be plasticized with a mixture of glycerol and PBSU. The blends have been prepared in a Haake mixer (64).

Processing properties, thermal behaviors, including melting and crystallization behavior, crystal structure, and compatibility of the blends, were investigated using DSC, wide-angle X-ray diffractometer, scanning electron microscopy (SEM) and thermal mechanical analysis (DMA).

The maximum and equilibrium torques decrease by increasing the glycerol contents and decreasing PBSU contents. The addition of starch and glycerol could lead to higher crystallinity and lower crystallization rate of PBSU, but would not change the crystal types and crystallite sizes of PBSU, as is suggested by DSC measurements (64).

A facile and effective way to fabricate PBSU carbon fiber composite foams with lightweight, high strength and improved conductive networks has been reported (65). The method involves the combination of solvent mixing, micro-injection molding and supercritical carbon dioxide foaming methods.

The experiments showed that the resulting composite foams possessed much higher electrical conductivity, suggesting that the introduction of the foaming technique could be beneficial for the formation of effective 3D conductivity networks. The composite foams presented a good compressive strength and a low density (65).

The flame retardancy of PBSU was improved by using isosorbide modified by sulfur, silicon or phosphorus as fire retardants. The syntheses of these fire retardants has been given in detail (66).

The TG results showed that all fire retardants exhibit thermal stabilities above 170°C and are suitable for the processing of PBSU. It was shown that (66):

1. Sulfur modified fire retardants showed no improvement of the PBSU flame retardancy,
2. Silicon modified fire retardants showed a slight improvement acting as char-forming agent when blending with intumescent fire retardant, and
3. The performances of phosphorus modified fire retardants depend on the oxidation degree of phosphorus.

4.1.11 *Dianhydrohexitol-Based Polymers*

The use of polymers from renewable 1,4,3,6-dianhydrohexitols has been reviewed (67, 68). Isosorbide, isomannide, and isoidide are basically stereo isomers from dianhydrohexitol compounds. 1,4,3,6-Dianhydrohexitol is shown in Figure 4.13.

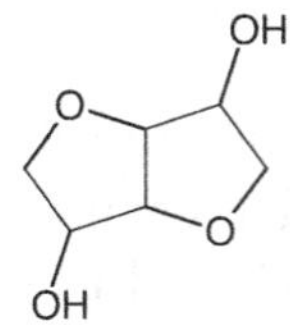

Figure 4.13 1,4,3,6-Dianhydrohexitol.

Isosorbide derivates are of considerable importance for the replacement of fossil resource-based products. Applications include building blocks for new polymers and functional materials, new organic solvents, for medical and pharmaceutical applications, as well as fuels (69). The derivatization of isosorbide is somewhat problematic because of the different configurations of the 2- and 5-positions and thus of the different reactivities of the hydroxyl groups.

Bis(2-hydroxyethyl)isosorbide is a biomass-derived glycol, which has been found to be a valuable monomer for a wide variety of polymeric materials (70). It may be prepared from isosorbide, which is derived from renewable resources such as sugars and starches. Glucose can be obtained from starch polymers. Isosorbide can be made from *D*-glucose by hydrogenation followed by an acid-catalyzed dehydration (70, 71). The synthesis of isosorbide from glucose is shown in Figure 4.14.

In general, hydroxyethylation processes with ethylene carbonate generally result in a mixture of products. The initially formed bis(2-hydroxyethyl)isosorbide may further react with ethylene carbonate to form higher hydroxy-terminated poly(ether)s. The formation of such higher poly(ether)s may be controlled by the amount of added ethylene carbonate and the nature of the catalyst system (70). Furthermore, the raw bis(2-hydroxyethyl)isosorbide may be purified by distillation.

Isosorbide is used for medical applications, such as a diuretic agent, and also can be used for the treatment of eye diseases to decrease the fluid pressure in the eye.

Isosorbide can be modified with a wide variety of chemicals to

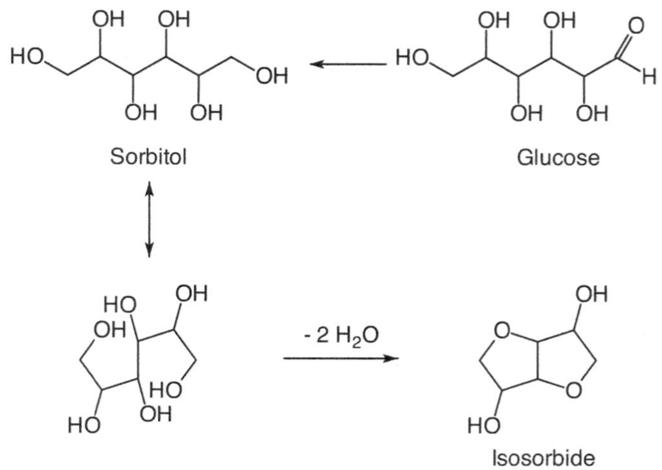

Figure 4.14 Synthesis of isosorbide from glucose.

get derivates (68). The reaction with ethylene carbonate to get bis(2-hydroxyethyl)isosorbide is shown in Figure 4.15.

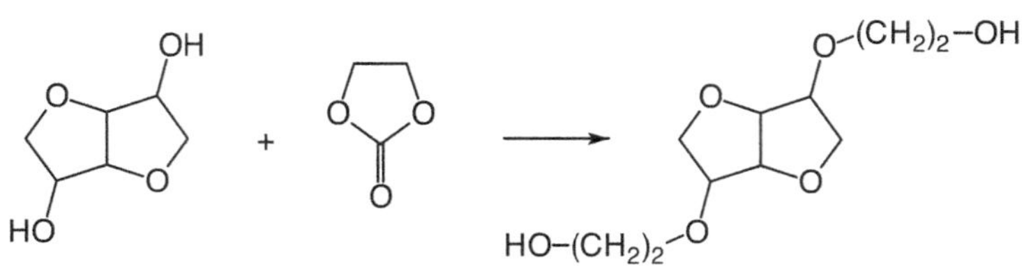

Figure 4.15 Synthesis of bis(2-hydroxyethyl)isosorbide.

Eventually, bis(2-hydroxyethyl)isosorbide may be chemically modified to form bisacrylate esters or bis-vinyl ethers, which can be used in photochemically or thermally cured vinyl resin compositions. Alternatively, this compound may be reacted with epichlorohydrin and bases to form bis(glycidyl ether) adducts which may be used in epoxy resins. Bis(2-hydroxyethyl)isosorbide will find use as a monomer in many polymeric compositions such as polyesters, poly(urethane)s and poly(ether sulfone)s.

Polyesters can be synthesized by the polycondensation dicarboxylic acid chlorides, as shown in Figure 4.16.

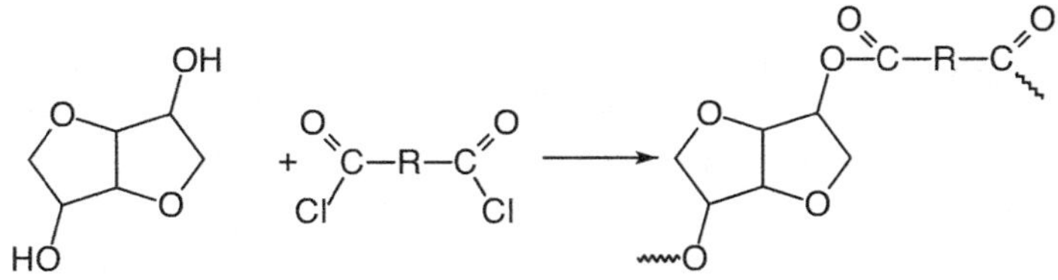

Figure 4.16 Synthesis of polyesters (68).

For technical purposes, a melt polymerization method is used, the dicarboxylic acid component being acids or esters (70). As catalysts, metal oxides or acetate salts may be used, such as manganese(II) acetate tetrahydrate and antimony(III) trioxide.

The glass transition temperature of isosorbide-based aliphatic polyesters depending on the number of methylene groups in the aliphatic acid is shown in Figure 4.17.

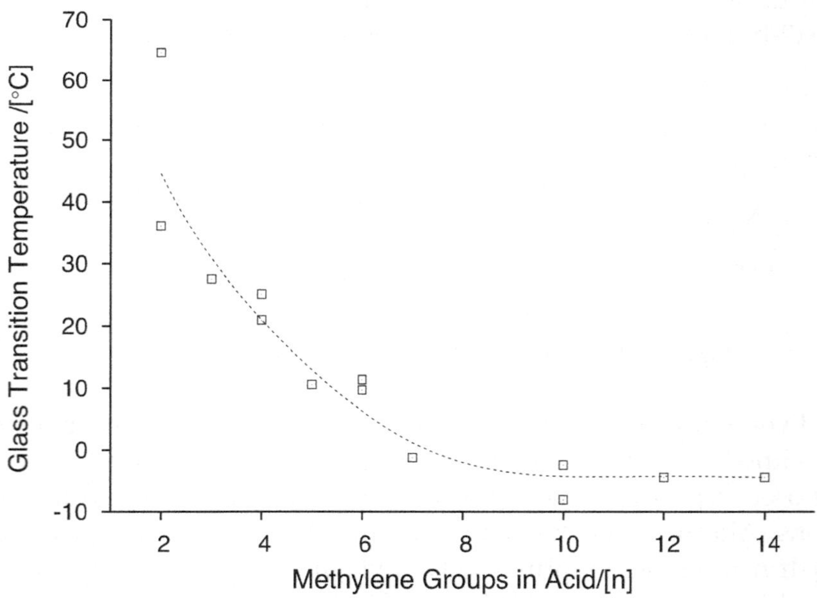

Figure 4.17 Glass transition temperature of isosorbide-based aliphatic polyesters depending on the number (n) of methylene groups in the aliphatic acid (67, 68, 72).

In contrast to aliphatic polyesters, for example, terephthalic-based polyesters show a much higher glass transition temperature of 155°C, or even higher, depending on the molar mass of the polyester (68, 73).

The Flory-Fox equation relates the number average molecular weight M_n to the glass transition temperature T_g of a polymer as (74)

$$T_g(M_n) = T_g(M_n = \infty) - \frac{K}{M_n} \quad . \tag{4.1}$$

A Flory-Fox plot of the glass transition temperatures of poly(isosorbide terephthalate) esters of different molar mass is shown in Figure 4.18.

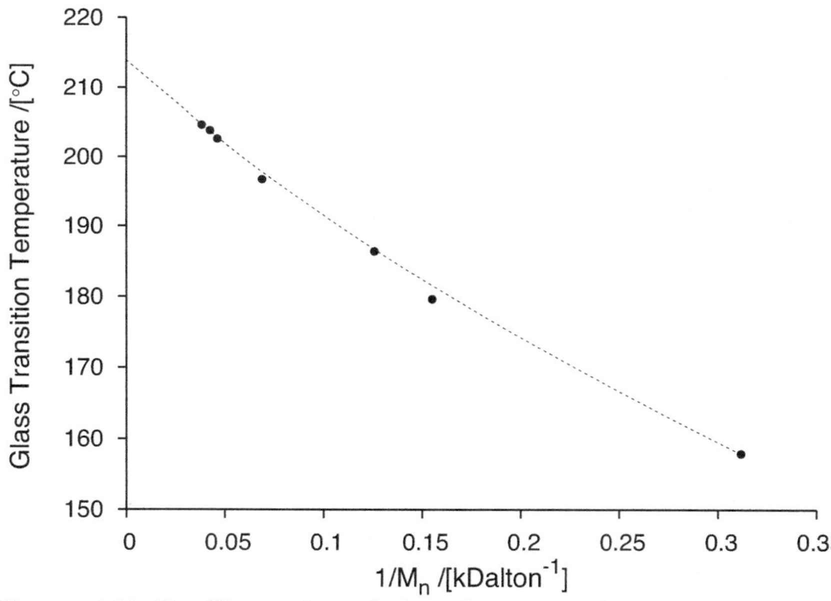

Figure 4.18 Fox-Flory plot of the glass transition temperatures of poly(isosorbide terephthalate) (68, 73).

The PBSU is restricted for wider applications due to its low thermal/mechanical properties and its unsatisfactory (bio)degradability (75). A series of PBSU copolyesters based on the carbohydrate-derived isosorbide (1,4:3,6-dianhydro-*D*-glucidol, isosorbide) were tested to be rigid and hydrophilic.

The results of this study clearly demonstrated that the presence of isosorbide enhances the glass transition temperature almost linearly and simultaneously promotes (neutral, acidic) hydrolytic and enzymatic degradations (with porcine pancreas) of the copolyesters (75).

The PBSU copolyesters that contain 20 *mol*% isosorbide display comparable hydrolytic and enzymatic degradation rates with those of copolyesters with 20 *mol*% adipic acid, but a substantially 23°C higher glass transition temperature value (75).

The synthesis and the properties of poly(1,4-cyclohexanedimethylene-*co*-isosorbide terephthalate) with an isosorbide content ranging from 16 *mol*% to 61 *mol*% relative to the total diol amount were investigated (76). Amorphous polymers were successfully synthesized without isosorbide ring hydration. The glass transition temperature linearly increased by 1.1°C for each molar percent of isosorbide included in the polyester; the melting temperature linearly decreased 2.1°C for semicrystalline samples (up to 25 *mol*% of isosorbide).

The molar masses of the semicrystalline samples could be easily increased using a high-temperature, solid-state polymerization method and a reduced viscosity higher than 70 $ml\,g^{-1}$. Semicrystalline samples exhibit outstanding impact resistance compared with commercial high performance copolyester, whereas amorphous copolyesters exhibit acceptable mechanical properties coupled with high thermal resistance (76).

4.1.12 *Aliphatic-Aromatic Copolyesters*

Copolymers of the poly(butylene adipate-*co*-terephthalate) type with a molar fraction of terephthalate of 42 *mol*%, biodegrade completely to form compost in twelve weeks, whereas products with 51 *mol*% of molar fraction of terephthalate show a percentage of biodegradation of less than 40% (77). This different behavior was attributed to the formation of a higher number of butylene terephthalate sequences with a length greater than or equal to 3, which are less easily biodegradable.

If it would be possible to maintain suitable biodegradation properties, an increase in the percentage of aromatic acid in the chain is desirable insofar as it would enable an increase in the melting point of the polyester. In this way important mechanical properties, such as ultimate strength and elastic modulus, are kept (78).

The polymerization of aliphatic monomers with aromatic monomers, such as terephthalic acid or their diester derivatives, such as dimethyl terephthalate (DMT), may improve the performance properties of aliphatic polyesters.

However, questions have been raised with regard to the complete biodegradability of aliphatic-aromatic copolyesters because aromatic copolyesters, such as poly(ethylene terephthalate) (PET), are resistant to microbial attack.

On the other hand, it has been established that aliphatic-aromatic copolyesters are indeed biodegradable and that the biodegradability of these copolyesters is related to the length of the aromatic sequence. Block copolyesters with relatively long aromatic sequences are not rapidly degraded by microorganisms.

To produce a terephtalate adipate copolyester, dimethyl terephthalate, adipic acid, 1,4-butanediol, and glycerol are mixed together with tetrabutyl orthotitanate as a catalyst (79). The reaction mixture is then heated to a temperature of 180°C for 6 *h* and at 240°C. The excess dihydroxy compound can be removed by distillation *in vacuo*. Then, hexamethylene diisocyanate is slowly added to the mixture. The diisocyanates act as chain extenders. From this material, a biodegradable polyester film can be produced. In addition, the syntheses of related copolyesters have been described in detail (80). It has been found that the addition of chalk can achieve an additional improvement in the biodegradability.

Monomers for aliphatic-aromatic copolyesters are summarized in Tables 4.3 and 4.4.

The polymer sold under the name Ecoflex® by BASF has a glass transition temperature of −33°C and a melting range of 105–115°C (33). Another soft aliphatic-aromatic copolyester is manufactured by Eastman Chemical Company and is sold under the trade name Eastar Bio. The aliphatic-aromatic copolyester manufactured by Eastman is a random copolymer derived from 1,4-butanediol, adipic acid, and DMT.

One particular grade of Eastar Bio is Eastar Bio 14766, with a glass transition temperature of −33°C and a melting point of 112°C. It has a tensile strength at break in the machine direction of 19 *MPa*, an elongation at break of 600%, and a tensile modulus of elasticity of 97 *MPa* (tangent). It has an Elmendorf tear strength of 282 *g*.

Table 4.3 Acids for aliphatic-aromatic copolyesters (84).

Aliphatic dicarboxylic acids
Malonic acid
Succinic acid
Glutaric acid
Adipic acid
Pimelic acid
Suberic acid
Azelaic acid
Sebacic acid
Maleic acid
Fumaric acid
2,2-Dimethyl glutaric acid
1,3-Cyclopentanedicarboxylic acid
1,3-Cyclohexanedicarboxylic acid
1,4-Cyclohexane-dicarboxylic acid
Diglycolic acid, itaconic acid
2,5-Norbornanedicarboxylic acid
4-(Hydroxymethyl)-cyclohexanecarboxylic acid
Hydroxypivalic acid
6-Hydroxyhexanoic acid
Glycolic acid
Lactic acid
Aromatic dicarboxylic acids
Terephthalic acid
2,6-Naphthalic acid
1,5-Naphthalic acid

Table 4.4 Alcohols for aliphatic-aromatic copolyesters (84).

Alcohols
Ethylene glycol
Diethylene glycol
Propylene glycol
1,3-Propylene glycol
2,2-Dimethyl-1,3-propylene glycol
1,3-Butylene glycol
1,4-Butylene glycol
1,5-Pentanediol
1,6-Hexanediol
2,2,4-Trimethyl-1,6-hexanediol
Thiodiethanol
1,3-Cyclohexanedimethanol
1,4-Cyclohexanedimethanol
2,2,4,4-Tetramethyl-1,3-cyclobutanediol
Triethylene glycol
Tetraethylene glycol

This test method, invented by Armin Elmendorf (81), is of value in ranking the relative tearing resistance of various plastic films and thin sheeting of comparable thickness (82, 83).

4.1.12.1 *Blends of Aliphatic-Aromatic Copolyesters with Poly(ethylene-co-vinyl acetate)*

Several blends of aliphatic-aromatic copolyesters with poly(ethylene-*co*-vinyl acetate) blends have higher melt strength than the aliphatic-aromatic copolyester as such and exhibit an increased melt strength and a better thermal processability (85). An example of a branched basic aliphatic-aromatic material is made from poly(tetramethylene adipate-*co*-terephthalate) by transesterification. As branching agents, pentaerythritol or pyromellitic dianhydride may be used. The poly(ethylene-*co*-vinyl acetate) resins used were from the DuPont Elvax® family.

Furthermore, the blends may contain biodegradable additives to enhance their disintegration and biodegradability in the environment. Examples of biodegradable additives include thermoplastic starch, microcrystalline cellulose, methyl cellulose, hydroxypropyl cellulose, PLA, poly(hydroxy butyrate), or PVA (85).

In addition, plasticizers may be added. Typical plasticizers include dioctyl adipate, phosphates, and diethyl phthalate. Preferably, the blend components are mixed by thermal compounding. Good mixing is desirable to help establish the integrity of the blends and is easily achieved in mixers with high shear and significant elongational flow, such as occurs with twin screw extruders. The high elongation stresses help to break the particles up, regardless of viscosity difference, and assure an uniform mix (85). Other blenders are a batch mixer, ribbon blender, Banbury mixer, or a roll mill.

4.1.12.2 Blends with Epoxy Containing Polymers

A copolymer that contains epoxy groups can be added to form biodegradable blends (86). It is preferable to use such copolymers based on styrene, acrylate, or methacrylate. In general, the compounds have two or more epoxy groups in the molecule. Furthermore, glycidyl acrylate and glycidyl methacrylate or epoxidized fatty acid esters are suitable.

The addition of branching agents in the form of master batches shows advantages with respect to the flow rate of the polyester mixtures. Branching agent compositions, shown in Table 4.5, are more useful as conventional branching agents, such as poly(lactide) and glycidyl methacrylate. These branching agent master batches also feature an excellent storage stability. As lubricants and antiblocking

Table 4.5 Branching agent compositions (86).

Acid components	Alcohol components
Adipic acid/terephthalic acid	1,4-butanediol
Adipic acid/terephthalic acid	1,3-propanediol
Succinic acid/terephthalic acid	1,4-butanediol
Succinic acid/terephthalic acid	1,3-propanediol
Sebacic acid/terephthalic acid	1,4-butanediol
Sebacic acid/terephthalic acid	1,3-propanediol
Azelaic acid/terephthalic acid	1,4-butanediol
Brassylic acid/terephthalic acid	1,4-butanediol

agents, calcium stearate or beeswax are used.

The fabrication of the polyester mixtures, including a lubricant masterbatch and a branching agent masterbatch, is done in a corotating, self-cleaning twin screw extruder (86).

The degradation rates of the compositions have been determined as follows (86): The samples were pressed at 190°C to give foils with a thickness of 30 μm. Each of these foils was cut into square pieces with an edge length of 20 cm. Then the initial weight of each of these foil pieces was measured. The foil pieces were placed on a soil-filled cloth in a cabinet under controlled atmospheric conditions for a period of four weeks. A temperature of 30°C was adjusted, a relative humidity of about 50%, and the foils were irradiated using a Heraeus SUNTEST accelerated irradiation equipment in the wavelength range of 300–800 nm with irradiation intensity of 765 $W\,m^{-2}$. The residue weight of the foil pieces was measured weekly.

A particular application for the biodegradable polyester mixtures is for mulch foils for agriculture. These mulch foils are laid to provide protection and to accelerate the growth of young seedlings in fields. After harvest, these mulch foils are left on the field or are plowed into the soil. A substantial biodegradation of these mulch foils has to occur by the start of sawing in the following year (86).

4.1.12.3 *Composites with Agar*

Agar particles can be reinforced butylene adipate-*co*-terephthalate polyesters. The agar has been obtained from red marine plant *Gelidium robustum*. The composites have been fabricated with extrusion and injection molding techniques. The effect of the agar content on the properties of the composites has been assessed (87).

The incorporation of agar particles into butylene adipate-*co*-terephthalate polyesters enhance the tensile strength and modulus accompanied by a reduced percentage of elongation at break.

Furthermore, a reduction in the mechanical loss factor is effected. In contrast, the situation is reversed for the storage and the loss modulus. Increasing agar content causes an increase in the melting temperature and the crystallization temperature.

The morphology of the biocomposites was analyzed by scanning electron microscopy, c.f. Figure 4.19.

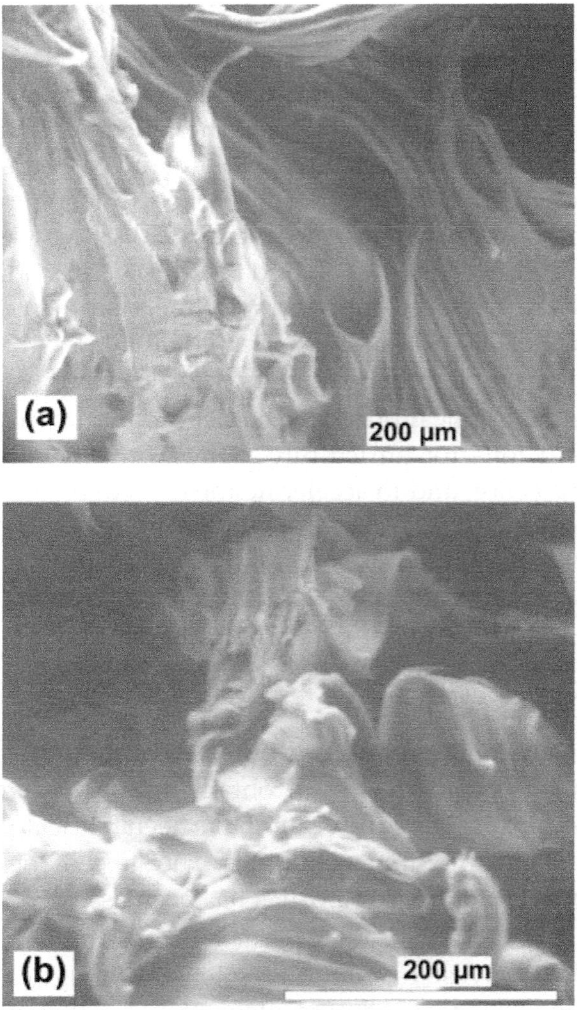

Figure 4.19 Scanning electron micrograph of the biocomposite: (a) 10% Agar, (b) 40% Agar. Reprinted from (87) with permission from Wiley.

4.1.13 *Succinate-Based Polyesters*

Another class of soft aliphatic polyesters are based on repeating succinate units such as PBSU, poly(butylene succinate adipate) (PBSA), and poly(ethylene succinate) (PES). Each of these succinate-based aliphatic polyesters are manufactured by Showa High Polymer Ltd. and are sold under the trade name Bionolle®. Glass transition temperatures (T_g) and melting points (T_m) of three types are shown in Table 4.6.

Table 4.6 Glass transition temperatures (T_g) and melting points (T_m) of Bionolle polymers (33).

Polymer	Trade name	T_g/[°C]	T_g/[°C]
PBSU	Bionolle 1001	−30	114
PBSA	Bionolle 3001	−35	95
PES	Bionolle 6000	−4	102

PBSU: Poly(butylene succinate)
PBSA: Poly(butylene succinate adipate)
PESU: Poly(ethylene succinate)

Applications for Bionolle include films, sheets, filaments, foam-molded products and foam-expanded products. Bionolle is biodegradable in compost, in moist soil, in water with activated sludge, and in seawater. PBSA degrades rapidly in a compost environment, similar to cellulose. PBSU degrades less rapidly, similar to a newspaper.

Bionolle is manufactured by a two-step process. In the first step, a low molecular weight hydroxyl-terminated aliphatic polyester prepolymer is made from a glycol and an aliphatic dicarboxylic acid. This polymerization is catalyzed by a titanium catalyst such as tetraisopropyltitanate, tetraisopropoxy titanium, dibutoxydiacetoacetoxy titanium, or tetrabutyltitanate. In the second step, a high molecular weight polyester is made by coupling the polyester prepolymer with a diisocyanate such as hexamethylene diisocyanate.

4.1.14 *Sebacate-Based Polyesters*

Biodegradable aliphatic-aromatic polyesters obtained from dicarboxylic acids and dialcohols are known in the literature and are commercially available. The presence of the aromatic component in the polyester chain is important to obtain polymers with sufficiently high melting points and with adequate crystallization rates (78).

Biodegradable polymers that can be prepared using an enzyme catalyst are exemplified by poly(propane-1,2-diol sebacate), poly(butane-1,3-diol-sebacate), poly(butane-2,3-diol sebacate), and poly(pentane-2,4-diol sebacate).

The enzyme catalyst used in the process of preparing the biodegradable polymer is preferably a lipase that is hydrolase. Particularly, *Candida antarctica* lipase B is preferable.

This enzyme includes the amino acid of serine-histidine-aspartame (Ser-His-Asp). The enzyme may be prepared as follows, or a commercial product may be used (88).

Preparation 4–2: *Candida* strains may be cultivated in a nutritive medium containing assimilable carbon and nitrogen sources, essential mineral matter, trace elements and the like under aerobic conditions, and the medium may be constructed in a conventional manner. After the cultivation, insoluble substances are removed by filtration or centrifuging to prepare a concentrated solution of liquid enzyme, and a culture solution may be subsequently evaporated or concentrated by reverse osmosis. The concentrated solution may be precipitated in a solvent capable of being mixed with salts or water, for example, ethanol, or may be dried in a conventional spray manner to prepare a solid enzyme preparation.

Since the biodegradable polymer is polymerized by using the enzyme in the course of the polycondensation, it is easily prepared and friendly to nature; and it is possible to control a molecular weight depending on a reaction time in the course of the polymerization. Therefore, the biodegradable polymer is usefully applied to control the dose of a drug in a drug release system and is useful as a supporter for regenerating a desired biological tissue of a patient in tissue engineering.

The method of polymerization is subsequently exemplified with poly(glycerol sebacate). Other sebacates may be prepared in the same way, by using different alcohols instead of glycerol (88):

Preparation 4–3: Glycerol is added to sebacic acid in equimolar amounts that is dissolved in toluene. The mixture is then stirred in an argon atmosphere at 55°C for 1 *h*. Afterwards, enzyme B is added as a catalyst in an amount of 5%. The mixture is allowed to react at 100°C for 72 *h*. The polycondensation is carried out using a Dean-Stark trap device. Eventually, the resulting product is dried for 24 *h* at reduced pressure.

The biodegradable polymers prepared were analyzed by a thermogravimetric analyzer. The temperatures at which a 10% weight loss was achieved are summarized in Table 4.7.

Table 4.7 Temperatures at 10% weight loss (88).

Polymer	Δw 10% at °C
Poly(glycerol sebacate)	193
Poly(1,2-propanediol sebacate)	233
Poly(1,3-butanediol sebacate)	324
Poly(2,3-butanediol sebacate)	257
Poly(2,4-pentanediol sebacate)	242

The compositions and properties of aromatic aliphatic polyesters prepared from terephthalic acid, sebacic acid (i.e., decanedioic acid), butanediol, and butylstannoic acid (as a catalyst) are shown in Table 4.8 and Table 4.9.

Table 4.8 Compositions of aromatic aliphatic polyesters (78).

No.	TPA/[g]	SA/[g]	BD/[g]	BSA/[g]	TAC/[%]
1	2890	3000	3500	6.1	53.5
2	2890	3000	3500	6.1	54
3	3476	3494	2411	0	60
4	2480	3400	3430	6.1	47
5	3294	3108	2923	0	54
6	3081	3277	2212	0	60
7	3858	3526	2071	0	66

No. Sample number
TPA Terephthalic acid
SA Sebacic acid
BD Butanediol
BSA Butylstannoic acid
TAC Total aromatic content

Table 4.9 Properties of aromatic aliphatic polyesters (78).

No.	ΔT_c /[°C]	ΔT_m /[°C]	US /[MPa]	BD /[%]
1	58	130	40	91
2	46	129	39	90
3	78	146	45	90
4	22	114	28	100
5	54	132	15	30
6	23	80	6.5	71
7	98	162	39.5	50

No. Sample number
T_c Crystallization temperature
T_m Melting temperature
US Ultimate strength
BD Biodegradation

Other suitable catalysts are orthobutyl titanate or triisopropyl aluminium (78). Besides sebacic acid, other dicarboxylic acids of natural origin have been proposed, such as azelaic acid and brassylic acid (89). These acids are shown in Figure 4.20.

Azelaic acid can be prepared by the oxidation of oleic acid. Sebacic acid can be produced from castor oil. Brassylic acid can be synthesized from erucic acid via an ozonolysis reaction and with the aid of microorganisms from tridecane (90, 91). The preparation of brassylic acid from tridecane occurs as follows (91):

Preparation 4–4: *Candida tropicalis* FERM-P No. 3,291 having the ability to produce a long-chain dicarboxylic acid from a straight-chain hydrocarbon is cultured in an incubator separately provided to obtain 120 l of seed broth containing 10–15 $g\,l^{-1}$ of cultured fungal body, and this seed broth was added into a reactor supplied with 1,200 l of a medium containing 240 l of n-tridecane.

This mixed solution is then adjusted to pH 5 and cultured at 32°C. for 12 h, during which period germfree air is supplied at a rate of 400 $l\,min^{-1}$. Since the pH of the medium tends to drop during culture, a 10 N KOH solution was added to maintain said pH at 5.0±0.1. After this 12 h culture, the pH of the medium was shifted to 7.0 and the culture was further continued for an additional 72 h.

After this 72 h culture, 1,200 l of fermentation broth was obtained con-

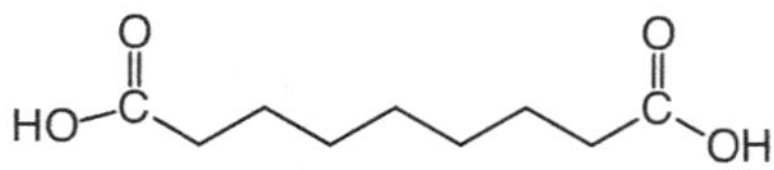

Azelaic acid

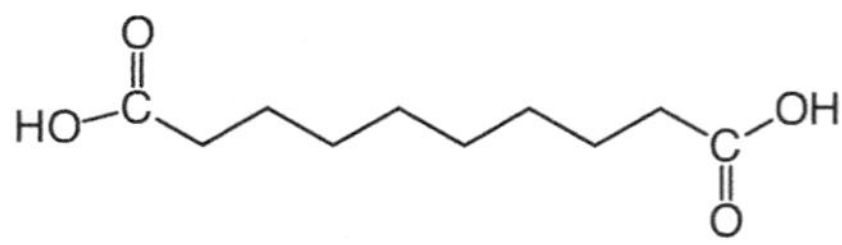

Sebacic acid

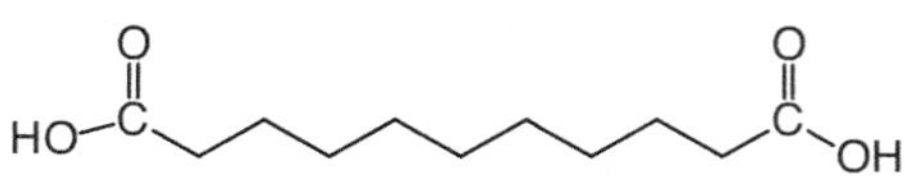

Brassylic acid

Figure 4.20 Dicarboxylic acids of natural origin.

taining 40 $g\,l^{-1}$ of brassylic acid, 8 $g\,l^{-1}$ of *n*-tridecane and 20 $g\,l^{-1}$ of cultured fungal body.

The population of other germs in the fermentation broth was less than 10 cells per *ml*.

In an improved process, the yield of the dicarboxylic acid can be increased by the addition of inducers, i.e., 0.1% H_2O_2 (92).

4.1.15 2,5-Furandicarboxylic Acid-Based Polyesters

Biodegradable polyesters can be obtained from aliphatic dicarboxylic acids, polyfunctional aromatic acids of renewable origin and particularly 2,5-furandicarboxylic acid, and diols (93).

The polyesters have properties and viscosity values that make them suitable, after adjusting their molecular weight, for use in numerous practical applications such as films, injection molded products, extrusion coatings, fibers, foams, thermoformed products, extruded profiles and sheets, extrusion blow molding, injection blow molding, rotomolding, and stretch blow molding.

Table 4.10 Diols (93).

Compound	Compound
1,2-Ethanediol	1,2-Propanediol
1,3-Propanediol	1,4-Butanediol
1,5-Pentanediol	1,6-Hexanediol
1,7-Heptanediol	1,8-Octanediol
1,9-Nonanediol	1,10-Decanediol
1,11-Undecanediol	1,12-Dodecanediol
1,13-Tridecanediol	1,4-Cyclohexanedimethanol
Propylene glycol	Neopentyl glycol
2-Methyl-1,3-propanediol	Dianhydrosorbitol
Dianhydromannitol	Dianhydroiditol
Cyclohexanediol	Cyclohexane-methanediol

Examples of diols that can be used are collected in Table 4.10.

It is possible to add chain extenders such as isocyanates and isocyanides, epoxides, and particularly polyepoxides, oxazolines or carbodiimides (93). Generally, such additives are used in percentages of 0.05% to 1.5%. Examples of epoxides are collected in Table 4.11.

4.1.16 Unsaturated Polyesters

4.1.16.1 Hemp Fibers for Unsaturated Polyester Resins

Nonwoven hemp mats and unsaturated polyester resin have been used for the fabrication of biocomposites (94). The amount of hemp fiber added has been optimized by mechanical measurements.

The effect of four surface treatments of industrial hemp fibers on mechanical and thermal properties of biocomposites was studied. The treatments done were alkali treatment, silane treatment, and UP matrix treatment; and a treatment with acrylonitrile of the hemp materials has been studied.

Both mechanical and thermal properties of the biocomposites were found to be improved due to the surface treatments. In order to get a reasonable balance in the properties, a hybrid composite of industrial hemp and glass fibers was tested (94).

ESEM micrographs of surface-treated hemp mat fibers are shown in Figure 4.21.

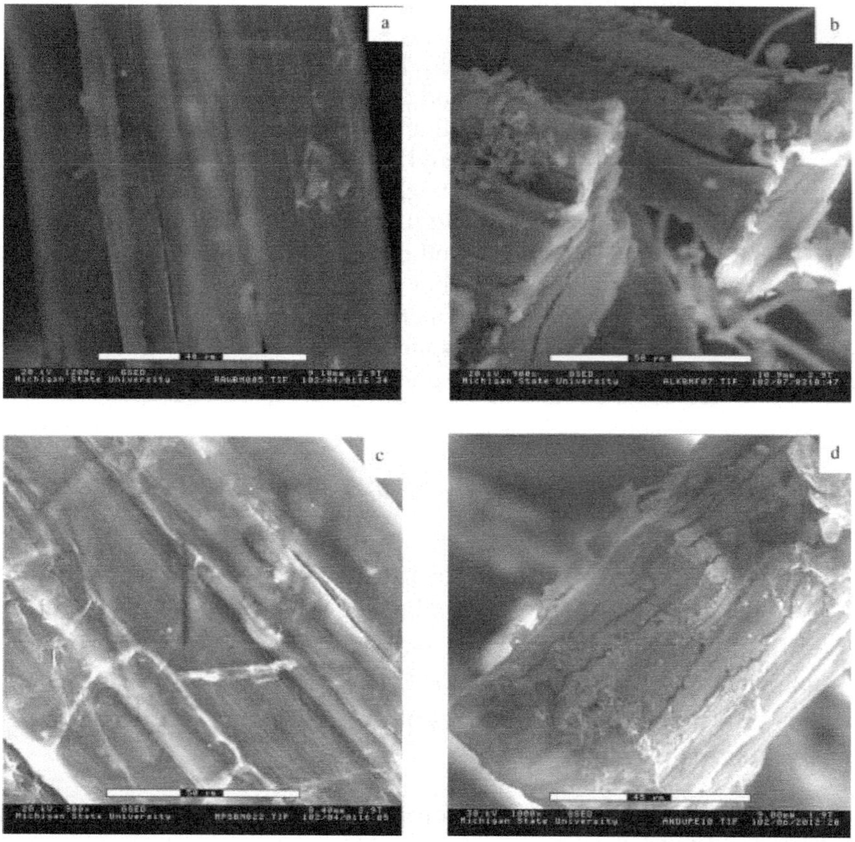

Figure 4.21 Micrographs of surface-treated hemp mat fibers: (a) untreated hemp mat, (b) alkali-treated hemp mat, (c) silane-treated hemp mat, (d) acrylonitrile-treated hemp mat. Reprinted from (94) with permission from Wiley.

Table 4.11 Epoxide additives (93).

Compound
Diethylene glycol diglycidyl ether
Polyethylene glycol diglycidyl ether
Glycerol polyglycidyl ether
Diglycerol polyglycidyl ether
1,2-Epoxybutane
Polyglycerol polyglycidyl ether
Isoprene diepoxide, and cycloaliphatic diepoxide
1,4-Cyclohexanedimethanol diglycidyl ether
Glycidyl 2-methylphenyl ether
Glycerol propoxylate triglycidyl ether
1,4-Butanediol diglycidyl ether
Sorbitol polyglycidyl ether
Glycerol diglycidyl ether
Tetraglycidyl ether of meta-xylenediamine
Diglycidyl ether of bisphenol A
Diglycidyl ether of bisphenol F

4.1.16.2 *Kenaf Fiber Reinforced Unsaturated Polyester Resins*

Kenaf fiber mat reinforced unsaturated polyester biocomposites have been prepared using a hand lay-up and a compression molding technique (95). A sodium hydroxide solution with three different soaking times ranging from 0 *h* to 6 *h* were used to improve the fiber-matrix adhesion properties of the kenaf fiber mat-unsaturated polyester composites.

The thermal stability and chemical changes of fiber constituents of the kenaf fiber mat-unsaturated polyester composites were characterized with TG and FTIR. A study of the mechanical properties revealed that strength and modulus properties were increased with the NaOH treatment. The incorporation of the alkalized kenaf fiber mat resulted in composites with better tensile and flexural properties. A 3 *h* soaking time treatment showed the best results. Further investigation on morphological properties was also done to correlate the efficiency of chemical treatment used to improve the mechanical properties of the kenaf fiber mat reinforced unsaturated polyester biocomposites (95).

4.1.16.3 *Reinforced Sandwich Biocomposites*

The tensile, flexural and low velocity impact response of regenerated biodegradable warp-knitted and nonwoven viscose fabric reinforced unsaturated polyester sandwich biocomposites was reported (96). Four different types of sandwich laminates comprised of different configurations as well as various weight fractions of warp knitted/nonwoven fabrics were manufactured by vacuum infusion molding.

The influence of warp-knitted/nonwoven ratios in the sandwich surface layers and core layer and lay-up configurations on the tensile, flexural and low velocity impact damage characteristics were investigated. From the experimental results it was observed that fiber lay-up configurations have a considerable effect on the mechanical properties of regenerated cellulose reinforced sandwich biocomposites.

The tensile strength was increased from 50 *MPa* for sandwich laminate sample SL1 (22% of reinforcement) to 160 *MPa* for the sandwich laminate sample SL4 (39.8% of reinforcement). A similar trend was observed for the impact behavior, showing the best penetration resistance at the SL4 laminate (96).

4.1.16.4 *Nanostructured Biocomposites*

Wood cellulose nanofibers were used to reinforce an unsaturated polyester matrix without the need for coupling agents or cellulose nanofibers surface modification (97). The nanostructured cellulose nanofibers network reinforcement strongly improves modulus and strength of unsaturated polyester but also ductility and toughness.

A template-based prepreg processing approach of industrial potential is adopted, which combines high cellulose nanofibers content (up to 45 *vol%*) with nanoscale cellulose nanofibers dispersion. The cellulose nanofibers/unsaturated polyester composites are subjected to moisture sorption, dynamic thermal analysis, tensile tests at different humidities, fracture toughness tests and fractography. The glass transition temperature increases substantially with the content of cellulose nanofibers.

The modulus and strength of the unsaturated polyester increase about 3 times at 45 *vol%* cellulose nanofibers whereas ductility and apparent fracture toughness are doubled (97).

4.1.17 *Sulfonated Polyesters*

It has been demonstrated that sulfonated terephthalate units contribute significantly to the biodegradability of the polymer (98). Biodegradable polymer blends manufactured from a hard biopolymer and a soft biopolymer have been described. Blends of hard and soft polymers may possess synergistic properties superior to those of either hard or soft polymers by themselves (98).

However, it has been stated that when even blends can reach a desirable balance of physical and thermal properties, the use of polymeric blends will complicate the processes used to produce the final products (99). Therefore, single products will be more desirable.

Sulfonated aliphatic-aromatic polyesters include polyesters derived from a mixture of aliphatic dicarboxylic acids and aromatic dicarboxylic acids and a sulfonated monomer, such as a salt of 5-sulfoisophthalic acid. Suitable dicarboxylic acids are summarized in Table 4.12.

Table 4.12 Dicarboxylic acid monomers and monomers suitable for sulfonated polyesters (99).

Aromatic acid	Aliphatic acid
Terephthalic acid	Oxalic acid
Isophthalic acid	Malonic acid
2,6-Napthalene dicarboxylic acid	Succinic acid
2,7-Naphthalenedicarboxylic acid	Glutaric acid
3,4´-Diphenyl ether dicarboxylic acid	Adipic acid
4,4´-Diphenyl ether dicarboxylic acid	Pimelic acid
3,4´-Diphenyl sulfide dicarboxylic acid	Suberic acid
4,4´-Diphenyl sulfide dicarboxylic acid	Azelaic acid
3,4´-Diphenyl sulfone dicarboxylic acid	Sebacic acid
4,4´-Diphenyl sulfone dicarboxylic acid	Brassylic acid
3,4´-Benzophenonedicarboxylic acid	Docosanedioic acid
4,4´-Benzophenonedicarboxylic acid	Undecanedioic acid
1,4-Naphthalene dicarboxylic acid	Tetracosanedioic acid
4,4´-Methylene bis(benzoic acid)	Hexadecanedioic acid

Basically, the sulfonated aliphatic-aromatic copolyesters should contain up to 4 *mol%* of sulfonate groups based on the total aliphatic dicarboxylic acid component.

The addition of a polyfunctional branching agent may be desirable when a higher resin melt viscosity is desired for specific end uses. Polyfunctional branching acid-based agents include trimellitic acid, pyromellitic acid, or the corresponding anhydrides, as well as citric acid. Polyfunctional alcohol-based branching agents may also be used, such as pentaerythritol, glycerol, and 2-(hydroxymethyl)-1,3-propanediol.

The sulfonated aliphatic-aromatic copolyesters exhibit improved thermal properties, in particular, a desirable balance of high temperature properties and improved compostability. Sulfonated polyesters can be prepared as follows (99):

Preparation 4–5: The following components are mixed together in a suitable glass flask:

Component	[*g*]
Dimethyl terephthalate	48.54
1,3-Propanediol	38.04
Isophthalate-3-sodium sulfonate	2.96
Dimethyl adipate	43.55
1,2,4,5-Benzenetetracarboxylic dianhydride	0.098
Titanium(IV) isopropoxide	0.0582

Then, the reaction mixture is stirred and heated to 180°C under a slow nitrogen purge. After reaching 180°C, the reaction mixture is heated to 200°C over 1.5 *h* with stirring under a slow nitrogen purge. The resulting reaction mixture is then stirred at 200°C for 1.0 *h* with a slight nitrogen purge. Then the reaction mixture is heated to 255°C over 1.0 *h* with stirring and a slow nitrogen purge. The resulting reaction mixture is stirred at 255°C under a slight nitrogen purge for 0.5 *h*.

Eventually, 21.35 *g* of a colorless distillate was collected over this heating cycle. The reaction mixture was then staged to full vacuum with stirring at 255°C. The resulting reaction mixture was stirred for 0.8 *h* under full vacuum, at a pressure of less than 100 *m Torr*. The vacuum was then released with nitrogen and the reaction mass allowed to cool to room temperature. An additional 8.39 *g* of distillate was recovered and 86.0 *g* of a solid product was recovered.

Several other methods for the preparation of sulfonated polyesters have been disclosed (100).

Sulfonated aliphatic-aromatic copolyetheresters have advantageous thermal properties. Here, in addition, poly(alkylene ether) glycols, such as poly(ethylene glycol), poly(trimethylene ether glycol), poly(tetramethylene ether)glycol, are used (100).

Sulfonated aliphatic-aromatic copolyetheresters have been found to be useful within a wide variety of shaped articles. These articles have advantageous thermal properties. In addition, the materials may be solution or melt processed to form coatings or films. Coatings may be produced by coating a substrate with a polymer followed by drying, by coextruding, or by melt coating. The coatings can be use as barriers to moisture, oxygen, carbon dioxide, etc., but also as adhesives (100).

Sulfonated aliphatic-aromatic copolyetheresters may also find utility as plastic containers. The preferred method for molding a container is stretch blow molding, which is generally used in the production of PET containers. Subsequently, a method of preparation is illustrated (100).

Preparation 4–6: To a 1 *l* glass flask was added 508 *g* bis(2-hydroxyethyl)terephthalate, 82 *g* dimethyl glutarate, 15 *g* sodium dimethyl 5-sulfoisophthalate, 2 *g* tris(2-hydroxyethyl)trimellitate, 42 *g* poly(ethylene glycol), 0.8 *g* sodium acetate, 0.2 *g* manganese(II) acetate tetrahydrate, and 0.2 *g* antimony(III) trioxide. The reaction mixture was stirred and heated to 180°C under a slow nitrogen purge. After achieving 180° C, the reaction mixture was heated to 200°C over 0.2 *h* with stirring under a slow nitrogen purge. The resulting reaction mixture was stirred at 200°C under a slight nitrogen purge for 1 *h*. The reaction mixture was then heated to 275°C over 1.3 *h* with stirring under a slight nitrogen purge. The resulting reaction mixture was stirred at 275°C for 1.2 *h* while under a slight nitrogen purge; 67.0 *g* of a colorless distillate was collected over this heating cycle. The reaction mixture was then staged to full vacuum with stirring at 275°C. The resulting reaction mixture was stirred for 3.2 *h* under full vacuum, at a pressure less than 100 *m Torr*. The vacuum was then released with nitrogen and the reaction mass allowed to cool to room temperature. An additional 68.2 *g* of distillate was recovered and 400 *g* of a solid product was recovered.

The introduction of a sulfonate functionality into some polyesters has been found to enhance the biodegradation rate of the materials. Shaped articles produced from sulfonated aromatic copolyesters containing hydroxyalkanoic acid residues have improved thermal properties (101).

Sulfonated polyesters also may be used for water-dispersible fibers (102). The fibers may contain a single sulfopolyester or a blend of a sulfopolyester with a water-dispersible polymer. The multicomponent fibers can be used to produce microdenier fibers. The biodegradability of the polymers have been assessed by the ASTM Standard Method D6340-98 (103).

4.2 Plant Oil-Based Biopolymers

Plant oils and their derivatives are attractive for building renewable polymers due to their renewability, worldwide availability, and their comparatively low price (104). In order to achieve a more sustainable production of polymers from these resources the recent methods of synthesis go in the direction of catalytic transformations. In particular, olefin metathesis and thiol-ene additions have been used for synthetic purposes. These methods have recently been reviewed (104–106).

From the chemical view, plant oils are triacylglycerols, i.e., triesters from longer chain acids. Fatty acids constitute some 95% of the total weight of triglycerides. Efficient routes to degradable and nondegradable renewable polymers from fatty acids have been summarized (107).

Fatty acid monomers esters can be easily obtained by the hydrolysis or alcoholysis of triglycerides. These compounds contribute to valuable renewable monomeric moieties for renewable polymers.

The double bonds that are present in some fatty acids may undergo olefin metathesis reactions. For example, the ethenolysis of fatty acid methyl esters in the presence of a Grubbs first generation catalyst was used to synthesize ω-unsaturated fatty acid methyl esters (108, 109).

The cross-metathesis reaction of fatty acid methyl esters with methyl acrylate has been studied (110). A wide variety of unsaturated and acid or hydroxyl functional monomers can be obtained in this way.

Also, the cross-metathesis reaction between oleyl alcohol and methyl acrylate has been studied (111). From these unsaturated fatty acid esters, several classes of polymers can be synthesized, such as polyolefins, poly(ester)s, and poly(ether)s.

Plant oils can be modified with acrylate groups. The crosslinking properties of acrylate triglycerides from high oleic sunflower oil have been investigated (112). The synthesis is shown in Figure 4.22.

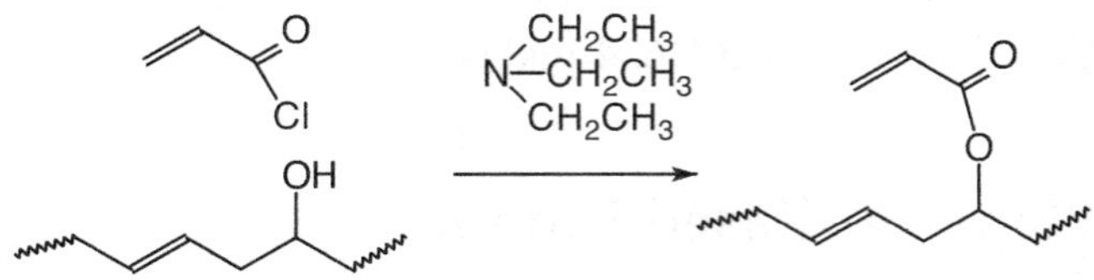

Figure 4.22 Introduction of acrylic groups in plant oils (113).

Polymers with bulky constituents, such as the long alkyl side-chains of fatty acid-derived polymers, typically exhibit poor mechanical performance due to lack of entanglements (114). Hydrogen bonding moieties were incorporated into the fatty acid-based midblock of a thermoplastic elastomeric triblock copolymer as a means of improving its mechanical behavior. Poly(styrene-*b*-(lauryl acrylate-*co*-acrylamide)-*b*-styrene), containing lauryl acrylate and the hydrogen bonding comonomer acrylamide, c.f. Figure 4.23, in the midblock, was synthesized via a reversible addition-fragmentation chain transfer polymerization (114).

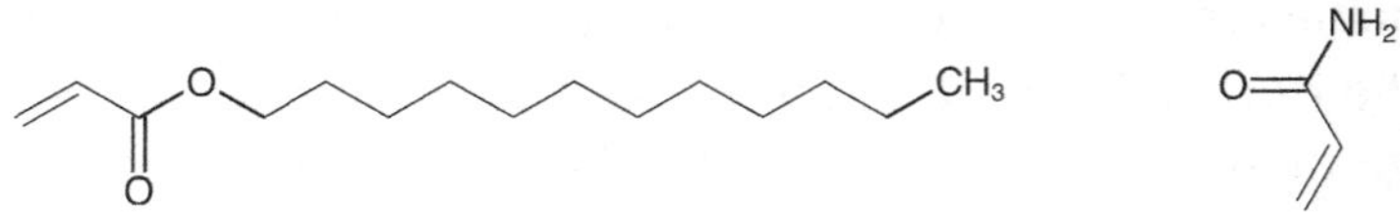

Lauryl acrylate

Acrylamide

Figure 4.23 Acrylic monomers.

The chemical and physical properties of triblock copolymers of varying composition were explored. FTIR analysis confirmed the formation of a transient network, which exhibited a reduction in the crosslinking density with increasing temperature, which is beneficial for high temperature melt processing.

The triblock copolymers exhibited spherical morphologies lacking a long-range order at room temperature, which were unaffected by the presence of acrylamide. Moreover, the order-disorder transition temperature was reduced with increasing acrylamide content, due to a reduction in the Flory-Huggins interaction parameter. The incorporation of acrylamide into the midblock greatly improved both the tensile strength and strain at break.

Therefore, the incorporation of a transient network into the midblock is an effective method of improving the mechanical properties of triblock copolymer-based thermoplastic elastomers containing bulky constituents (114).

The thiol-ene click reaction for the synthesis of plant oil-derived polymers has been reviewed (115). The thiol-ene click reaction is a chain reaction with the basic steps shown in Eq. 4.2.

$$\begin{aligned} \mathrm{R{-}SH + J\cdot} &\rightarrow \mathrm{R{-}S\cdot + HJ} \\ \mathrm{R{-}S\cdot + CH_2{=}CH{-}R} &\rightarrow \mathrm{R{-}S{-}CH_2{-}CH\cdot{-}R} \\ \mathrm{R{-}S{-}CH_2{-}CH\cdot{-}R + R{-}SH} &\rightarrow \mathrm{R{-}S{-}CH_2{-}CH_2{-}R + R{-}S\cdot} \end{aligned} \tag{4.2}$$

For example, mercapto acetate can be added onto the unsaturated moieties (116). Two oleic acid molecules can also be linked together using thioacetate. The reaction is shown in Figure 4.24.

Mercapto oleic esters can also be prepared by the reaction with potassium thioacetate followed by deacetylation or directly from the unsaturated esters by radical addition of thioacetic acid (117).

Unsaturated moieties in a polymer can be modified with the thiol-ene (click) reaction. This is of interest because some monomers are not readily polymerizable to give the desired polymer. In fact, post-polymerization modification of a polymer backbone is an important strategy in polymer science.

4.2.1 *Water Treatment*

Conventional water treatment processes include coagulation and flocculation, especially that which is chemically assisted, for both drinking water and wastewater. The removal capacity of these

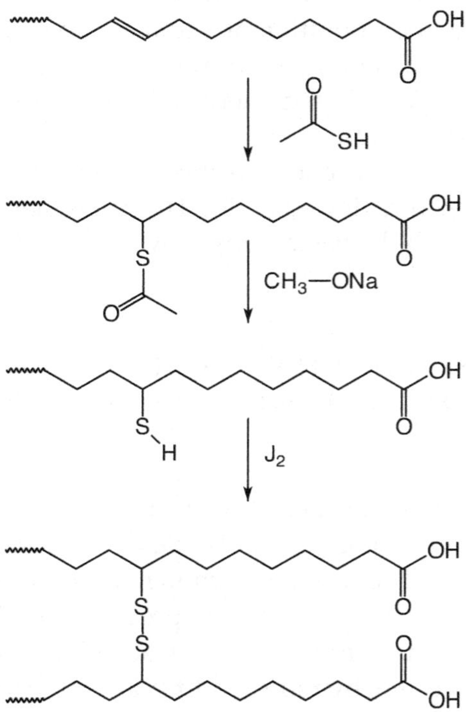

Figure 4.24 Linking reaction of oleic acid (118, 119).

operations is effective and with a speed that could be used especially with waters that have high concentrations of organic matter; for this purpose, reagents have been used, such as aluminum sulfate, which is usually obtained from the reaction of aluminum hydroxide with sulfuric acid. Another reagent used is ferric chloride, which is obtained by the reaction of chlorine gas on heated iron. A reagent used in recent years is the aluminum polychloride obtained by the reaction of aluminum with hydrochloric acid in aqueous solution.

The use of plant biopolymers for water treatment and water desalination has been detailed in a monograph (120).

The action of promoting the removal of particles in water requires coagulant substances, which destabilize the equilibrium of the mixture in aqueous solution. This is needed to perform the coagulation and flocculation operations in the treatment of drinking water and wastewater; especially for the removal of solids with diameters smaller than 0.2 *mm*.

This operation is carried out with chemical compounds capable of breaking the ionic stability of a mixture and segregating the solids that cannot be separated without this operation. The plants tested for use as coagulants or flocculants have had a traditional use, which indicates their ability to carry out the separation of solids (120).

4.2.2 *Plant Oils with Acrylic Moities*

Acrylated oils can be obtained by converting the triglyceride double bonds into an epoxide. Afterwards, the epoxy groups are opened with acrylic acid to get hydroxy acrylated oils (113, 121–123).

Alternatively, acrylic moieties can be introduced directly into the unsaturated fatty oil groups through a bromonium cyclic intermediate (112). *N*-Bromosuccinimide is used for the bromination. The bromine moiety is removed from the nitrogen by an heterolytic fission as a bromonium ion and a succinimide anion is formed. The bromonium ion is attacked by the π-bonds of the alkene. At the same stage, the succinimide anion is protonated by the acidic solvent used. This gives a weak nucleophile succinimide and a carboxylate anion that attacks the carbonium ions (124). The reaction is shown in Figure 4.25.

Eventually, a bromoacrylate ester is obtained that can be polymerized and also imparts flame resistance by the bromine groups.

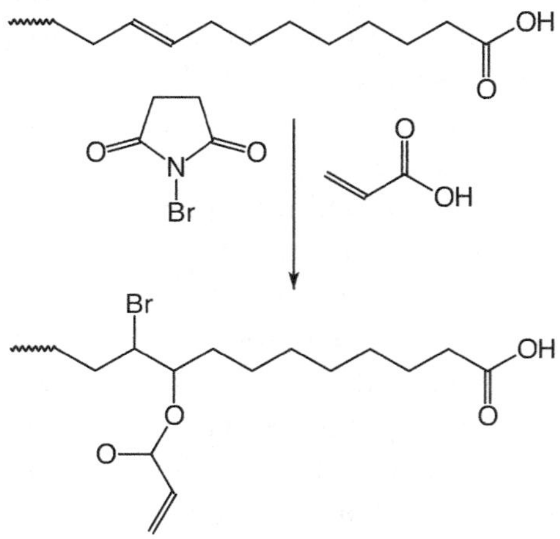

Figure 4.25 Formation of a bromoacrylated adduct (124).

However, bromine-based flame retardant resins release hydrogen bromide during combustion, which is toxic and corrosive (125, 126).

4.2.3 *Plant Oils with Phosphorus Moities*

The concept of sustainable development requires fire retardant technologies to be developed that have a minimum impact on health and the environment throughout the life-cycle of the fire-resistant material: Starting from its synthesis, via fabrication, use, and recycling to its final disposal. Therefore, the search for new environmentally friendly flame retardant polymeric materials is of great current interest.

In contrast to bromine, phosphorus-based polymers are an effective class of flame retardant materials (127). In particular, they exhibit a good flame retardant performance and are thus preferred in comparison to halogenated flame retardants for environmental and health reasons (128).

Acyclic diene metathesis (ADMET) polymerization is a valuable tool for the synthesis of polymers that eventually may contain a wide variety of functional groups (129, 130). ADMET is a special type of olefin metathesis that can be used to polymerize terminal dienes into polyenes (131), as shown in Figure 4.26.

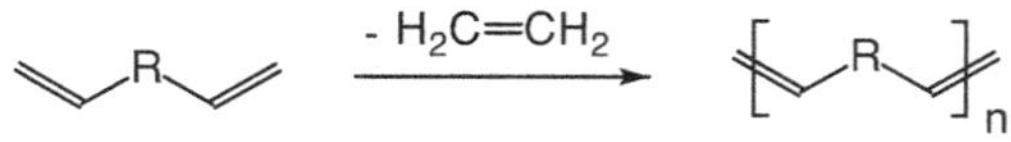

Figure 4.26 Schematic of ADMET polymerization.

The formation of unsaturated polyphosphoesters via acyclic diene metathesis polymerization has been described (132). The ADMET copolymerization of a phosphorus-based α,ω-diene with different amounts of a castor oil-derived diene has been exemplified. These polymer types show a good flame retardancy. The synthesis of a di-10-undecenoylglycerol mixture is shown in Figure 4.27.

The photoperoxidation of high oleic sunflower oil with singlet oxygen is an efficient reaction for the synthesis of functionalized triglyceride derivatives. In this reaction, high oleic sunflower oil and oxygen are the only reactants, which makes this transformation very attractive from an environmental point of view (113).

4.2.3.1 *Acrylated Soybean Oil Coating*

The combination of biomass and ultraviolet curing technology can reduce the dependence on petroleum products and emissions of volatile organic compounds (133).

An environment friendly UV curable coating with good performances was successfully prepared from acrylated soybean oil (133). Acrylated soybean oil was obtained at low temperature without catalyst and inhibitor by reacting soybean oil-based polyol with acryloyl chloride and characterized by FTIR spectroscopy and NMR. It was found that the UV-curable oligomer of acrylated soybean oil had a lower viscosity and a better ductility compared to the acrylated epoxidized soybean oil.

The properties of UV-cured films prepared by acrylated soybean oil were studied and compared with films from acrylated soybean oil. The results showed that the acrylated soybean oil-based films were significantly improved in the adhesion and impact resistance. Furthermore, the addition of styrene could enhance the hardness and impact resistance of films (133).

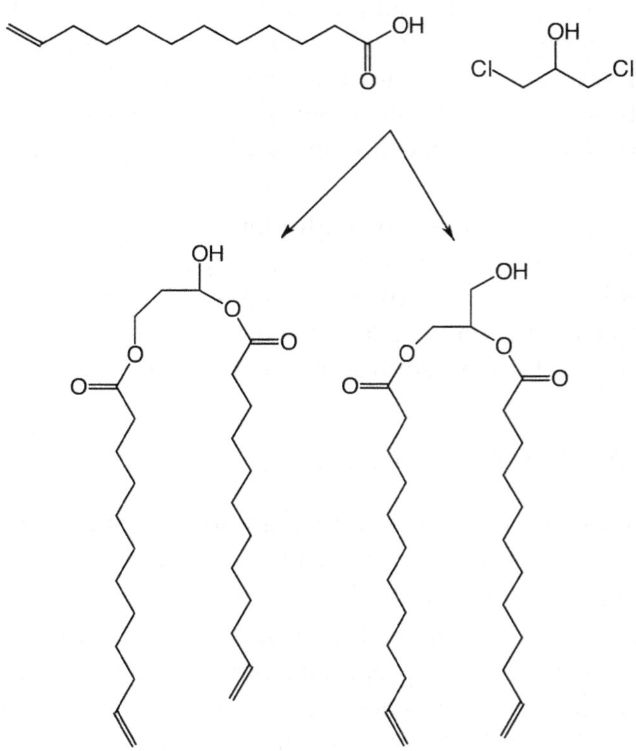

Figure 4.27 Synthesis of 1,3-di-10-undecenoylglycerol and 1,2-di-10-undecenoylglycerol (113).

4.2.4 *Vanillin-Based Monomers and Polymers*

Renewable monomers derived from vanillin and fatty acids have been studied with various polymerization techniques such as ADMET, thiol-ene addition, polycondensation (134), and Schiff base click reaction (135). The synthesis of such monomers is shown schematically in Figure 4.28.

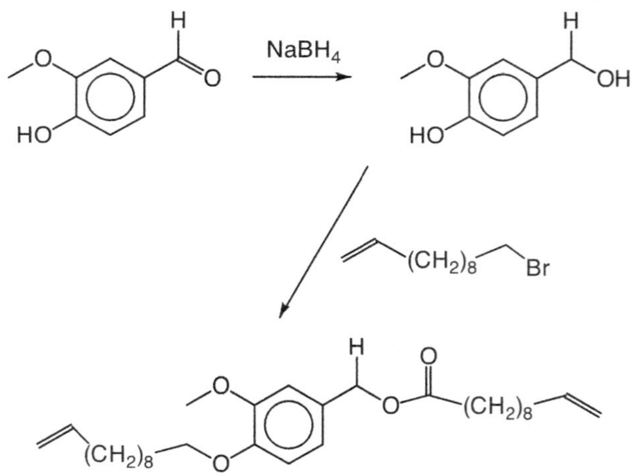

Figure 4.28 Synthesis of monomers derived from vanillin and fatty acids (134).

Vanillin is 4-hydroxy-3-methoxybenzaldehyde. It is shown in Figure 4.29.

4.2.4.1 *Vanillin Acrylate*

The synthesis of vanillin acrylate (4-Formyl-2-methoxyphenylacrylate) has been described as (135):

Preparation 4–7: In a 250 *ml* two-neck flask fitted with argon balloon, 8 *g* of vanillin (0.052 *mol*) was dissolved in 100 *ml* dry CH_2Cl_2 and strongly stirred and 10.52 *g* (0.1 *mol*) of triethylamine was added. The reaction mixture was cooled in an ice bath to 0°C to 5°C. Then, 5.4 *g* (0.059 *mol*) of acryloyl chloride was added dropwise. The yellowish suspension was stirred at 5°C for 1 *h* then was stirred at room temperature overnight. The precipitate was filtered and solvent was evaporated under reduced

Vanillin Vanillin acrylate

Figure 4.29 Vanillin compounds.

pressure. The product was extracted by CH_2Cl_2 and washed three times with distilled water, once with sodium carbonate and once with 0.1 *M* HCl. The organic phase was dried with $MgSO_4$ overnight, then filtered, and the product was distilled using an oil pump at 100°C. The yield was 85%.

Vanillin acrylate acts as the hydrophobic molecule when copolymerized with *N*-isopropylacrylamide in different mole ratios (5, 10 and 15 *mol*%) by a random free-radical polymerization using 2,2′-azobisisobutyronitrile as initiator (135).

4.2.4.2 *Vanillin Acrylate Grafting*

The functionality with the aldehyde group in the polymer main chain was a promising start to a grafting reaction with an amino compound to produce a familiar Schiff base by click reaction. The grafting process was done at room temperature in a mixture of ethannol and water at a ratio of 1:1 and a pH of 11 (135).

4.2.4.3 *Methacrylated Vanillyl Alcohol*

Methacrylated vanillyl alcohol was synthesized using vanillyl alcohol and methacrylic anhydride in the absence of solvents (136). The resulting methacrylated vanillyl alcohol was used as a sustainable comonomer to replace styrene in a maleinated acrylated epoxidized soybean oil resin to produce novel thermosets by free-radical polymerization.

The influence of methacrylated vanillyl alcohol loading on the viscosity, gelation time, curing extent, and thermomechanical and tensile properties of the maleinated acrylated epoxidized soybean oil methacrylated vanillyl alcohol thermoset was investigated. The synthesized methacrylated vanillyl alcohol exhibited a very low volatility, which is beneficial for the development of construction materials with low or zero emission. The viscosity of the maleinated acrylated epoxidized soybean oil/methacrylated vanillyl alcohol system can be tailored to meet the commercial requirements. Increasing the methacrylated vanillyl alcohol content accelerates the crosslinking reaction rate and improves thermal and mechanical properties of the maleinated acrylated epoxidized soybean oil/methacrylated vanillyl alcohol system. The glass transition temperature increases with increasing methacrylated vanillyl alcohol content. Soxhlet extraction experiments showed that more than 90% of the components were incorporated into the crosslinking network. So, the here developed methacrylated vanillyl alcohol monomer showed promising properties for use as an effective, green comonomer to replace styrene (136).

4.2.4.4 *Acyclic Diene Metathesis Polymerization*

An ADMET polymerization resulted in high molecular weights of up to 50 $k\,D$. In contrast, the thiol-ene polyaddition and the polycondensation reactions yielded polymers with molecular weights of 15 $k\,D$ and 17 $k\,D$, respectively. Thermoplastic polymers with melting points of 16–78°C were obtained (134).

4.2.4.5 *Poly(ether) Esters*

Bio-based polyether/esters have been synthesized by a one-pot polymerization reaction of ricinoleic acid, vanillic acid and ethylene carbonate (137). In particular, ethylene carbonate selectively reacts with the phenolic group of vanillic acid to obtain *in-situ* methyl 4-(2-hydroxyethoxy)-3-methoxybenzoate, c.f. Figure 4.30, which is suitable for subsequent copolymerization with ricinoleic acid. The procedure was carried out in a single step without any solvent.

The chemical structure and thermal properties of the new materials were studied in order to explore relationships between composition and final performances. The combination of 4-(2-hydroxyethoxy)-3-methoxybenzoate, bearing a rigid aromatic struc-

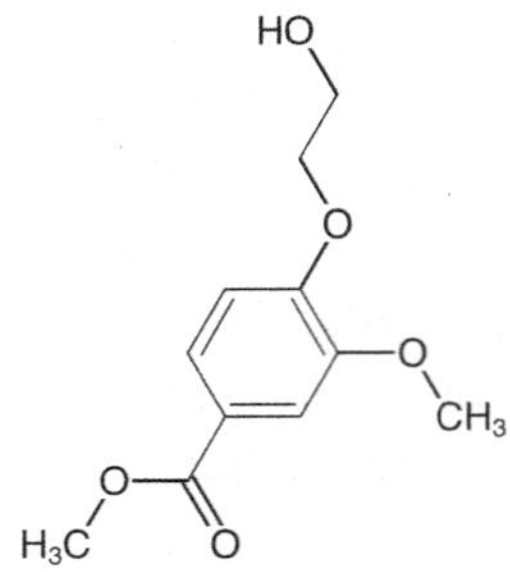

Figure 4.30 Methyl 4-(2-hydroxyethoxy)-3-methoxybenzoate.

ture, with ricinoleic acid, characterized by high flexibility and potential biocidal activity, allows the production of a novel class of fully bio-based aliphatic-aromatic polymers presenting tuneable thermal properties and suitable for a range of applications, for example, in active packaging and biomedical fields (137).

4.2.4.6 *Poly(imide)s*

An asymmetric bio-based diamine, 2-methoxy-1-(4′-aminophenoxy)-4-((4′-aminophenoxy)methyl)benzene (MAAMB), containing phenoxyl-benzyloxyl linkage and a pendent methoxyl group, was synthesized starting from the aromatic, lignin-derivative vanillin through three steps (138).

Lignin-derivative vanillin was reacted with hydroquinone and hydrazine to get a bio-based asymmetric diamine. Then the diamines were reacted with terephthalic anhydride to obtain asymmetric poly(imide)s (PIs).

Two series of PI films were prepared by condensing MAAMB or an analogous, symmetric diamine, 1,4-bis(4-aminophenoxy)benzene (1,4,4-APB), with four different dianhydride 4,4′-hexafluoroisopropylidene diphthalic dianhydride, 3,3′,4,4′-oxydiphthalic anhydride, 3,3′,4,4′-benzophenone dianhydride, and 3,3′,4,4′-biphenyl dianhydride, respectively, via classic two-step procedure. Comparative studies between the MAAMB-resulting PI series (APIs) and the 1,4,4-APB-resulting PI series (SPIs) showed that the introduction of benzyl linkage and pendent methoxyl group in MAAMB imparts APIs with better solubility in organic solvents and higher optical transparence with 63.7–89.2% at 500 *nm*.

The API series retains good thermal properties with T_g ranging from 241°C to 262°C, T_5 ranging from 448°C to 478°C, and favorable mechanical properties with tensile strengths ranging from 82.2–92.7 *MPa* and tensile modulus ranging from 2.27–2.53 *GPa*. In addition, APIs exhibit a lower water absorption (0.71–1.01%) and surface energy (31.35–42.03 $mN\,m^{-1}$) than those (0.75–1.25%) and (34.06–44.35 $mN\,m^{-1}$) of the analogous SPIs. The study indicated that the renewable vanillin could be employed to prepare bio-based poly(imide)s with balanced comprehensive properties (138).

4.2.5 *Soybean Oil Epoxidized Acrylate*

Two nanocomposite inks, consisting of a plant oil-based biopolymer resin, i.e., soybean oil epoxidized acrylate and silanized nano-hydroxyapatite particles with and without diluent hydroxyethyl acrylate, were extruded using direct ink writing to fabricate filaments of differing volume fractions (139).

Hydroxyethyl acrylate was hypothesized to lower the viscosity of the ink, improve the dispersion of the silanized nano-hydroxyapatite and the extrudability, and therefore result in better relative improvements in mechanical properties over those of the base resin.

TG, SEM, and uniaxial tensile tests were conducted. The strength, toughness, and stiffness of both resin types were improved by the addition of silanized nano-hydroxyapatite. The hydroxyethyl acrylate improved the dispersion and the extrudability (139). This resulted in a better particle alignment and a relatively greater augmentation of the mechanical properties.

The experimental data for both biopolymer resins were used to calibrate 3D representative volume element finite element micromechanical models, while the experimental data for the nanocomposite filaments were employed to validate these models. The Young's moduli from the computational models were also compared with the experimental data and with analytical predictions calculated from the Mori-Tanaka method (139).

The Mori-Tanaka model can predict the elastic stiffness tensor of a composite material with assumed elliptical inclusions and interaction among inclusions and matrix using (140, 141):

$$C_c = C_m + \phi_i \frac{(C_i - C_m)A_m}{(1 - \phi_i)I + \phi_i A_i} \tag{4.3}$$

Here, C_c, C_m,and C_i are the stiffness tensors of the composite, matrix, and inclusion respectively, ϕ_i is the inclusion volume fraction, I is the fourth order unit tensor, and A_i is the dilute mechanical strain concentration.

The predictions of the 3D representative volume element models correlated well with the experimental data while, for higher volume fractions, the Mori-Tanaka results diverged, similar to previous reports (139).

4.2.6 *Vegetable Oil Thermosets*

The majority of biodegradable matrices are thermoplastic; however, research in thermoset matrices is on the way with some industrial players producing thermoset matrices with precursors from plant sources (142). The process of preparation and manufacturing of green thermoset matrices has been discussed in a monograph (142).

Also, the recent developments in vegetable oil-based thermosets prepared by a variety of polymerization methods have been reviewed (143, 144). The obtained thermosets exhibit a wide range of thermomechanical properties from soft and flexible rubbers to rigid and hard plastics. Some of the thermosets have properties comparable to petroleum-based analogs and show promise as replacements, providing possible solutions to environmental and energy concerns.

A study has been presented that describes the use of an epoxy functional fatty acid, 9,10-epoxy-18-hydroxyoctadecanoic acid, c.f. Figure 4.31, extracted from birch (*Betula pendula*) outer bark to produce renewable thermosets (145).

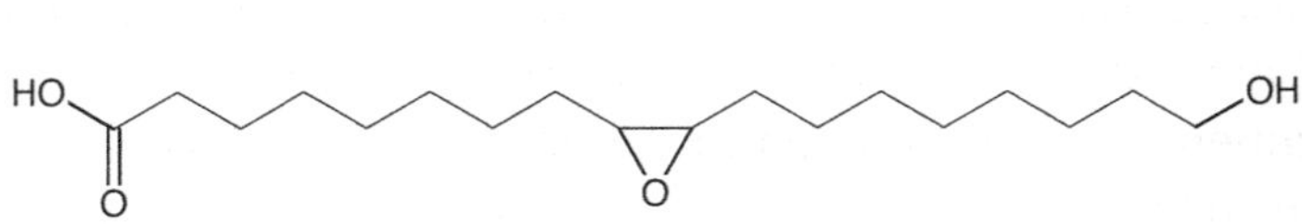

Figure 4.31 9,10-Epoxy-18-hydroxyoctadecanoic acid.

An epoxy fatty acid was polymerized by enzyme-catalyzed polycondensation utilizing *Candida antarctica* lipase B to form oligomers with degrees lower than polymerization of up to 9. The crosslinked polyesters form insoluble, amorphous, and transparent films. So, it was demonstrated that thermoset films with tailored properties can be effectively made using forest products.

Bio-based thermoset polymers were produced from epoxidized waste vegetable oils cured with anhydrides at different molar ratios (146). The properties of these materials were compared to analogues produced with neat oil and diglycidyl ether of bisphenol A as a feedstock.

The thermal stability proved to be affected by the molar ratio, and the use of feedstock from waste oil resulted in having no effect on this property. A dynamic mechanical analysis has shown that higher concentrations of anhydride enhance the storage modulus, glass transition temperature (up to 42.5°C) and crosslinking density. The frying process proved to play a minor role in tuning the dynamic mechanical properties (146).

However, the contribution of the anhydride was demonstrated to be significant enough to mitigate the losses caused by the waste oil, as shown statistically in a design of experiment study. All formulations were chemically resistant to aqueous, organic and acidic media. The identification of the effects of critical parameters on the properties of waste vegetable oil-based thermosets enables the further production of polymers from waste streams (146).

An aza-Michael addition on acrylated linseed oil was performed to synthetize bio-based bulk thermosets without any catalyst (147). First, the acrylation of epoxidized linseed oil allowed obtainment of acrylate functions with vicinal hydroxyl groups, which enhanced the reactivity of acrylates. The autocatalytic effect of hydroxyl groups on acrylate monomers was highlighted by kinetic studies monitored by NMR and FTIR analysis on model molecules.

Then, Priamine 1071, amine terminated poly(propylene oxide) and *m*-xylylenediamine (MXDA) were used as crosslinking agents with acrylated linseed oil. The curing kinetics were studied by DSC analyses to compare the reactivity of these structures.

Priamine 1071 showed the highest reactivity. Curing at room temperature was performed and a high conversion was reached. Two enthalpies were observed with MXDA and only one at high temperature for PPO-based materials. Thermosets with a large range of mechanical properties were finally obtained from soft materials with PPO-diamine to hard materials with MXDA (147).

The bio-based monomer 2,5-furandicarboxylic acid has been extended into epoxy thermosets (148). The study presents the synthesis of diallyl furan-2,5-dicarboxylate followed by an epoxidation of the allyls to form diglycidyl furan-2,5-dicarboxylate. Diglycidyl furan-2,5-dicarboxylate was then copolymerized in both stoichiometric and off-stoichiometric ratios with epoxidized fatty methyl esters to form a range of thermosets.

The crosslinking reaction was either a thermally or UV-induced cationic polymerization utilizing onium salt initiators. The reactivity was studied by DSC and real-time FTIR analysis. Structure-property relationships of the final thermosets were obtained by DMA that revealed a possibility to tune the properties over a wide range. In addition, thermosets were made from diglycidyl bisphenol A with epoxidized fatty methyl esters made for comparative purposes (148). Diglycidyl bisphenol A is shown in Figure 4.32.

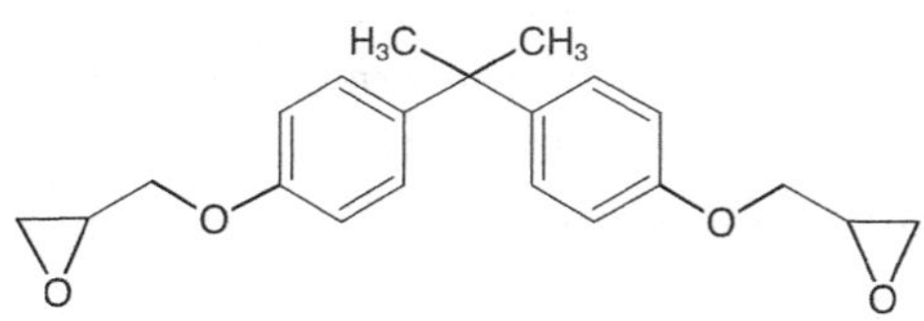

Figure 4.32 Diglycidyl bisphenol A.

4.3 Poly(amide)s

4.3.1 Soy-Based Bioplastic and Chopped Industrial Hemp

Soy-based bioplastics have been prepared by cooking with a plasticizer and blending with a biodegradable poly(ester amide) (149). The biocomposites thus prepared were blended with chopped industrial hemp fiber via twin screw extrusion and also by an injection molding process.

The tensile strength and modulus, flexural strength and modulus, impact strength and heat deflection temperature of the fiber reinforced biocomposites were found to be significantly improved.

In addition, the fracture surfaces showed no traces of the matrix resin on the fiber surface. This suggests a rather poor interfacial adhesion (149).

4.3.2 *Soybean-Based Composites*

Soy proteins are complex macromolecules containing some 20 amino acids (150). By extrusion cooking, soy protein polymers can be converted into biodegradable plastics. However, the main disadvantage of soy protein plastic comes from its low strength and high moisture absorption.

The simplest and most effective way to improve the properties is to combine soy protein with a biodegradable polymer to form soy-based biopolymers. The other method is to use functional monomers or oligomers to modify the soy protein during the processing of soy protein with an extruder. Casting methods or pretreatment methods with soy protein to form articles for applications can be used.

Fiber reinforced composites can increase the scope of these materials and will increase applications in various arenas such as the automotive and packaging industries. Natural fibers, such as kenaf, flax, jute, hemp, and sisal reinforced composites, have become attractive, because natural fibers have the advantages of low cost, low density, acceptable specific strength properties, ease of separation, carbon dioxide sequestration and biodegradability.

Also, agricultural plant materials, such as corn stalk, rice stalk, wheat straw and grass, have been investigated as a potential resource for natural fibers since they are inexpensive, economically friendly, sustainable, recyclable and biodegradable, and thus are attractive materials with commercial viability and environmental acceptability.

In general, such agro-based materials are only used as feed for livestock and not as load-bearing materials. Therefore, grass reinforced composites have an excellent potential to be used as fibers. There are plenty of grass resources. Elephant grass-based biocomposites are of interest for automotive applications.

Biocomposites are formed through the combination of natural cellulose fibers with other resources such as biopolymers, resins, or binders based on renewable raw materials. The goal is to combine the materials in such a way that a synergism between the components results in a new material that is much better than the individual components.

The properties of plant fibers can be modified through physical and chemical technologies to improve performance of the final biocomposite. Some of the plant fibers with suitable properties for making biocomposites are summarized in Table 4.13

Table 4.13 Plant fibers for biocomposites (150).

Fiber Material		
Hemp	Kenaf	Coir
Henequen	Jute	Flax
Sisal	Banana	Kapok

The chemical modification of a soy protein can be made by the reaction with some functional groups such as carboxyl or amino groups in the protein. Also, a low molecular weight PCL/hexamethylene diisocyanate (HDI) prepolymer can be used for the modification (150).

The addition of salts, e.g., sodium bicarbonate or sodium carbonate (SC), into the formulation of bioplastic materials may alter their hydrophilic character to a significant extent (151). Soy protein isolate is a byproduct of the soybean oil industry, which, when properly blended with glycerol, can be further processed through a lab-scale injection molding device. A maximum in the water uptake of around 2250% or 2500% is obtained for bioplastics obtained when either sodium bicarbonate or sodium carbonate content is around 1%, respectively. Thus, they exceed the limit to be considered superabsorbent materials. Regarding their mechanical properties, a higher presence of sodium bicarbonate within the soy protein isolate/glycerol matrix provides materials with a higher extensibility and lower Young's modulus. A higher water uptake is observed after replacing sodium bicarbonate for sodium carbonate in the formulation, probably related to a higher alkalinization of the material as well as to the high hydrophilicity of the carbonate (151).

Soy protein-based bioplastics have been developed with a lignocellulosic fiber as additive by injection molding (152). The mechanical and absorption properties of the bioplastics reinforced with lignocellulose (0.1%, 1.0%, 5.0%) have been studied, as well as their microstructure. Also, the effect produced by changing the mold temperature (70°C, 90°C, 110°C, and 13°C) was checked (152).

The so-obtained results confirmed an improvement of the mechanical properties of these bioplastics, depending on the amount of fiber incorporated (152).

By the reaction between the amino acid in the protein, and the HDI modified PCL, a urea-urethane linkage in the protein and PCL prepolymer complex is formed. This results in an increase in toughness and water resistance of the modified soy protein (153).

A vinyl polymer modified soy protein, using vinyl monomers for grafting, like methyl methacrylate, ethyl acrylate, butyl acrylate and hexyl acrylate, does not exhibit improved mechanical properties (154).

Enzyme modification is another method to enhance the properties of soy protein (155). The influence of enzyme modification on mechanical properties shows a significant decreased of the tensile strength; however, the elongation increases.

Starch can be used as a filler material to modify soy, a protein plastic, since it forms hydrogen bonding or polar interaction with the soy protein (156).

The degradability of nanocomposites derived from crosslinked soy protein modified with organoclay has been studied (157).

A soy protein isolate has been crosslinked with furfural, and then blended with an organoclay (Cloisite 30B). The crosslinking process has been ascertained from FTIR data. The degradation pattern of the crosslinked product has been examined using thermogravimetry. The kinetic parameters were assessed by a LOTUS software. From the kinetic data the mechanism of degradation of the nanocomposites was asserted (157).

Soybean oil-based poly(urethane) (PU)–poly(acrylonitrile) interpenetrating polymer networks were synthesized by the reaction of a soybean oil-based polyol, isophorone diisocyanate, c.f. Figure 4.33, and acrylonitrile (158).

The formation of the interpenetrating polymer networks were confirmed by FTIR and DSC studies. The mechanical properties,

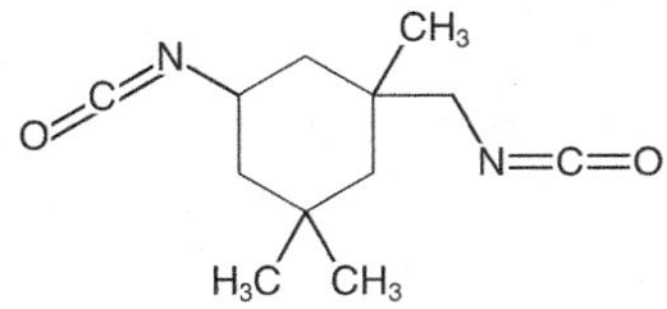

Figure 4.33 Isophorone diisocyanate.

such as tensile strength, Young's modulus, hardness, and abrasion resistance, were found to be increased with the increase of PU content in the interpenetrating polymer network. The interpenetrating polymer networks showed a higher thermal stability and a lesser burning rate than native PU. In addition, the interpenetrating polymer network composition showed a higher hydrophobic nature than PU, as studied by a contact angle measurement (158).

Modified or unmodified natural oils and compatibilizers can be used with a biorenewable thermoplastic such as PLA in order to produce an improved thermoplastic material (159–161). Modified natural oils include acrylated epoxidized soybean oil and unmodified soybean oil.

Acrylated epoxidized soybean oil is a multifunctional, commercially available modified soybean oil produced through the epoxidation of soybean oil followed by the ring-opening of the epoxide groups with acrylic acid. Through this process, the carbon-carbon double bonds on the soybean oil structure are functionalized with both acrylate and hydroxyl groups. Acrylated epoxidized soybean oil is a reactive oil, as the acrylate groups polymerize at elevated temperatures in the presence of air (without the addition of a radical initiator), in stark contrast to unmodified soybean oil, which is a relatively non-reactive oil. Prior studies have utilized acrylated epoxidized soybean oil and other functionalized vegetable oils as monomers for the preparation of highly crosslinked thermosets.

The compatibilizers may be star polymer compatibilizers. An example of a star polymer compatibilizer is shown in Figure 4.34.

The morphology of a PLA oil blend consists of oil droplets dispersed throughout the PLA matrix.

The addition of the PLA star polymer to the PLA oil blend results

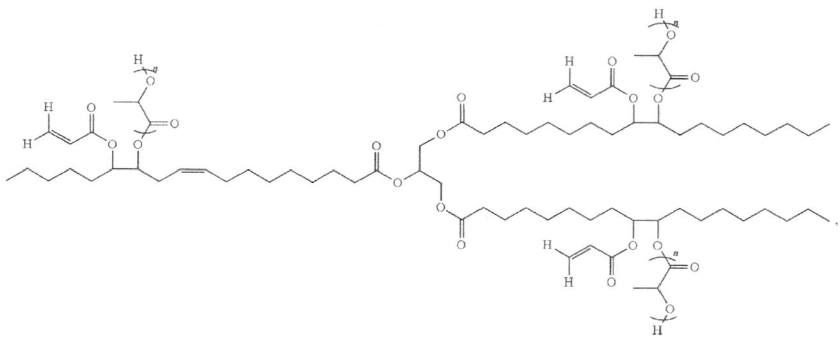

Figure 4.34 Star polymer compatibilizer (161); n is between 10 and 10,000.

in a reduction in the average droplet diameter and an increased reproducibility of the measured tensile parameters, c.f. Figure 4.35.

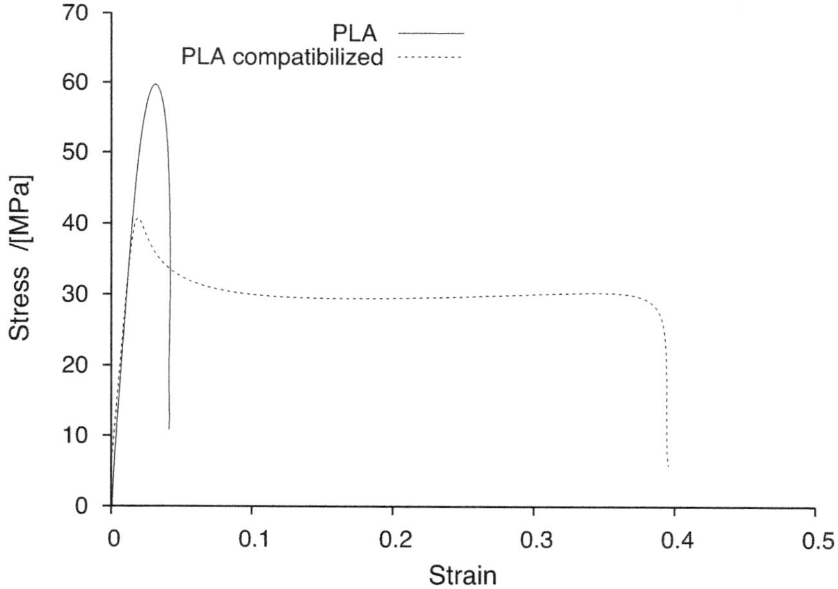

Figure 4.35 Stress-strain curves (160).

The so improved thermoplastic may have improved tensile properties with only little plasticization (161).

Trade names appearing in the references are shown in Table 4.14.

Table 4.14 Trade names in references.

Trade name Description	Supplier
Acrawax®	Lonza AG
Amide wax (99, 101)	
Acronal® 4F	BASF AG
Poly(*n*-butylacrylate) (99, 101)	
Acronal®	BASF
Acrylic resins (99, 101)	
Adcote® 313	Morton International
Polyethyleneimine (99, 101)	
Adcote® 50T4983	Morton International
Ethylene acrylic acid dispersion (99, 101)	
Adcote® 50T4990	Morton International
Ethylene acrylic acid dispersion (99, 101)	
Airflex® (Series)	Air Products and Chemicals Inc.
Vinyl acetate/ethylene copolymer emulsions (99, 101)	
Aquathane® 97949	Reichhold
Nonsulfonated urethane dispersion (99, 101)	
Aquathane® 97959	Reichhold
Nonsulfonated urethane dispersion (99, 101)	
Avicel®	FMC BioPolymer
Microcrystalline cellulose (78, 89)	
Biocycle®	PHB Ind.
Poly(hydroxybutyrate) (86)	
Biomax®	DuPont
Sulfonated aliphatic-aromatic copolyesters (99, 101)	
Bionolle®	Showa Highpolymer Co.
Poly(butylene succinate) (99, 101)	
Biopol®	Zeneca
Biodegradable hot melt adhesive (PHV/B) (29, 86, 99, 101)	
Bynel® (Series)	DuPont
Anhydride modified ethylene vinyl acetate resin, adhesion promoter (85, 99, 101)	
Carboset® CR-760	B.F. Goodrich Co.
Anionic acrylate-styrene dispersion (99, 101)	
Celvol® (Series)	Celanese
Poly(vinyl alcohol) (150)	

Table 4.14 (cont) Trade names in references.

Trade name Description	Supplier
Chinet®	Huhtamaki
Disposable paper plates (101)	
Citroflex® A-4	Morflex Inc.
Acetyltri-*n*-butyl citrate (99, 101)	
Claytone®	Claytone
Organophilic bentonite (101)	
Cymel® (Series)	Cytec
Amino resins (99, 101)	
Dispercoll® (Series)	Bayer Corp.
Sulfonated polyester urethane dispersions (99, 101)	
Eastar® Bio	Eastman Chemical Company
Compostable copolyester (85, 86, 99, 101, 102)	
EcoFoam®	National Starch and Chemical Co.
Biodegradable materials for use in packaging (99, 101)	
EcoPLA®	Cargill Dow Co.
Poly(lactide) (99, 101)	
Edenol®	Cognis
Epoxidized soybean oil (86)	
Elvax® (Series)	DuPont
Ethylene/vinyl acetate copolymers (85)	
Enmat®	TianAn
Polyhydroxybutyrate-*co*-valerate (86)	
EnPol®	IRe Chemical Ltd.
Poly(1,4-butylene succinate) (99, 101)	
EnviroFil®	ENPAC
Modified starch (99, 101)	
Escorez™ (Series)	ExxonMobil Chemical Co.
Tackifying resins (EVA) (99, 101)	
Flexthane® (Series)	Air Products and Chemicals, Inc.
Nonsulfonated urethane dispersions (99, 101)	
Hycar® (Series)	Lubrizol Advanced Materials Inc.; B.F. Goodrich Co.
Amine-terminated butadiene-acrylonitrile (99, 101)	
Irganox® (Series)	Ciba-Geigy AG
Hindered phenols, polymerization inhibitor (101)	
Jagotex® KEA	Jager Co.
Acrylic dispersions (99, 101)	

Table 4.14 (cont) Trade names in references.

Trade name Description	Supplier
Joncryl® ADR 4368	Johnson Polymer
Copolymers from styrene and of glycidyl (meth)acrylates (86)	
Kevlar®	DuPont
Aramid (121)	
Kynar®	Arkema Inc.
Poly(vinylidene fluoride) (99, 101)	
Luphen® D 200 A	BASF
Nonsulfonated urethane dispersions (99, 101)	
Merginat® ESBO	Hobum Oleochemicals GmbH
Epoxidized soybean oil (86)	
MicroMid® (Series)	Union Camp Corp.
Poly(amide) dispersions (99, 101)	
Myvacet® (Series)	Kerry Group Services Ltd.
Acetylated monoglycerides of modified fats (99, 101)	
Myvaplex® 600	Kerry Group Services Ltd.
2,3-Dihydroxypropyl octadecanoate (glycerol monostearate) (99, 101)	
Myverol®	Kerry Group Services Ltd.
Propylene glycol monostearate (99, 101)	
NatureWorks™ PLA	Cargill Dow
Poly(lactate) (86)	
NeoRez® (Series)	Zeneca Resins
Nonsulfonated urethane dispersions (99, 101)	
Papermatch®	A. Schulman, Inc.
Masterbatch to impart paper-like properties to polymer films (99, 101)	
Piccotex® LC-55WK	Hercules Inc.
Styrene resin emulsion (99, 101)	
Pluriol® 1500	BASF
Poly(ethylene glycol) (86)	
Polectron® 430	ISP Corp.
Vinylpyrrolidone/styrene copolymer emulsion fluid (99, 101)	
Quilastic® (Series)	Merquinsa
Nonsulfonated urethane dispersions (99, 101)	
Resyn® (Series)	ICI
Vinyl acetate homopolymer dispersions (99, 101)	
Rhoplex® (Series)	Evonik; Rohm & Haas
Acrylic latex (99, 101)	

Table 4.14 (cont) Trade names in references.

Trade name Description	Supplier
Rucoflex®	Bayer
Polyester polyol (99, 101)	
Sancure®	Lubrizol Advanced Materials Inc.
Nonsulfonated urethane dispersions (99, 101)	
Supercoat™	Imerys Pigments Inc.
Calcium carbonate powder (101)	
Synthemul® (Series)	Reichhhold
Carboxylated acrylic copolymer (99, 101)	
Tacolyn® 5001	Hercules Inc.
Styrene resin dispersion (99, 101)	
Tecoflex® (Series)	Lubrizol Advanced Materials Inc.
Urethane elastomers (86, 99, 101, 102)	
Texacar®	Texaco Chemical Co.
Ethylene and propylene carbonates (70)	
Texigel® (Series)	Scott Bader Inc.
Silanated anionic acrylate-styrene polymer dispersions (99, 101)	
Tronox® 470	Tranox Ltd.
Titanium dioxide pigment (99, 101)	
Tween® 20	Uniqema
Sorbitan monolaurate (99, 101)	
Tween® 40	Uniqema
Sorbitan monopalmitate (99, 101)	
Tween® 60	Uniqema
Sorbitan monostearate (99, 101)	
Tween® 85	Uniqema
Sorbitan monooleate (99, 101)	
Tyzor® PC-42	DuPont
Titanium complex, adhesion promoter (100, 101)	
Vancryl® (Series)	Air Products & Chemicals Co.
Anionic acrylate-styrene dispersions (99, 101)	
Vycar®	Lubrizol Advanced Materials Inc.
Poly(vinylidene chloride) emulsion (99, 101)	

References

1. J.M. Nduko and S. Taguchi, Microbial production and properties of LA-based polymers and oligomers from renewable feedstock in Z. Fang, R.L. Smith, Jr, and X.-F. Tian, eds., *Production of Materials from Sustainable Biomass Resources*, pp. 361–390. Springer Singapore, Singapore, 2019.
2. S.K. Bhatia, P. Wadhwa, R.K. Bhatia, S.K.S. Patel, and Y.-H. Yang, Strategy for biosynthesis of polyhydroxyalkonates polymers/copolymers and their application in drug delivery in V.C. Kalia, ed., *Biotechnological Applications of Polyhydroxyalkanoates*, pp. 13–34. Springer Singapore, Singapore, 2019.
3. T. Lebarbé, L. Maisonneuve, T.H. Nga Nguyen, B. Gadenne, C. Alfos, and H. Cramail, *Polymer Chemistry*, Vol. 3, p. 2842, 2012.
4. V.K. Thakur and M.K. Thakur, eds., *Functional Biopolymers*, Springer Series on Polymer and Composite Materials, Springer Nature, Urdorf, Switzerland, 2018.
5. M. von Czapiewski and M.A.R. Meier, *Macromolecular Chemistry and Physics*, Vol. 218, p. 1700153, 2017.
6. A. Fleckhaus, G. Klaassen, and M. Rüsch gen. Klaas, *Journal of Polymer Science Part A: Polymer Chemistry*, Vol. 55, p. 3352, 2017.
7. A. Fleckhaus, P.A. Fokou, G. Klaassen, and M. Rüsch gen. Klaas, *European Journal of Lipid Science and Technology*, Vol. 121, p. 1700429, 2019.
8. K. Hu, D. Zhao, G. Wu, and J. Ma, *RSC Adv.*, Vol. 5, p. 85996, 2015.
9. F.A. Ferraz, A.S. Muniz, A.R.d. Santos Oliveira, and M.A. Ferreira César-Oliveira, *Journal of Polymer Science Part A: Polymer Chemistry*, Vol. 56, p. 1039, 2018.
10. C. Valverde, G. Lligadas, J.C. Ronda, M. Galià, and V. Cádiz, *Polymer Degradation and Stability*, Vol. 155, p. 84, 2018.
11. J. Fang, X. Sui, Y. Li, and C. Chen, *Polym. Chem.*, Vol. 9, p. 4143, 2018.
12. J. Gao, B. Yang, and C. Chen, *Journal of Catalysis*, Vol. 369, p. 233, 2019.
13. O. Lamarzelle, G. Hibert, S. Lecommandoux, E. Grau, and H. Cramail, *Polym. Chem.*, Vol. 8, p. 3438, 2017.
14. A. Celli, P. Marchese, L. Sisti, D. Dumand, S. Sullalti, and G. Totaro, *Polymer International*, Vol. 62, p. 1210, December 2012.
15. E. Akaraonye, T. Keshavarz, and I. Roy, *Journal of Chemical Technology & Biotechnology*, Vol. 85, p. 732, June 2010.
16. J.-P.M. Autran, Biodegradable polyester blend compositions and methods of making the same, US Patent 7 265 188, assigned to The Procter & Gamble Company (Cincinnati, OH), September 4, 2007.
17. W. Blunt, A. Lagassé, Z. Jin, C. Dartiailh, R. Sparling, D.J. Gapes, D.B. Levin, and N. Cicek, *Process Biochemistry*, Vol. 82, p. 19, 2019.

18. H. Verhoogt, B. Ramsay, and B. Favis, *Polymer*, Vol. 35, p. 5155, November 1994.
19. J. Walker and J.R. Whitton, Extraction of poly(β-hydroxy butyric acid), US Patent 4 358 583, assigned to Imperial Chemical Industries PLC (London, GB2), November 9, 1982.
20. Y. Doi, Random copolymer comprising d-3-hydroxybutyrate units and d-3-hydroxyvalerate, and process for production thereof, US Patent 4 997 909, assigned to Mitsubishi Gas Chemical Company, Inc. (Tokyo, JP), March 5, 1991.
21. M. Lemoigne, *Bull. Soc. Chim. Biol.*, Vol. 8, p. 770, 1926.
22. A. Steinbüchel and H.E. Valentin, *FEMS Microbiology Letters*, Vol. 128, p. 219, 1995.
23. E. Shimamura, K. Kasuya, G. Kobayashi, T. Shiotani, Y. Shima, and Y. Doi, *Macromolecules*, Vol. 27, p. 878, May 1994.
24. Y. Doi, S. Kitamura, and H. Abe, *Macromolecules*, Vol. 28, p. 4822, July 1995.
25. Y. Hori, T. Nishishita, and A. Yamaguchi, Process for producing poly(3-hydroxybutyric acid), US Patent 5 430 125, assigned to Takasago International Corporation (Tokyo, JP), July 4, 1995.
26. J. Fang, M.J.-L. Tschan, T. Roisnel, X. Trivelli, R.M. Gauvin, C.M. Thomas, and L. Maron, *Polymer Chemistry*, Vol. 4, p. 360, 2013.
27. F. Hempel, A.S. Bozarth, N. Lindenkamp, A. Klingl, S. Zauner, U. Linne, A. Steinbüchel, and U.G. Maier, *Microbial Cell Factories*, Vol. 10, p. 81, 2011.
28. C. Nawrath, Y. Poirier, and C. Somerville, *Proceedings of the National Academy of Sciences*, Vol. 91, p. 12760, 1994.
29. K.J. Gruys, T.A. Mitsky, G.M. Kishore, S.C. Slater, S.R. Padgette, and D.M. Stark, Methods of optimizing substrate pools and biosynthesis of poly-β-hydroxybutyrate-*co*-poly-β-hydroxyvalerate in bacteria and plants, US Patent 5 958 745, assigned to Monsanto Company (St. Louis, MO), September 28, 1999.
30. N.S. Elsayed, K.M. Aboshanab, M.A. Yassien, and N.H. Hassouna, *Molecular Biology Reports*, Vol. 46, p. 3357, Jun 2019.
31. J. Kobayashi and A. Kondo, *Microbial Cell Factories*, Vol. 18, p. 40, 2019.
32. J.-J. Gu, Y. Zhou, Y.-B. Tong, J.-J. Lu, and B.-C. Ye, *Biotechnology and Applied Biochemistry*, Vol. 65, p. 622, 2018.
33. K. Khemani, P.J. Andersen, S.K. Hodson, and H. Schmidt, Biodegradable polymer films and sheets suitable for use as laminate coatings as well as wraps and other packaging materials, US Patent 6 573 340, assigned to Biotec Biologische Naturverpackungen GmbH & Co. KG (DE), June 3, 2003.
34. P. Lemechko, M.L. Fellic, and S. Bruzaud, *International Journal of Biological Macromolecules*, Vol. 128, p. 429, 2019.

35. J.H. Wang and D.M. Schertz, Grafted biodegradable polymer blend compositions, US Patent 7 053 151, assigned to Kimberly-Clark Worldwide, Inc. (Neenah, WI), May 30, 2006.
36. S.K. Bhatia, P. Wadhwa, J.W. Hong, Y.G. Hong, J.-M. Jeon, E.S. Lee, and Y.-H. Yang, *International Journal of Biological Macromolecules*, Vol. 123, p. 117, 2019.
37. A. Svendsen, *Biochimica et Biophysica Acta (BBA) - Protein Structure and Molecular Enzymology*, Vol. 1543, p. 223, 2000. Protein engineering of enzymes.
38. Wikipedia contributors, Lipase — Wikipedia, the free encyclopedia, https://en.wikipedia.org/w/index.php?title=Lipase&oldid=894978757, 2019. [Online; accessed 11-June-2019].
39. G. Chen, G. Zhang, S. Park, and S. Lee, *Applied Microbiology and Biotechnology*, Vol. 57, p. 50, October 2001.
40. S.H. Lee, D.H. Oh, W.S. Ahn, Y. Lee, J.-I. Choi, and S.Y. Lee, *Biotechnology and Bioengineering*, Vol. 67, p. 240, January 2000.
41. X. Qu, Q. Wu, K. Zhang, and G. Chen, *Biomaterials*, Vol. 27, March 2006.
42. Y.-G. Hong, Y.-M. Moon, J.-W. Hong, T.-R. Choi, H.-R. Jung, S.-Y. Yang, D.-W. Jang, Y.-R. Park, C.J. Brigham, J.-S. Kim, Y.K. Lee, and Y.H. Yang, *Journal of Microbiology and Biotechnology*, Vol. 29, p. 382, 2019.
43. J. Zhou, X. Ma, J. Li, and L. Zhu, *Cellulose*, Vol. 26, p. 979, Jan 2019.
44. S. Baidurah, P. Murugan, K.Y. Sen, Y. Furuyama, M. Nonome, K. Sudesh, and Y. Ishida, *Journal of Analytical and Applied Pyrolysis*, Vol. 137, p. 146 , 2019.
45. D. Battegazzore, A. Noori, and A. Frache, *Polymer Composites*, Vol. in press, p. N.A., 2019.
46. A. Dufresne and E. Samain, *Macromolecules*, Vol. 31, p. 6426, September 1998.
47. D. Dubief, E. Samain, and A. Dufresne, *Macromolecules*, Vol. 32, p. 5765, September 1999.
48. N. Kurth, E. Renard, F. Brachet, D. Robic, P. Guerin, and R. Bourbouze, *Polymer*, Vol. 43, p. 1095, February 2002.
49. A. Dufresne and M. Vincendon, *Macromolecules*, Vol. 33, p. 2998, April 2000.
50. B. Hazer, E. Akyol, T. Şanal, S. Guillaume, B. Çakmakli, and A. Steinbuchel, *Polymer Bulletin*, Vol. 76, p. 919, Feb 2019.
51. D.F. Gilmore, S. Antoun, R.W. Lenz, and R.C. Fuller, *Journal of Environmental Polymer Degradation*, Vol. 1, p. 269, October 1993.
52. T.W. Federle, M.A. Barlaz, C.A. Pettigrew, K.M. Kerr, J.J. Kemper, B.A. Nuck, and L.A. Schechtman, *Biomacromolecules*, Vol. 3, p. 813, July 2002.

53. C. Zhang, R. Jia, Y. Dong, and L. Zhao, *Environmental Science and Pollution Research*, Vol. 26, p. 15518, May 2019.
54. Wikipedia contributors, Buprofezin — Wikipedia, the free encyclopedia, https://en.wikipedia.org/w/index.php?title=Buprofezin&oldid=833019432, 2018. [Online; accessed 12-June-2019].
55. S. Muñoz-Guerra, M. García-Alvarez, and J.A. Portilla-Arias, *Journal of Renewable Materials*, Vol. 1, p. 42, January 2013.
56. H. Wang, X. Zhuang, X. Wang, C. Li, Z. Li, W. Kang, Y. Yin, M.D. Guiver, and B. Cheng, *ACS Applied Materials & Interfaces*, Vol. 11, p. 21865, 2019.
57. J.J. Feminus, R. Manikandan, S.S. Narayanan, and P.N. Deepa, *Journal of Chemical Sciences*, Vol. 131, p. 11, Jan 2019.
58. A. Gamarra-Montes, B. Missagia, J. Morató, and S. Muñoz-Guerra, *Polymers*, Vol. 10, p. 21, December 2017.
59. M. Biagiotti, G. Borghese, P. Francescato, C.F. Morelli, A.M. Albertini, T. Bavaro, D. Ubiali, R. Mendichi, and G. Speranza, *RSC Adv.*, Vol. 6, p. 43954, 2016.
60. S. Velmathi, R. Nagahata, J.-i. Sugiyama, and K. Takeuchi, *Macromolecular Rapid Communications*, Vol. 26, p. 1163, July 2005.
61. M.S. Nikolic and J. Djonlagic, *Polymer Degradation and Stability*, Vol. 74, p. 263, January 2001.
62. V. Tserki, P. Matzinos, E. Pavlidou, and C. Panayiotou, *Polymer Degradation and Stability*, Vol. 91, p. 377, February 2006.
63. D.P.R. Kint, A. Alla, E. Deloret, J.L. Campos, and S. Muñoz-Guerra, *Polymer*, Vol. 44, p. 1321, March 2003.
64. W. Wang, G. Zhang, W. Zhang, W. Guo, and J. Wang, *Journal of Polymers and the Environment*, Vol. 21, p. 46, March 2013.
65. T. Kuang, J. Ju, Z. Yang, L. Geng, and X. Peng, *Composites Science and Technology*, Vol. 159, p. 171, 2018.
66. C. Hu, S. Bourbigot, T. Delaunay, M. Collinet, S. Marcille, and G. Fontaine, *Polymer Degradation and Stability*, Vol. 164, p. 9, 2019.
67. D. Braun and M. Bergmann, *Journal für Praktische Chemie/Chemiker-Zeitung*, Vol. 334, p. 298, 1992.
68. F. Fenouillot, A. Rousseau, G. Colomines, R. Saint-Loup, and J.-P. Pascault, *Progress in Polymer Science*, Vol. 35, p. 578, May 2010.
69. M. Rose and R. Palkovits, *ChemSusChem*, Vol. 5, p. 167, January 2012.
70. R. Hayes and C. Brandenburg, Bis(2-hydroxyethyl isosorbide); preparation, polymers derived therefrom, and enduses thereby, US Patent 6 608 167, assigned to E. I. du Pont de Nemours and Company (Wilmington, DE), August 19, 2003.
71. G. Flèche and M. Huchette, *Starch - Stärke*, Vol. 38, p. 26, 1986.
72. M. Okada, K. Tsunoda, K. Tachikawa, and K. Aoi, *Journal of Applied Polymer Science*, Vol. 77, p. 338, July 2000.

73. R. Storbeck, M. Rehahn, and M. Ballauff, *Die Makromolekulare Chemie*, Vol. 194, p. 53, January 1993.
74. T.G. Fox and P.J. Flory, *Journal of Applied Physics*, Vol. 21, p. 581, 1950.
75. J. Qi, J. Wu, J. Chen, and H. Wang, *Polymer Degradation and Stability*, Vol. 160, p. 229, 2019.
76. S. Legrand, N. Jacquel, H. Amedro, R. Saint-Loup, J.-P. Pascault, A. Rousseau, and F. Fenouillot, *European Polymer Journal*, Vol. 115, p. 22 , 2019.
77. U. Witt, M. Yamamoto, U. Seeliger, R.-J. Müller, and V. Warzelhan, *Angewandte Chemie International Edition*, Vol. 38, p. 1438, 1999.
78. C. Bastioli, T. Milizia, G. Floridi, A.S. Lallaro, G. Cella, and M. Tosin, Biodegradable aliphatic-aromatic copolyester, US Patent 8 193 300, assigned to Novamont S.p.A. (Novara, IT), June 5, 2012.
79. F. Freese, R. Loos, J. Keck, J. Auffermann, and X. Yang, Biodegradable polyester film, US Patent Application 20 120 288 650, assigned to BASF, Ludwigshafen (DE), November 15, 2012.
80. k.O. Siegenthaler and J. Auffermann, Item produced via injection molding, US Patent Application 20 130 004 759, assigned to BASF Ludwigshafen (DE), January 3, 2013.
81. A. Elmendorf, Tearing tester, US Patent 1 423 841, July 25, 1922.
82. Plastics – Film and sheeting – Determination of tear resistance – Part 2: Elmendorf method, ISO Standard 6383-2, International Organization for Standardization, Geneva, Switzerland, 1983.
83. Standard test method for propagation tear resistance of plastic film and thin sheeting by pendulum method, ASTM Standard D1922-09, ASTM International, West Conshohocken, PA, 2011. DOI:10.1520/D1922-09.
84. R.-Y. Wu, C.-W. Chu, L.-L. Chang, Y.-H. Tang, and P.-S. Lai, Method for preparing a biodegradable copolyester, US Patent 7 129 301, assigned to Far Eastern Textile Ltd. (Taipei, TW), October 31, 2006.
85. M.D. Shelby, A.J. Matosky, C.M. Tanner, and M.E. Donelson, Blends of aliphatic-aromatic copolyesters with ethylene-vinyl acetate copolymers, US Patent 7 241 838, assigned to Eastman Chemical Company (Kingsport, TN), July 10, 2007.
86. U. Seeliger, G. Skupin, and D. Starke, Biologically-degradable polyester mixture, US Patent 8 003 731, assigned to BASF SE (Ludwigshafen, DE), August 23, 2011.
87. T. Madera-Santana, M. Misra, L. Drzal, D. Robledo, and Y. Freile-Pelegrin, *Polymer Engineering & Science*, Vol. 49, p. 1117, June 2009.
88. I.S. Choi and K.R. Yoon, Process of preparing a biodegradable polymer using an enzyme catalyst and a biodegradable polymer prepared through the process, US Patent 7 642 075, assigned to Korea

Advanced Institute of Science and Technology (Taejon, KR), January 5, 2010.

89. C. Bastioli, T. Milizia, G. Floridi, A.S. Lallaro, G. Cella, and M. Tosin, Biodegradable aliphatic-aromatic polyesters, US Patent 8 193 298, assigned to Novamont S.p.A. (Novara, IT), June 5, 2012.
90. H.J. Neischlag, I.A. Wolff, T.C. Manley, and R.J. Holland, *Industrial & Engineering Chemistry Product Research and Development*, Vol. 6, p. 120, June 1967.
91. K. Kato and N. Uemura, Process for the preparation of long-chain dicarboxylic acids by fermentation, US Patent 4 339 536, assigned to Nippon Mining Co., Ltd. (Tokyo, JP), July 13, 1982.
92. A. Kamal, S. Gurrala, and W.N. Fadnavis, Process for preparing long-chain dicarboxylic acids, US Patent 8 383 373, assigned to Council of Scientific & Industrial Research (New Delhi, IN), February 26, 2013.
93. C. Bastioli, G. Borsotti, L. Capuzzi, and R. Vallero, Aliphatic-aromatic biodegradable polyester, assigned to Novamont S.p.A. (Novara, IT), June 11, 2019.
94. G. Mehta, L.T. Drzal, A.K. Mohanty, and M. Misra, *Journal of Applied Polymer Science*, Vol. 99, p. 1055, February 2006.
95. W. Chee Sai, M.B.A. Bakar, M.N. Masri, M.A. Sulaiman, M.H.M. Amini, and M.Z. Mohd, *Int. J. Cur. Res. Eng. Sci. Tech. ISSN*, Vol. 2581, p. 5, 2018.
96. M. Skrifvars, H. Dhakal, Z. Zhang, J. Gentilcore, and D. Åkesson, *Composites Part A: Applied Science and Manufacturing*, Vol. 121, p. 196, 2019.
97. F. Ansari, M. Skrifvars, and L. Berglund, *Composites Science and Technology*, Vol. 117, p. 298, 2015.
98. K. Khemani, P.J. Andersen, H. Schmidt, and S.K. Hodson, Biodegradable polymer films and sheets suitable for use as laminate coatings as well as wraps and other packaging materials, WO Patent 0 216 468, assigned to Khashoggi E. Industries, February 28, 2002.
99. R.M. Gohil, N.M. Hasty, R.A. Hayes, J.V. Kurian, Y. Liang, E.J. Stancik, M. Strukelj, and S.Y.-T. Tseng, Aliphatic-aromatic polyesters, and articles made therefrom, US Patent 7 888 405, assigned to E. I. du Pont de Nemours and Company (Wilmington, DE), February 15, 2011.
100. R.A. Hayes, Sulfonated aliphatic-aromatic copolyetheresters, US Patent 7 625 994, assigned to E.I. du Pont de Nemours and Company (Wilmington, DE), December 1, 2009.
101. R.A. Hayes, Sulfonated aromatic copolyesters containing hydroxyalkanoic acid groups and shaped articles produced therefrom, US

Patent 7 358 325, assigned to E. I. du Pont de Nemours and Company (Wilmington, DE), April 15, 2008.

102. W.A. Haile, S.E. George, W.R. Hale, and W.L. Jenkins, Water-dispersible and multicomponent fibers from sulfopolyesters, US Patent 8 444 896, assigned to Eastman Chemical Company (Kingsport, TN), May 21, 2013.
103. Standard test methods for determining aerobic biodegradation of radiolabeled plastic materials in an aqueous or compost environment, ASTM Standard D6340-98, ASTM International, West Conshohocken, PA, 2007.
104. L. Montero de Espinosa and M.A. Meier, *European Polymer Journal*, Vol. 47, p. 837, May 2011.
105. R. Malacea and P.H. Dixneuf, Alkene metathesis and renewable materials: Selective transformations of plant oils in V. Dragutan, A. Demonceau, I. Dragutan, and E. Finkelshtein, eds., *Green Metathesis Chemistry*, NATO Science for Peace and Security Series A: Chemistry and Biology, pp. 185–206. Springer Netherlands, 2010.
106. J. Shakina and N.B.M. Eben, *International Journal on Recent Advancement in Biotechnology & Nanotechnology*, Vol. 2, 2019.
107. O. Türünç, *Efficient Routes to Degradable and non-Degradable Renewable Polymers from Fatty Acids*. PhD thesis, Fakultät für Chemie und Biowissenschaften Karlsruher Institut für Technologie, University of Karlsruhe, Germany, 2012.
108. S. Warwel, F. Brüse, C. Demes, M. Kunz, et al., *Chemosphere*, Vol. 43, p. 39, 2001.
109. C. Thurier, C. Fischmeister, C. Bruneau, H. Olivier-Bourbigou, and P.H. Dixneuf, *ChemSusChem*, Vol. 1, p. 118, 2008.
110. A. Rybak and M.A.R. Meier, *Green Chemistry*, Vol. 9, p. 1356, 2007.
111. A. Rybak and M.A.R. Meier, *Green Chemistry*, Vol. 10, p. 1099, 2008.
112. L.M. de Espinosa, J.C. Ronda, M. Galià, and V. Cádiz, *Journal of Polymer Science Part A: Polymer Chemistry*, Vol. 47, p. 1159, February 2009.
113. L. Montero de Espinosa Meléndez, *Plant oils as renewable precursos of thermosetting and flame retardant polymers*. PhD thesis, Universitat Rovira i Virgili, Tarragona, Spain, 2010.
114. W. Ding and M.L. Robertson, *European Polymer Journal*, Vol. 113, p. 411, 2019.
115. O. Türünç and M.A.R. Meier, *European Journal of Lipid Science and Technology*, Vol. 115, p. 41, 2013.
116. N.H. Koenig and D. Swern, *Journal of the American Chemical Society*, Vol. 79, p. 362, January 1957.
117. F.D. Gunstone, M.G. Hussain, and D.M. Smith, *Chemistry and Physics of Lipids*, Vol. 13, p. 71, August 1974.

118. N.H. Koenig, G.S. Sasin, and D. Swern, *Journal of Organic Chemistry*, Vol. 23, p. 1525, October 1958.
119. A.W. Schwab, L.E. Gast, and J.C. Cowan, *Journal of the American Oil Chemists Society*, Vol. 45, p. 461, June 1968.
120. J. Epalza, J. Jaramillo, and O. Guarín, Extraction and use of plant biopolymers for water treatment in M. Eyvaz and E. Yüksel, eds., *Desalination and Water Treatment*. IntechOpen, London, UK, 2018.
121. R. Wool, S. Kusefoglu, G. Palmese, S. Khot, and R. Zhao, High modulus polymers and composites from plant oils, US Patent 6 121 398, assigned to University of Delaware (Newark, DE), September 19, 2000.
122. A. Guo, Y. Cho, and Z.S. Petrovic, *Journal of Polymer Science Part A: Polymer Chemistry*, Vol. 38, p. 3900, 2000.
123. J. La Scala and R.P. Wool, *Polymer*, Vol. 46, p. 61, January 2005.
124. T. Eren and S.H. Küsefoğlu, *Journal of Applied Polymer Science*, Vol. 91, p. 2700, February 2004.
125. L.M. de Espinosa, M.A.R. Meier, J.C. Ronda, M. Galià, and V. Cádiz, *Journal of Polymer Science Part A: Polymer Chemistry*, Vol. 48, p. 1649, April 2010.
126. A. Horrocks, *Fire Retardant Materials*, CRC Press, Boca Raton, FL; Woodhead Publishing, Cambridge, England, 2001.
127. L.A. Mercado, G. Ribera, M. Galià, and V. Cádiz, *Journal of Polymer Science Part A: Polymer Chemistry*, Vol. 44, p. 1676, March 2006.
128. S.-Y. Lu and I. Hamerton, *Progress in Polymer Science*, Vol. 27, p. 1661, October 2002.
129. D.J. Valenti and K.B. Wagener, *Macromolecules*, Vol. 31, p. 2764, May 1998.
130. T.W. Baughman and K.B. Wagener, *Recent Advances in ADMET Polymerization*, Springer-Verlag, February 2005.
131. K.B. Wagener, J.M. Boncella, and J.G. Nel, *Macromolecules*, Vol. 24, p. 2649, May 1991.
132. F. Marsico, M. Wagner, K. Landfester, and F.R. Wurm, *Macromolecules*, Vol. 45, p. 8511, November 2012.
133. H. Liu, W. Lu, and S. Liu, *Journal of Applied Polymer Science*, Vol. 135, p. 45698, 2018.
134. M. Firdaus and M.A.R. Meier, *European Polymer Journal*, Vol. 49, p. 156, January 2013.
135. M.S.A. Abdelaty and D. Kuckling, *Open Journal of Organic Polymer Materials*, Vol. 8, p. 15, 2018.
136. Y. Zhang, V.K. Thakur, Y. Li, T.F. Garrison, Z. Gao, J. Gu, and M.R. Kessler, *Macromolecular Materials and Engineering*, Vol. 303, p. 1700278, 2018.

137. C. Gioia, M.B. Banella, G. Totaro, M. Vannini, P. Marchese, M. Colonna, L. Sisti, and A. Celli, *Journal of Renewable Materials*, Vol. 6, p. 126, 2018.

138. Z. Wang, Y. Li, T. Zhu, L. Xiong, F. Liu, and H. Qi, *Polymer Degradation and Stability*, 2019. in press.

139. A. Bahmani, P.A. Comeau, J. Montesano, and T.L. Willett, *Materials & Design*, Vol. 174, p. 107790, 2019.

140. T. Mori and K. Tanaka, *Acta Metallurgica*, Vol. 21, p. 571, 1973.

141. M. Mirkhalaf and B. Ashrafi, *Materials Today Communications*, Vol. 13, p. 144, 2017.

142. S. Rwahwire, B. Tomkova, A.P. Periyasamy, and B.M. Kale, Green thermoset reinforced biocomposites in G. Koronis and A. Silva, eds., *Green Composites for Automotive Applications*, Woodhead Publishing Series in Composites Science and Engineering, chapter 3, pp. 61 – 80. Woodhead Publishing, 2019.

143. Y. Xia, R.L. Quirino, and R.C. Larock, *Journal of Renewable Materials*, Vol. 1, p. 3, January 2013.

144. S.S. Panda, B.P. Panda, S.K. Nayak, and S. Mohanty, *Polymer-Plastics Technology and Engineering*, Vol. 57, p. 500, 2018.

145. A. Rüdiger, P. Hendil-Forssell, C. Hedfors, M. Martinelle, S. Trey, and M. Johansson, *Journal of Renewable Materials*, Vol. 1, p. 124, April 2013.

146. F.C. Fernandes, K. Kirwan, P.R. Wilson, and S.R. Coles, *Green Materials*, Vol. 6, p. 38, 2018.

147. Y. Ecochard, R. Auvergne, B. Boutevin, and S. Caillol, *European Journal of Lipid Science and Technology*, Vol. in press, p. 0, 2019. in press.

148. S. Nameer, D.B. Larsen, J.Ø. Duus, A.E. Daugaard, and M. Johansson, *ACS Sustainable Chemistry & Engineering*, Vol. 6, p. 9442, 2018.

149. A.K. Mohanty, P. Tummala, W. Liu, M. Misra, P.V. Mulukutla, and L.T. Drzal, *Journal of Polymers and the Environment*, Vol. 13, p. 279, July 2005.

150. L.T. Drzal, A.K. Mohanty, W. Liu, K. Thayer, and M. Misra, Cellulosic biomass soy flour based biocomposites and process for manufacturing thereof, US Patent 7 576 147, assigned to Board of Trustees of Michigan State University (East Lansing, MI), August 18, 2009.

151. L. Fernández-Espada, C. Bengoechea, J.A. Sandía, F. Cordobés, and A. Guerrero, *Journal of Applied Polymer Science*, Vol. 136, p. 47012, 2019.

152. S. Gamero, M. Jiménez-Rosado, A. Romero, C. Bengoechea, and A. Guerrero, *Journal of Polymers and the Environment*, Vol. 27, p. 1285, June 2019.

153. Y.T.P. Ly, L.A. Johnson, and J. Jane, Soy protein as biopolymer in D.L. Kaplan, ed., *Biopolymers from renewable resources*, pp. 144–176. Springer, Berlin, 1998.
154. Y. Chen, L. Zhang, and L. Du, *Industrial & Engineering Chemistry Research*, Vol. 42, p. 6786, December 2003.
155. R. Kumar, V. Choudhary, S. Mishra, I. Varma, and B. Mattiason, *Industrial Crops and Products*, Vol. 16, p. 155, November 2002.
156. J.U. Otaigbe, H. Goel, T. Babcock, and J. Jane, *Journal of Elastomers and Plastics*, Vol. 31, p. 56, 1999.
157. A. Sasmal, P.L. Nayak, and S. Sasmal, *Polymer-Plastics Technology and Engineering*, Vol. 48, p. 905, August 2009.
158. S. Das, H. Kalita, S. Mohanty, and S.K. Nayak, *Advances in Polymer Technology*, Vol. 37, p. 1050, 2018.
159. J. Wen and R.X. Zhuo, *Macromolecular Rapid Communications*, Vol. 19, p. 641, 1998.
160. S.C. Mauck, S. Wang, W. Ding, B.J. Rohde, C.K. Fortune, G. Yang, S.-K. Ahn, and M.L. Robertson, *Macromolecules*, Vol. 49, p. 1605, 2016.
161. M. Robertson, S. Mauck, and S. Wang, Biorenewable blends of polylactide and acrylated epoxidized soybean oil compatibilized by a polylactide star polymer, US Patent Application 20 190 010 326, assigned to University of Houston System, Houston (TX), January 10, 2019.

5

Carbohydrate-Related Polymers

5.1 Starch

Starch molecules appear in two forms: The substantially linear amylose polymer and the highly branched amylopectin polymer (1).

These two forms of starch have very different properties, probably due to the ease of association of the hydroxyl groups among different molecules. The molecular structure of amylose is essentially linear with two to five relatively long branches. The average degree of polymerization of the branches is about 350 monomer units.

Under conditions that provide sufficient freedom of molecular movements, primarily by dilution with suitable solvents, and in some instances, dilution coupled with heating, the linear amylose chains can be oriented into preferentially parallel alignments such that the hydroxyl groups on one chain are in close proximity with those on the adjacent chains. The alignment of neighboring amylose molecules is believed to facilitate intermolecular hydrogen bonding. Consequently, the amylose molecules form strong aggregates (1).

In contrast, the molecular structure of amylopectin is highly branched via 1,6-α linkages. The average degree of polymerization of the branches is about 25 monomer units. Due to the highly branched structure, the amylopectin molecules cannot move as freely and do not align and associate as readily (1).

5.1.1 Starch Modification

Aspects of genetic engineering and also green applications of starch polymers have been described in monographs (2–4).

Starch can be modified in various ways, which are classified by the nature of action. These include (1):

- Physical modifications,
- Chemical modifications, and
- Biological modifications.

Physical modifications of the starch may be intramolecular or intermolecular modifications. Intramolecular modifications include reduced molecular weight or molecular weight distribution, and changes in the polymer chain conformation. Intermolecular modifications include melting or disordering the starch molecules, and reducing crystallinity, crystallite size, and granular size. Physical modifications may be achieved by input of energy such as thermal, mechanical, thermomechanical, electromagnetic, and ultrasonic energy. Other methods for modification occur by pressure, moisture, or fractionation.

Chemical modifications of starch typically include acid or alkali hydrolysis and oxidative chain scission to reduce the molecular weight and the molecular weight distribution. Suitable compounds for chemical modification of starch include organic acids such as citric acid, acetic acid, glycolic acid, and adipic acid. Furthermore, inorganic acids can be used, such as hydrochloric acid, sulfuric acid, nitric acid, phosphoric acid, boric acid, and partial salts of polybasic acids. Suitable bases are sodium hydroxide, potassium hydroxide and ammonia. Oxidizing agents are hydrogen peroxide, benzoyl peroxide, ammonium persulfate, potassium permanganate, sodium bicarbonate, or hypochloric acid salts.

Chemical modifications may also include the derivatization of starch by reaction of its hydroxyl groups with alkylene oxides. Other suitable substances are those that form ether linkages, ester linkages, urethane linkages, carbamate linkages, or isocyanate linkages. The degree of substitution of the chemically modified starch varies from 0.05 to 0.2.

Biological modifications of starch include bacterial digestion of the carbohydrate bonds, or enzymatic hydrolysis using enzymes such as amylase or amylopectase.

Recent research has focused on identifying genes and proteins that are required for the production of starch polymers (5). Four core groups of genes are required for the biosynthesis of starch, i.e.:

1. ADP-glucose pyrophosphorylases,
2. Starch synthases,
3. Starch-branching enzymes, and
4. Starch-debranching enzymes.

Phosphorylases are enzymes that catalyze the addition of a phosphate group from an inorganic phosphate to an acceptor. Starch synthase is an enzyme that catalyzes the polymerization of glucose and related compounds.

Recent research and knowledge has been sampled about the functional significance of the regulation of biosynthetic enzymes for starch; it has been concluded that it could be possible to alter the biosynthetic process (5).

A dynamic mediatory role between starch synthesis and degradation has been ascribed to starch phosphorylase (6). Actually, plant starch phosphorylase has been rather believed to be involved in the degradation reactions. The various biochemical properties have been reviewed. However, there are different varieties of starch phosphorylase. The importance of plastidial and cytoplasmic starch phosphorylase has been discussed as well as various biotechnological aspects.

α-Glucan phosphorylase can be used for the production of glucose-1-phosphate for the synthesis of amylose and a variety of other glucans (7, 8).

Glucans are polysaccharides in which the saccharide components consist only of *D*-glucose. Examples of glucans include starch and cellulose, and the like. Starch is an α-glucan in which the monomeric components are linked by α-glucoside bonds.

Glucans are useful as raw materials for starch processing industries, dietary compositions, compositions for food additives, adhesive compositions, inclusion compounds and absorbing compounds, medical and cosmetic compositions, film-type product compositions, and as an alternative for starches used in biodegradable plastics. Methods for the production of glucan have been described (7). The reaction starts from glucose and inorganic phosphate, such as sodium dihydrogen phosphate, and uses glucan phosphorylase to produce glucans. The reaction is conducted at 40–70°C.

Hulless barley, with its unique nutritional value and potential health benefits, has increasingly attracted attention in recent years (9). However, the transcription dynamics during hulless barley grain development is not well understood.

The transcriptome changes during barley grain development were investigated using illumina paired-end RNA sequencing. Two datasets of the developing grain transcriptomes from two barley landraces with the differential seed starch synthesis traits were generated, and comparative transcriptome approach in both genotypes was performed. The results showed that 38 differentially expressed genes were found co-modulated in both genotypes during the barley grain development. Of those, the proteins encoded by most of those differentially expressed genes were found, such as α-amylase-related proteins, lipid-transfer protein, homeodomain leucine zipper, and others.

Moreover, six co-expression modules associated with specific biological processes or pathways were identified by consensus co-expression network. Significantly enriched pathways of those module genes showed differences in both genotypes. These results will expand our understanding of the complex molecular mechanism of starch synthesis during barley grain development (9).

5.1.1.1 Starch-Based Flame Retardants

The flame retardancy of bio-based polymers, the flame retardancy of natural fiber composites, and the synthesis and efficiency of bio-based flame retardants have been described in a monograph (10).

A flame retardant thermoplastic starch material has been described. The composition contains 100 parts by weight of starch, 5 to 75 parts by weight of a plasticizer, and 5 to 30 parts by weight of an organic phosphonate flame retardant (11). Examples of the phosphonate flame retardant are Chemguard-1045, triphenyl phosphate, and resorcinol bis(dixylenyl phosphate).

Ammonium starch phosphate carbamates with varying degrees of substitution were analyzed by differential scanning calorimetry (DSC) and thermogravimetry (TG) coupled with Fourier transform infrared spectroscopy (FTIR) (12).

The data were analyzed with respect to the structural features of starch phosphate carbamates and their potential flame-retardant properties.

It turned out that charring of starch phosphate carbamates and the formation of polyphosphate in the condensed phase increased significantly in the case of rising degree of substitution of starch phosphate carbamates.

Thus, temperature mass losses and amounts of evolved decomposition products decreased noticeably in comparison to native starch. The activation temperatures of starch phosphate carbamates markedly decreased with an increasing degree of substitution and were considerably lower than decomposition temperatures of native starch.

As a byproduct of starch phosphate carbamates the degradation of ammonia was identified, which contributes as inert gas to flame extinguishment.

The results suggest that starch phosphate carbamates represent promising candidates for a new generation of sustainable and environmentally friendly flame retardants based on renewable resources (12).

A flame retardant foaming material based on corn starch was prepared by a mold pressing method (13). The starch was crosslinked using $Na_3(PO_3)_3$ and became gelatinized. Initiated by using $(NH_4)_2S_2O_8$, the starch was grafted with methyl acrylate and vinyl acetate by a free-radical polymerization.

The starch was mixed with a foaming regent composed of $NaHCO_3$ and azodicarbonamide, and molded at 135°C and 9 *MPa* for 10 *min*. A foaming material with a uniform honeycomblike structure and relatively high strength was prepared.

The composition showed an excellent flame resistance of V-0 rating. $Na_3(PO_3)_3$ acted both as crosslinking agent and as flame retardant, which effectively simplified the formulation. When the material was burnt, a compact charred crust was formed on its surface, which prevented the permeation of oxygen and the release of combustible organic compounds (13).

5.1.2 *Reactive Dye Removal*

A starch-montmorillonite/polyaniline nanocomposite was synthesized by a chemical oxidative polymerization method of aniline in the presence of starch-montmorillonite nanocomposite dispersion (14).

The prepared ternary nanocomposite was characterized using FTIR, X-ray diffraction (XRD), scanning electron microscopy (SEM), TG and transmission electron microscopy (TEM) techniques. XRD patterns combined with TEM results confirmed the intercalation of montmorillonite in the starch matrix. SEM micrographs revealed the growth of polyaniline over the surface of the starch-montmorillonite nanocomposite. The starch-montmorillonite/polyaniline nanocomposite was used for the adsorption of a reactive dye. Batch removal experiment results showed complete removal of dye in a very short contact time. Further investigations indicated that the removal mechanism was based on both the adsorption and electrostatic attraction between nanocomposite and dye molecules. The experimental data were well fitted to the Langmuir isotherm and pseudo-second-order kinetic model.

The adsorption capacity of reactive dye on starch-montmorillonite/polyaniline nanocomposite was 91.74 $mg\,g^{-1}$. The study results demonstrated the effectiveness of the hybrid system as an efficient adsorbent for removal of reactive dyes from textile effluents (14).

A natural polymeric adsorbing material, γ-cyclodextrin immobilized starch, was prepared by introducing a host functional molecule γ-cyclodextrin into starch, using epichlorohydrin as a crosslinking agent and sodium hydroxide as a catalyst, via the aqueous phase synthesis method (15).

The γ-cyclodextrin content in the compositions was up to 8.15%, as determined by the bromocresol green hyperchromic spectrophotometric method. The structures of CS-γCD were characterized by FTIR, ^{13}C-nuclear magnetic resonance spectroscopy (NMR), SEM, XRD and gel permeation chromatography.

The adsorption performances and kinetics of three different kinds of dyestuff were investigated, including methylene blue, methyl purple, and congo red, c.f. Figure 5.1.

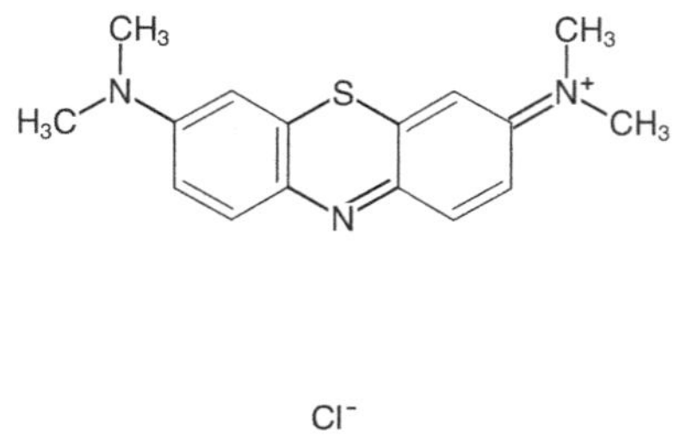

Figure 5.1 Methylene blue.

The results showed the adsorbing capacities of the γ-cyclodextrin immobilized starch to three kinds of dyes had obviously increased compared with native starch and diatomite, and the γ-cyclodextrin immobilized starch was also found to have a significant advantage in adsorption of macromolecule dye, such as CR, over other adsorbing materials (15).

The adsorption behaviors of γ-cyclodextrin, c.f. Figure 5.2, immobilized starch on the three dyestuffs could be better described by a Langmuir model (R2>0.99), but also had relatively high correlation to the Freundlich model.

The pseudo-second-order model could better describe the adsorption behaviors. The γ-cyclodextrin immobilized starch could be enzymolized, while its degradation rate was much lower than native starch, which indicated that CS-γCD had a relatively higher structural stability and longer service life. These characters could be beneficial for the popularization and application of γ-cyclodextrin immobilized starch in the fields of wastewater treatment and environmental remediation (15).

5.1.3 *Starch Granules*

Starch granules exhibit hydrophilic properties and strong intermolecular association via hydrogen bonding due to the hydroxyl groups on the granule surface. The hydrophilicity and thermal

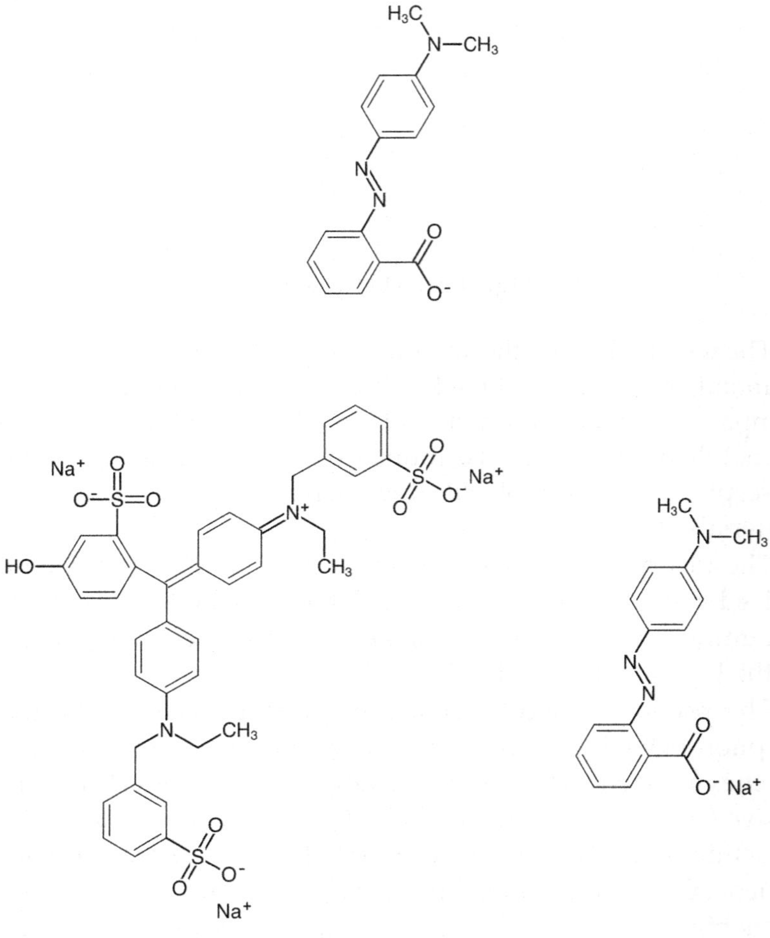

Figure 5.1 (cont) Methyl purple.

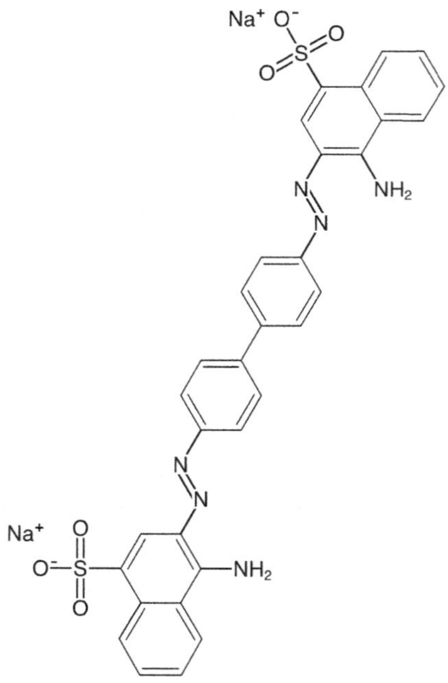

Figure 5.1 (cont) Congo red.

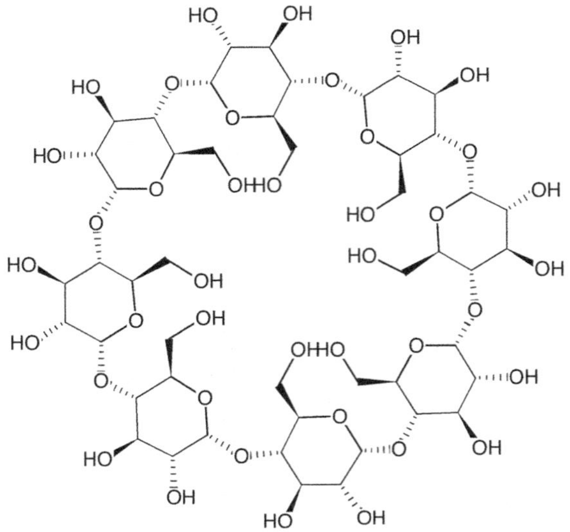

Figure 5.2 γ-Cyclodextrin.

sensitivity render the starch polymer unsuitable for thermoplastic applications (16).

Therefore, blends with other polymers have been proposed to optimize the properties of starch. However, most of such polymers are significantly more expensive than starch. This fact tends to increase the cost of such polymer blends in comparison to pure starch. Furthermore, most of such compositions will not significantly alter the mechanical properties of these blends.

Blends of thermoplastic corn starch and chitosan were obtained by melt extrusion (17). The effects of the incorporation of chitosan in the thermoplastic corn starch matrix and polymer interaction on morphology and thermal and mechanical properties were investigated. Possible interactions between the starch molecules and thermoplastic chitosan were assessed by XRD and FTIR techniques. SEM showed a homogeneous fracture surface without the presence of starch granules or chitosan aggregates. Although the incorporation of thermoplastic chitosan caused a decrease in both tensile strength and stiffness, films with better extensibility and thermal stability were produced (17).

Granular starches grafted with poly(caprolactone) (PCL) were obtained using *N*-methylimidazole as a catalyst (18). *N*-Methylimidazole is shown in Figure 5.3.

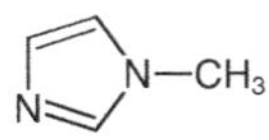

Figure 5.3 *N*-Methylimidazole.

The effect of the starch/monomer ratio and catalyst content was studied to obtain different levels of grafted PCL. The highest grafting percentage (76%) and addition (43%) were achieved for reactions with a starch/monomer ratio of 50/50 and 25% catalyst. The grafting of PCL on the starch granule was verified by the emergence of the carbonyl group in the FTIR spectra and the increased diameter of the grafted starch granule.

Thermoplastic starch from ungrafted starch and grafted starch was obtained by mixing ungrafted or grafted starch granules with water, glycerol, or sorbitol in a mixer. Ungrafted starch and grafted starch behave as plastic materials, and their mechanical properties depend on the type of plasticizer used. Materials with glycerol as the plasticizer exhibited less rigidity. The presence of starch grafted PCL results in a dramatic increase in the elongation of the thermoplastic material. The starch present in the ungrafted starch or grafted starch was completely biodegraded while the grafted PCL was partially biodegraded after the enzymatic degradation of the materials (18).

A quantitative trait locus was identified on the group 4 chromosomes of *Aegilops peregrina* and bread wheat that controls the B-type starch granule content (19). Also, a candidate gene was identified by fine-mapping of *Aegilops peregrina* and its function was confirmed using wheat targeting induced local lesions in genomes (TILLING) mutants. This gene is orthologous to the floury endosperm 6 (FLO6) gene of rice and barley and the PTST2 gene of *Arabidopsis*. In Triticeae endosperm, a reduction in the gene dose of functional FLO6 alleles results in a reduction, or loss, of B-granules. This occurs due to the repression of granule initiation in late grain development, but has no deleterious impact on the synthesis of A-granules. The complete absence of functional FLO6, however, results in reduced numbers of normal A-type and B-type granules and the production of highly-abnormal granules that vary in size and shape. This polymorphous starch seen in a wheat flo6 triple mutant is similar to that observed in the barley mutant Franubet. Analysis of Franubet (fractured Nubet) starch suggests that the mutant A-granules are not fractured but compound, due to stimulation of granule initiation in plastids during early grain development. Thus, in different situations in Triticeae, FLO6 either stimulates or represses granule initiation (19).

5.1.4 *Baked Foams*

The issues of starch-based composite foams have been presented in a review (20). Many techniques, including extrusion, hot-mold baking/compression, microwave heating, freeze-drying/solvent exchange, and supercritical fluid extrusion, can be used to produce starch foams with different cellular structures and properties.

Biodegradable trays from cassava starch, sugarcane fibers and sodium montmorillonite have been developed by a baking process (21). A nanoclay dispersion, resulting in an exfoliated structure, was obtained. The physical properties, the photodegradation and biodegradation of these trays have been studied. The formulations resulted in well-shaped trays with densities in the range of 0.19–0.30 $g\,cm^{-3}$.

The addition of the fibers and sodium montmorillonite reduce the density and the rigidity. UV stability tests of the trays for 336 *h* showed that a sample with 20% fiber and 5% nanoclay had the highest loss in stress at break of 91%. The results of degradation studies are shown in Table 5.1.

Table 5.1 Degradation studies (21).

		Weight loss/[%] after		
Fiber/[%]	MMT/[%]	30 *d*	60 *d*	90 *d*
0	0	59.3	77.0	85.5
20	0	48.0	76.0	82.7
0	5	41.7	66.3	75.0
20	5	40.3	71.7	78.0

MMT Sodium montmorillonite

At the end of the degradation experiment, i.e., after 90 *d*, all of the samples showed an irregular texture with large pores and cracks. Their surfaces exhibited dense microbial colonization (21).

The effect of the addition of sugarcane bagasse and asparagus peel fiber on the physical and mechanical properties of baked foams based on oca starch was investigated (22).

Low concentrations of fiber reduce the density of the foams and the addition of fiber does not improve the flexural strength of the foams, but generates harder and deformable trays. In contrast, high concentrations of sugarcane bagasse fiber generate more compact trays with a lower water absorption capacity. Foams with asparagus peel fiber showed higher rates of thermal degradation than the control samples, but not so extensive as to affect their applicability (22).

Starch was isolated from three Andean native crops, such as arracacha (*Arracacia xanthorrhiza*), oca (*Oxalis tuberosa*), and sweet potato (*Ipomoea batatas*), for its use as a raw material for the production of foam trays (23). The starches were characterized in their proximal composition, crystallinity, microstructure and thermal properties.

The sweet potato starch showed the highest amylose content of 42.65% and the lowest protein content of 0.30%. The highest crystallinity of sweet potato starch caused larger values of onset temperature, peak temperature, and conclusion temperature, than arracacha and oca starch. The produced foam trays showed a good appearance, adequate expansion, and low density. However, all foam trays showed a water absorption capacity of greater than 50%, which was related to their porosity and low density (23).

Malt bagasse is a byproduct of the brewing industry (24). This compound was used to produce biodegradable foam trays made from cassava starch using a baking process. The foams were prepared with different concentrations of malt bagasse (0–20%). The foams were characterized according to their microstructure, physical and mechanical properties. The trays produced with 5% and 10% malt bagasse had the highest production yields of 100%, the trays' thicknesses ranged from 2.16 *mm* to 2.24 *mm* and the densities ranged from 0.415 to 0.450 $g\,cm^{-3}$.

A good distribution of the malt bagasse throughout the polymeric matrix was found, with an amorphous structure. The variation in the relative humidity from 33% to 58% did not affect the mechanical properties of the produced foams, but the stress at break decreased and the strain at break increased when the trays were stored at 90% relative humidity. As demonstrated by the sorption isotherm data, the incorporation of malt bagasse at 10% resulted in a decrease in the hygroscopicity of starch foams. The addition of bagasse at concentrations up to 15% decreased the initial water adsorption rate of the trays. The produced trays could be an alternative for packing dry foods, thus reducing the use of petroleum-based materials (24).

Also, biodegradable foam trays were developed based on cassava bagasse and poly(vinyl alcohol) (PVA) (25). The trays were produced by baking mixtures of 90–100% of cassava bagasse and 0–10% of PVA. All these formulations were able to form well-shaped and homogeneous trays with a good appearance. The incorporation of PVA improved the production yield and reduced the water sorption capacity of the produced materials, mainly at low levels. Also, the addition of PVA resulted in an increase in stress and strain at break values. The most positive effects of PVA were obtained when this polymer was used at lower levels of 2.5% and 5.0%.

Baked foams based on cassava starch coated with PVA have a higher degree of hydrolysis of 98% (26). A decrease in the water absorption capacity of 50% was observed when the coated samples were compared with the control sample after 30 *min* of immersion in water.

5.1.5 Starch Composite Foam

At the moment, mainstream bio-foam materials consist of two kinds of materials, poly(lactic acid) (PLA) foam materials and starch foam materials (27). A deficiency of heat resistance exists in the former, which is incapable of meeting the requirements of the environmental tests for container transportation.

To match the mechanical function of current PLA foam materials, the applications of PLA foam materials in the industry are mostly focused on the market for single-use disposable food trays. Nowadays, starch is one of the most abundant renewable resources in the world, with the characteristics of large output, diversified supply, low prices and superior biodegradability and antistaticity, etc., which is appropriate for developing competitive new generation, low carbon, lightweight, and environmentally friendly materials for packaging or transportation.

However, since ordinary starch is not easily foamed, the main technology of current starch foam material products comes from the patented technology of high-content, straight-chain starch (gene-modified corn starch) of Warner-Lambert, at an extremely expensive price. This technology mainly uses water as a foaming agent, with the deficiency of difficulty of controlling the manufacturing process and having a narrow process window. It is mostly employed as loose filler with poor properties, and, therefore, its application is difficult to expand.

A modified starch composition has been described, consisting of starch with a terminal siloxane having 100 parts by weight; water having 30–70 parts by weight; and a polyol having 5–35 parts by weight.

The starch may comprise corn starch, tapioca starch or potato starch, or even an esterified starch or an etherified starch.

The siloxane connected to the terminal of the starch may be tetraethoxysilane, 3-aminopropyltriethoxysilane, γ-glycidoxypropyltrimethoxysilane, γ-methacryloxypropyltrimethoxysilane, vinyltrimethoxysilane monohydrochloride, or tetramethyl orthosilicate.

The polyol may be glycerol, sorbitol, poly(ethylene glycol) (PEG), or combinations thereof. PEG should have a weight average molecular weight ranging from about 200 *D* to 6,000 *D*.

A method of preparation may run as follows (27):

Preparation 5–1: First, 100 *phr* of corn starch, 50 *phr* of water and 25 *phr* of glycerol were poured into a kneading machine and stirred for 5 *min* to 20 *min* at 65–95°C. Then, 2 *phr* of a modifier of γ-glycidoxypropyltrimethoxysilane was added to the kneading machine and kneaded for 5–25 *min*. After kneading, the kneaded matter was conducted into a single-screw granulator at 85–130°C to prepare the modified thermoplastic starch particles. Next, 30% of modified thermoplastic starch and 70% of poly(butylene adipate-co-terephthalate) were processed in a twin screw extruder and granulated with a screw speed of 30–250 *rpm* at 100–190°C to form pellets of modified starch glue.

After injection molding of the pellets, a supercritical foaming process was conducted on the resulting square specimen at a pressure of 800–5,000 *psi* and a temperature of 80–120°C.

The starch composite foam material was thus prepared. The density, foaming ratio, hardness, permanent compression set value and rebound resilience of the starch composite foam material were tested.

Several other examples of preparation have been presented in the study (27).

Here, a competitive starch-based biodegradable foam material with high starch content could be synthesized through the functionalized modification of industrial starch and glue technology, giving it an appropriate melting strength and chain entanglement, and using a green foam-processing technology (using supercritical fluid for forming micropores) to develop the starch-based microporous foam material with characteristics such as cell size uniformity, lively elasticity, flexibility, and superior buffering. The method is capable for application in the fields of fitness equipment and disposable materials (27).

A technique of preparing starch-based foam by steam has been widely used in the food and packaging industries, but many aspects of the process are far from being fully understood, in particular the multi-functions of water (28).

The relationship between phase transition and foaming behavior of starch-based material was established by investigating the effects of water content on melting temperature, crystallinity, foaming process, cell structure as well as mechanical properties.

It was found that there was a critical point of water content, between 16% and 18%, where the expending ratio changed significantly since the cell structure changed from open to closed. The lower the water content, the higher the melting temperature of starch-based materials, which resulted in lower melting strength and open cell structure.

Conversely, a higher water content and a lower melting temperature resulted in higher melting strength and closed cell structure. The closed cell structure prevented moisture evaporation during the foaming process and kept the lower melting temperature, which resulted in the shrinking of the foam when the material was cooled down due to a reduced pressure in the cell. The water contained in the open cell was evaporated mostly during the foaming process, which resulted in higher melting temperature and stable rigid foam structure, with unit density of about 26 $kg\,cm^{-3}$ (28).

The water forms hydrogen bonds with a hydroxyl chain in starch, and the bonded water is more stable under high temperature. Therefore, the bonded water is a good plasticizer but an inefficient blowing agent for preparing starch-based foam. The compressive strength of the starch-based foams increased from about 42 to 67 *kPa* with decreasing water content from about 15 to 10%, which could be explained by their thermal properties and cell structures (28).

Single-use food packaging made from expanded poly(styrene) (PS) is generally discarded after use, and, since it is a difficult material to recycle, generates a large amount of waste. Expanded PS can be replaced by thermoplastic starch foams, which are made from renewable sources and are biodegradable (29).

However, thermoplastic starch foams are hydrophilic and absorb large amounts of water, which makes them difficult to use. An alternative to this problem is to chemically modify the starch to make it more hydrophobic. Two modifications for potato starch were evaluated: acetylation and esterification with maleic anhydride (29).

Foams with 6.67%, 13.34%, 20% and 26.67% (modified starch weight/total starch paste mass) of modified starch were prepared. While thermoplastic starch foams without modified starch absorbed 75 *g* water/100 *g* solids, foams with 13% acetylated starch and 20% esterified starch presented lower absorption results (42 *g* and 45 *g* water/100 *g* solids, respectively), which is an improvement in the water resistance (29).

Starch foams from wheat, pea and potato starches were prepared by reactive extrusion with different concentrations of starch, glycerol, gluten, and sodium bicarbonate (30).

The density, the expansion ratio, the elasticity and the dynamic damping properties of these foams were determined. Foams made from wheat starch exhibited the best mechanical properties, especially during aging. Using the ratios of wheat starch, glycerol, gluten, and sodium bicarbonate that gave the best results (100/46/25/1), the extrusion parameters were adjusted and higher and more stable expansion (expansion ratio = 9.1) was obtained when the screw speed was as low (300 *rpm*) as the highest input rate could afford (21 $kg\,h^{-1}$) (30).

The agro-industrial wastes, sugarcane bagasse and asparagus peel, were used to enhance the properties of biodegradable foam trays based on sweet potato starch-based foam trays (31). Starch sugarcane bagasse and starch asparagus peel trays containing different concentrations of sugarcane bagasse and asparagus peel (0%–40%) were prepared. Their microstructure and the physical, thermal, and mechanical properties were assessed.

The addition of fiber wastes allowed obtaining a yellowish foam tray with a lower luminosity and a higher porosity, mechanical resistance, deformability, and better ability to absorb water as compared to the sweet potato starch foam trays without fibers. The addition of sugarcane bagasse yielded foam trays less porous, with lower water absorption capacity and greater tensile strength than the addition of asparagus peel (31). Higher concentrations of asparagus peel fibers (greater than 30%) are generating more extendible foam trays. The addition of fibrous wastes improved the thermal stability of the sweet potato starch foam trays (31).

The chemical modification of starch, i.e., via oxidation, can improve its expansion capability and, thus, reduce the density of the resulting foam (32). How the oxidation of potato starch with different concentrations of hydrogen peroxide changes its chemical structure was investigated.

The intrinsic viscosity of the starch solutions exhibited a linear decrease as the degree of oxidation increased, indicating that the molecular weight of the starch is reduced by oxidation because of chain scission. FTIR analysis revealed the decrease in the number of C–H and CO bonds caused by the conversion of CH_2–OH–6 into carboxyl groups in the oxidation reaction. In addition, NMR measurements showed a reduction in the signal assigned to the OH–6 hydroxyl groups, confirming their transformation into carboxyl groups.

The densities of the foams prepared with the modified starches decreased as the oxidant concentration increased from 308 $kg\,m^{-3}$ for native starch to 142 $kg\,m^{-3}$ for starch oxidized with 25% H_2O_2 (OS25). However, the starch prepared with the highest peroxide concentration showed a high density of 258 $kg\,m^{-3}$ (OS70) (32).

The dependence of the starch paste viscosity on the oxidant concentration, which results from a balance between the decrease in molecular weight and the increase in attractive chain interactions, was analyzed and an adequate interpretation for this behavior was provided. The internal bond strength of the foams exhibited the same trend as the density as a function of oxidant concentration (from 0.641 in native starch to 0.439 in OS25) as expected, because a lower density implies less mechanical cohesion within the foam (32).

5.1.6 *High Starch Polymer*

A substantially biodegradable high starch composition has been developed. A PVA is made compatible with starch. Optionally, an elasticity enhancer and a stabilizing agent may be added.

The elasticity enhancer is made from a mixture of polyols, polyglycol and water. Suitable polyols include glycol, propanediol, glycerol, or sorbitol.

The stabilizing agent is synthesized from PVA, ethylene vinyl alcohol, ethylene acrylic acid and poly(methyl methacrylate) (PMMA). An example of a formulation of a stabilizing agent is given in Table 5.2.

Table 5.2 Stabilizing agent (33).

Component	Amount/[%]
Poly(vinyl alcohol)	23.7
Ethylene vinyl alcohol	24.8
Ethylene acrylic acid	23.8
Poly(methyl methacrylate)	27.7

Three examples of high starch compositions are given in Table 5.3.

Table 5.3 Starch compositions (33).

Component	Amount/[g]		
	(1)	(2)	(3)
Maize starch	45	53	35
Oxidized starch	0	0	20
Poly(vinyl alcohol)	28	22	20
Glycerol	12	11	11
Polyglycol	5	6	6
Water	10	8	8
Stabilizing agent	0	4	4

Starch and PVA are mixed and heated. During the mixing and heating process the starch molecular structures are randomized. The randomized starch molecules, the PVA, and the other ingredients undergo a reaction.

This composition is then extruded, at a temperature that is above the crystallization temperature of the starch molecules, to form the first stage product, the pellets. Eventually, the pellets can then be extruded into films using a traditional plastic film extruder (33).

The starch composition has high starch content so it is cost efficient. The inexpensiveness of the product makes it easy for large-scale commercialization.

The films manufactured have comparable strength, tearing resistance and puncture resistance to that of traditional polyester film. It is a suitable alternative and substitute for oil polymer plastic film and can be used in a wide range of domestic and commercial applications. It is 100% biodegradable in a natural environment in a very short period of time. It dissolves in hot water almost immediately. It does not dissolve in cold water.

The manufacturing process uses a relatively simple technology. It does not require capital-intensive equipment. The high starch polymer film can be made by using conventional plastic film extruders. The entire operation is cost-effective and makes the large-scale industrial production readily feasible (33).

5.1.7 Destructurization of Natural Starch

For destructurization, the starch is typically heated above its softening and melting temperature under a pressurized condition. Melting and disordering of the molecular structure of the starch granules takes place and a destructurized starch is obtained.

Chemical or enzymatic agents may also be used to destructurize, oxidize, or derivatize the starch. Modified starches have been used to make biodegradable plastics, wherein the modified starch is blended as an additive or the minor component with petroleum-based or synthetic polymers.

However, when the modified starch is processed by itself or as the major component in a blend with other materials using conventional thermoplastic processing techniques, such as molding or extrusion, the finished parts tend to have a high incidence of defects. Moreover, the modified starch has been found to have a poor melt extensibility. Consequently, it cannot be successfully processed by uniaxial or biaxial extensional processes into fibers, films, foams or the like.

5.1.8 *Melt Processable Starch*

The relationships between the structure, the process, and the properties of biodegradable multiphase systems based on plasticized starch, i.e., thermoplastic starch, have been reviewed (34). Such multiphase materials are blends of plasticized starches and other biodegradable materials, such as PCL, PLA, poly(β-hydroxyalkanoate)s, poly(esteramide)s and others. The materials can be tailored as soft or rigid. A critical issue is the compatibility between these hetero-materials. The composites show some interesting properties and have applications in a wide variety of fields such as packaging, sports, catering, agriculture and gardening, or hygiene.

In addition, newer aspects, such as the effects of sustainable materials based on starch on the macro- or nanostructure and subsequent processing, thermomechanical properties and performance properties of plasticized starch polymers have been examined (35). Specific structures and the resulting properties are controlled by many specific factors such as filler shape, size and surface chemistry, processing conditions and environmental aging. In the case of nanosized biocomposites, the interfacial interactions are extremely important to the final nanostructures and performance of these materials.

A melt processable composition can be prepared by mixing starch urea, sucrose, and free water (1). The mixture is manually stirred to form a slurry. Poly(acrylamide) (PAAm) is dissolved in water solution. An aliquot of the solution is added to the slurry. Water in the slurry is then evaporated until the weight percent of PAAm in the final mixture is 0.2%. In another example, the PAAm was completely omitted. Typical compositions are shown in Table 5.4.

A wide variety of other examples have been given elsewhere (36).

It is generally known that melt shear viscosity is a material property useful for evaluating melt processability of the material in traditional thermoplastic processes such as injection molding or extrusion. For conventional fiber spinning of thermoplastics such as polyolefins, polyamides and polyesters, there is a strong correlation

Table 5.4 Typical compositions (1).

Formulation →	A	B
Ingredient	Amount/[%]	
Starch	45	45
Urea	40.5	40.5
Sucrose	4.5	4.5
Water	9.8	9.8
Poly(acrylamide) 15 *k D*	0.2	0.0
Properties		
Shear viscosity *Pa s*	0.65	1.35
Extensional viscosity *Pa s*	1863.2	43.02

between shear viscosity and extensional viscosity of these conventional thermoplastic materials and blends thereof.

Thus, the spinnability of the material can be determined simply by the melt shear viscosity, even though the spinnability is a property controlled primarily by melt extensional viscosity. Since the correlation is quite robust, the fiber industry has relied on the melt shear viscosity in selecting and formulating melt spinnable materials. The melt extensional viscosity has rarely been used as an industrial screening tool.

It is therefore surprising to find that the starch compositions shown in Table 5.4 do not exhibit such a correlation between shear and extensional viscosities. Specifically, when a high polymer is added to a starch composition, the shear viscosity of the composition remains relatively unchanged, or even decreases slightly.

Based on conventional wisdom, such a starch composition would exhibit decreased melt processability and would not be suitable for melt extensional processes. However, the results reveal that the starch composition shows a significant increase in extensional viscosity when even a small amount of high polymer is added. Consequently, the starch composition is expected to have an enhanced melt extensibility and is suitable for melt extensional processes such as blow molding, spun bonding, blown film molding, or foam molding (1).

A controlled esterification of starch was used to replace the –OH moieties with bio-derived medium chain fatty acids, and the changes in the polymer structure and properties for material applications (37). The esterification was conducted via a homogeneous esterification process using an activated lauric acid in the presence of a base catalyst. The degree of esterification through the replacement of hydroxyl groups of starch was estimated using elemental analysis and proton NMR. The effect of the modification on the structural and material properties of the modified starch polymer could be elucidated by evaluating the changes in morphology, network thermal stability, hydrophobicity, solubility profile, and thermal transition events. SEM imaging revealed structural changes ranging from surface roughness to complete disruption, depending on the degree of substitution.

Because of the esterification of starch, the resulting polymers become melt processable thermoplastic that forms a transparent film with an elastic storage modulus of up to 226 *MPa* at room temperature. This shows that the starch-fatty acid polymer can be used for various industrial and advanced material applications without any other plasticizers or modifiers. The final material is completely bio-based, and is expected to be biodegradable in the environment (37).

5.1.9 *Starch-Based Aerogels*

Starch-based aerogels are a class of advanced biomaterials with a low density and a high specific surface area (38). The commonly used procedures for the fabrication of such aerogels are starch gelatinization, retrogradation, organic solvent exchange, and supercritical carbon dioxide drying. The so-produced starch-based aerogels are nanoporous and exhibit a lot of potential applications.

The production, the properties, and various applications of starch-based aerogels have been reviewed (38). A comparative analysis between starch-based aerogels and other similar porous materials was done to show any advantages of these products. It could be shown that the production of starch-based aerogels is more environmentally friendly in comparison to other similar porous materials.

Starch-based aerogels can be used for the encapsulation and the controlled release of bioactive compounds, biomedical uses and tissue engineering as packaging and thermal materials, and as templates for the synthesis of functional materials (38).

5.1.10 *Spinning Processes for Starch*

5.1.10.1 *Wet-Spinning*

Previous attempts to produce starch fibers relate principally to wet-spinning processes. For example, a starch/solvent colloidal suspension can be extruded from a spinneret into a coagulating bath. This process relies on the marked tendency of amylose to align and form strongly associated aggregates to provide strength and integrity to the final fiber (1).

Any amylopectin present is tolerated as an impurity that adversely affects the fiber spinning process and the strength of the final product. Since natural starch is rich in amylopectin, pretreating of natural starch is done to obtain the amylose-rich portion, which is highly desirable for fiber spinning (1).

5.1.10.2 *Electrospinning*

Electrospinning has become an increasingly attractive technique to produce microscale and nanoscale fibers from bio-based polymers, including starch (39). A comprehensive review of the recent attempts made on electrospinning of starch and its potential applications in biomedical and tissue engineering has been presented (40). Also, the recent trends in the fabrication of the pristine/modified starch-based composite scaffolds by electrospinning along with their possible applications has been summarized (41). Apart from the electrospinning technique, the review also provided comprehensive information on various other techniques employed in the fabrication of the starch-based nanofibers.

Electrospun starch nanofibers have a high porosity, extremely high specific surface area, and biocompatible, biodegradable, and bioabsorbable properties. Therefore, starch nanofibers have a great potential in pharmaceutical applications, including drug delivery, wound dressing, and tissue engineering. The recent developments in the synthesis of electrospun starch fibers from common starch, modified starch, and hybrids with other polymers have been reviewed (42).

In comparison to their synthetic counterparts, electrospun fibers from bio-based polymers generally have an inferior mechanical strength. The aim of one study was to enhance the mechanical strength of electrospun starch-based fibers by using nanocellulose as a reinforcing filler and cationic starch as a binding agent (39).

The potential ionic bonding between cationic starch and anionic nanocellulose and the hydrogen bonding, as well as the macromolecular entanglement between these three components, were expected to improve the compatibility of the polymers and the mechanical strength of their composite fibers. The rheological properties of the ternary systems were studied and correlated with their electrospinnability and the tensile strength of the composite fiber mats.

The results suggested that nanocellulose-cationic starch at its percolation concentration of ca. 2% of starch has the most profound influence on the fiber strength, and the ratios of cationic starch and anionic nanocellulose of 1:2 and 1:1 improved the intermolecular ionic bonding for reinforcing the fiber structure (39).

Potatoes are a viable source of starch. The preparation and characterization of native starch's microfiber by electro-wet-spinning technique has been tested (43).

The physical and chemical properties of membranes conformed by potatoes starch microfiber were tested. The parameters that were changed for each treatment were as follows (43):

1. The amount of potential applied to the solution,
2. The distance between the needle and the collector, and
3. The rate of injection of the solution.

The microfibers were analyzed in order to determine their structural properties and thus define the range of application. SEM was used to determine the morphologies and diameters of microfibers, FTIR was used to determine the chemical changes, TG and DSC to obtain the thermal transitions and the corresponding temperatures. In summary, potatoes starch microfibers were obtained with average diameters of 15, 17, 23 and 25 μm, the fibers presented a degradation temperature of 304°C, indicating that the fibers are available with diameters of small scale, with good thermal properties (43).

Through starch phosphorylation and solution aging treatments, it was aimed to produce electrospun fibers derived from native and anionic corn starches, modified with sodium tripolyphosphate, with amylose contents of less than 70% (44). The fibers of native and anionic corn starches were prepared by electrospinning of starch solutions dissolved in aqueous 75% v/v formic acid. The effects of the aging (24 *h*, 48 *h*, and 72 *h*) on the rheology and electrical conductivity of the starch solutions, as well as the material properties (size distribution, morphology, and infrared spectrum) of the resulting electrospun fibers, were evaluated.

Fibers produced from Hylon VII and Hylon V starches showed homogeneous morphologies, whereas the fibers from regular corn starches exhibited droplets and had heterogeneous morphologies, with varying diameters of 70 *nm* to 264 *nm*. Both native and anionic corn starches, with amylose contents of less than 70%, produced smooth continuous fibers. The electrospun corn starch fibers potentially can be used as carriers for the encapsulation of active components in food and packaging applications (44).

A method of making a concentric multilayered starch-formate fiber has been presented. The method consists of the following steps (45):

1. Providing a first spinning dope for forming at least one layer of the fiber; the first spinning dope comprises a solution or dispersion of starch in a solvent comprising at least 50% by volume of formic acid,
2. Providing one or more additional spinning dopes for forming at least one additional layer within said fiber, and
3. Co-electrospinning the spinning dopes through multi-axial capillaries to produce a starch-formate concentric multilayered fiber.

At ambient temperatures during the solution preparation and electrospinning process, nanofibers with the diameters of about 200 *nm* could be produced. Formic acid plays a dual role in the simultaneous destruction of the starch granular structure: the esterification of starch to starch-formate and as dispersing solvent for the electrospinning process.

Rheological measurements evidenced the complete starch granule destruction in pure formic acid solutions at ambient temperatures, while the destruction was only partial for aqueous dispersions of starch-formate in formic acid. Final fibrous mat showed decreased crystallinity and improved mechanical properties, thereby exhibiting an economic and ecological biomaterial for use in many industries, including, but not limited to the food packaging, nutraceutical and pharmaceutical industries (45).

5.1.10.3 Centrifugally Spun Fibers

A method for controlling the surface morphology of centrifugally spun starch-based fibers has been presented (46). This can be done by adjusting the ratio of amylopectin/amylose in the starch materials and using a hot blast temperature. The effects of hot blast temperature, amylopectin, and amylose on fiber surface morphologies were investigated.

SEM was used to characterize the morphology of the prepared fibers. The results showed that fibers with burr-shaped nanostructures on the surface can be fabricated by adding amylopectin to the starches and are promoted by increasing the hot blast temperature. However, amylose in the starch compositions smooths the fiber surfaces. XRD showed that the fibers are amorphous. FTIR spectroscopy analysis indicated that some physicochemical changes occur during centrifugal spinning (46).

5.1.11 Pre-gelled Starch Suspensions

Ingredients of a pre-gelled starch composition are shown in Table 5.5.

Table 5.5 Pre-gelled starch composition (47).

Component	Amount/[*g*]		
	(1)	(2)	(3)
3% Potato starch gel			16.3
5% Potato starch gel	31.5	5	
15% Corn starch gel		19.5	
Dry corn starch	18		5.9
Dry wood flour (60 mesh soft wood)	6		
Dry wood flour (80 mesh soft wood)		0.125	14
Baking powder		0.125	1
Glycerol			1

Baking powder is added to elevate the number of open cells in the final structure by introducing a source of carbon dioxide released by heat and water. Glycerol is added to produce a product that releases better from the mold and to produce a smoother surface on the finished product.

Several other modified compositions have been presented, using the addition of bentonite clay and recycled paper pulp (47). The starch-wood flour mixture, with any included additives, is added to the pre-gelled starch and mixed until a homogeneous mixture is generated. Varying amounts of additional water can be added to facilitate different types of molding, since the form of the pre-molded (green) product is dependent on the mold, heating rate and drying/melt time (47).

If the product is to be molded by classic injection methods the material must be rather thin. The material can also be rolled into green sheets and molded, extruded and made into dry pellets for other processes.

Using a microstructural engineering approach, a variety of articles can be produced, including plates, cups, cartons, and other types of containers and articles having mechanical properties substantially similar or even superior to their counterparts made from conventional materials such as paper, poly(styrene) foam, plastic, metal and glass. The articles can also be made at a fraction of the cost of their conventional counterparts. The minimal cost is a result of the relatively inexpensive aggregate which typically comprises a large percentage of the mixture, and the minimum processing energy required (47).

5.1.12 *Processing of Natural Starch*

Attempts have been made to process natural starch on standard equipment and existing technology known in the plastic industry. Since natural starch generally has a granular structure, it needs to be destructurized and modified before it can be melt processed like a thermoplastic material (1).

5.1.13 *Granular Starch as Additive to Conventional Polymers*

Granular starch can be added to a polymer in order to decrease the cost of the base polymer. Clearly, the use of granular starch makes the derivative more biodegradable (48).

The fabrication involves mixing the starch with a primary polymer and a compatibilizer that has grafting compounds attached to it. These grafting compounds covalently bond to the hydroxyl groups located on granular starch. Starch hydroxyl groups are susceptible to binding by these grafting compounds. Because the starch is chemically grafted to the compatibilizer, which in turn physically interacts with the polymer, the resulting mixture has substantially the same physical properties as the polymer in its pure form. Up to 30% of the mixture may be granular starch.

Furthermore, glycerol is not added to the mixture. This reduces the water absorbency of the final product (48). An example of a grafting compound is maleic anhydride (MA). An example of a good compatibilizer is poly(ethylene) (PE) containing approximately 5% MA, which is commercially available and only slightly more expensive than pure poly(ethylene).

When PE-*g*-MA is added to low density poly(ethylene) (LDPE)/corn starch blends, the tensile strength and elongation at break of the blends are improved, and the improvement was even more pronounced at higher starch contents.

Because the starch may be granular wheat or rice starch, this greatly reduces the cost of manufacturing the polymer. In addition, it greatly enhances the biodegradability of the mixture.

Polymers having grafting compounds already attached to them are presently commercially available and are only slightly more expensive than the unaltered polymers. This difference in cost is more than compensated for by the use of inexpensive starch. The compatibilized polymer blend has very similar mechanical properties to pure polymer (48).

The miscibility between granular corn starch and LDPE is improved with the addition of a commercially available compatibilizer, PE-*g*-MA. Results from DSC, SEM and tensile properties suggest that the improved compatibility was attributed to the chemical reaction between hydroxyl groups in starch and anhydride groups in PE-*g*-MA, and the physical interaction between the PE in PE-*g*-MA and LDPE. Therefore it is possible to blend a high percentage of granular corn starch with LDPE while still showing comparable tensile properties (48).

Polymers such as petroleum-based plastics can be altered by the incorporation of carbohydrates to increase their biodegradability (49). For example, in a synthetic resin, particles of biodegradable substances and a substance that can be auto-oxidized are introduced. During the processing the starch granules are preserved and thus can be found in the final product. Such a polymer, when it contacts a transition metallic salt, auto-oxidizes to generate a peroxide or a hydroperoxide.

The autoxidation of fats can be catalyzed by transition metals such as iron (50). In a soil burial situation, such metals are available. The autoxidation of the fat soluble transition metal salt generates peroxides which initiate the breakdown of the polymeric molecules. Fat-soluble transition metal salts can be preformed and included in plastics.

A biodegradable plastic blend consists of both a biodegradable material and a non-biodegradable polymer such as a poly(propylene) (PP), poly(urethane) (PU), PE, PP, or poly(carbonate). These materials are treated (49):

1. Under heat, pressure and reagents to break down the polymers, and
2. By adding an oxidizing agent to them.

This treatment forms reactive groups for bonding (49):

1. On the biodegradable material, in the case of the carbohydrates aldehyde, or hydroxyl groups in the case of proteins and urea amine groups, and
2. On the non-biodegradable polymers' aldehyde groups, hydroxyl groups, or alkyl and benzyl groups.

The procedure is now illustrated by an example. Wheat starch and PS are blended together. To this blend, citric acid and sodium bicarbonate are added. Citric acid acts as the oxidizing agent and sodium bicarbonate is the substance for degrading the starch. This blend is extrusion-cooked at a temperature of 140°C and a pressure of approximately 20 *M Pa*.

Heat and pressure results in the release of carbon dioxide as a foaming agent, degradative oxidation of the methylene groups in PS to form aldehyde groups which react with groups on the starch, and the formation of sodium hydroxide from the bicarbonate that degrade the starch and thus form aldehydes such as formaldehyde or hydroxyl groups that react with the styrene moieties. The resulting product has the appearance of an original PS (49). The amounts of ingredients are given in Table 5.6.

Table 5.6 Typical compositions (49).

Formulation →	A	B
Ingredient	Amount/[parts]	
Wheat starch	66	20
Wheat protein	0	1
Poly(styrene)	27	79
Citric acid	3	1.6
Sodium bicarbonate	6	4.7

A biodegradable polymer prepared in this way has several advantages: The biodegradable polymer retains its physical characteristics with a large percentage of carbohydrate or protein or other biodegradable material added. The biodegradable polymer effectively degrades when discarded. The process permits the inclusion of a large amount of carbohydrate. Moreover, the biodegradable polymer is less expensive than other biodegradable polymers (49).

The incorporation of platelet-like starch nanocrystals into natural rubber may remarkably enhance the barrier performances of, for example, water vapor and oxygen (51, 52).

5.2 Cellulose

Cellulose is a naturally occurring polysaccharide composed of β-linked D-glucose. The type of linkage is its only difference to starch, which is α-linked. However, this makes a great difference to the degradability, in particular digestibility. Humans cannot digest cellulose and use a form of it in food, but they can digest starch, whereas cows can digest cellulose. The basic structure of cellulose is shown in Figure 5.4.

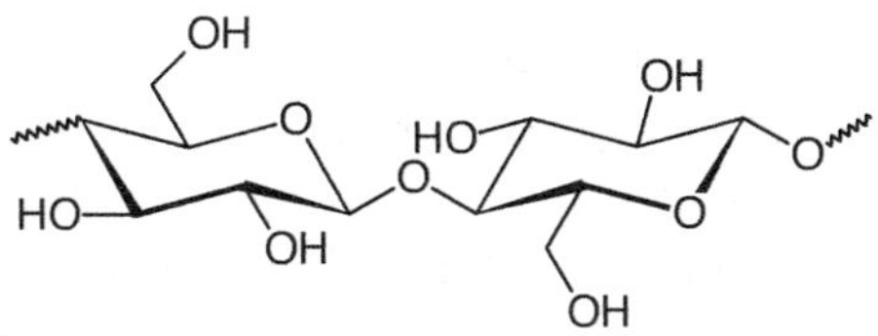

Figure 5.4 Basic structure of cellulose.

The aspects of molecular biology of cellulose degradation have been reviewed (53). In industrial uses, cellulose is mainly used for paperboard and paper. For special purposes, cellulose is mostly not used as such, but is modified into cellulose ethers and esters, as well, in compositions.

Bacterial cellulose is a special kind of cellulose which is produced by microorganisms. It has unique properties. The biosynthesis and purification of bacterial cellulose have been reviewed (54).

5.2.1 *Liquid Crystalline Derivatives*

One of the general methods for improving the properties of a polymer is the introduction of a fiber structure into a polymer by blending. Conventionally, this method includes the blending of an inorganic material, such as a glass or carbon fiber, or by blending with an organic polymer fiber.

In the same way, the properties of a biodegradable polymer can also be improved by blending it with fiber materials. However, if the blended fiber is not biodegradable, recycling becomes difficult, resulting in an increased burden on the environment (55).

Compositions from PLA, which is a biodegradable non-liquid-crystalline polymer, and a biodegradable liquid crystalline polymer have been described. The liquid crystalline polymer may result in an oriented polymer molecule structure in a molded article, which improves the mechanical strength of the molded article.

Biodegradable liquid crystalline polymers derived from cellulosics have been described (56, 57). Such polymers also include the butyric acid ester of hydroxyethyl cellulose, the butyl ether of hydroxypropyl cellulose, and cellulose butyl ether (55). This type of liquid crystalline polymer can be derived from waste paper. The butyric acid ester is prepared as follows (55):

Preparation 5–2: First, 39.3 *g* of butyric acid was placed in a four-neck separable flask equipped with a refluxing device and a stirrer. Nitrogen gas was bubbled therein for one hour to replace the dissolved oxygen. Thereto, 50 *ml* of trifluoroacetic anhydride was added gradually under a nitrogen gas flow. The temperature was elevated gradually to 50°C by a water bath, and the activation was conducted for 30 *min*. Therein 4.6 *g* of dried solid hydroxyethyl cellulose was added gradually through an inlet.

The polymer was dissolved quickly. Under these conditions, the reaction was continued further for 5 *h*. After the reaction, the reaction mixture was poured into a large amount of water to precipitate a butyric acid ester of hydroxyethyl cellulose. The obtained polymer was dried and purified by repetition of redissolution in acetone and reprecipitation in water. The polymer yield was 6.1 *g*.

The obtained polymer (M_w: 351 *D*) was confirmed to be the intended substance by infrared spectroscopy and NMR and found to have a liquid crystal transition temperature of 135°C by thermal analysis.

In the same way as in the example above, a cellulose material separated from waste paper has been used in place of the industrial cellulose as the source material (55).

Further methods for recycling the materials obtained as described above have been reported. The steps of recycling include shredding into chips and treatment with fuming hydrochloric acid. The glucose chemically separated from the decomposition product can be used as a source material for synthesis of a glucose-sebacic acid copolymer (55).

A low-cost approach was developed to prepare carboxylated cellulose nanocrystals from pulp with high yield at mild reaction conditions (50°C, 1% sulfuric acid medium) using potassium permanganate and oxalic acid as the oxidizing and reducing agents, respectively (58). The oxidant dosage in this strategy is much lower than that in a conventional 2,2,6,6-tetramethylpiperidine-*N*-oxyl (TEMPO) method, and the yield of carboxylated cellulose nanocrystals can reach as high as 68.0%, with a carboxylate content of 1.58 $mmol\ g^{-1}$.

In this reaction system, the presence of the oxalic acid can complex with Mn^{3+} to form a complex with oxalic acid and so prevent the Mn^{3+} from being reduced to Mn^{2+}, leading to the strong oxidizing capacity of the reaction system maintained for a longer time.

Atomic force microscopy (AFM) analysis showed that rod-like cellulose nanocrystals were obtained with an average size of 10–22 *nm* in diameter and 150–300 *nm* in length. The crystal structure of the prepared carboxylated cellulose nanocrystals was nearly unchanged, and the crystallinity was 89.2% based on wide-angle X-ray diffractometer analysis. Of particular interest, carboxylated cellulose nanocrystals suspension with high concentration (>6%) also exhibited the same intriguing chiral nematic liquid crystalline self-assembly behaviors as sulfate cellulose nanocrystals prepared by traditional H_2SO_4 hydrolysis method. The study provided an efficient and cost-effective way to fabricate carboxylated cellulose nanocrystals, leading to great potential applications in constructing advanced functional material (58).

5.2.2 *Cellulose Fibers*

The research on cellulose nanofibers (59) and nanocomposites has been reviewed (60). Some mechanical properties of cellulose in comparison to other engineering materials are given in Table 5.7.

Table 5.7 Moduli of engineering materials (60–62).

Material	Modulus /[GPa]	Density /[$g\,cm^{-3}$]	Specific modulus /[$GPa\,g^{-1}cm^3$]
Aluminium	69	2.7	26
Steel	200	7.8	26
Glass	69	2.5	28
Crystalline cellulose	138	1.5	92

A thermoplastic polymer composition reinforced with fibers such as cellulose or other fillers, particularly from natural sources, and a process for manufacturing the composition has been described (63). The polymer is extruded with a salt which reduces the melting point, and is pelletized. The pellets are then extruded again with the filler. The composition with the filler can then be melted at the reduced melting temperature to manufacture an article.

The thermal degradation of natural fibers results in poor physico-mechanical properties and in the discoloration of the fibers. Therefore, reinforcing high temperature melting thermoplastics which melt above 200°C with natural fibers has turned out to be a challenging task (63).

Lithium chloride, lithium bromide, and copper chloride were found to be effective in reducing the melting temperature of polyamides. The reduction of the melting temperature of Nylon 6 by the addition of lithium chloride is shown in Table 5.8.

Table 5.8 Melting temperature of Nylon 6 in the presence of lithium chloride (63).

Lithium chloride/[%]	Melting Temperature/[°C]
0	223
2.5	208
3.0	199
3.5	194
5	191

The addition of salt to the polymer affects the crystallization properties of the polymer. This change can be noticed by looking at the morphology of the polymer-salt and comparing it to the morphology of the polymer. The injection molded samples were microtomed onto a glass slide to give a sample of thickness 20 μ.

The area to be microtomed was chosen far from the surface of the injection molded coupon, deep into it. Confocal laser scanning microscopy was used to examine the surface morphology of the microtomed sample. A laser light of wavelength 488 *nm* was used to scan the surfaces of the microtomed samples. The advantage of using the confocal laser scanning microscopy is that the sample can be sectioned in steps along the thickness using the laser light and each section can be imaged. In all cases the lithium modified Nylon-6 possesses a different morphology than its unmodified precursor. The results are shown in Figure 5.5.

The preparation, properties and applications of polysaccharide nanocrystals in advanced functional nanomaterials have been reviewed (64). Cellulose-based nanocrystals decompose and undergo a carbonization reaction over 250°C, whereas starch nanocrystals exhibit a lower thermal decomposition temperature. Polysaccharide-based nanocrystals exhibit specific properties, especially bioactivity, biocompatibility, biodegradability, and low-toxic-

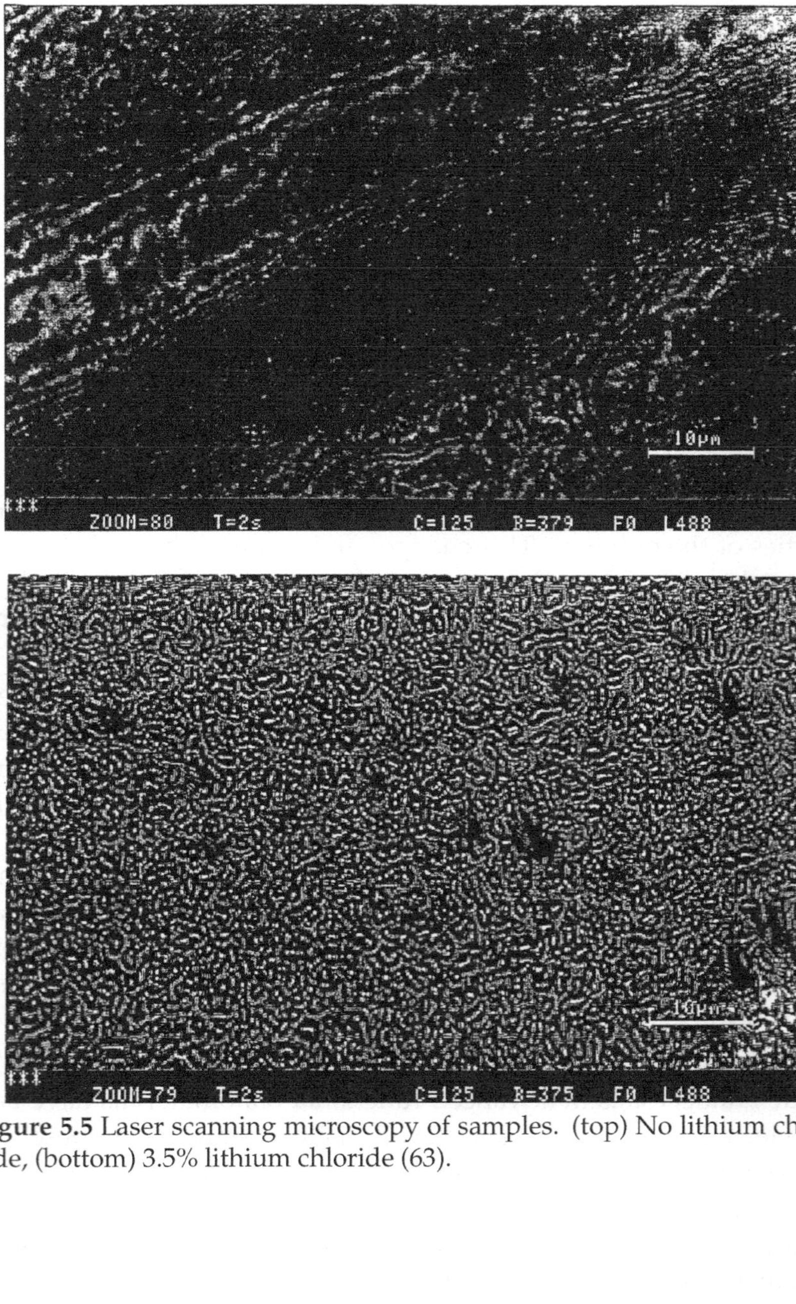

Figure 5.5 Laser scanning microscopy of samples. (top) No lithium chloride, (bottom) 3.5% lithium chloride (63).

ity suitable particularly for pharmaceutical and biomedical applications.

Cellulose nanofibers are conventionally used together with carbon fibers and glass fibers, as reinforcing materials for the fiber reinforced composite materials (65).

On the other hand, cellulose nanofibers are advantageous in that they not only exert excellent reinforcing effects as compared with other reinforcing fibers, but also environmental load during production is low since they are produced from cellulose derived from plants, bacteria, or the like, and further combustion residues do not remain at the time of disposal, unlike glass fibers. Therefore, attention has been given to fiber reinforced composite materials using cellulose nanofibers as reinforcing fibers in recent years (65).

Cellulose nanofibers can be produced by several methods such as mechanical shearing using a homogenizer or by treating a solution containing an ionic liquid (65).

Cellulose nanoparticles from a vegetable source, i.e., cellulose fiber, have been evaluated for their use as reinforcement of polymeric matrixes, e.g., biodegradable films (66). Cellulose nanoparticles have numerous advantages: They are inexpensive and biodegradable, and they originate from renewable sources.

Here, cellulose nanofibers were isolated from banana peel by chemical methods, i.e., alkaline treatment and bleaching followed by acid hydrolysis with 0.1, 1, or 10% (v/v) H_2SO_4, and mechanical methods such as high-pressure homogenization treatments.

AFM analysis showed that all treatments effectively isolated banana fibers at the nanometer scale (average diameter of 3.72 *nm*).

The cellulose nanofibers had high crystallinity values from 63.1 to 66.4%, which indicated they could be good reinforcing agents. FTIR results confirmed that the chemical and mechanical treatments removed the amorphous fractions. Regarding cytotoxicity, low cellulose nanofiber concentrations of 50–500 $\mu g\ ml^{-1}$ did not cause cell death, but cellulose nanofibers at concentrations above 1000 $\mu g\ ml^{-1}$ significantly decreased the cell viability.

The use of different sulfuric acid concentrations provided more detailed knowledge of the treatment methods and cellulose nanofiber features, which could help to improve the cellulose nanofiber production process. The combination of chemical and mechanical treatments proved to be an efficient strategy to prepare cellulose nanofibers from banana peels as a potential reinforcing agent of polymeric matrixes, e.g., food packaging (66).

The cytotoxicity of cellulose nanofibers was assessed on Caco-2 cell line (67). The Caco-2 cell line is a continuous line of heterogeneous human epithelial colorectal adenocarcinoma cells (68). Cellulose nanofibers were not cytotoxic at 50–2000 $\mu g\, ml^{-1}$ (67). However, cellulose nanofibers above 2000 $\mu g\, ml^{-1}$ significantly decreased the cell viability.

When wood cellulose fibers are oxidized with NaClO and catalytic amounts of TEMPO and NaBr in water at a pH of 10, significant amounts of sodium carboxylate groups are introduced into the oxidized celluloses (69). The original fibrous morphologies and cellulose-I crystal structures are unchanged by oxidation. The TEMPO-oxidized cellulose fibers can be converted into partially fibrillated nanonetworks, completely individualized cellulose nanofibers with high aspect ratios, and needle-like cellulose nanocrystals with low aspect ratios by controlling the conditions of mechanical disintegration in water (69).

The catalytic oxidation of cellulose using TEMPO under aqueous conditions at room temperature is a position-selective and efficient chemical modification (70). TEMPO-mediated oxidation of plant cellulose fibers, followed by gentle mechanical disintegration of the oxidized celluloses in water, results in the formation of TEMPO-oxidized cellulose nanofibers with homogeneous widths of around 3 *nm* and high aspect ratios.

The huge presence of carboxylic groups in TEMPO-oxidized cellulose nanofibers indicates that this material could interact with cationic species, such as metal ions, in aqueous solution (71). However, the contact between nanofibers and water solutions requires a 3D structure to entrap and retain the nanofibers.

In this sense, two different 3D structures were synthesized: cellulose nanofiber-calcium alginate beads and cellulose nanofiber aerogels (71). After the synthesis and characterization of 3D structures, batch sorption studies were performed by using these sorbents to study their ability for metal removal. Equilibrium data fitted very well with Langmuir and Freundlich isotherm models in the studied concentration range of copper(II) ions and confirmed that the copper sorption is a favorable process. Both synthetized materials were found to be effective for Cu(II) removal, and maximum sorption capacity was higher for cellulose nanofiber aerogels than cellulose nanofiber-calcium alginate beads (71). Thus, it could be demonstrated that the synthetized 3D cellulose nanofiber aerogel structure is an efficient sorbent for copper ion removal from aqueous solutions, and the use of this synthetized structure for environmental decontamination opens a great opportunity for cellulose nanofiber applications.

Cellulose nanofibers were derived from waste pineapple leaves with the aim of developing a nanocomposite with high impact strength and transparency (72). First, cellulose nanofibers were prepared using an acid-base treatment and TEMPO-mediated oxidation. Then, the surface was modified with polymerized methyl methacrylate (MMA) using an eco-friendly method in an aqueous system. The MMA-modified cellulose nanofiber showed an increased contact angle from 12.02° to 57.45°, and the surface hydrophobicity improved the interfacial compatibility between the cellulose nanofiber and PMMA matrix. At 1–3% of the modified cellulose nanofiber, the impact strength of the nanocomposites was significantly improved by 2.7% to 22.9%.

The eco-friendly modification allowed cellulose nanofibers to be dispersed homogeneously in the nanocomposite, showing an excellent light transmittance. In summary, this eco-friendly modification can replace a conventional solvent-based modification, allowing modified cellulose nanofiber to effectively reinforce the PMMA nanocomposite (72).

A green process to produce cellulose nanofibers has been proposed (73). Cellulose nanofibers are usually produced via mechanical, enzymatic, and/or chemical treatment such as TEMPO-mediated oxidation of cellulose fibers, but this method involves high energy consumption, which limits the commercialization of the cellulose nanofiber. Moreover, an expensive effluent treatment system is required to complete the cellulose nanofiber manufacturing process. In this context, a novel process using a green method was developed to improve cellulose nanofiber production. The cellulose, sourced from eucalyptus was modified by adding dimethyl carbonate in ethanolic potassium hydroxide medium. The effect of the reaction temperature at 4°C, 25°C, and 40°C was evaluated and the obtained cellulose carbonate was characterized by several techniques, including 13C cross-polarization magic angle spinning NMR and X-ray photoelectron spectroscopy.

After the chemical step, cellulose nanofiber was manufactured with a supermasscolloider ultrafine friction grinder. The resultant cellulose nanofiber suspension was characterized in terms of fibrillation yield, transparency, rheological behavior, morphological features, and quality index. The here described chemical approach for the production of cellulose nanofiber seems to hold promise not only for its green features but also for its lesser and cleaner effluent discharge and low cost of reagents (73).

A nanofluidic membrane has been reported consisting of densely packed cellulose nanofibers directly derived from wood (74). Numerous nanochannels were produced among an expansive array of one-dimensional cellulose nanofibers. The abundant functional groups of cellulose enable a facile tuning of the surface charge density via chemical modification. The nanofiber-nanofiber spacing can also be tuned from ≈ 2 to ≈ 20 *nm* by structural engineering. The surface-charge-governed ionic transport region shows a high ionic conductivity plateau of $\approx 2\ mS\,cm^{-1}$.

The nanofluidic membrane also exhibits excellent mechanical flexibility, demonstrating stable performance even when the membrane is folded 150°. Combining the inherent advantages of cellulose, this class of membrane offers an environmentally responsible strategy for flexible and printable nanofluidic applications (74).

Bacterial cellulose is an appealing biomass for developing functional electrodes with multiple components along their crosslinked nanofibers (75). A bacterial cellulose-directed ternary composite was fabricated through the successive layer-by-layer assembly of poly(pyrrole) (PPY) and a bimetallic hydroxide (NiMn-LDH). In this strategy, both *in-situ* layer-by-layer depositions have been realized at mild conditions of less than 25°C. By tailoring the feeding weights of PPY/bacterial cellulose in NiMn-LDH precursor solutions, ternary composites with a maximum loading of NiMn-LDH has been achieved with the highest specific surface.

The optimized bacterial cellulose-based electrode material delivers a specific capacity of 653 $C\,g^{-1}$ at a current density of 1.0 $C\,g^{-1}$. Hybrid supercapacitor assembled by NiMn-LDH/PPY/bacterial cellulose as cathode and Fe_3O_4 multiwall carbon nanotubes as anode showed a stable performance with a high energy density of 29.8 $Wh\,kg^{-1}$ at a power density of 299.0 $W\,kg^{-1}$ (75).

5.2.3 *Cellulose Nanopapers*

Cellulose nanofibers show a high water retention that leads to exorbitant time consumption in the dewatering of cellulose nanofiber suspensions (76). This has become a bottleneck, which is restricting the commercialization of cellulose-nanofiber-derived products such as nanopapers and cellulose nanofiber reinforced paper sheets.

An eco-friendly water-based approach was suggested that involves the use of sonication energy and lactic acid to modify the surface of the cellulose nanofibers. The suggested modification resulted in a rapid water drainage, and dewatering was completed in 10 *min*. With unmodified cellulose nanofibers, it took around 45 *min* (76).

A completely water-based approach has been reported to prepare water resistant modified cellulose nanopapers (77). Here, lactic acid in an aqueous medium was attached on a cellulose nanofiber's surface using ultrasonication. Then the material was oligomerized by compression molding under high temperature and pressure to obtain the modified nanopapers with enhanced mechanical properties.

The modified nanopapers showed an increase of 32% in the elastic modulus and 30% in the yield strength over conventional nanopapers. Additionally, the modified nanopaper was hydrophobic in nature and had superior storage modulus under moist conditions. The storage modulus of wet modified nanopaper was three times (2.4 *GPa*) that of the reference nanopapers (0.8 *GPa*) after 1 *h* immersion in water. Finally, the thermal stability of the modified nanopaper was also higher than reference nanopaper. Thus, the material is 100% bio-based (77).

In another study, water-based PU was dispersed along with cellulose nanofibers in water and hybrid nanopapers were prepared by draining water under vacuum followed by forced drying (78). These hybrid nanopapers have a gradient interpenetrating structure with PU concentrated towards one side and cellulose nanofibers towards the other. This could be confirmed by SEM scanning electron microscopy, X-ray photoelectron spectroscopy and contact angle measurements.

Because of this structure, the nanopapers are water resistant on one surface, the PU rich side, but hydrophilic on the other side, the cellulose rich side. This makes them stereoselectively water resistant. When wetted with water on the PU side, the hybrid nanopaper with 10% PU is able to retain 65% modulus. On the other hand, the reference retains only 10% of the modulus. Similar results could be seen with regard to the tensile and the yield strength. Additionally, the hybrid nanopapers have a higher elongation and an improved thermal stability. This material could be relevant to applications such as flexible electronics and transparent displays (78).

A green method has been developed to prepare hybrid lignocellulosic nanopapers by combining wood nanofibers and cellulose nanofibers (79). The wood nanofibers and cellulose nanofibers behave synergistically to compensate for the drawbacks of each other, resulting in enhanced hybrid nanopapers. The draining time of hybrid nanopapers was improved by up to 75% over cellulose nanofiber nanopaper. The mechanical properties, modulus, strength and elongation were improved up to 35%, 90% and 180%, respectively, over a wood nanofiber nanopaper. The stress-strain curves of the dry nanopapers are shown in Figure 5.6.

In addition, the water resistance of hybrid nanopapers was considerably improved, with a water contact angle of 95°. The neat cellulose nanofiber nanopaper had a contact angle of 52°. The morphology of nanopapers indicated that lignin acts as a matrix, which binds the nanofibers together and makes them impervious to external environmental factors such as high humidity. The here reported hybrid nanopapers are 100% bio-based and can be prepared by a simple and environmentally friendly processing route. Such hybrid nanopapers can be used in novel applications such as gas barrier membranes and printable electronics (79).

The preparation of nanofibers obtained from low-cost agro-industrial byproducts, such as soybean hulls and pods, and their use as nanoreinforcers of soy protein films has been described (80).

From hulls, white and stable in-time aqueous dispersions of cellulose nanofibers were obtained that were about 20 *nm* wide and 500 *nm* long with a hydrodynamic diameter of 269 *nm*, showing a marked increase in the crystalline index as compared to raw material.

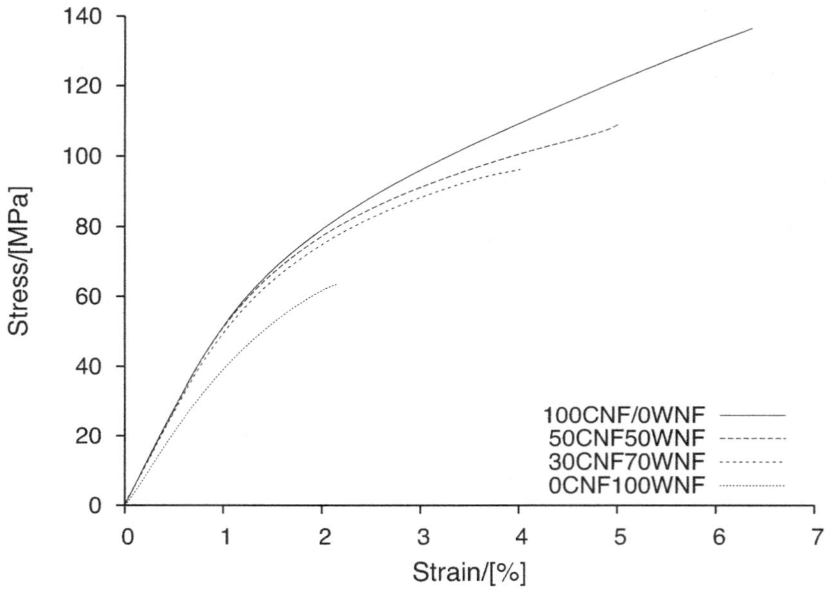

Figure 5.6 Stress-strain curves of the dry nanopapers (79).

Filmogenic solutions of soy proteins were prepared using glycerol as a plasticizer and different amounts of cellulose nanofibers of 0% 10% 20%, and 40% in relation to the amount of soy protein. The rheological behavior of filmogenic solutions containing nanofibers suggested an interaction between the matrix materials and the reinforcements. Films were prepared by casting of filmogenic solutions (80).

As a result, films were obtained with a homogeneous surface having a large number of nanofibers, showing increasing opacity in samples with larger amounts of cellulose nanofibers. The addition of nanofibers decreases the susceptibility of the film to water. A significant decrease in the total soluble matter and swelling in water was observed with an increase of the contact angle. Also, the mechanical properties were improved by obtaining more rigid films as the number of nanofibers increased (80).

5.2.4 *Modified Cellulose Fibers*

A method has been presented in which cotton fibers are treated with a combination of an alkyl ketene dimer, modified corn starch and a cationic retention agent in order to get a cellulose material which is both hydrophobic and biodegradable (81).

Alkyl ketene dimers are typically utilized for sizing in the pulp and paper industry. The term *sizing* means to reduce the tendency of the dry material to absorb water. Alkyl ketene dimers are synthesized from fatty acids, as shown in Figure 5.7.

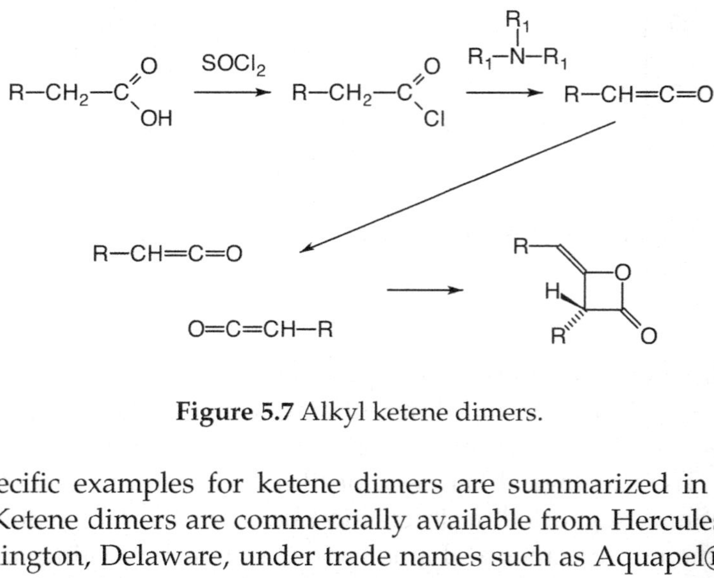

Figure 5.7 Alkyl ketene dimers.

Specific examples for ketene dimers are summarized in Table 5.9. Ketene dimers are commercially available from Hercules Inc., Wilmington, Delaware, under trade names such as Aquapel® and Hercon® (82), or NovaSize® AKD 3016 made by Georgia-Pacific Corp. (83).

The reaction with ketene dimers effects the formation of bridges that narrow the pores of the cotton fiber so that the passage of water across the fiber is greatly delayed. The main stages of preparation are (81):

1. Preparing an aqueous alkaline emulsion of alkyl ketene dimer with cationic starch and a retention agent,
2. Saturating a basic laminar cellulose material with the above emulsion, and

3. Drying the saturated laminar cellulose material at elevated temperature, thus accelerating the reaction between the alkyl ketene dimer and cellulose.

A specific example of preparation is given subsequently (81):

Preparation 5–3: An emulsion is prepared by dispersing the alkyl ketene dimer in a cationic starch/water solution in a concentration ranging from 2.5–4.5% at 70°C. The emulsification is carried out under a high pressure of 20 MPa. A solution of sodium bicarbonate is added to the vessel with water at 45°C to achieve a pH of around 7.6. Then, to the emulsion a retention agent is added, e.g. Kymene® 557 (Hercules Inc.), an adipic acid epoxypropyl diethylenetriamine copolymer, in a concentration of 4.0 g per kilogram of textile material that will be processed. Single samples of a textile material made of cotton fibers and having a woof of 72 $g\,m^{-2}$ are first submerged in the above bath and then dried at a temperature of 65–160°C.

Table 5.9 Alkyl ketene dimers (85).

Component	
Octyl ketene dimer	Decyl ketene dimer
Dodecyl ketene dimer	Tetradecyl ketene dimer
Hexadecyl ketene dimer	Octadecyl ketene dimer
Eicosyl ketene dimer	Docosyl ketene dimer
Tetracosyl ketene dimer	Phenyl ketene dimer
Benzyl β-naphthyl ketene dimer	Cyclohexyl ketene dimer
Montanic acid ketene dimer	Naphthenic acid ketene dimer
10-Decylenic acid ketene dimer	10-Dodecylenic acid ketene dimer
Palmitoleic acid ketene dimer	Oleic acid ketene dimer
Ricinoleic acid ketene dimer	Petroselinic acid ketene dimer
Vaccenic acid ketene dimer	Linoleic acid ketene dimer
Linolenic acid ketene dimer	Eleostearic acid ketene dimer
Licanic acid ketene dimer	Parinaric acid ketene dimer
Tariric acid ketene dimer	Gadoleic acid ketene dimer
Arachidonic acid ketene dimer	Cetoleic acid ketene dimer
Erucic acid ketene dimer	Selacholeic acid ketene dimer
Coconut oil ketene dimer	Babassu oil ketene dimer
Palm kernel oil ketene dimer	Palm oil ketene dimer
Olive oil ketene dimer	Peanut oil ketene dimer
Rape oil ketene dimer	Beef tallow ketene dimer

Composites from natural cellulose fibers or other fibrous components contain conventional polyolefin or poly(vinyl chloride) fibers. Molded composite structures may contain up to 50% of binder polymer. One of the limitations of conventional articles is their lack of biodegradability when composted, which is due to the nature of the binder polymer used (84).

Instead of the conventional binder polymers described above, biodegradable polymers, such as aliphatic-aromatic copolyesters, may be advantageously used. A preferred aliphatic dicarboxylic acid is adipic acid and terephthalic acid can be used as aromatic dicarboxylic acid. As a glycol component, 1,4-butanediol has been suggested. The natural cellulose fibers are selected from hemp, sisal, flax, kenaf, cotton, jute, or coconut (84). Commercial natural cellulose containing fibers are summarized in Table 5.10.

Table 5.10 Commercial natural cellulose containing fibers (84, 86).

Name	Type	Botanical name	Growing area
Cotton	Seed Hair	*Gossypium* sp.	United States, Asia, Africa
Coir	Seed Hair	*Cocos nucifera*	Tropics, India, Mexico
Flax	Bast	*Linum usitatissimum*	North and South Temperate Zones
Hemp	Bast	*Cannabis sativa*	China
Jute	Bast	*Corchorus capsularis*	India
Kenaf	Bast	*Hibiscus cannabinus*	India, Iran, South America
Sisal	Leaf	*Agave sisalana*	Haiti, Java, Mexico, South Africa

The world production of cellulosic fibers is given in Table 5.11. In contrast, protein fibers play a comparatively minor role.

Composite structures containing natural cellulose fibers have been developed for use in the automotive industry for door panels and acoustic insulation.

Cellulose microfibrils have been investigated as nanofillers for polymers in order to increase their biodegradability (88). The anaerobic degradation of cellulose microfibril films extracted from banana and plantain plant rachis residues has been studied. Samples were exposed to burial tests in nature compost for 14 *d*. The degradation

Table 5.11 World production of cellulosic fibers (87).

Name	Amount/[$Mt\,a^{-1}$]
Cotton	25
Coir	0.45
Flax	0.08
Hemp	0.09
Jute	2.7
Kenaf	0.5
Sisal	0.3
Ramie	0.15
Abaca	0.08

process was assessed by optical microscopy, mechanical tests, viscosity measurements, ATR-FTIR, diffraction and thermogravimetry. For cellulose microfibril films extracted from banana the biodegradability was found to be higher than for plantain films.

The production of nanocomposites of PLA grafted on surface-modified cellulose nanofibers (TEMPO-oxidized cellulose nanofibers) was reported using an an improved process (89).

The surface-modified cellulose nanofibers forming the backbone of the composite matrix were prepared from sugarcane bagasse bleached pulp. The formation of ester bonds between the L-lactide and carboxylate groups on the surfaces of the modified cellulose nanofibers were enhanced. The carbonyl groups of PLA were covalently linked to the surface functional groups of bagasse bleached pulp.

The compositions had a high crystallinity (>76%) and a small crystallite size (3.8 *nm*) (89).

5.3 Cellulose Ethers

In a recent study, the etherification of cellulose was performed using a mixture of ionic liquids playing roles in both cellulose dissolution and catalysis (90). The effects of the reaction time and the ratio of these ionic liquids in the mixture were investigated. The cellulose etherification was performed in these ionic liquid mixtures. The proportion of propoxy cellulose exceeded 2.5 after 24 *h*.

Cellulose ethers, such as hydroxyethyl cellulose or hydroxypropyl cellulose, are water-soluble cellulose derivatives which may be produced by etherification of alkali cellulose with epoxides, e.g., ethylene oxide or propylene oxide (91). Cellulose is made alkaline in sodium hydroxide solution and subsequently reacted with the corresponding alkylene oxides.

Water-soluble cellulose ethers have been used for pharmaceutical products, binders for foods, thickeners for various solvents, and binders for extrusion molding and ceramic-forming materials. A method for producing a highly viscous water-soluble cellulose ether has been shown (92). The method for producing a water-soluble cellulose ether consists of the steps of (92):

1. Bringing cellulose pulp into contact with an alkali metal hydroxide solution to obtain an alkali cellulose mixture,
2. Draining the alkali cellulose mixture to obtain alkali cellulose,
3. Reacting the alkali cellulose with an etherifying agent to obtain a water-soluble cellulose ether mixture,
4. Washing and draining the water-soluble cellulose ether mixture to obtain a first moist cellulose ether,
5. Mixing the first moist cellulose ether with water to obtain a second moist cellulose ether,
6. Coarsely pulverizing the second moist cellulose ether to obtain a coarsely pulverized cellulose ether,
7. Cooling the coarsely pulverized cellulose ether, and
8. Drying and pulverizing the coarsely pulverized cellulose ether to obtain a water-soluble cellulose ether.

In detail, the preparation may run as follows (92):

Preparation 5–4: A pulp sheet having an intrinsic viscosity of 1,800 $ml\ g^{-1}$ and a thickness of 1.5 *mm* was immersed in a 49% aqueous NaOH solution of 39°C for 31 *s*. The mass ratio of the 49% aqueous NaOH solution in the alkali cellulose mixture to the solid component in pulp was 200. Then, the pulp sheet was pressed to remove an excess 49% aqueous NaOH solution to obtain alkali cellulose. The mass ratio of the NaOH component in the alkali cellulose to the solid component in pulp was 1.06. Next, 17.9 *kg* of the alkali cellulose was placed in an internal-stirring pressure-resistant reactor with a jacket, and vacuumed and purged with nitrogen to thoroughly remove the oxygen in the reactor. Then, the inside of the reactor was stirred while adjusting the inside temperature of the reactor to 60°C. Subsequently, 2.2 NaOH of dimethyl ether was added thereto, while the inside of the temperature was adjusted to 60°C. After the addition of dimethyl ether, while increasing the inside temperature of the reactor from 60°C to 80°C, methyl chloride was added in such an amount as to make a molar ratio of (methyl chloride) to (NaOH component in alkali cellulose) of 1.3, and propylene oxide was added in such an amount as to make a mass ratio of (propylene oxide) to (solid component in pulp) of 0.26. After the addition of methyl chloride and propylene oxide, the inside temperature of the reactor was increased from 80°C to 90°C, and the reaction was continued at 90°C for another 20 *min*. The gas in the reactor was then discharged, and crude hydroxypropyl methyl cellulose was taken out of the reactor. The crude hydroxypropyl methyl cellulose was dispersed in hot water of 95°C, and then drained to obtain a first moist cellulose ether in a block shape. The first moist cellulose ether had a temperature of 85°C and a water content of 50%. While the first moist cellulose ether was stirred in a batch-type ProShear mixer equipped with a spray nozzle and having a jacket temperature of 90°C, water of 85°C was continuously supplied from the spray nozzle over 5 *min* thereto in such an amount that a second moist cellulose ether had a water content of 80%. The mixing with stirring was continued for another 5 *min* to obtain a second moist cellulose ether. The second moist cellulose ether had a temperature of 85°C and a water content of 80%. The second moist cellulose ether had an average particle size of 15 *mm*, which was separately measured as the particle size at an integrated value of 50% in a mass-based cumulative particle size distribution as determined from the ratios of particles passing through five sieves having different openings (22.4 *mm*, 16.0 *mm*, 11.2 *mm*, 8.0 *mm* and 4.0 *mm*). Several grams of sample were collected from each of five different positions in the surface part or from each of five different positions in the center part of the second moist cellulose ether, and an average of the water contents at the five positions was calculated. The absolute value of the difference in (average) water content between the surface part to the center part was 3.5%, and the water content ratio of the surface part to the center part was 1.05:1.00. Successively, the second moist cellulose ether was coarsely pulverized with a feather mill (manufactured by Hosokawa Micron Corporation) to obtain a coarsely pulverized cellulose ether. The coarsely pulverized cellulose ether had a temperature of 60°C and a water content of 78%. Several grams of sample were collected from each of five different positions in the surface part or from each of five different positions in the center part of the coarsely pulverized cellulose ether, and an average of the water contents at the five positions was calculated. Subsequently, the absolute value of the difference in (average) water content between the surface part to the center part was calculated to be 0.5%, and the water content ratio of the surface part to the center part was 1.01:1.00. The coarsely pulverized cellulose ether had an average particle size of 8 *mm*, which was separately measured as the particle size at an integrated value of 50% in a mass-based cumulative particle size distribution determined from the ratios of particles passing through five sieves having different openings (16.0 *mm*, 11.2 *mm*, 8.0 *mm*, 4.0 *mm* and 2.0 *mm*). The residence time in the pulverizer during the coarse pulverization was about 1 *s*, the pulverizing blade of the coarse pulverizer was a knife blade, and the circumferential speed thereof was 79 $m\,s^{-1}$. The obtained coarsely pulverized cellulose ether was cooled to 15°C by being mixed and granulated in a batch-type ProShear mixer having a jacket temperature of 5° C., and was then introduced into an Ultra Rotor IIS impact mill (manufactured by Altenburger Maschinen Jäckering) which was driven at a pulverizing blade tip-circumferential speed of 108 $m\,s^{-1}$ and was subjected to addition of a high temperature gas of 120°C containing nitrogen at a speed of 800 $m^3\,h^{-1}$. Thus, the coarsely pulverized cellulose ether was dried and pulverized concurrently to obtain a hydroxypropyl methyl cellulose powder. The hydroxypropyl methyl cellulose powder had a degree of substitution (DS) of methoxy group of 1.80 and a molar substitution (MS) of hydroxypropoxy group of 0.16. A viscosity at 20°C of a 2% aqueous solution of the hydroxypropyl methyl cellulose powder was 110,000 *mPa s*. A loose bulk density of the hydroxypropyl methyl cellulose powder was 0.48 $g\,ml^{-1}$, as determined with a powder characteristic evaluation apparatus, Powder Tester PT-S, manufactured by Hosokawa Micron Corporation. An average particle size of the hydroxypropyl methyl cellulose powder was 59 μm, which was determined as the particle size at an integrated value of 50% on a volume-based cumulative distribution curve obtained from the measurement by a laser diffraction particle size distribution analyzer, Mastersizer 3000 (manufactured by Malvern) using a dry method. The number of undissolved fibers, each fiber having a dimension of 8 to 200 μm, was 720, which was determined at 25° C in 2 *ml* of a 0.1% aqueous solution of the hydroxypropyl methyl cellulose powder by using a Multisizer 3 (manufactured by Beckmann Coulter Inc.).

5.4 Nonionic Cellulose Ethers

The current literature suggests that nonionic cellulose ethers adsorb on cement. To clarify whether such interaction indeed occurs, the solution depletion of nonionic cellulose ethers from cement slurries was investigated (93).

It was found that the retained amounts of nonionic cellulose ethers on cement are generally quite low, decrease with increasing w/c ratio and are independent of cement composition and surface area. Moreover, D-glucose, cellulose's monomeric unit, was not retained by silica as model substrate, and the ζ-potential of the silica suspension was unchanged upon nonionic cellulose ether addition. So, it was concluded that nonionic cellulose ethers do not adsorb on cement (93).

A phase-transfer method has been described to synthesize hydrophobically modified hydroxyethyl cellulose with 1-bromododecane (94). The synthesis of hydroxyethyl cellulose lauryl ether occurred by an etherification reaction between hydroxyethyl cellulose and 1-bromododecane using water and tetramethylene oxide as the solvent and tetrabutylammonium bromide as the phase-transfer agent.

A series of hydroxyethyl cellulose lauryl ethers with different substitution degrees were obtained by changing the reaction temperature and the volume ratio of hydroxyethyl cellulose aqueous solution to tetramethylene oxide. The molecular structures of obtained hydroxyethyl cellulose lauryl ethers were confirmed by FTIR and H NMR. The properties of the hydroxyethyl cellulose lauryl ethers were tested by particle size analysis, rheological testing and pyrene probe method. Hydroxyethyl cellulose lauryl ethers with DS range of 0.21–1.95 showed typical properties of amphiphilic polymers and self-assembled into aggregates in aqueous solution because of hydrophobic interaction. The sizes of aggregates and critical aggregate concentrations were found to correlate with the molecular structure of the polymers, and were varied in the range of 2–4 μm and 0.019–0.088 $mg\ ml^{-1}$, respectively. The apparent viscosity under low shear and thickening nature of hydroxyethyl cellulose lauryl ethers modified by moderately grafting lauryl group was increased in contrast to a native hydroxyethyl cellulose. Therefore, hydroxyethyl cellulose lauryl ethers can be applied as hydrophobic active compounds for coating, and possibly in many other areas (94).

5.5 Cellulose Esters

Cellulose can be reacted with lactones to result in cellulose hydroxy carboxylic acid esters (91). Cellulose can be activated with lye and subsequently reacted with lactones in suspension. There is no need to use specific cellulose solvents. In this way, polymeric cellulose esters with side chains are obtained. The cellulose esters are thermoplastic and biodegradable.

Cellulose fibers could be chemically functionalized under mild conditions by introducing palmitate groups in order to produce a hydrophobic cellulose-based ester (95). The effectiveness and the degree of functionalization were evaluated by FTIR spectroscopy and TG. The structural integrity of the products could be confirmed by XRD measurements.

The prepared materials were tested in water remediation processes with fatty oils mixtures from both animal (cod liver oil) and vegetable (sunflower oil) sources. Kinetic measurements, rationalized by means of the most common models, showed that the best performances were reached with the sunflower oil. These results are proving to be a promising application of such green materials as a sustainable alternative tool for oil-water remediation processes (95).

Reactive polysaccharide-based nanoparticles with activated *N*-hydroxysuccinimide ester moieties were prepared by a two-step approach (96):

1. Homogeneous synthesis of activated cellulose ester derivatives, followed by
2. Nano-self-assembling of these derivatives into spherical polysaccharide nanoparticles.

Four different conversions of cellulose acetate phthalate into cellulose acetate (*N*-succinimidyl)phthalate (*N*-hydroxysuccinimide-cellulose acetate phthalate) were studied (96). The carboxyl groups of cellulose acetate phthalate were converted into *N*-hydroxysuccinimide ester in good yield by using *N,N*′-disuccinimidyl carbonate as reagent. Spherical nanoparticles with diameters in the range of 200–400 *nm* were obtained by self-assembling the *N*-hydroxysuccinimide-cellulose acetate phthalate derivatives.

The *N*-hydroxysuccinimide moieties were found to be stable in aqueous media and remained reactive. Reactive polysaccharide-nanoparticles were used for the immobilization of an amine-functionalized fluorescent dye. Under basic conditions, an amide bond was formed, which resulted in high binding efficiencies of up to 90%. In contrast, only weak non-covalent interaction with the dye was observed for nanoparticles derived from cellulose acetate phthalate and cellulose acetate. These *N*-hydroxysuccinimide-activated polysaccharide-nanoparticles have a great potential for applications in drug delivery and bioimaging (96).

5.6 Cellulose Ether Esters

Various esters of cellulose ethers are useful as enteric polymers for pharmaceutical dosage forms such as methyl cellulose phthalate, hydroxypropyl methyl cellulose phthalate, methyl cellulose succinate, or hydroxypropyl methyl cellulose acetate succinate (97).

Cellulose ether esters can be synthesized by the reaction of alkylene oxides with alkaline-activated cellulose and also by grafting of the cellulose ether with lactones (91).

By the combination of the ether and the ester functionality and a proper selection of the degree of substitution, an increased range of property profiles can be achieved.

The polymers are injection moldable and thermoplastic with a melting point varying between 80°C to 250°C, depending on the degree of substitution and ratio of alkylene oxide and lactone. In contrast, pure hydroxyalkyl celluloses are thermoplastic only at high degrees of substitution of greater than 4 and exhibit melting points of 180–200°C. An example for the preparation of a cellulose ether ester is as follows (91):

Preparation 5–5: First, 1 *kg* cotton linters is stirred in 20 *kg* 20% NaOH for 30 *min* at 25°C and the aqueous NaOH solution is subsequently centrifuged off. Then the filter cake is stirred in 9.6 *l* methanol/water with a ratio of 85:15 and centrifuged, stirred again in 8 *kg* methanol and centrifuged again. Eventually, the alkali cellulose is stirred in 7 *kg* dimethyl sulfoxide (DMSO) and centrifuged, and ca. 2.6 *kg* DMSO-moist alkali cellulose remains with an alkali content of some 7%. Then, 52.4 *g* of the DMSO-moist alkali cellulose is stirred in 700 *g* DMSO with 62 *g* ε-caprolactone and 23.9 *g* ethylene oxide in a 0.5 *M Pa* nitrogen atmosphere at 140°C for 1.5 *h*. The reaction solution is subsequently poured into 5 *l* water and the product obtained is filtered off. The cellulose hydroxyalkyl hydroxy carboxylic acid ester thus obtained is purified of DMSO by washing with water and subsequently dried. The yield is 80 *g*, which is 77% of the theoretical value with a softening point of 165°C.

The dependence of the softening point of the cellulose ether esters versus composition is shown in Table 5.12.

Table 5.12 Softening point versus composition (91).

CE	CA	EO	PO	SP
37.89	44.83	17.28	0	165
36.34	41.75	0	21.91	200
31.79	61.95	6.26	0	200

CE Cellulose /[%]
CA ε-Caprolactone /[%]
EO Ethylene oxide /[%]
PO Propylene oxide /[%]
SP Softening point /[°C]

Hydroxyalkyl celluloses with degrees of substitution of less than 1 may be reacted in conventional organic solvents, such as dioxan, dimethylacetamide, *N*-methylpyrrolidone or *tert*-butyl alcohol, with glycolide or lactide, the cyclic dimer of lactic acid, to form water-soluble hydroxyalkyl cellulose-2-hydroxy carboxylic acid esters (98). So, specific solvent systems for cellulose, e.g., *N*-methylmorpholine oxide or dimethylacetamide/lithium chloride, are not needed. Despite their high total degree of substitution, such hydroxyalkyl cellulose-2-hydroxy carboxylic acid esters biodegrade in aqueous solution markedly more quickly than alkyl and hydroxyalkyl cellulose ethers with comparable total degrees of substitution.

Suitable 2-hydroxy carboxylic acid derivatives are the cyclic dimers glycolide, *L*-lactide, *D*-lactide and meso-lactide, as well as lactic acid esters and lactic acid oligomers (98).

Recently, it has been found that sodium diacetate is as effective as sodium acetate as reaction catalyst for esterifying cellulose ethers and here much less heat is generated when sodium diacetate is dissolved in acetic acid instead of sodium acetate under comparable conditions (97).

A process for preparing an esterified cellulose ether contains the step of esterifying a cellulose ether with (97):

1. An aliphatic monocarboxylic acid anhydride, or
2. A dicarboxylic acid anhydride, or

3. A combination of an aliphatic monocarboxylic acid anhydride and a dicarboxylic acid anhydride.

The cellulose ether used as a starting material in the process is preferably an alkyl cellulose, hydroxyalkyl cellulose or hydroxyalkyl alkylcellulose (97).

5.7 Lignin

Lignin is the only large-volume renewable aromatic feedstock, which is found in most terrestrial plants typically in a range of 15–30% by dry weight and 40% by energy (99, 100)

So, lignins are considered the main aromatic renewable resource (101). They represent an excellent alternative feedstock for the elaboration of chemicals and polymers.

Lignin is a highly abundant biopolymeric material which, along with cellulose, constitutes one of the major components in structural cell walls of higher vascular plants. Large quantities of lignin are available from numerous pulping processes such as paper and biorefinery industries. Lignin extraction from lignocellulosic biomass, such as wood, is a key point for its extensive use in industrial applications.

The exploration of effective approaches for the valorization of lignin to valuable products has attracted broad interest (102). Fully unlocking the potential of the world's most abundant resource of bio-aromatics could improve the profitability and carbon efficiency of the entire biorefinery process, thus accelerating the replacement of fossil resources with bioresources in our society.

One of the major problems still remaining is its unclearly defined structure and its versatility according to the origin, separation and fragmentation processes, which mainly limits its utilization (101). While currently often used as a filler or additive, lignin is rarely exploited as a raw material for chemical production. However, it may be an excellent candidate for chemical modifications and reactions due to its highly functional character, i.e., rich in phenolic and aliphatic hydroxyl groups, for the development of new bio-based materials.

During recent decades, numerous research efforts driven by the chemical modification of lignin have resulted in significant studies. Details of different chemical modifications of lignins have been reviewed. These can be classified into three groups (101):

1. Lignin fragmentation into phenolic or other aromatic compounds for fine chemistry,
2. Synthesis of new chemical active sites to impart new reactivity to lignin, and
3. Functionalization of hydroxyl groups to enhance their reactivity.

Also, the major achievements and remaining challenges for lignin modifications and its uses as a macromer for polymer synthesis were discussed with emphasis on the most promising and relevant applications (101).

Actually, lignin is a component of all vascular plants, found mostly between cellular structures but also within the cells and in the cell walls (103). It functions to regulate the transport of liquid in the plant (in part by reinforcing cell walls and keeping them from collapsing and in part by regulating the flow of liquid) and enables the plant to grow tall and compete for sunshine.

Lignin is a complex, amorphous, three-dimensional polymer having a structure based on phenylpropane. In the natural unprocessed form, the molecular structure of lignin varies according to the source and is so complex and varied that its molecular structure has never been completely described. Lignin is mostly composed from *p*-coumaryl alcohol, coniferyl alcohol, and sinapyl alcohol. These compounds are shown in Figure 5.8.

These monomers can be biosynthesized in plants via the shikimic acid pathway. *p*-Coumaryl alcohol is a component of grass and forage-type lignin. Coniferyl alcohol is the predominant lignin monomer found in softwoods.

Methods for the isolation of lignin polymers, and for use of the isolated lignin polymers, have been described (104). Lignin containing compositions can be used in carbon fiber composites, resins, adhesive binders and coatings, PU-based foams, rubbers and elastomers, plastics, films, paints, nutritional supplements, food and beverage additives. Lignin polymers may be produced at low cost, and may be used as substitutes for feedstocks originating from fossil fuel or petrochemical sources in the manufacture of various products.

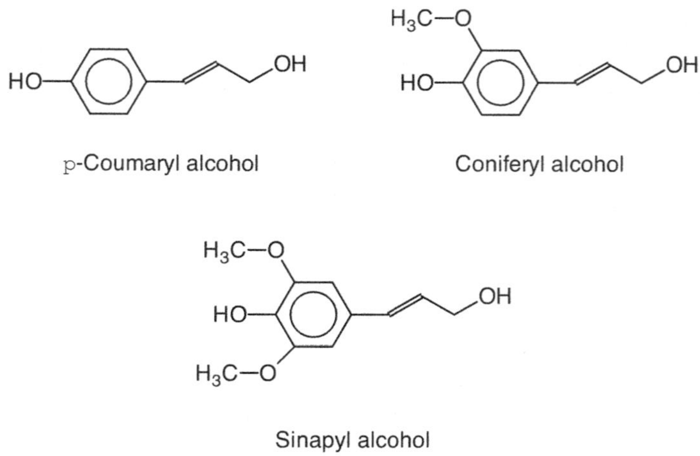

Figure 5.8 Monomers for lignin.

Lignin-based PU can be formulated by using the lignin directly as a polyol precursor or together with other polyols. The efficient functionalization of lignin with diisocyanates also allows the formulation of a high quality PU resin (104).

Phenolic resins can also be formulated from lignin. High-grade lignin can be either directly blended with the phenol resin or incorporated into the resin at high ratios by condensation or derivatization with phenol or formaldehyde (104).

Modified lignin polymers, i.e., alkoxy lignin sulfonates, have been described that may be incorporated into a cleaning composition (103). Lignin sulfonate is modified to produce a modified lignin polymer having both an anionic and alkoxy containing substituent. 2-Hydroxybutyl lignin can be prepared by reacting 1,2-epoxybutane with lignin sulfonate. The modified lignin is isolated by precipitation in ethanol.

5.7.1 Lignocellulose Biorefinery

Lignocellulose biorefinery includes both process engineering and biotechnology tools for the processing of lignocellulosic biomass. These methods are used for the manufacturing of bio-based products such as biofuels, biochemicals, and biomaterials (105).

Lignocellulose biorefinery offers a clear value proposition; however, success at an industrial level has not been shown for the commercial production of renewable chemicals and fuels. This is because of high capital and operating expenditures, irregularities in biomass supply chain, technical process immaturity, and scale-up challenges. As a result, the commercial production of biochemicals and biofuels is still lagging behind.

However, efforts are underway in specialty chemical producing companies for bulk production of compounds with succinic acid, cellulosic ethanol, 1,3-propanediol, 1,4-butanediol, ethylene, cellulosic ethanol, or farnesene (105).

Lignin gas been exploitated as a resource for chemicals. Three technological biorefinery aspects are used for the synthesis of chemicals from lignin (106):

1. Lignocellulose fractionation,
2. Lignin depolymerization, and
3. An upgrading towards targeted chemicals.

The research that has been devoted to each of these three interconnected biorefinery aspects has been reviewed (106).

The effective extraction and separation of lignin with a high purity and less condensed structure from lignocellulose is crucial for lignin valorization (107). Emerging technologies are ionizing and non-ionizing radiation, pulsed electrical field, ultrasound and high pressure technologies.

A key objective during the fractionation and depolymerization is to minimize the lignin condensation, i.e., the formation of resistive carbon-carbon linkages. During fractionation, this can be achieved by either preserving the native lignin structure or by tolerating the depolymerization of the lignin polymer, but preventing a condensation reaction through chemical quenching or physical removal of certain reactive intermediates. The latter strategy is commonly applied in the lignin depolymerization stage, while an alternative approach is to augment the relative rate of depolymerization vs. condensation by enhancing the reactivity of the lignin structure towards depolymerization.

Because depolymerized lignins often consist of a complex mixture of various compounds, an upgrading of the raw product mixture through convergent transformations can be a promising approach to decrease the complexity of the compounds. This particular upgrading approach is termed funneling. It includes both chemocatalytic and biological strategies (106).

Deep eutectic solvents showed potential as solvents for biomass pretreatment in terms of lignin removal and sugar release after an enzymatic saccharification. The recycling and reusability of deep eutectic solvents were also evaluated. The results from this study prove that deep eutectic solvents synthesized from renewable resources are promising solvents for biorefineries (108).

A hydrothermally pretreated rice straw using various pretreatment temperatures was anaerobically digested in a whole slurry (109). The results of these studies indicated that a promoting pretreatment temperature significantly deconstructed the rice straw and facilitated the conversion of insoluble fractions to soluble fractions. Although 306.6 $ml\,g^{-1}$ of biogas was maximally yielded in hydrothermally pretreated rice straw-90 and hydrothermally pretreated rice straw-180, respectively, via digestion in whole slurry, it was only a 3% increase compared to the unpretreated rice straw. Hydrothermally pretreated rice straw-210 yielded 208 $ml\,g^{-1}$ biogas, which was a 30% reduction with longer lag period of 19.8 *d*, suggesting that serious inhibition reactions happened.

Through slightly increasing the organic loading, more serious acidification and reduction on biogas yield, especially at higher pretreatment temperatures, indicated the soluble fractions controlled digestion performances. A Pearson correlation analysis suggested that a negative relationship existed between the yield of methane and the soluble fractions, including soluble carbohydrates, formic acid and furfural. A hydrothermal pretreatment, especially at higher temperature, did not improve the anaerobic digestion (109).

5.7.2 *Acid Hydrolysis*

The acid hydrolysis method uses dilute acid hydrolysis processes (0–5% HCl, H_2SO_4 or H_3PO_4, 120°C–240°C) (110, 111) and also concentrated acid hydrolysis processes (concentrated HCl, H_2SO_4 or HF, 18°C–25°C) (112).

These methods intend to depolymerize hemicellulose and cellulose while also affecting lignin. During acid hydrolysis, low amounts of lignin could be removed in forms of soluble fragments.

The soluble lignin fragments are usually highly reactive under the hydrolysis conditions, leading to undesired condensation reactions. The introduction of a flow-through reactor design can separate lignin from the heating zone in time and restrict condensation reactions (113–115)

However, a flow-through reactor suffers from high water consumption, resulting in a low sugar concentration for downstream conversion. The key to solving this problem is to overcome the solid loading limitation and to increase the mass and heat transfer.

The hydrolysis with concentrated mineral acid, e.g., 72% H_2SO_4 at room temperatures, aims to convert carbohydrates in biomass into soluble sugars, where lignin also undergoes a severe degradation and repolymerization reactions. The so separated lignin is highly condensed and is known as Klason lignin. The calculation of Klason lignin is commonly used to evaluate lignin content in biomass (116). However, due to its highly condensed structure, the selective conversion of Klason lignin to fuels and/or chemicals is difficult.

5.7.3 *Alkaline Hydrolysis*

An alkaline hydrolysis pretreatment is similar to that of an alkaline pulping process, which uses alkaline compounds, e.g. NaOH, $Ca(OH)_2$ and NH_3, to solubilize or depolymerize lignin and make lignin extractable from a biomass matrix (117). The main features of an alkaline pretreatment are that it selectively removes lignin without degrading carbohydrates, and increases the porosity and the surface area, thereby enhancing enzymatic hydrolysis.

Partial hydrolysis of ether linkages serves to yield lignin oligomers, phenolic dimers, and monomers. Due to the low reaction severity, lignin is usually separated with low condensation structure. However, the delignification efficiency of alkaline pretreatments is relatively low. Ammonia or its aqueous solutions are extensively used as alkaline reagents for biomass pretreatment with several well-known technologies such as ammonia fiber explosion (118), anhydrous ammonia pretreatment (119), and ammonia recycled percolation (120). An ammonia-induced biomass pretreatment may alter the cellulose crystalline structure, break lignin-carbohydrate complex bonds, and also partially cleave the lignin ether linkages. Thus, it facilitates the lignin extraction from a biomass matrix. The advantage of these technologies is that ammonia can be easily recovered after pretreatment due to its high volatility.

5.7.4 *Enzymatic Hydrolysis*

Enzymatic hydrolysis of pretreated biomass uses cellulases and hemicellulases to degrade polysaccharides in biomass to soluble oligosaccharides or monosaccharides. This method can generate lignin-enriched solid residues (119, 121).

The direct enzymatic hydrolysis of biomass is hindered by the rigid and recalcitrant structure of lignocellulose, and thus an additional physical or chemical pretreatment is usually needed (110). The molecular weight of lignin is a fundamental property that influences the recalcitrance of biomass and the valorization of lignin. The determination of the molecular weight of lignin in native biomass is dependent on the bioresources used and the isolation and purification procedures employed.

Commonly used pretreatment methods are ball-milling, hot water pretreatment, dilute acid or alkaline hydrolysis, ammonia fiber explosion (118), and mechanical refining (106). A pretreatment step intends to reduce the recalcitrance of raw biomass and to increase the digestibility of plant polysaccharides by enzymes, while lignin is not isolated directly at this stage.

After enzymatic hydrolysis, solid residues with relatively low lignin purity are obtained. Enzymatic hydrolysis is usually carried out under mild conditions, so, the purity and the structural modification of lignin mainly depends on the pretreatment method used. To further increase the purity of lignin, a post-treatment, such as a mild acid hydrolysis or an organic solvent extraction, can be used to remove residual carbohydrates, protein, and/or ash in the lignin-enriched solid residue (122)

5.7.5 *Reductive or Oxidative Fractionation*

Reductive or oxidative catalytic fractionation reactions use reductive reagents, such as H_2 or oxidants (e.g., O_2, O_3, H_2O_2), to promote the cleavage of ether linkages and lignin-carbohydrate complex bonds (123, 124)

These processes usually use redox catalysts, and in some cases they are combined with solvolytic extraction or acidic/alkali hydrolysis (125). The isolated lignin products from these processes are complicated and diverse in structure, including lignin oligomers along with high molecular weight components, highly depolymerized and deoxygenated lignin oil and/or lignin-derived acids.

Solvolysis biomass fractionation, which is similar to organosolv pulping, refers to the use of organic solvents (e.g., alcohols, ethers, ketones, esters, acids) or their water-mixtures to treat biomass feedstock, and are often combined with redox catalysts or mineral acids/alkalis (108, 126, 127). Since organic media exhibit a higher lignin solubility than water, the lignin extraction efficiency is high.

By using catalysts, lignin in an organic medium can undergo depolymerization and/or condensation reactions. In some cases, the organic medium can react with the lignin component to modify the lignin structures. For example, low molecular weight alcohol solvents can etherify with lignin at the α-position of alkyl chains to generate alkoxy groups (128, 129). The structure of lignin that is obtained from solvolysis fractionation varies depending on the organic medium employed, catalyst, and the severity of the pretreatment method.

5.7.6 *Combined Pretreatment Methods*

Mechanocatalytic fractionation methods have been investigated (130). Mechanocatalysis is an efficient method for deep depolymerization of lignocellulosic materials. It renders the full conversion of lignocellulose into water-soluble products, a mixture of oligosaccharides and lignin oligomers. In an aqueous monophasic system, a further saccharification of these products affords high yields of fermentable sugars in addition to a lignin precipitate. The saccharification of water-soluble lignocellulose in a biphasic mixture of water and 2-methyltetrahydrofuran was used in order to prevent lignin fragments from being recondensed.

The lignin stream extracted from the biphasic system has a molecular weight much lower than that of the lignin precipitate obtained from monophasic aqueous saccharification, and that of lignin isolated by an organosolv process with no added acid catalyst. The organosolv process is a pulping technique that uses an organic solvent to solubilize lignin and hemicellulose (131).

Because of the lower molecular weight, low carbohydrate content, and high solubility in organic solvents, the catalytic downstream processing of the extracted lignin leads to higher yields of liquid products, e.g., 2-methoxyphenols and 2,6-dimethoxyphenols (130). After a subsequent acid catalyzed post-hydrolysis, high yields of monosaccharides together with lignin precipitate can be obtained. Due to the mild pretreatment conditions, the structure of lignin isolated by this method is similar to that of ethanosolv lignin, which only has a limited condensed structure.

Another alternative combined pretreatment method uses deacetylation and mechanical refining (132, 133). This method relies on the use of dilute alkaline extraction of biomass to remove acetyl groups in the feedstock, followed by mechanical disc refining to overcome the recalcitrance of lignocellulose. Most of the lignin in the biomass is solubilized under alkaline conditions. After acid precipitation, solid lignin can be recovered. The isolated lignin also shows a less condensed structure, which is suitable for a subsequent conversion.

Other combined techniques are the use of concentrated phosphoric acid with hydrogen peroxide (134), inorganic salts with calcium peroxide (135) or with deep eutectic solvents (127), deep eutectic solvent with microwave (126), and electrolysis combined with ultrasonication (136).

5.7.7 *Pyrolysis*

Pyrolysis is a thermochemical decomposition process that rapidly heats the lignin to elevated temperatures of 450°C–650°C in the absence of oxygen (137). After rapidly quenching the pyrolysis vapor, up to 75% of the feedstock can be converted into bio-oil, the remainder being non-condensable gases and solid char byproducts.

Bio-oil is an unstable mixture of a lot of oxygenated organic compounds, including monolignols, phenols, such as phenol, syringol, guaiacol, and catechol, benzene and its alkyl substituted derivatives, ketones, aldehydes, and acids (138, 139). So, a bio-oil cannot be directly used as a fuel due to its high oxygen content, reduced volatility, high viscosity, acidity, and cold flow problems. Therefore, upgrading technologies using a hydrodeoxygenation process are needed.

In-situ upgrading of bio-oil during pyrolysis by addition of specific catalysts is a good strategy (140).

The incondensable gases generated from pyrolysis of lignin include CO and CO_2 and some low molecular hydrocarbons (CH_4, C_2H_4, C_2H_2, C_3H_6, etc.) which are mainly derived from reformation of –CO and –COOH functional groups or of lignin alkyl chains. The solid residue left after pyrolysis is known as biochar, which has an aromatic polycyclic structure formed from intramolecular and intermolecular rearrangement of lignin degradation intermediates that generally contain benzene rings (141). Biochar can be modified to become a biofertilizer, catalyst support, and an adsorbent.

The yield of bio-oil mainly depends on the heating rate, particle size, temperature, and the used catalyst. The heating rate has significant effects on the bio-oil yield. A high heating rate is beneficial for the production of bio-oil compounds whereas a low heating rate leads to greater char formation. Using a high heating rate, various types of chemical bonds can be broken nearly simultaneously to degrade lignin mainly into volatile compounds, which then can reduce the production of stable oligomers that are prone to formation of char via rearrangement reactions. The particle size of the raw material can also indirectly influence the heating rate during the pyrolysis due to the heat transfer changes. Also, the final temperature is a critical factor that can uniquely affect the product distribution in pyrolysis.

The addition of catalysts can enhance the efficiency of pyrolysis and can tune product distribution. Acidic zeolites, inorganic salts, and some Lewis acids (142–144) have been used as catalysts for lignin pyrolysis.

The catalytic activity of a Lewis acid arises from two aspects (145):

1. The reaction of the Lewis acid with water or alcohol to generate a Brönsted acid, which then promotes lignin ether bonds hydrolysis.
2. The metal center of a Lewis acid can interact with oxygens in ether bonds and promote the depolymerization of lignin.

5.7.8 *Acidic Conversion*

Various types of acids have been tested in lignin conversion. These types are including mineral acids (146–148), zeolites (149–151), organic acids (152–154), and acidic ionic liquids (155).

5.7.9 *Reductive Conversion*

As mentioned above, pyrolysis, as well as acid or base hydrolysis, can be used for lignin depolymerization. However, the generated products have a high oxygen content. These oxygen-containing compounds have a strong tendency to polymerize and are responsible for the thermal and chemical instability of lignin depolymerized fragments (106, 156, 157).

Therefore, the high oxygen-containing products must be refined before utilization. Reductive conversion of lignin, especially hydrodeoxygenation of lignin, with hydrogen as a co-reactant under thermal and catalytic conditions, is currently one of the most popular and efficient strategies for lignin valorization. Lignin hydrodeoxygenation integrates lignin depolymerization and deoxygenation in one process, resulting in production of components with improved stability and heating value that are suitable for drop-in biofuel usage (158–162). The lignin hydrodeoxygenation process involves multiple reactions such as hydrogenolysis, hydrocracking, hydrogenation, dehydration, decarboxylation, isomerization, repolymerization, dimerization, etc. The desired hydrodeoxygenation process should selectively break lignin into its monomers or dimers and simultaneously remove oxygen in the form of water with the minimum loss of carbon.

Catalysts play a crucial role in lignin conversion by hydrodeoxygenation. Both homogeneous and heterogeneous catalysts made from noble metals have been tested. Also, the application of bimetallic catalysts is a promising option for the catalytic conversion of lignin, since bimetallic catalysts usually exhibit enhanced catalytic performances as compared with their parent metals.

5.7.10 *Oxidative Conversion*

Oxygen or peroxide chemicals for lignin oxidative conversion have been developed to generate acids or aldehydes because of their economic feasibility (163–165).

The mechanism of lignin depolymerization is dependent on the nature of the oxidants and the catalysts. Oxidants for lignin conversion are often hydrogen peroxide, ozone, and peroxyacids of chlorine, including chlorine dioxide, while electrophilic, radical and nucleophilic mechanisms are involved (160, 163, 166).

When oxygen is used as an oxidant, radical and electrophilic mechanisms occur in the lignin degradation, with vanillaldehyde/acid, syringaldehyde/acid, and *p*-hydrobenzaldehyde/acid as the dominant products.

However, for hydrogen peroxide, though generating similar reactive species with O_2, the mechanisms usually follow radical and nucleophilic pathways, and the main products are aldehydes, acids, and quinines. With respect to economic feasibility and operability, oxygen is the most promising oxidant.

Metal oxides are the most promising catalysts for oxidative biomass conversion because of their high activity, low cost, and convenient manipulation (161, 166, 167).

5.8 Biodegradable Nanocomposites

The state-of-the-art in the field of biodegradable nanocomposites has been reviewed (168, 169). Biodegradable bio-based nanocomposites can be made from cellulosic materials, corn-derived plastics, and poly(hydroxyalkanoate)s. The latter materials are fabricated from bacterial resources.

Nanocomposites can be produced in combination with nanoclays for a wide field of applications. Even low loadings of nanoclays with high aspect ratios to biopolymers can have a profound enhancing effect over the material properties (170). Such nanocomposites have an improved strength and stiffness, a low gas vapor permeability, a low coefficient of thermal expansion, and an increased heat deflection temperature. However, some technical difficulties also emerge in this field, but these can be circumvented (168).

The preparative techniques for the production of bionanocomposites are based largely on existing techniques for the processing of ordinary plastics. The processes can be classified into three main categories (169):

1. Solution intercalation,
2. *In-situ* intercalative polymerization, and
3. Melt intercalation.

5.8.1 Oxidation of Spruce and Pulps

The TEMPO-mediated oxidation of cellulose has been studied (171). TEMPO is shown in Figure 5.9.

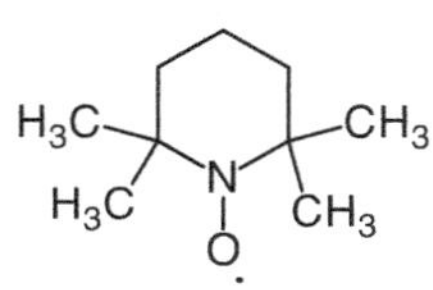

Figure 5.9 2,2,6,6-Tetramethylpiperidine-1-oxyl (TEMPO).

TEMPO is a stable radical that can be used for a wide variety of oxidation reactions in organic chemistry. Softwood and hardwood pulp fibers were suspended in water and oxidized to various extents at pH 10 and 22°C using sodium hypochlorite in the presence of TEMPO radicals and sodium bromide. This reaction composition seems to be the most efficient one for the creation of both surface carboxyl and aldehyde groups.

Important relationships between formation of these functional groups and the fibrillation yield, light transmittance of the water dispersions and degree of polymerization of the oxidized softwood and hardwood pulps were established (171). The dependence of the yield of fibrillation on the content of carboxylate in TEMPO-oxidized pulps is shown in Figure 5.10.

A birefringence test confirmed the presence of nanofibers with diameters in the range of 1.6–3.8 *nm*. The oxidation of spruce is more efficient than the oxidation of eucalyptus. Spruce yields higher carboxylate contents and nanofiber yields.

In summary, the TEMPO-mediated oxidation is a promising technique applicable for a variety of cellulosic materials. A high reaction rate and yield, high selectivity and a low degradation of the substrate is obtained. Both hardwood and softwood can be used for the preparation of nanofibers with different properties depending on the carboxylate contents (171).

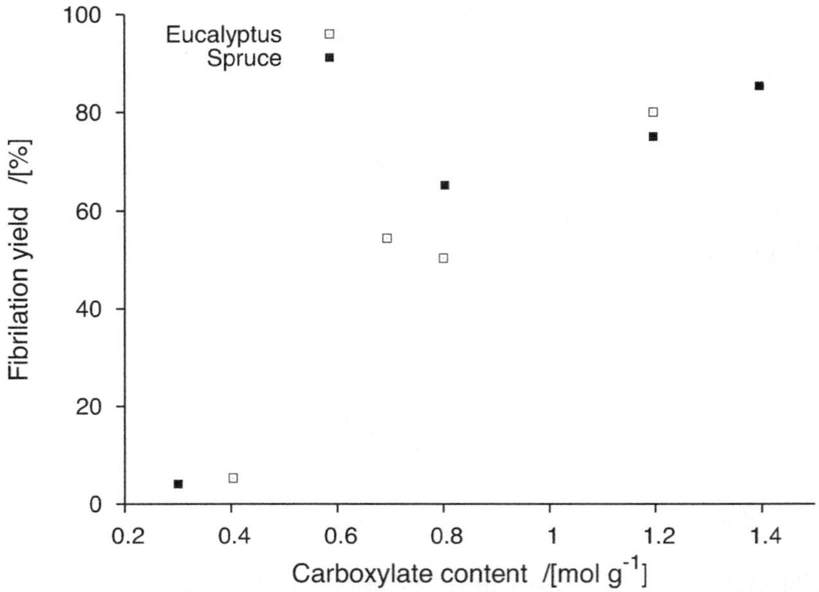

Figure 5.10 Fibrilation yield against carboxylate content of oxidized pulps (171).

5.8.2 Modified Cellulose Nanofibers

The effect of chemical modification of cellulose nanofibers on the properties of PLA nanocomposites has been investigated (172). Acetylated nanofibers, with a degree of substitution 1.07, were isolated from acetylated kenaf fibers by a mechanical procedure. The acetylated nanofibers were found to be more hydrophobic. Both crystallinity and thermal stability of acetylated nanofibers were lower in comparison to the non-acetylated nanofibers.

Nanocomposites were prepared by premixing PLA master batches, with both acetylated nanofibers and non-acetylated nanofibers. These master batches were eventually diluted to their final concentrations of 5% in the course of the extrusion.

The tensile properties and the dynamic mechanical properties were found to be enhanced for both acetylated and non-acetylated nanocomposites in comparison to the neat PLA matrix. The storage modulus increased slightly for the acetylated nanocomposites in comparison to the non-acetylated nanocomposites (172).

5.8.3 *Bio-based Epoxy Nanocomposites*

Bio-based fiber reinforced resins were prepared using bacterial cellulose as the matrix reinforcing phase and diglycidyl ether diphenolate-ethyl ester as the bio-based epoxy resin (173).

Bacterial cellulose mats were prepared by the static cultivation of the strain *Gluconacetobacter xylinus* ATCC 700178 in Hestrin-Schramm medium augmented with mannitol in sterile containers. After freeze-drying, the surface of the bacterial cellulose matrix fibers was modified to introduce trimethylsilyl moieties. Reinforced nanofiber networks were fabricated by impregnation of bacterial cellulose trimethylsilyl matrixes with the resin mixture followed by hot-pressing and curing.

The bacterial cellulose network was effective in reinforcing the epoxy resin matrix. The Young's modulus increased from 1.22 GPA for the neat diglycidyl ether diphenolate-ethyl thermoset to 8.8 for the composite with 30% bacterial cellulose with trimethylsilyl moieties (173).

Also, bio-based epoxy nanocomposites can be reinforced with organo montmorillonite clay and carbon fibers obtained from poly(acrylonitrile) (174). To get the organically modified clay into the glassy bio-based epoxy networks, a sonication technique was used. In this way, clay nanoplatelets were obtained that were homogeneously dispersed and completely exfoliated in the matrix.

The mechanical properties showed an increase of 0.9 *GPa* for the storage modulus at 30°C when 5% of exfoliated clay nanoplatelets were added. In contrast, the glass transition temperature decreases with the addition of the organoclay nanoplatelets (174).

5.9 Chitin

In 1995 the first *in-vitro* synthesis of chitin was succeeded by a ring-opening polyaddition of a chitobiose oxazoline monomer, c.f., Figure 5.11, catalyzed by chitinase (175–178).

The aminodisaccharide in Figure 5.11 is obtained by reacting di-*N*-acetylchito-biosehexacetate with acetyl chloride in the presence of a base to chlorinate the acetate derivative and reacting the resultant chlorinated compound with acetonitrile in the presence

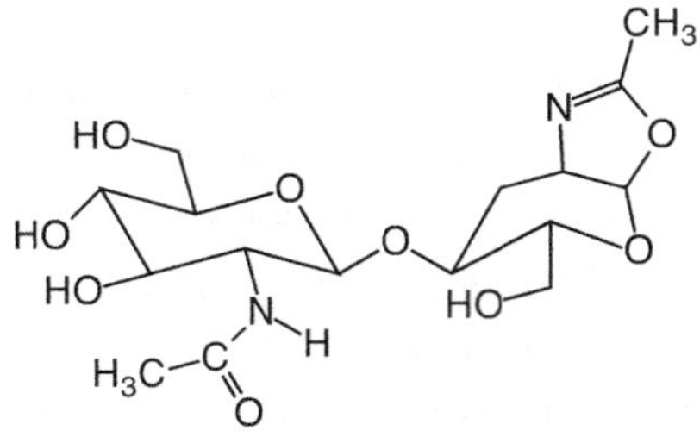

Figure 5.11 Chitobiose oxazoline monomer (178).

of tetraethylammonium chloride, and further deprotecting the hydroxyl group.

The reaction is done at a pH of 9–11. There, the hydrolysis of the product chitin is much suppressed. The degree of polymerization of synthetic chitin is some 10–20 depending on the reaction conditions (179). Peracetyloxazolines can be obtained from peracetyl saccharides (180). A facile preparation method of chitin cellulose composite films has been described (181). Here, the ionic liquids 1-allyl-3-methylimidazolium bromide and 1-butyl-3-methylimidazolium chloride are used. These compounds are shown in Figure 5.12.

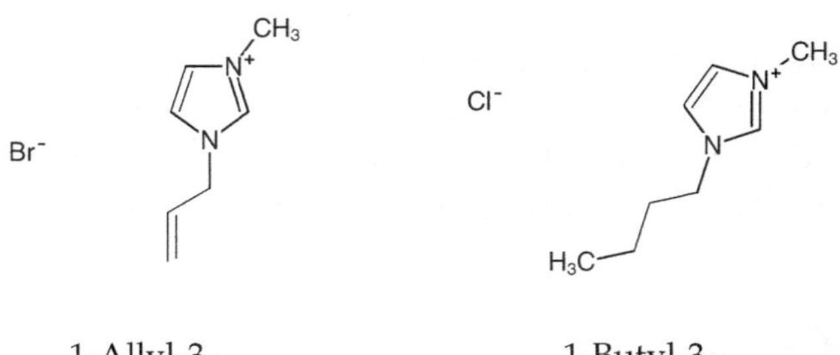

1-Allyl-3-methylimidazolium bromide | 1-Butyl-3-methylimidazolium chloride

Figure 5.12 Ionic liquids.

The former liquid dissolves chitin and the latter liquid dissolves cellulose.

First, solutions of chitin in 1-allyl-3-methylimidazolium bromide and cellulose in 1-butyl-3-methylimidazolium chloride are separately prepared by heating each mixture up to 100°C for 24 *h*. Then, the homogeneous mixture of the two solutions is casted onto a glass plate, followed by standing at room temperature for 2 *h*. Eventually, the material is subjected to successive Soxhlet extraction, first with ethanol for 12 *h* and then with water for 12 *h*. Finally, the residue is dried at room temperature to give the desired composite film (179). The thermal stability of the produced films are comparable to their constituents, chitin and cellulose (181).

The Pickering emulsion polymerization of styrene was investigated using self-assembled chitin nanofibers as stabilizers to produce chitin nanofiber-based composite particles (182). These were further converted into hollow particles. Prior to the emulsion polymerization, the chitin nanofibers were maleylated by the reaction with maleic anhydride in the presence of perchloric acid. After styrene was added to a dispersion of the produced anionic chitin nanofibers in aqueous ammonia, a mixture was ultrasonicated to give an emulsion, in which styrene droplets were stably surrounded by chitin nanofibers. Then, a radical polymerization was done in the presence of potassium persulfate as initiator to produce the composite particles.

The particle sizes became smaller when the amounts of chitin nanofibers increased. Hollow particles were prepared by solubilizing out the PS cores with toluene (182).

The modification of chitin and chitosan were successfully done by directly grafting poly(3-hexylthiophene) onto their surfaces using simple oxidative polymerization with $FeCl_3$ (183). The thermal stability and crystallinity of grafted chitin and chitosan changed upon grafting with poly(3-hexylthiophene).

The buildup of a $\pi - \pi^*$ structure from the poly(3-hexylthiophene) on the surface of chitin and chitosan resulted in the appearance of UV-Vis absorption and fluorescence emission peaks in the range from 500 poly(3-hexylthiophene) to 600 *nm*. The introduction of poly(3-hexylthiophene) onto the surface of chitin and chitosan significantly improved the electrical properties of chitin and chitosan with the increase in conductivity from 10^{-9} to 10^{-7} $S\,cm^{-1}$. The usual behavior of hydrophilic surface of chitin and chitosan turned hydrophobic with water contact angle of 97.7° and 107.0°, respectively, in the presence of poly(3-hexylthiophene) (183).

Also, a copolymer of 3-hexylthiophene and fluorene was directly grafted onto chitin and chitosan using $FeCl_3$ as an oxidant (184). UV-Vis absorption peaks of the chitin/chitosan grafted with 3-hexylthiophene and fluorene copolymer were increasingly blue-shifted upon increasing the fluorene content and the red-shifted emission of the grafted chitin/chitosan were controlled by varying the monomer feed of the copolymer units.

Trade names appearing in the references are shown in Table 5.13.

Table 5.13 Trade names in references.

Trade name Description	Supplier
Amres®	Georgia-Pacific Chemicals LLC
Polyamide-epichlorohydrin resin (83)	
Aquapel®	Hercules Inc.
Ketene dimer (82)	
Brij® (Series)	ICI Surfactants
Ethoxylated fatty alcohols (104)	
Bufloc® (Series)	Buckman Laboratories
Cationic coagulants (83)	
Eastar® Bio	Eastman Chemical Company
Compostable copolyester (84)	
Hercon®	Hercules Inc.
Ketene dimer (82)	
Kymene® 557H	Hercules Inc.
Adipic acid-diethylenetriamine copolymer, reaction product with epichlorohydrin, Wet strength resin (83)	
Lodyne®	Ciba-Geigy Corp.
Fluorochemical surfactant (83)	
Lubrol™ PX	Various
Fatty alcohol polyoxyethylene ether (104)	
Nonidet® P40	Roche
Octylphenolpoly(ethylenglycolether), nonionic detergent (104)	
Novamax™	Georgia-Pacific Chemicals
Neutral rosin sizing agent (83)	
Novasize® AKD 3016	Georgia-Pacific Corp.
Alkyl-ketene-dimer (AKD) emulsion (83)	
Organosolv®	Natural Pulping AG
Aqueous ethanol solvent (104)	
Primacor® (Series)	Dow
Ethylene acrylic acid copolymers (83)	
Teflon®	DuPont
Tetrafluoro polymer (83)	
Triton® X (Series)	Union Carbide Corp. (Rohm & Haas)
Poly(alkylene oxide), nonionic surfactants (104)	
Triton™ X-100	Dow
Hydrophilic poly(ethylene oxide) (104)	

References

1. V.A. Bailey, L.N. Mackey, and P.D. Trokhan, Fiber comprising unmodified and/or modified starch and a crosslinking agent, US Patent 7 938 908, assigned to The Procter & Gamble Company (Cincinnati, OH), May 10, 2011.
2. P. Halley and L. Avérous, eds., *Starch Polymers: From Genetic Engineering to Green Applications*, Elsevier, Burlington, MA, 2014.
3. M.T.P.S. Clerici and M. Schmiele, eds., *Starches for Food Application: Chemical, Technological and Health Properties*, Academic Press, 2018.
4. S. Marques, A.D. Moreno, M. Ballesteros, and F. Gírio, Starch biomass for biofuels, biomaterials, and chemicals in S. Vaz Jr., ed., *Biomass and Green Chemistry: Building a Renewable Pathway*, pp. 69–94. Springer International Publishing, Cham, 2018.
5. M.K. Morell and A.M. Myers, *Current Opinion in Plant Biology*, Vol. 8, p. 204, April 2005.
6. R. Tiwari, *Biotechnology and Molecular Biology Reviews*, Vol. 7, September 2012.
7. K. Fujii, Y. Terada, M. Yanase, K. Odan, H. Takata, T. Takaha, T. Kuriki, and S. Okada, Production method and preparation method of glucans, US Patent 7 229 801, assigned to Ezaki Glico Co., Ltd. (Osaka, JP), June 12, 2007.
8. R.S. Rathore, N. Garg, S. Garg, and A. Kumar, *Critical Reviews in Biotechnology*, Vol. 29, p. 214, 2009.
9. Y. Tang, X. Zeng, Y. Wang, L. Bai, Q. Xu, Z. Wei, H. Yuan, and T. Nyima, *Functional & Integrative Genomics*, Vol. 17, p. 107, Jan 2017.
10. R. Sonnier, A. Taguet, L. Ferry, and J.-M. Lopez-Cuesta, *Towards Bio-based Flame Retardant Polymers*, SpringerBriefs in Molecular Science, Springer, 2018.
11. S.-J. Liao, C.-J. Chang, Y.-C. Chung, F.-M. Chien, and C.-M. Chen, Flame-retardant thermoplastic starch material, flame-retardant thermoplastic starch-based bio-composite, and method for manufacturing the same, US Patent 9 127 156, assigned to Industrial Technology Research Institute (Chutung, Hsinchu, TW), September 8, 2015.
12. L. Passauer, *Carbohydrate Polymers*, Vol. 211, p. 69, 2019.
13. Y. Wang, Q. Su, H. Wang, X. Zhao, and S. Liang, *Journal of Applied Polymer Science*, Vol. 136, p. 47193, 2019.
14. A. Olad and F.F. Azhar, *Fibers and Polymers*, Vol. 15, p. 1321, Jun 2014.
15. Z. Hao, Z. Yi, C. Bowen, L. Yaxing, and Z. Sheng, *Polish Journal of Environmental Studies*, Vol. 28, p. 1701, 2019.
16. R. Narayan, S. Balakrishnan, Y. Nabar, J.-M. Raquez, and P. Dubois, Starch-polyester biodegradable graft copolymers and a method of preparation thereof, US Patent 7 985 794, assigned to Board of

Trustees of Michigan State University (East Lansing, MI), July 26, 2011.

17. J.F. Mendes, R.T. Paschoalin, V.B. Carmona, A.R.S. Neto, A.C.P. Marques, J.M. Marconcini, L.H.C. Mattoso, E.S. Medeiros, and J.E. Oliveira, *Carbohydrate Polymers*, Vol. 137, p. 452, 2016.
18. Z.B. Cuevas-Carballo, S. Duarte-Aranda, and G. Canché-Escamilla, *International Journal of Polymer Science*, Vol. 2017, p. 1, 2017.
19. T. Chia, M. Chirico, R. King, R. Ramirez-Gonzalez, B. Saccomanno, D. Seung, J. Simmonds, M. Trick, C. Uauy, T. Verhoeven, and K. Trafford, *bioRxiv*, pp. 1–31, 2019.
20. N. Soykeabkaew, C. Thanomsilp, and O. Suwantong, *Composites Part A: Applied Science and Manufacturing*, Vol. 78, p. 246, 2015.
21. A.E.S. Vercelheze, A.L.M. Oliveira, M.I. Rezende, C.M. Muller, F. Yamashita, and S. Mali, *Journal of Polymers and the Environment*, Vol. 21, p. 266, 2013.
22. J. Cruz-Tirado, R. Siche, A. Cabanillas, L. Díaz-Sánchez, R. Vejarano, and D.R. Tapia-Blácido, *Procedia Engineering*, Vol. 200, p. 178, 2017. 3rd International Conference on Natural Fibers: Advanced Materials for a Greener World, ICNF 2017, 21-23 June 2017, Braga, Portugal.
23. J. Cruz-Tirado, R. Vejarano, D.R. Tapia-Blácido, G. Barraza-Jáuregui, and R. Siche, *International Journal of Biological Macromolecules*, Vol. 125, p. 800, 2019.
24. L.R.P.F. Mello and S. Mali, *Industrial Crops and Products*, Vol. 55, p. 187, 2014.
25. F. Debiagi, B.M. Marim, and S. Mali, *Journal of Polymers and the Environment*, Vol. 23, p. 269, 2015.
26. F.A. de Carvalho, A.P. Bilck, F. Yamashita, and S. Mali, *Journal of Polymers and the Environment*, Vol. 26, p. 1445, 2018.
27. S.-J. Liao, C.-J. Chang, Y.-P. Liu, S.-J. Liou, Y.-C. Chung, and C.-M. Chen, Modified starch compositions, starch composite foam materials and method for preparing the starch composite foam material, US Patent 9 850 362, assigned to Industrial Technology Research Institute (Hsinchu, TW), December 26, 2017.
28. L. Meng, H. Liu, L. Yu, Q. Duan, L. Chen, F. Liu, Z. Shao, K. Shi, and X. Lin, *Industrial Crops and Products*, Vol. 134, p. 43, 2019.
29. B.F. Bergel, S.D. Osorio, L.M. da Luz, and R.M.C. Santana, *Carbohydrate Polymers*, Vol. 200, p. 106 , 2018.
30. A. Georges, C. Lacoste, and E. Damien, *Industrial Crops and Products*, Vol. 115, p. 306, 2018.
31. J.P. Cruz-Tirado, R. Vejarano, D.R. Tapia-Blácido, L.M. Angelats-Silva, and R. Siche, *Packaging Technology and Science*, Vol. 32, p. 227, 2019.
32. J.V. Barbosa, J. Martins, L. Carvalho, M.M.S.M. Bastos, and F.a.D. Magalhães, *Industrial Crops and Products*, Vol. 137, p. 428, 2019.

33. C.Y. Ding, Y.H. Zheng, and X. Zhang, Substantially completely biodegradable high starch polymer, US Patent 7 939 599, assigned to Biostarch Technology PTE Ltd. (Singapore, SG) and Beijing Jinbaodi Bio-Environmental Technology Ltd. (Beijing, CN), May 10, 2011.
34. L. Avérous, *Journal of Macromolecular Science, Part C: Polymer Reviews*, Vol. 44, p. 231, December 2004.
35. L. Avérous and P.J. Halley, *Biofuels, Bioproducts and Biorefining*, Vol. 3, p. 329, May 2009.
36. V.A. Bailey, L.N. Mackey, and P.D. Trokhan, Starch fiber, US Patent 7 704 328, assigned to The Procter & Gamble Company (Cincinnati, OH), April 27, 2010.
37. E. Ojogbo, R. Blanchard, and T. Mekonnen, *Journal of Polymer Science Part A: Polymer Chemistry*, Vol. 56, p. 2611, 2018.
38. F. Zhu, *Trends in Food Science & Technology*, Vol. 89, p. 1, 2019.
39. H. Wang, L. Kong, and G.R. Ziegler, *Food Hydrocolloids*, Vol. 90, p. 90, 2019.
40. T. Hemamalini and V.R.G. Dev, *International Journal of Biological Macromolecules*, Vol. 106, p. 712, 2018.
41. R. Ashraf, H.S. Sofi, A. Malik, M.A. Beigh, R. Hamid, and F.A. Sheikh, *Applied Biochemistry and Biotechnology*, Vol. 187, p. 47, January 2019.
42. G. Liu, Z. Gu, Y. Hong, L. Cheng, and C. Li, *Journal of Controlled Release*, Vol. 252, p. 95, 2017.
43. W. Cárdenas, E.Y. Gómez-Pachon, E. Muñoz, and R. Vera-Graziano, *IOP Conference Series: Materials Science and Engineering*, Vol. 138, p. 012001, July 2016.
44. L.M. Fonseca, J.P. de Oliveira, P.D. de Oliveira, E. da Rosa Zavareze, A.R.G. Dias, and L.-T. Lim, *Food Research International*, Vol. 116, p. 1318, 2019.
45. E. Zussman and A. Lancuski, High-amylose starch-formate electrospun fibers, US Patent Application 20 180 044 818, assigned to Nanospun Technologies Ltd., February 15, 2018.
46. X. Li, T. Hou, Y. Lu, and B. Yang, *Journal of Applied Polymer Science*, Vol. 135, p. 45810, 2018.
47. J.A. Bowden and C.C. Johnston, Biodegradable or compostable containers, US Patent 7 967 904, assigned to New Ice Limited (Isle of Man), June 28, 2011.
48. Z. Sun, W. Liu, and Y.-J. Wang, Biodegradable materials from starch-grafted polymers, US Patent 7 608 649, assigned to Cereplast, Inc. (Los Angeles, CA), October 27, 2009.
49. R. Chinnaswamy and M.A. Hanna, Biodegradable polymers, US Patent 5 496 895, assigned to The Board of Regents of the University of Nebraska (Lincoln, NE), March 5, 1996.

50. G.J.L. Griffin, Biodegradable synthetic resin sheet material containing starch and a fatty material, US Patent 4 016 117, assigned to Coloroll Limited (London, EN), April 5, 1977.
51. H. Angellier, S. Molina-Boisseau, L. Lebrun, and A. Dufresne, *Macromolecules*, Vol. 38, p. 3783, 2005.
52. H. Angellier, S. Molina-Boisseau, and A. Dufresne, *Macromolecular Symposia*, Vol. 233, p. 132, 2006.
53. P. Beguin, *Annual Review of Microbiology*, Vol. 44, p. 219, October 1990.
54. E. Trovatti, *Journal of Renewable Materials*, Vol. 1, p. 28, January 2013.
55. H. Maehara, Resin composition with biodegradable non-liquid-crystalline and liquid-crystalline polymers, US Patent 7 608 593, assigned to Canon Kabushiki Kaisha (Tokyo, JP), October 27, 2009.
56. S.L. Tseng, G.V. Laivins, and D.G. Gray, *Macromolecules*, Vol. 15, p. 1262, September 1982.
57. D.G. Gray, *J. Appl. Polym. Sci., Appl. Polym. Symp.*, Vol. 37, p. 179, 1983.
58. L. Zhou, N. Li, J. Shu, Y. Liu, K. Wang, X. Cui, Y. Yuan, B. Ding, Y. Geng, Z. Wang, Y. Duan, and J. Zhang, *ACS Sustainable Chemistry & Engineering*, Vol. 6, p. 12403, 2018.
59. Y.L. Hsieh, Cellulose nanofibers: Electrospinning and nanocellulose self-assemblies in A. Netravali, ed., *Advanced Green Composites*, chapter 4, pp. 67–95. Wiley-Scrivener, Hoboken, NJ, 2018.
60. S.J. Eichhorn, A. Dufresne, M. Aranguren, N.E. Marcovich, J.R. Capadona, S.J. Rowan, C. Weder, W. Thielemans, M. Roman, S. Renneckar, and et al., *Journal of Materials Science*, Vol. 45, p. 1, January 2010.
61. I. Sakurada, Y. Nukushina, and T. Ito, *Journal of Polymer Science*, Vol. 57, p. 651, March 1962.
62. M.F. Ashby, *Engineering Materials 1: An Introduction to Their Properties and Applications*, Butterworth-Heinemann, Oxford, Boston, 1996.
63. A.K. Mohanty, P. Tummala, M. Misra, and L.T. Drzal, Filler reinforced thermoplastic compositions and process for manufacture, US Patent 7 582 241, assigned to Board of Trustees of Michigan State University (East Lansing, MI), September 1, 2009.
64. N. Lin, J. Huang, and A. Dufresne, *Nanoscale*, Vol. 4, p. 3274, 2012.
65. M. Wada, A. Yoshimura, and Y. Amano, Cellulose nanofibers, method for producing same, aqueous dispersion using cellulose nanofibers, and fiber-reinforced composite material, US Patent 9 951 192, assigned to Nitto Boseki Co., Ltd. (Fukushima, JP), April 24, 2018.
66. H. Tibolla, F.M. Pelissari, J.T. Martins, A.A. Vicente, and F.C. Menegalli, *Food Hydrocolloids*, Vol. 75, p. 192, 2018.
67. H. Tibolla, F. Pelissari, J. Martins, E. Lanzoni, A. Vicente, F. Menegalli, and R. Cunha, *Carbohydrate Polymers*, Vol. 207, p. 169, 2019.

68. Wikipedia contributors, Caco-2 — Wikipedia, the free encyclopedia, https://en.wikipedia.org/w/index.php?title=Caco-2&oldid=893805679, 2019. [Online; accessed 30-May-2019].
69. A. Isogai and Y. Zhou, *Current Opinion in Solid State and Materials Science*, 2019.
70. A. Isogai and L. Bergström, *Current Opinion in Green and Sustainable Chemistry*, Vol. 12, p. 15, 2018. Green and Sustainable Nanotechnology / Selected papers from the ACS Symposium on Green chemistry synthesis of nanomaterials and manufacturing.
71. N. Fiol, M.G. Vásquez, M. Pereira, Q. Tarrés, P. Mutjé, and M. Delgado-Aguilar, *Cellulose*, Vol. 26, p. 903, Jan 2019.
72. Y.F. Shih, M.Y. Chou, H.Y. Lian, L.R. Hsu, and S.M. Chen-Wei, *Express Polymer Letters*, Vol. 12, 2018.
73. R. Khiari, F. Rol, M.-C. Brochier Salon, J. Bras, and M.N. Belgacem, *ACS Sustainable Chemistry & Engineering*, Vol. 7, p. 8155, 2019.
74. T. Li, S.X. Li, W. Kong, C. Chen, E. Hitz, C. Jia, J. Dai, X. Zhang, R. Briber, Z. Siwy, M. Reed, and L. Hu, *Science Advances*, Vol. 5, 2019.
75. Y. Yuan, J. Zhou, M.I. Rafiq, S. Dai, J. Tang, and W. Tang, *Electrochimica Acta*, Vol. 295, p. 82, 2019.
76. J. Sethi, K. Oksman, M. Illikainen, and J.A. Sirviö, *Carbohydrate Polymers*, Vol. 197, p. 92 , 2018.
77. J. Sethi, M. Farooq, S. Sain, M. Sain, J.A. Sirviö, M. Illikainen, and K. Oksman, *Cellulose*, Vol. 25, p. 259, 2018.
78. J. Sethi, M. Farooq, M. Österberg, M. Illikainen, and J.A. Sirviö, *Carbohydrate Polymers*, Vol. 199, p. 286, 2018.
79. J. Sethi, M. Visanko, M. Österberg, and J.A. Sirviö, *Carbohydrate Polymers*, Vol. 203, p. 148, 2019.
80. A. González, G. Gastelú, G.N. Barrera, P.D. Ribotta, and C.I. Álvarez Igarzabal, *Food Hydrocolloids*, Vol. 89, p. 758, 2019.
81. P. Aldrett, Substantially hydrophobic and biodegradable laminar cellulose material, its manufacturing method, and substantially biodegradable disposable diapers made of said material, US Patent 5 163 931, November 17, 1992.
82. D.A. Jones and N.C. Clungeon, Soft absorbent paper product containing deactivated ketene dimer agents, US Patent 6 458 243, assigned to Kimberly Clark Worldwide Inc. (Neenah, WI), October 1, 2002.
83. D.M. Bhat and D.W. Sumnicht, Multi-ply paper product with moisture strike through resistance and method of making the same, US Patent 8 216 424, assigned to Georgia-Pacific Consumer Products LP (Atlanta, GA), July 10, 2012.
84. F.W. Williams and E.E. Ellery, Composite structures, US Patent 6 872 674, assigned to Eastman Chemical Company (Kingsport, TN), March 29, 2005.

85. P.H. Aldrich and D.H. Dumas, Ketene dimer modified water-dispersible thermosettable cationic resins, US Patent 3 922 243, assigned to Hercules Incorporated (Wilmington, DE), November 25, 1975.
86. L. Rebenfeld, Fibers (survey) in J.I. Kroschwitz and M.H. Grant, eds., *Kirk Othmer, Encyclopedia of Chemical Technology*, Vol. 10, pp. 539–558. John Wiley & Sons, Ltd, New York, 4th edition, 1993.
87. C. Johansson, J. Bras, I. Mondragon, P. Nechita, D. Plackett, P. Šimon, D.G. Svetec, S. Virtanen, M.G. Baschetti, C. Breen, et al., *BioResources*, Vol. 7, p. 2506, 2012.
88. C. Castro, R. Zuluaga, L. Vélez, A. Retegi, I. Mondragon, and P. Gañán, *Journal of Polymers and the Environment*, Vol. 20, p. 774, September 2012.
89. C. Chuensangjun, T. Kitaoka, Y. Chisti, and S. Sirisansaneeyakul, *European Polymer Journal*, Vol. 114, p. 308, 2019.
90. T. Kakibe, S. Nakamura, K. Amakuni, and H. Kishi, *Australian Journal of Chemistry*, Vol. 72, p. 101, 2019.
91. J. Kalbe, H.P. Muller, R. Koch, J. Engelhardt, W. Koch, K. Szablikowski, and G. Weber, Thermoplastic cellulose ether ester graft copolymers and process for their production, US Patent 5 466 794, assigned to Wolff Walsrode Aktiengesellschaft (Walsrode, DE), November 14, 1995.
92. A. Kitamura and M. Yoshida, Water-soluble cellulose ether and method for producing the same, US Patent Application 20 190 010 252, assigned to Shin-Etsu Chemical Co., Ltd., Tokyo (JP), January 10, 2019.
93. T. Hurnaus and J. Plank, *Construction and Building Materials*, Vol. 195, p. 44, 2019.
94. X. Wang, M. Wu, B. Zhang, Y. Zhang, C. Hu, L. Shi, Y. Lv, and R. Ran, *Colloids and Surfaces A: Physicochemical and Engineering Aspects*, Vol. 562, p. 383, 2019.
95. L. Lavagna, R. Nisticò, S. Musso, and M. Pavese, *Wood Science and Technology*, Vol. 53, p. 249, January 2019.
96. P. Schulze, M. Gericke, and T. Heinze, *Cellulose*, Vol. 26, p. 475, January 2019.
97. R.B. Appeil, Process for preparing an ester of a cellulose ether in the presence of acetic acid and a reaction catalyst, US Patent Application 20 180 273 646, assigned to Dow Global Technologies LLC, September 27, 2018.
98. J. Simon, H.-P. Muller, R. Koch, V. Muller, J. Engelhardt, K. Szablikowski, and W. Koch, Water-soluble biodegradable hydroxyalkyl cellulose-2-hydroxycarboxylic acid esters which can flocculate, US Patent 6 559 300, assigned to Wolff Walsrode AG (Walsrode, DE), May 6, 2003.

99. J.S. Lupoi, S. Singh, R. Parthasarathi, B.A. Simmons, and R.J. Henry, *Renewable and Sustainable Energy Reviews*, Vol. 49, p. 871, 2015.
100. S. Gillet, M. Aguedo, L. Petitjean, A.R.C. Morais, A.M. da Costa Lopes, R.M. Łukasik, and P.T. Anastas, *Green Chem.*, Vol. 19, p. 4200, 2017.
101. S. Laurichesse and L. Avérous, *Progress in Polymer Science*, Vol. 39, p. 1266, 2014. Topical Issue on Biomaterials.
102. H. Wang, Y. Pu, A. Ragauskas, and B. Yang, *Bioresource Technology*, Vol. 271, p. 449, 2019.
103. Y. Gizaw, B.A.J. Hubesch, J.S. Dupont, X.J. Wang, and L.A. Zannoni, Modified lignin biopolymer useful in cleaning compositions, US Patent 8 075 637, assigned to The Procter & Gamble Company (Cincinnati, OH), December 13, 2011.
104. H. Belanger, R.L. Prestidge, T.J. Lough, and J.D. Watson, Methods for isolating and harvesting lignin and isolated lignin preparations produced using the methods, US Patent 8 053 566, assigned to Vertichem Corporation (Toronto, Ontario, CA), November 8, 2011.
105. A.K. Chandel, V.K. Garlapati, A.K. Singh, F.A.F. Antunes, and S.S. da Silva, *Bioresource Technology*, Vol. 264, p. 370, 2018.
106. W. Schutyser, T. Renders, S. Van den Bosch, S.-F. Koelewijn, G.T. Beckham, and B.F. Sels, *Chem. Soc. Rev.*, Vol. 47, p. 852, 2018.
107. S.S. Hassan, G.A. Williams, and A.K. Jaiswal, *Bioresource Technology*, Vol. 262, p. 310, 2018.
108. K.H. Kim, T. Dutta, J. Sun, B. Simmons, and S. Singh, *Green Chem.*, Vol. 20, p. 809, 2018.
109. D. Wang, F. Shen, G. Yang, Y. Zhang, S. Deng, J. Zhang, Y. Zeng, T. Luo, and Z. Mei, *Bioresource Technology*, Vol. 249, p. 117, 2018.
110. V.A. Thomas, B.S. Donohoe, M. Li, Y. Pu, A.J. Ragauskas, R. Kumar, T.Y. Nguyen, C.M. Cai, and C.E. Wyman, *Biotechnology for Biofuels*, Vol. 10, p. 252, 2017.
111. W. Liu, W. Chen, Q. Hou, S. Wang, and F. Liu, *RSC Advances*, Vol. 8, p. 10207, 2018.
112. A. Shahbazi and B. Zhang, Dilute and concentrated acid hydrolysis of lignocellulosic biomass in K. Waldron, ed., *Bioalcohol Production*, Woodhead Publishing Series in Energy, chapter 5, pp. 143–158. Woodhead Publishing, 2010.
113. D.D. Laskar, J. Zeng, L. Yan, S. Chen, and B. Yang, *Industrial Crops and Products*, Vol. 50, p. 391 , 2013.
114. L. Yan, L. Zhang, and B. Yang, *Biotechnology for Biofuels*, Vol. 7, p. 76, 2014.
115. S. Bhagia, H. Li, X. Gao, R. Kumar, and C.E. Wyman, *Biotechnology for Biofuels*, Vol. 9, p. 245, 2016.
116. A.B. Ibáñez and S. Bauer, *Biomass and Bioenergy*, Vol. 68, p. 75 , 2014.

117. J.S. Kim, Y. Lee, and T.H. Kim, *Bioresource Technology*, Vol. 199, p. 42, 2016. Pretreatment of Biomass.
118. M.W. Lau, C. Gunawan, and B.E. Dale, *Biotechnology for Biofuels*, Vol. 2, p. 30, 2009.
119. C.G. Yoo, C.W. Lee, and T.H. Kim, *Applied Biochemistry and Biotechnology*, Vol. 174, p. 2639, Dec 2014.
120. T.H. Kim, R. Gupta, and Y.Y. Lee, Pretreatment of biomass by aqueous ammonia for bioethanol production in J.R. Mielenz, ed., *Biofuels: Methods and Protocols*, pp. 79–91. Humana Press, Totowa, NJ, 2009.
121. B. Jiang, T. Cao, F. Gu, W. Wu, and Y. Jin, *ACS Sustainable Chemistry & Engineering*, Vol. 5, p. 342, 2017.
122. A. Tolbert, H. Akinosho, R. Khunsupat, A.K. Naskar, and A.J. Ragauskas, *Biofuels, Bioproducts and Biorefining*, Vol. 8, p. 836, 2014.
123. T.-Q. Yuan, S.-N. Sun, F. Xu, and R.-C. Sun, *Journal of Agricultural and Food Chemistry*, Vol. 59, p. 10604, 2011.
124. M.V. Galkin and J.S.M. Samec, *ChemSusChem*, Vol. 9, p. 1544, 2016.
125. M.V. Galkin, A.T. Smit, E. Subbotina, K.A. Artemenko, J. Bergquist, W.J.J. Huijgen, and J.S.M. Samec, *ChemSusChem*, Vol. 9, p. 3280, 2016.
126. Z. Chen and C. Wan, *Bioresource Technology*, Vol. 250, p. 532, 2018.
127. Y.-L. Loow, T.Y. Wu, G.H. Yang, L.Y. Ang, E.K. New, L.F. Siow, J.M. Jahim, A.W. Mohammad, and W.H. Teoh, *Bioresource Technology*, Vol. 249, p. 818, 2018.
128. C.S. Lancefield, I. Panovic, P.J. Deuss, K. Barta, and N.J. Westwood, *Green Chemistry*, Vol. 19, p. 202, 2017.
129. T. Renders, S. Van den Bosch, S.-F. Koelewijn, W. Schutyser, and B.F. Sels, *Energy Environ. Sci.*, Vol. 10, p. 1551, 2017.
130. G. Calvaruso, M.T. Clough, and R. Rinaldi, *Green Chem.*, Vol. 19, p. 2803, 2017.
131. Wikipedia contributors, Organosolv — Wikipedia, the free encyclopedia, https://en.wikipedia.org/w/index.php?title=Organosolv&oldid=843856778, 2018. [Online; accessed 24-May-2019].
132. X. Chen, E. Kuhn, E.W. Jennings, R. Nelson, L. Tao, M. Zhang, and M.P. Tucker, *Energy Environ. Sci.*, Vol. 9, p. 1237, 2016.
133. X. Chen, W. Wang, P. Ciesielski, O. Trass, S. Park, L. Tao, and M.P. Tucker, *ACS Sustainable Chemistry & Engineering*, Vol. 4, p. 324, 2016.
134. Q. Wang, Z. Wang, F. Shen, J. Hu, F. Sun, L. Lin, G. Yang, Y. Zhang, and S. Deng, *Bioresource Technology*, Vol. 166, p. 420, 2014.
135. R.G. Saratale, H.S. Shin, G.S. Ghodake, G. Kumar, M.K. Oh, and G.D. Saratale, *Bioresource Technology*, Vol. 258, p. 26, 2018.
136. G. Kumar, P. Sivagurunathan, G. Zhen, T. Kobayashi, S.-H. Kim, and K. Xu, *Bioresource Technology*, Vol. 245, p. 196, 2017.

137. C. Zhao, E. Jiang, and A. Chen, *Journal of the Energy Institute*, Vol. 90, p. 902, 2017.
138. M. Saidi, F. Samimi, D. Karimipourfard, T. Nimmanwudipong, B.C. Gates, and M.R. Rahimpour, *Energy & Environmental Science*, Vol. 7, p. 103, 2014.
139. L. Fan, Y. Zhang, S. Liu, N. Zhou, P. Chen, Y. Cheng, M. Addy, Q. Lu, M.M. Omar, Y. Liu, Y. Wang, L. Dai, E. Anderson, P. Peng, H. Lei, and R. Ruan, *Bioresource Technology*, Vol. 241, p. 1118, 2017.
140. L. Fan, P. Chen, N. Zhou, S. Liu, Y. Zhang, Y. Liu, Y. Wang, M.M. Omar, P. Peng, M. Addy, Y. Cheng, and R. Ruan, *Bioresource Technology*, Vol. 247, p. 851, 2018.
141. F.-X. Collard and J. Blin, *Renewable and Sustainable Energy Reviews*, Vol. 38, p. 594, 2014.
142. X. Huang, J. Zhu, T.I. Korányi, M.D. Boot, and E.J.M. Hensen, *ChemSusChem*, Vol. 9, p. 3262, 2016.
143. R. Jastrzebski, S. Constant, C.S. Lancefield, N.J. Westwood, B.M. Weckhuysen, and P.C.A. Bruijnincx, *ChemSusChem*, Vol. 9, p. 2074, 2016.
144. H. Wang, H. Wang, E. Kuhn, M.P. Tucker, and B. Yang, *ChemSusChem*, Vol. 11, p. 285, 2018.
145. H. Wang, Y. Duan, Q. Zhang, and B. Yang, *ChemSusChem*, Vol. 11, p. 2562, 2018.
146. L. Zhang, Y. Tian, S. Zhu, and C. Gao, The financing innovation pattern based on supply chain finance – A case study in Chinese steel industry, in R. Zhang, Z. Zhang, K. Liu, and J. Zhang, eds., *LISS 2013*, pp. 203–208, Berlin, Heidelberg, 2015. Springer Berlin Heidelberg.
147. A.W. Pelzer, M.R. Sturgeon, A.J. Yanez, G. Chupka, M.H. O'Brien, R. Katahira, R.D. Cortright, L. Woods, G.T. Beckham, and L.J. Broadbelt, *ACS Sustainable Chemistry & Engineering*, Vol. 3, p. 1339, 2015.
148. X. Wan, D. Tian, F. Shen, J. Hu, G. Yang, Y. Zhang, S. Deng, J. Zhang, and Y. Zeng, *Energy & Fuels*, Vol. 32, p. 5218, 2018.
149. A.K. Deepa and P.L. Dhepe, *ACS Catalysis*, Vol. 5, p. 365, 2015.
150. B. Güvenatam, E.H. Heeres, E.A. Pidko, and E.J. Hensen, *Catalysis Today*, Vol. 259, p. 460, 2016. Catalytic Science and Technology in Sustainable Energy.
151. J.A. Santana, W.S. Carvalho, and C.H. Ataíde, *Industrial Crops and Products*, Vol. 111, p. 126, 2018.
152. L. Chen, J. Dou, Q. Ma, N. Li, R. Wu, H. Bian, D.J. Yelle, T. Vuorinen, S. Fu, X. Pan, et al., *Science Advances*, Vol. 3, p. e1701735, 2017.
153. M. Oregui-Bengoechea, I. Gandarias, P.L. Arias, and T. Barth, *ChemSusChem*, Vol. 10, p. 754, 2017.
154. L. Shao, Q. Zhang, T. You, X. Zhang, and F. Xu, *Bioresource Technology*, Vol. 264, p. 238, 2018.

155. B.G. Janesko, *Phys. Chem. Chem. Phys.*, Vol. 16, p. 5423, 2014.
156. S. Van den Bosch, W. Schutyser, R. Vanholme, T. Driessen, S.-F. Koelewijn, T. Renders, B. De Meester, W.J.J. Huijgen, W. Dehaen, C.M. Courtin, B. Lagrain, W. Boerjan, and B.F. Sels, *Energy Environ. Sci.*, Vol. 8, p. 1748, 2015.
157. L. Shuai, M.T. Amiri, Y.M. Questell-Santiago, F. Héroguel, Y. Li, H. Kim, R. Meilan, C. Chapple, J. Ralph, and J.S. Luterbacher, *Science*, Vol. 354, p. 329, 2016.
158. H. Wang, L. Zhang, T. Deng, H. Ruan, X. Hou, J.R. Cort, and B. Yang, *Green Chem.*, Vol. 18, p. 2802, 2016.
159. H. Wang, H. Ruan, M. Feng, Y. Qin, H. Job, L. Luo, C. Wang, M.H. Engelhard, E. Kuhn, X. Chen, M.P. Tucker, and B. Yang, *ChemSusChem*, Vol. 10, p. 1846, 2017.
160. W. He, W. Gao, and P. Fatehi, *ACS Sustainable Chemistry & Engineering*, Vol. 5, p. 10597, 2017.
161. L.-P. Xiao, S. Wang, H. Li, Z. Li, Z.-J. Shi, L. Xiao, R.-C. Sun, Y. Fang, and G. Song, *ACS Catalysis*, Vol. 7, p. 7535, 2017.
162. S. Wang, W. Gao, H. Li, L.-P. Xiao, R.-C. Sun, and G. Song, *ChemSusChem*, Vol. 11, p. 2114, 2018.
163. R. Prado, A. Brandt, X. Erdocia, J. Hallet, T. Welton, and J. Labidi, *Green Chem.*, Vol. 18, p. 834, 2016.
164. R. Gao, Y. Li, H. Kim, J.K. Mobley, and J. Ralph, *ChemSusChem*, Vol. 11, p. 2045, 2018.
165. S. Yao, J.K. Mobley, J. Ralph, M. Crocker, S. Parkin, J.P. Selegue, and M.S. Meier, *ACS Sustainable Chemistry & Engineering*, Vol. 6, p. 5990, 2018.
166. R. Ma, Y. Xu, and X. Zhang, *ChemSusChem*, Vol. 8, p. 24, 2015.
167. R. Behling, S. Valange, and G. Chatel, *Green Chem.*, Vol. 18, p. 1839, 2016.
168. J.K. Pandey, A.P. Kumar, M. Misra, A.K. Mohanty, L.T. Drzal, and R. Palsingh, *Journal of Nanoscience and Nanotechnology*, Vol. 5, p. 497, April 2005.
169. R. Zhao, P. Torley, and P.J. Halley, *Journal of Materials Science*, Vol. 43, p. 3058, May 2008.
170. J.M. Lagaron and A. Lopez-Rubio, *Trends in Food Science & Technology*, Vol. 22, p. 611, November 2011.
171. G. Rodionova, T. Saito, M. Lenes, Ø. Eriksen, Ø. Gregersen, R. Kuramae, and A. Isogai, *Journal of Polymers and the Environment*, Vol. 21, p. 207, March 2013.
172. M. Jonoobi, A.P. Mathew, M.M. Abdi, M.D.˙ Makinejad, and K. Oksman, *Journal of Polymers and the Environment*, Vol. 20, p. 991, 2012.

173. L. Yue, F. Liu, S. Mekala, A. Patel, R.A. Gross, and I. Manas-Zloczower, *ACS Sustainable Chemistry & Engineering*, Vol. 7, p. 5986, 2019.
174. H. Miyagawa, R.J. Jurek, A.K. Mohanty, M. Misra, and L.T. Drzal, *Composites Part A: Applied Science and Manufacturing*, Vol. 37, p. 54, January 2006.
175. T. Kiyosada, E. Takada, S. Shoda, and S. Kobayashi, *Polymer Preprints Japan*, Vol. 44, p. 660, 1995.
176. S. Kobayashi, T. Kiyosada, and S.-I. Shoda, *Journal of the American Chemical Society*, Vol. 118, p. 13113, January 1996.
177. J. Sakamoto, T. Watanabe, Y. Ariga, and S. Kobayashi, *Chemistry Letters, Japan*, Vol. 30, p. 1180, 2001.
178. H. Kobayashi, A. Matsuda, T. Ikoma, and J. Tanaka, Production of aminodisaccharide and chitin or its analogue polysaccharides, JPH Patent Application 093 088, assigned to Shin-Etsu Chemical Co., January 7, 1997.
179. S. Kobayashi, *Proceedings of the Japan Academy, Series B*, Vol. 83, p. 215, 2007.
180. M. Colon, J.T. Davis, J.R. Rasmussen, M. Borowski, B.Y. Wan, and S. Hirani, Oligosaccharide oxazolines, oligosaccharide conjugates and methods of preparation thereof, US Patent 5 241 072, assigned to Genzyne Corporation, August 31, 1993.
181. J.-I. Kadokawa, K. Hirohama, S. Mine, T. Kato, and K. Yamamoto, *Journal of Polymers and the Environment*, Vol. 20, p. 37, March 2012.
182. S. Noguchi, K. Sato, K. Yamamoto, and J.-I. Kadokawa, *International Journal of Biological Macromolecules*, Vol. 126, p. 187, 2019.
183. T.A.P. Hai and R. Sugimoto, *Applied Surface Science*, Vol. 434, p. 188, 2018.
184. T.A.P. Hai and R. Sugimoto, *RSC Advances*, Vol. 8, p. 7005, 2018.

6

Other Polymer Types

6.1 Terpenes

Terpenes, terpenoids, and rosin are being widely used as renewable chemicals; however, for a long time they were ignored as biomass for the synthesis of green polymers, primarily due to the difficulty of precisely controlling their molecular structures (1).

However, there seems to be a change in judging the situation due to the development of new techniques such as highly efficient post-polymerization modification techniques such as *click chemistry*. The recent state of the field of terpenes has been described in a chapter of a monograph. Also, chemical reactions of these compounds, such as isomerization, oxidation, polymerization, dehydrogenation, and hydrogenation, have been detailed (2).

Terpenes are based on isoprene units. Some elementary terpenes are shown in Figure 6.1.

6.1.1 Grafted Terpene

Rosin esters have been grafted onto poly(caprolactone) (PCL) side chains by a combination of a ring-opening polymerization and click chemistry (3). The rosin moieties can be effectively grafted onto each repeat unit of the caprolactone. This synthesis does not require a sophisticated purification of the raw renewable biomass.

The bulky hydrophenanthrene group of the rosin increases the glass transition temperature of the modified PCL by greater than 100°C (3).

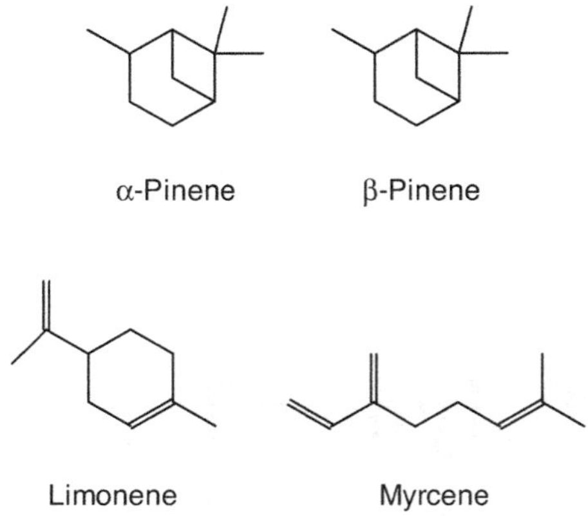

Figure 6.1 Elementary terpenes.

Furthermore, the hydrocarbon nature of rosin structures imparts an excellent hydrophobicity to the PCL with a contact angle close to that of poly(styrene), thus effecting a very low water absorption. The graft copolymers maintain the full degradability and a good biocompatibility.

6.1.2 *Thiol-ene Additions*

The addition of thiols to terpenes, e.g., limonene and pinene, has been studied (4). This is a simple approach to synthesize a wide range of alcohol or ester functionalized renewable monomers. The endocyclic and exocyclic double bonds of the limonene monomers exhibit different reactivities. Monoaddition or diaddition products can be obtained by varying the thiol feed ratio. Similarly, pinene derivates can be prepared.

Long-chain diesters and diols have been synthesized using castor oil (4). These are suitable comonomers that give polycondensates number average molecular weights of up to 25 $k\,D$.

The synthesis of five-membered cyclic carbonates from bio-based limonene oxide and CO_2 catalyzed by commercially available inexpensive, tetrabutylammonium halides has been described (5).

The cycloaddition of CO_2 with a commercial limonene oxide mixture of cis/trans-isomers (40:60) is highly stereoselective. The trans-isomer exhibits a considerably higher conversion than the cis-isomer. Therefore, a stereoselective method of (R)-(+)-limonene epoxidation was performed to achieve a significantly higher yield of the trans-isomer (87 ± 2%) than cis-isomer, which leads to high conversion and yield of the corresponding cyclic carbonates. The epoxidation of (R)-(+)-limonene is shown in Figure 6.2

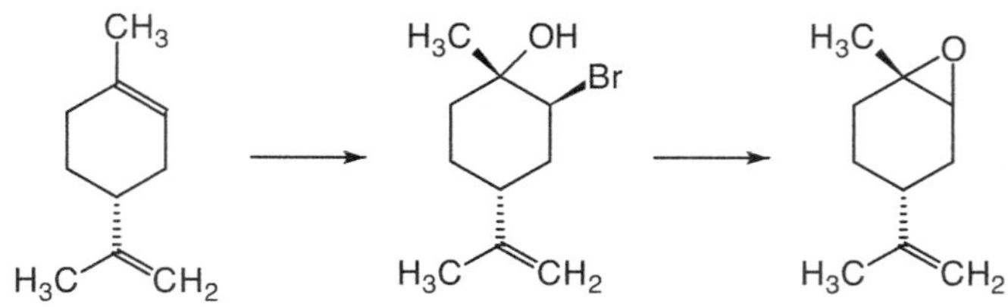

Figure 6.2 Epoxidation of (R)-(+)-limonene (5).

The catalytic effect of halide anions and the influence of operational reaction parameters such as temperature, pressure, and catalyst amount were studied. A high conversion (87%) was obtained after 20 *h* at 120°C, 40 *bar* CO_2, using 6 *mol*% tetrabutylammonium chloride as catalyst.

A detailed study of the reaction kinetics revealed the reaction to be first-order in epoxide (limonene oxide), CO_2 and catalyst concentrations. The activation energy of the reaction was calculated to be 64 $kJ\,mol^{-1}$. A high positive value of the Gibbs free energy and the negative value of the activation entropy obtained indicated that the reaction was endergonic and kinetically controlled in nature (5).

6.1.3 *Pinenes*

Pinenes are a group of monoterpenes containing a double bond (6). They are very suitable renewable building blocks for a variety of sustainable polymers and materials. Their abundance from mainly non-edible parts of plants, as well as the feasibility of isolating them, render these molecules unique amongst the variety of biomass that is utilizable for novel materials. Accordingly, their use for the synthesis of bio-based polymers has been investigated intensively, and strong progress has been made in the last few years (6).

Pinenes are very interesting, because they can be easily isolated by steam-distillation from the resins of coniferous trees. By isomerization into the α- form and optionally some further functionalization, these modified pinenes may undergo a common cationic polymerization (7).

Also, β-pinene may undergo a cationic polymerization. The mechanism of the cationic polymerization of β-pinene is shown in Figure 6.3.

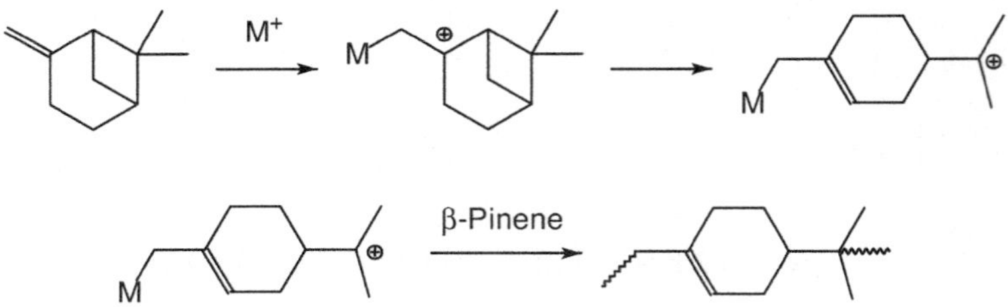

Figure 6.3 Cationic polymerization of β-pinene (1).

The cationic polymerization of β-pinene is initiated with a hydrogen chloride 2-chloroethyl vinyl ether adduct and isopropoxytitanium trichloride together with tetra-n-butylammonium chloride at a temperature below −40°C (8).

These catalysts deliver a rather low molecular weight polymer. High molecular weight polymers can be obtained with special catalyst systems such as diethyl aluminum dichloride (9).

Also, a Schiff-based nickel complex is suitable for the cationic polymerization of β-pinene (10). This nickel complex is shown in Figure 6.4.

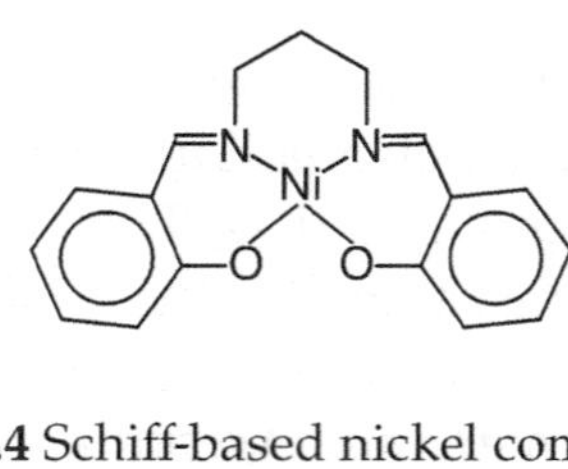

Figure 6.4 Schiff-based nickel complex (10).

A series of copolymers of β-pinene have been described using isobutylene as comonomer (11). Random copolymers of β-pinene and isobutylene are of interest because a wide variety of physical properties may be tailored, since isobutylene is a soft monomeric unit, whereas β-pinene contributes as hard units. It has been demonstrated that perfectly random copolymers of β-pinene and isobutylene with narrow molecular weight distributions can be synthesized by the living cationic copolymerization.

There have been reports in which β-pinene was copolymerized by a radical copolymerization, reversible addition-fragmentation chain transfer (RAFT). As comonomers, methyl acrylate or *n*-butyl acrylate have been used (12, 13).

In comparison to other controlled radical polymerization techniques, RAFT is more tolerant to a wide range of monomers and reaction conditions.

6.1.3.1 *Limonene*

The unsaturated double bonds in limonene can be epoxidized (14). Carbon dioxide and limonene oxide can be used to get a cyclic limonene dicarbonate. The reaction is shown in Figure 6.5. This compound is useful to produce linear poly(carbonate)s or poly(urethane)s (PUs).

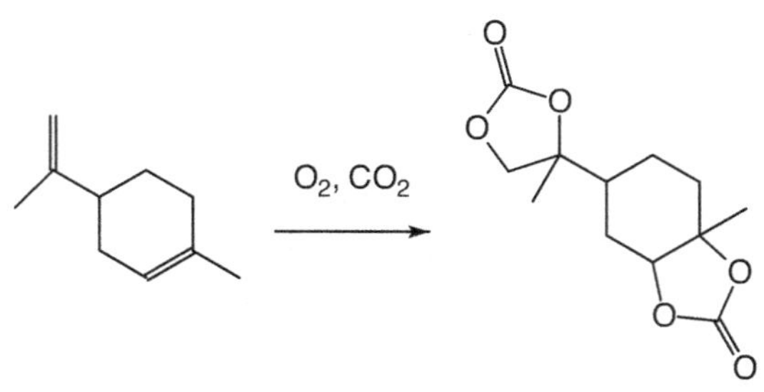

Figure 6.5 Synthesis of limonene dicarbonate.

In contrast to conventional plant oil-based cyclic carbonates, terpene-based cyclic carbonates achieve a much higher CO_2 fixation and do not contain ester groups. Actually, for PU formation, the absence of ester groups is essential to prevent side reactions during amine curing, such as ester cleavage and amide formation. Such reaction would impair the network formation and cause the emission of low molecular weight polyols. The latter compounds could result in a plasticization of the PUs.

Limonene oxide may also undergo an alternating copolymerization with carbon dioxide. As catalysts, β-diiminate zinc acetate compounds are used (15). The reaction is shown in Figure 6.6.

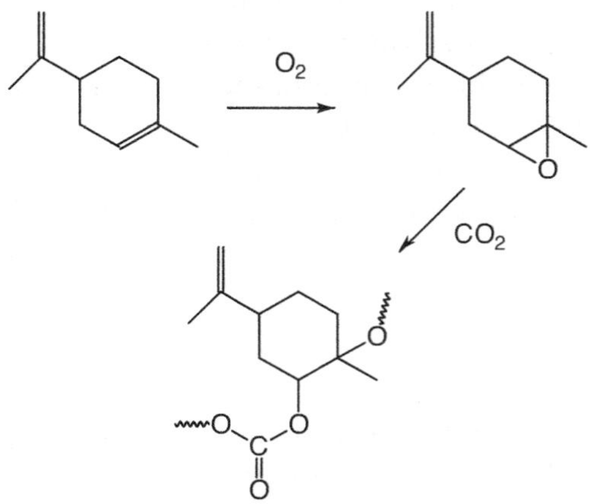

Figure 6.6 Alternating copolymerization limonene oxide with carbon dioxide (15).

6.1.3.2 Myrcene

3-Methylenecyclopentene can be produced by a ring-closing metathesis of myrcene. From this monomer, a high molecular weight polymer could be obtained by the combination of a ring-closing metathesis reaction and a cationic polymerization (16). The synthesis is shown in Figure 6.7.

Elastomers based on biological monomers can be obtained by the stereoselective coordination polymerization of β-myrcene (17). The combination of a neodymium pre-catalyst with triisobutyl aluminum in the presence of a boron activator results in a good yield of polymer, but there is evidence for crosslinking since the material displays a low solubility.

1,4-Poly(myrcene) was synthesized by anionic polymerization of β-myrcene and was subjected to photochemical functionalization

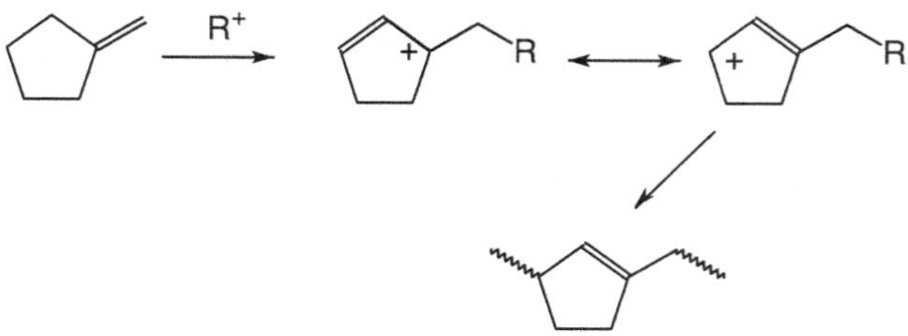

Figure 6.7 Synthesis of poly(3-methylenecyclopentene) (16).

with various thiols, i.e., methyl thioglycolate, methyl 3-mercaptopropionate, butyl 3-mercaptopropionate, ethyl 2-mercaptopropionate and 2-methyl-2-propanethiol, c.f. Figure 6.8, using benzophenone/UV light as the radical source (18).

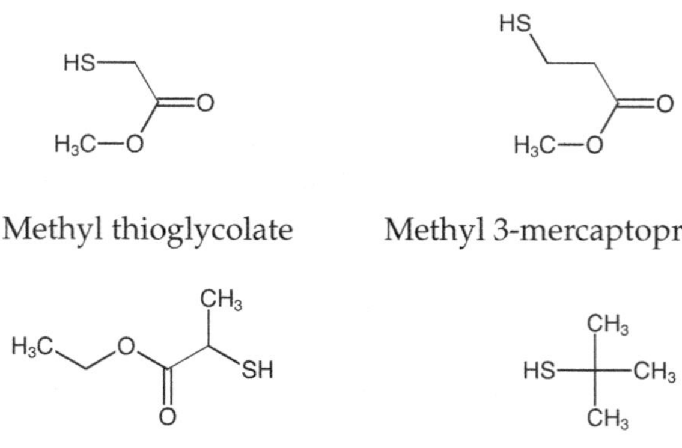

Methyl thioglycolate   Methyl 3-mercaptopropionate

Ethyl 2-mercaptopropionate   2-Methyl-2-propanethiol

Figure 6.8 Thiols.

The yield of thiol addition to the trisubstituted double bonds of 1,4-poly(myrcene) decreased in the order 1° thiol (ca 95%) > 2° thiol (ca 80%) > 3° thiol (<5%), due to the reversibility of the thiol-ene reaction (18). Remarkably, thiol addition to the side chain double bonds was 8–10 times (1° thiol) or 24 times (2° thiol) faster than to the main chain double bonds, which can be explained by the different accessibility of the double bonds and steric hindrance.

Despite the use of a 10-fold excess of thiol with respect to myrcene units, the thiol-ene addition was accompanied by chain coupling reactions, which in the extreme case of 3° thiol (or in the absence of thiol) resulted in the formation of an insoluble crosslinked material. As an example, a methyl-thioglycolate-functionalized 1,4-poly(myrcene) was saponified/crosslinked to give submicron polyelectrolyte particles in dilute alkaline solution (18).

6.2 Poly(urethane)s

Commercially available PUs are synthesized by the polyaddition of diisocyanates with polyols and low molecular weight chain extenders (19). A new approach to poly(urethane)s synthesis could be realized via non-isocyanate routes. Negative impacts of petroleum-based chemicals on the environment and human health, as well as gradual reduction of fuel-based resources, which leads to an increase in their prices, are an incentive to look for bio-based components derived from renewable resources for the synthesis of PUs.

The possibilities have been reviewed of using renewable natural resources as green intermediates to obtain non-isocyanate poly(urethane)s by polyaddition of bis(cyclic carbonate)s and di- or polyamines. The possibility of using carbon dioxide, plant oils, fatty acids and biomass-derived platform chemicals in non-isocyanate poly(urethane) synthesis has been exhaustively commented in a review (19). Also, the use of vegetable oils for the preparation of PUs has been reviewed (20).

The preparation of PU foam directly from various kinds of industrial lignin and molasses, which has scarcely been utilized, has been described (21). A one-step reaction using hydroxyl groups of plant materials as an active site makes it possible to produce a wide variety of PUs such as foams, sheets, gels and composite matrix. By controlling the conditions of reaction and the combination of plant components, mechanical and thermal properties can be varied in a wide range. Furthermore, industrial products using PUs derived from lignin and saccharides are competitive with those from petroleum in both functional properties and production price (21).

Bio-based polyols with high functionality have been synthesized by ring-opening epoxidized sucrose esters of soybean oil with methanol under acidic conditions, and were subsequently formulated with blocked isocyanates to form one package PUs (22). Bio-based PU coatings were prepared by formulating the polyols with blocked poly(isocyanate)s based on isophorone diisocyanate and hexamethylene diisocyanate (HDI).

The coatings can be applied to steel substrates. A commercial soybean oil polyol was used and was found to have a lower modulus and a lower crosslink density and glass transition temperature compared to coatings formulated with sucrose soyate polyols (22).

Naturally occurring oleic acid and undecylenic acid derivatives can be used to synthesize semicrystalline PUs via an AB-type polycondensation. The AB-type monomers were self-polycondensed in two ways (23):

1. The hydroxyl-acyl azide approach, and
2. The melt transurethane method.

Reasonably high molar mass PUs can be obtained. These PUs exhibit a fair thermal stability with no significant weight loss below 200°C. PUs based on undecylenic acid show a thermoplastic behavior with well-identified melting transitions. PUs prepared from undecylenic acid using different methods show different glass and melting transitions.

Two glass transition temperatures are ascribed due to soft and hard segments in the case of PUs synthesized from oleic acid. Thus, a phase-separated morphology is suggested (23).

A bio-based non-isocyanate urethane could be obtained by the reaction of a cyclic carbonate synthesized from a modified linseed oil and an alkylated phenolic polyamine from cashew nut shell liquid (24). The incorporation of the functional cyclic carbonate moieties into the triglyceride units of the oil was done by reacting epoxidized linseed oil with carbon dioxide in the presence of a catalyst. The reaction is shown schematically in Figure 6.9.

6.2.1 *Poly(ester urethane)s*

An aliphatic poly(ester urethane) based on poly(lactic acid) (PLA) and poly(ethylene succinate) has been prepared by a chain extension reaction of a PLA diol and a poly(ethylene succinate) diol with HDI (25).

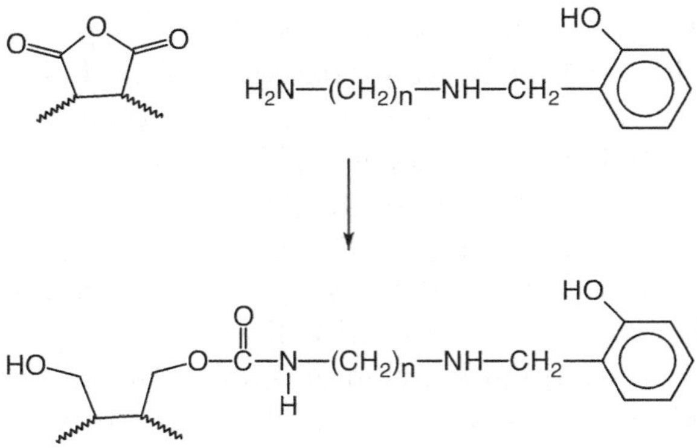

Figure 6.9 Reaction of a cyclic carbonate with amine (24).

The end-capped polymeric diols were obtained by the direct polycondensation of *l*-lactic acid in the presence of 1,4-butanediol and by the condensation of succinic acid with excessive ethylene glycol, respectively. Polymers with high molecular weights of more than 200 *k D* were obtained. The poly(ester urethane) exhibits excellent tensile properties (25).

The radical addition of 2-mercaptoethanol into oleic acid from rapeseed oil, has been achieved by the photochemical generation of radicals without any photoinitiator (26). The 2-mercaptoethanol is photochemically grafted onto vinyl groups of rapeseed oil by a thiol-ene addition to yield polyols. The higher the content of polyunsaturated fatty chains, the lower the rate of 2-mercaptoethanol grafting was found.

Eventually, the functionalized rapeseed oil was used to synthesize PUs with 1,6-hexamethylene diisocyanate and methylene diphenyl-4,4´-diisocyanate. The thermal properties of elastomeric products were found to be similar to those from a commercial product, Desmophen® 1150 (26).

6.3 Cationic Lipopolymers

Cationic lipopolymers may be prepared by reacting a diacrylate monomer with a poly(ethyleneimine), as shown in Figure 6.10. The

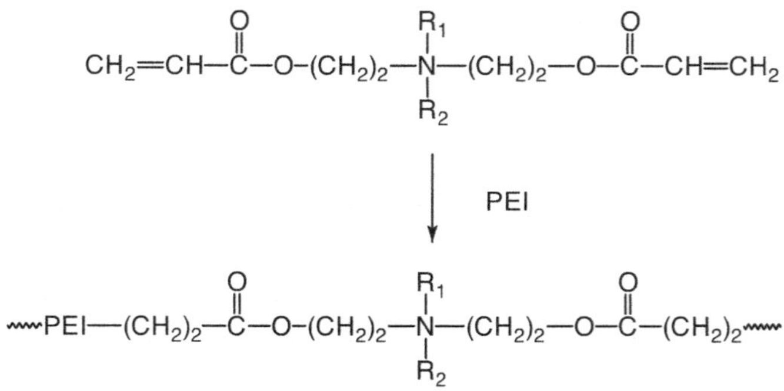

Figure 6.10 Cationic lipopolymers (27).

reaction between the poly(ethyleneimine) and the diacrylate(III) runs via a Michael reaction (27).

A wide variety of polymers may be made by varying the molecular weight and structure of the poly(ethyleneimine), the size and type of the pendant groups on the diacrylate, and the ratio of diacrylate to the poly(ethyleneimine). In addition, mixtures of different diacrylates and different poly(ethyleneimine)s may be used (27).

The cationic lipopolymer is biodegradable, the polymers may form complexes with biomolecules and thus are useful as carriers for the delivery of biomolecules to cells. Examples of biomolecules that form complexes with cationic lipopolymers include nucleic acids, proteins, peptides, lipids, and carbohydrates.

6.4 Plastics from Bacteria

A wide variety of microorganisms are known that can produce interesting products for natural polymers. These are mostly poly(β-hydroxybutyrate), but also copolymers with different alkyl moieties at the β-position.

6.4.1 *Poly(β-hydroxyalkanoate)s*

Poly(β-hydroxyalkanoate)s (PHAs) are of general interest for their use as natural, biodegradable, and biocompatible plastics for applications such as surgical sutures or packaging containers (28).

For industrial applications, the controlled incorporation of repeating units with different chain lengths into other copolymers is desirable. Thus, polyesters can be produced with tailored properties.

Such copolymers can be produced under controlled growth conditions as a defined mixture of substrates for a certain type of microorganism. In this way, a well-defined and reproducible copolymer will be formed (28). PHAs can be produced by both wild-type and recombinant bacteria. The production of medium-chain-length PHAs occurs in *Pseudomonas* (29).

Since a wide variety of bacteria accumulate PHAs as intracellular storage material, this natural polymer is considered a potential substitute for petrochemical plastics (30). Medium-chain-length types are semicrystalline elastomers with a low melting point, low tensile strength and high elongation to break. Therefore, these materials can be used as a biodegradable rubber.

Tropical mangrove and marine ecosystems from India have been screened for promising bacteria, that have the capability of accumulating high amounts of PHA (31). From 866 bacterial cultures isolated from this region, 337 cultures were scored positive for the PHA production. Amongst these isolates, seven cultures accumulated more than 1 $g\,l^{-1}$ in the culture broth. The thin-layer chromatograms of the methyl esters of PHA from the seven cultures showed varied profiles. The sediment samples also showed the presence of PHA with five different monomeric units.

Other microbial plastics are polymers from lactic acid, succinic acid, poly(ethylene) ethylene from ethanol and its polymer poly(ethylene), 1,3-propanediol, as well as poly(*p*-phenylene) (29).

An overview of the development and production of various important natural PHAs has been presented (32). This will give an insight into the recent advances and provide indicators for the future direction of engineering microorganisms as plastic cell factories. On the basis of our current understanding of PHA biosynthesis systems, the recent advances in the approaches adopted for strain development in the production of non-natural polyesters have been discussed, notably 2-hydroxycarboxylic acid-containing polymers, with particular reference to systems metabolic engineering strategies (32).

An overview of the main PHAs used in medical applications has been given (33). In this review, it has been demonstrated that the properties of PHAs can be controlled both by varying the concentration of units in the copolymer and the substrate for the PHA production.

Another way of controlling the mechanical properties of PHAs is varying the 3HV content, such as in the case of P(3HB-co-3HV). A higher 3HV content in the structure of this polyester will lead to a lower cristallinity and, therefore, to greater flexibility, strength and elongation at break (33).

6.4.1.1 *Biodegradability of Poly(hydroxyalkanoate)s*

Poly(β-hydroxyalkanoate)s (PHAs) can be degraded by microorganisms that produce extracellular PHA depolymerase. The degradability of PHA films produced by *Halomonas hydrothermalis* was evaluated using Jatropha biodiesel byproduct as a substrate. Poly(3-hydroxybutyrate) (PHB) films were subjected to the degradation in soil and also compared with the corresponding synthetic acrylate polymer with blends of the synthetic acrylate polymer and PHB (34). After 50 *d*, a weight loss of 60% in the PHB film was observed. However, no degradation was observed in the synthetic film.

6.4.1.2 *Poly(β-hydroxyalkanoate)*

Poly(3-hydroxybutyrate)s (PHBs) belong to the family of PHAs (28). These polymers can be formed as intracellular inclusions under unbalanced growth conditions.

The high production cost of PHB, particularly its substrate cost, limits its industrial applications (35). Reducing the substrate cost by utilizing cheaper carbon source enables improving its economic viability. Cane final molasses is considered potential feedstock due to its abundance, low cost, and high sugar content.

However, sucrose in molasses requires pretreatments to convert to its monomers, fructose and glucose, for the bacteria consumption during the fermentation process (35, 36).

The main components of molasses and sugar content in various pretreated molasses have been analyzed. The PHB content and cell dry weight of *Cupriavidus necator* biomass fermented utilizing glucose, untreated molasses and pretreated molasses as carbon sources were compared. The molasses consisted of 34.49 ± 0.13% sucrose, 5.93 ± 0.07% glucose and 5.00 ± 0.04% fructose respectively. The hydrothermal acid pretreated molasses (in an oil bath) contains the highest fructose and glucose content, which are 21.31 ± 0.04% and 20.32 ± 0.06% respectively.

The hydrothermal acid pretreated molasses contained the highest fructose and glucose content and the cell dry weight and the PHB produced by *C. necator* using hydrothermal acid pretreated molasses was the highest at 2.86 ± 0.82 *g* and 27%, respectively (35).

6.5 Bio-based Epoxy Resins

In epoxy materials, some of the conventional diglycidyl ether of bisphenol F has been replaced with functionalized vegetable oils, such as epoxidized linseed oil and epoxidized soybean oil. The materials have been cured using an anhydride-based curing agent.

The modified materials have been demonstrated to form a bio-based epoxy material with a high elastic modulus and a high glass transition temperature. The Izod impact strength and the fracture toughness were significantly improved by the addition of the functionalized vegetable oils as a phase-separated morphology was formed (37).

Vegetable oil-, furan-, lignin-, rosin-, vanillin-, and itaconic acid-derived bio-based epoxy resins are used in various applications, including paints, coatings, adhesives, and biomedicine. The utilization of natural resources for the synthesis of bio-based epoxy and its curing agents have been reviewed (38).

High molecular weight epoxy resins could be obtained by the reaction of epoxidized soybean oil and bisphenol A using a modern and pro-ecological modification of the well-known method of the synthesis of epoxy resins, called the *epoxy fusion process* (39).

The accelerated hydrolytic degradation of bio-based epoxy resins containing ester linkages was investigated (40). Epoxidized bio-based molecules were utilized as sustainable replacements for the diglycidyl ether of bisphenol A as an epoxy monomer, including epoxidized vanillic acid derived from lignin, epoxidized plant-based phenolic acids, such as epoxidized salicylic acid, 4-hydroxybenzoic acid, and epoxidized soybean oil. All bio-based epoxy monomers contain esters in contrast to diglycidyl ether of bisphenol A, which contains no ester. The epoxidized molecules were cured through the reaction with an anhydride curing agent.

The epoxy resins exhibited comparable glass transition temperatures to that of the diglycidyl ether of bisphenol A-based epoxy resin. All the bio-based epoxy resins underwent a rapid degradation in a basic solution as compared to the conventional diglycidyl ether of bisphenol A-based epoxy resin. The epoxidized salicylic acid- and epoxidized soybean oil-based epoxy resins exhibited the fastest degradation rates, whereas 4-hydroxybenzoic acid-based epoxy resins and epoxidized vanillic acid-based epoxy resins exhibited more moderate degradation rates.

These variations in degradation rate were attributed to differences in the epoxide content, the monomer structure, degree of hydrophilicity, crosslink density, and proximity to glass transition temperature (40).

6.5.1 Poloxamers

Poloxamers are block copolymers from ethylene oxide and propylene oxide. Poloxamers are also known as Pluronic® (41). Poloxamers can be subdivided into 3 categories, namely, emulsion forming, micelle forming, and water-soluble poloxamers.

Various factors which determine the poloxamer characteristics and behavior are the molecular weight, the ratio of poly(phenylene oxide) to poly(ethylene oxide), temperature conditions, concentration, and the presence of ionic materials. There is thus a wide range of characteristics in existing commercially available poloxamers which can be exploited in formulating the compositions of the present disclosure, especially where the composition further includes a medicinal agent and is utilized for drug delivery purposes (42).

Particularly useful for these purposes are triblock copolymers of the formula $HO(C_2H_{40})_a(C_3H_{60})_b(C_2H_{40})_cH$, wherein *a* and *c* are independent from 1–150 units and *b* ranges from 10–200 units, with the overall molecular weight ranging from 1–50 *k D*.

A biocompatible, biodegradable polymer can be produced as follows. A poloxamer (Pluronic® F68) (43) is dried *in vacuo* at about 105°C for 14 *h*. Afterwards, glycolide and then ε-caprolactone are added. As a catalyst, stannous octoate is used (42).

The reaction is allowed to proceed at 178°C for some 4 *h*. After completition of the reaction, the bioabsorbable polymer is extruded and allowed to cool for a minimum of 16 *h*. The resulting bioabsorbable terpolymer contains about 40% Pluronic® F68, 51% of caprolactoyl groups, 9% of glycoyl groups, and less than 1% of residual caprolactone monomer (42). The resulting composition may be utilized to form medical devices, drug delivery devices, or coatings for other medical devices.

A rotational rheometer was adopted to study the rheological properties of the mixtures of agarose/poloxamer 407 gels with different concentrations at various frequencies, strain rates and temperatures (44).

The addition of poloxamer 407 decreased the gel stiffness by an order of magnitude. For the combinations in this study, the increase in agarose concentration would increase both the storage modulus and loss modulus of the gel mixtures. The variation in poloxamer 407 concentration (2.5% to 10%) minimally changes the composite moduli. These agarose/poloxamer 407 gel mixtures also exhibited a shear thinning behavior (44).

However, the addition of poloxamer 407 (2.5%–10%) to agarose gel only has a very small effect on the thermomechanical properties of agarose gels. The overall transition temperature for these gel mixtures is governed by poloxamer 407 melting point, where the phase change starts around 55°C and the gels completely collapse at the melting temperature of agarose (44).

6.6 Phosphate-Containing Polymers

Polymers with phosphate linkages include poly(phosphate)s, poly(phosphonate)s, and poly(phosphite)s.

The respective structures of these three classes of compounds, each having a different side chain connected to the phosphorus atom, are shown in Figure 6.11.

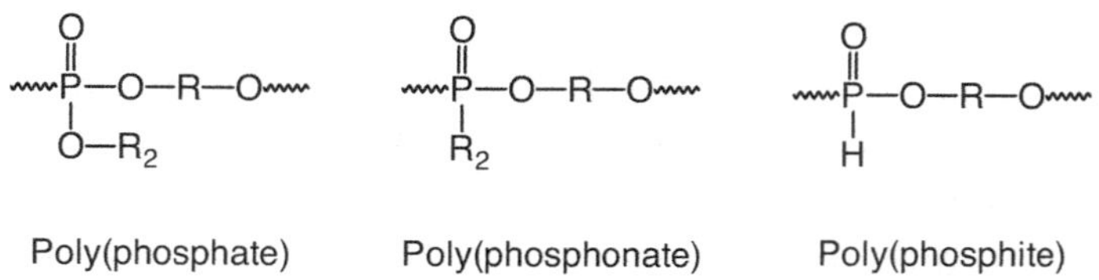

Figure 6.11 Polymers with phosphate linkages.

The versatility of these polymers comes from the versatility of the phosphorus atom, which is known for a multiplicity of reactions. Its bonding can involve the $3p$ orbitals or various $3s - 3p$ hybrids. Also, *spd* hybrids are possible because of the accessible *d* orbitals. Thus, the physicochemical properties of the poly(phosphoester)s can be readily changed by varying either the R or R_2 group.

The P–O–C group in the backbone lowers the glass transition temperature of the polymer and may confer solubility in common organic solvents, which is desirable for easy characterization and processing.

The biodegradability of the polymer is due primarily to the physiologically labile phosphoester bond in the backbone of the polymer. By manipulating the backbone or the side chain, a wide range of biodegradation rates are attainable.

Because phosphorus can be pentavalent, drug molecules or other biologically active substances can be chemically linked to the polymer. For example, drugs with carboxyl groups may be coupled to the phosphorus via an ester bond which is hydrolyzable (45).

The rate of release of the therapeutic agent is dependent on the hydrolytic cleavage of the polymer therapeutic agent conjugate. Alternatively, therapeutic agents containing two hydroxyl groups can be directly incorporated into the backbone of the polymers.

Four different methods have been used for the synthesis of phosphorus-containing polymers, including (46):

1. Bulk polycondensation,
2. Solution polymerization,
3. Interfacial polycondensation, and
4. Ring-opening polymerization.

In melt polycondensation the phosphonic or phosphoric dichloride is mixed with the diol in the absence of solvent. A Lewis acid catalyst, such as $FeCl_3$ or $MgCl_2$, is added and the mixture is heated *in vacuo* or in nitrogen blanket in order to remove the hydrochloric acid that is formed.

These admittedly vigorous conditions may lead to chain acidolysis or hydrolysis reactions. Undesirable side reactions such as crosslinking may take place. On the positive side, melt polycondensation avoids solvents and large amounts of other additives, thus making purification more straightforward. Melt polycondensation can also provide polymers of reasonable molecular weight.

Interfacial polycondensation can yield polymers with high molecular weights at high reaction rates. Since the interfacial technique is a non-equilibrium method, the critical dependence of high molecular weight on exact stoichiometric equivalence between diol and dichloridate inherent in bulk and solution methods is removed. The limitation of this method is the hydrolysis of the acid chloride in the alkaline aqueous phase.

Ring-opening polymerization of phosphorus-containing monomers is particularly useful for the synthesis of high molecular weight polymers (47, 48).

Polymer compositions with pendant 5-fluorouracil have been prepared. 5-Fluorouracil is heated with 1,1,1,3,3,3-hexamethyldisilazane, which results in the formation of 2,4-bis-*o*-trimethylsilyl-5-fluorouracil. The reactants are shown in Figure 6.12 and the reaction is shown in Figure 6.13.

A polymer-drug conjugate is obtained by the reaction of a chlorinated poly(phosphoester) with the 5-fluorouracil derivative in the presence of a stoichiometric amount of pyridine (46).

A mixture of 5-fluorouracil (7 *m mol*) and 1,1,1,3,3,3-hexamethyldisilazane (30 *ml*) was heated at reflux temperature for 20 hours in the presence of a catalytic amount of ammonium sulfate to derivatize the 5-fluorouracil. Evaporation of the mixture under reduced pressure resulted in the formation of 2,4-bis-*o*-trimethylsilyl-5-fluorouracil. To obtain the final polymer-drug conjugate, the chlorinated poly(phosphoester) (5 *g*) in methylene chloride (20 *ml*) was reacted with the 5-fluorouracil derivative in the presence of a stoichiometric amount of pyridine. After stirring for 18 hours at room temperature, 15 *ml* of methanol was added. After evaporation of the solvent,

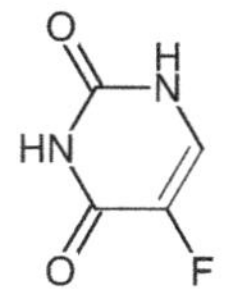

5-Fluorouracil

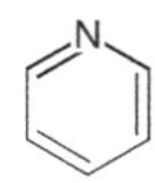

Pyridine

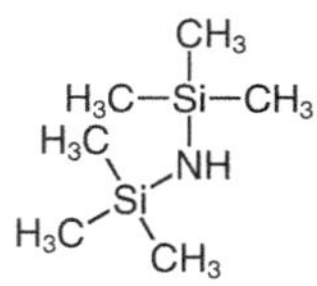

1,1,1,3,3,3-Hexamethyldisilazane

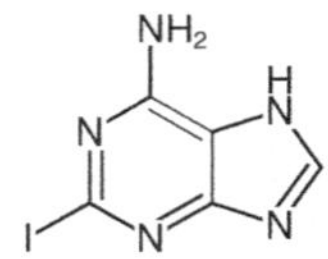

Iodoaminopurine

Figure 6.12 Reactants for polymer-drug conjugate (46).

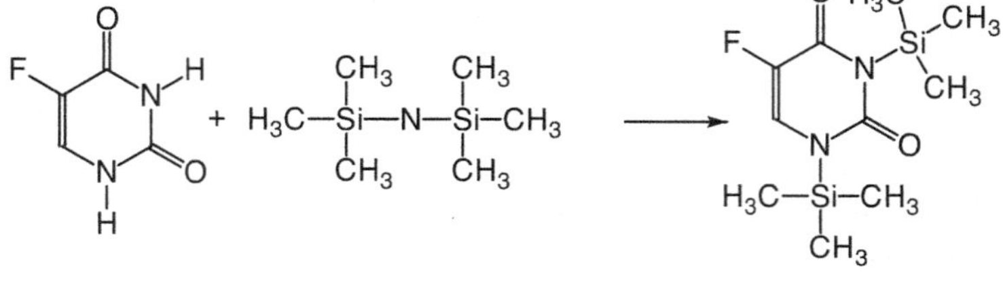

Figure 6.13 Synthesis of a 5-fluorouracil bonded to poly(phosphoester) (46).

the residue was redissolved in dimethyl formamide and repeatedly precipitated into acetone. Linking of iodoaminopurine (IAP) to the polymer can be achieved in a similar manner by taking advantage of the facile reaction between the primary amine of the drug and the chlorine in the side chain of the polymer.

Also, copolymers have been prepared. These copolymers contain both a phosphorous-containing moiety and a phosphor-free moiety, e.g., a lactic acid-based moiety.

Preparation 6–1: 2-Methoxy-2-oxo-1,3,2-dioxaphospholane is synthesized by a modified Edmundson method (45, 49–51). Phosphorus tinchloride is allowed to react with ethylene glycol dry 1,2-dichloroethane. The 2-chloro-1,3,2-dioxaphospholane is oxidized to give 2-chloro-2-oxo-1,3,2-dioxaphospholane. Finally, this intermediate is reacted with methanol to give the desired 2-methoxy-2-oxo-1,3,2-dioxaphospholane. The synthesis is shown in Figure 6.14.

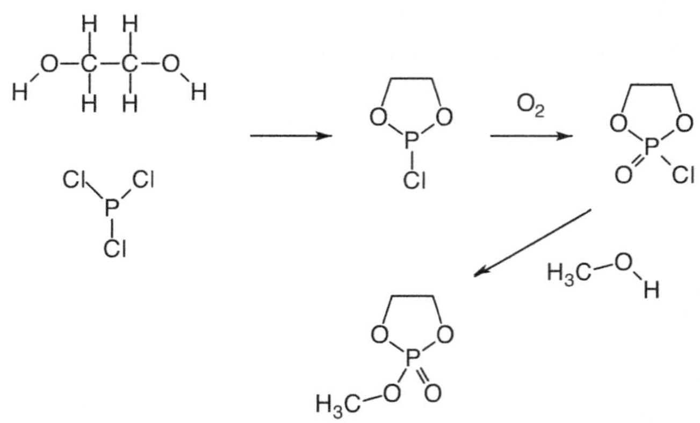

Figure 6.14 Synthesis of 2-methoxy-2-oxo-1,3,2-dioxaphospholane (45).

Furthermore, the synthesis of poly(*D*,*L*-lactide-*co*-ethylene methyl phosphate) is shown in Figure 6.15.

The lactide and the phosphate are copolymerized in bulk under a dry argon atmosphere with aluminum triisopropyloxide as initiator (45). A plot of water absorption of poly(*D*,*L*-lactide-*co*-ethylene methyl phosphate) as a function of ethylene methyl phosphate content is shown in Figure 6.16. Other polymers reported are shown in Table 6.1.

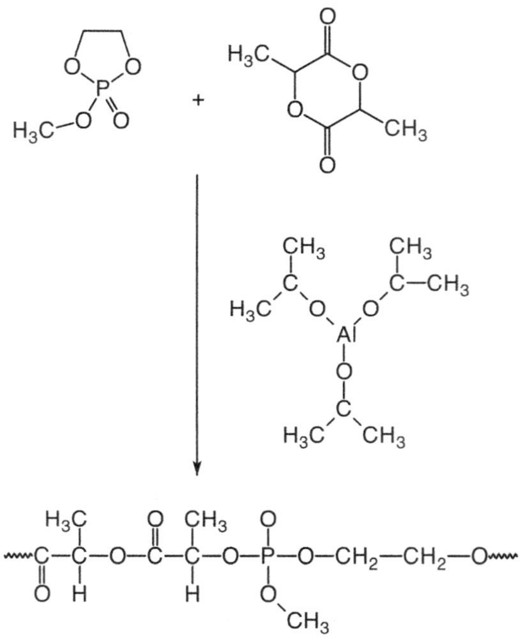

Figure 6.15 Synthesis of poly(*D,L*-lactide-*co*-ethylene methyl phosphate) (45).

Table 6.1 Phosphorus-containing polymers (45).

Polymer
Poly(*D,L*-lactide-*co*-ethylene methyl phosphate)
Poly(*D,L*-lactide-*b*-ethylene methyl phosphate-*b*-*D,L*-lactide)
Poly(*L*-lactide-*b*-ethylene methyl phosphate-*b*-*L*-lactide)
Poly(ε-caprolactone-*co*-ethylene-ethylphosphate)
Poly(lactide-*co*-ethylene ethyl phosphate)

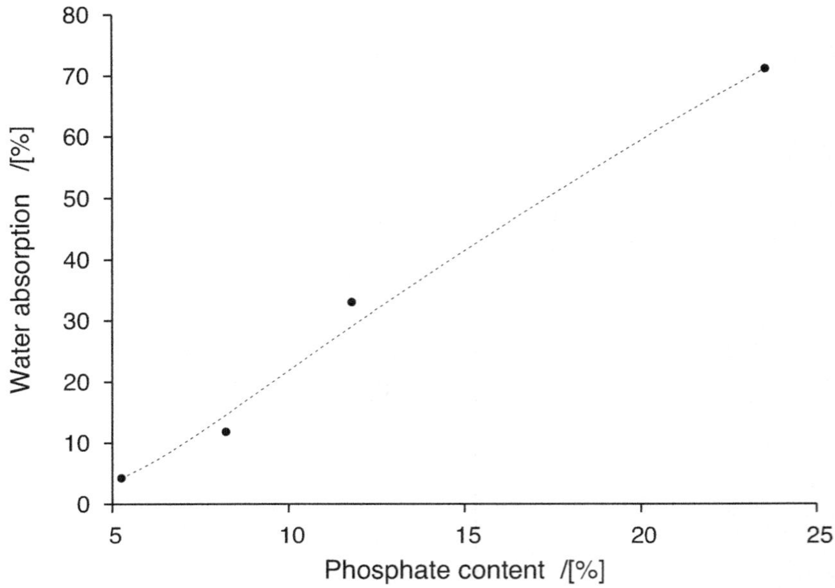

Figure 6.16 Water absorption as a function of ethylene methyl phosphate content (45).

Phosphonate-functionalized poly(β-amino ester) macromers were efficiently synthesized through an aza-Michael addition of various diacrylates, such as 1,6-hexanediol diacrylate, poly(ethylene glycol) diacrylate, 1,4-butanediol diacrylate, 1,6-hexanediol ethoxylate diacrylate, and triethylene glycol diacrylate, with a phosphonate-containing primary amine (diethyl 2-aminoethylphosphonate) without any catalyst (52). Some of these compounds are shown in Figure 6.17.

The replacement of diethyl 2-aminoethylphosphonate with propyl amine served as control. The macromers, whose molecular weight is ca. 1000–4000 *Da*, as confirmed by both gel permeation chromatography and 1H-nuclear magnetic resonance spectroscopy, are photopolymerized to give biodegradable gels.

The degradation behavior and cell interaction of these gels were studied. The degradation rates of the gels can be varied by the choice of the starting acrylates and the acrylate:amine ratio. Furthermore, the gels showed a slightly higher degradability than the propyl amine-based analogs (the controls). Except for the triethylene glycol diacrylate and poly(ethylene glycol) diacrylate-based ones, all

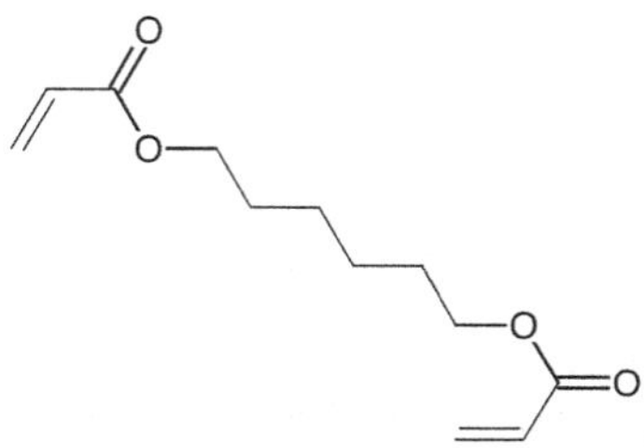

1,6-Hexanediol diacrylate

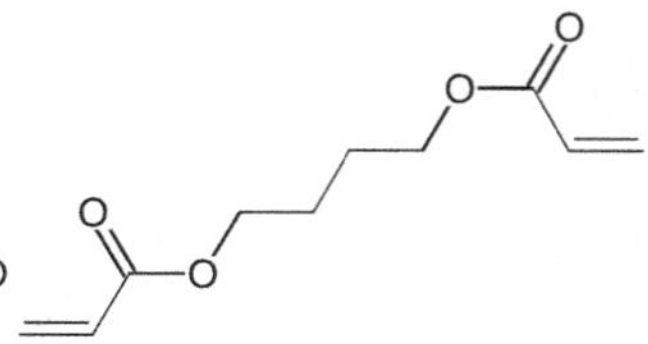

1,4-Butanediol diacrylate

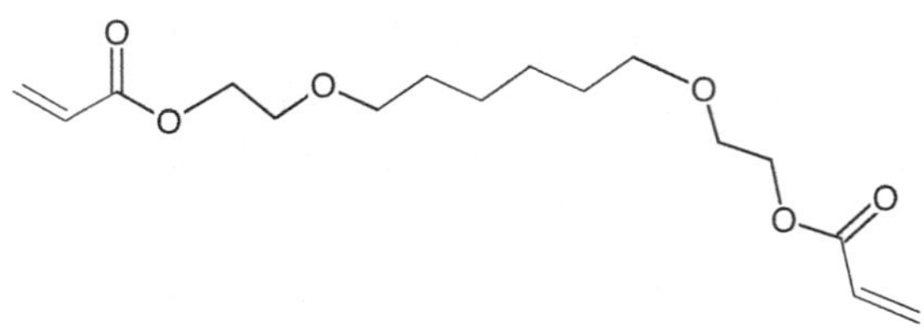

1,6-Hexanediol ethoxylate diacrylate

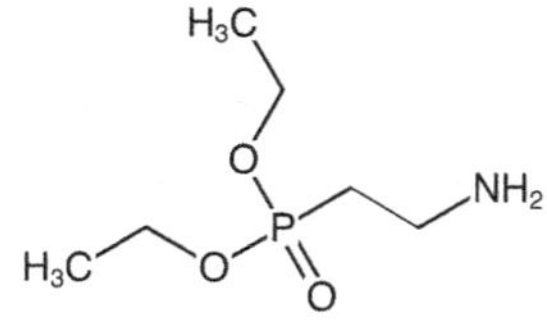

Diethyl 2-aminoethylphosphonate

Figure 6.17 Acrylate and amino compounds.

phosphonate-functionalized poly(β-amino ester) gels supported the attachment of a larger number of SaOS-2 cells than nonphosphonated ones, and the best film was found to be the one based on 1,6-hexanediol ethoxylate diacrylate, diethyl 2-aminoethylphosphonate with balanced hydrophilicity. Degradation products of these films have no significant cytotoxicity except for 1,6-hexanediol diacrylate-propyl amine (52).

6.7 Polyketals

Polyketals can be used for treating human patients. The treatment consists of administering in the form of soluble polymers, colloids, particles, gels, solid items, fibers, films, etc. Methods have been disclosed for the treatment of lung cancer (53). The lung cancer can be small cell carcinoma of the lung or non-small cell carcinoma of the lung. The methods include locally administering to the subject a therapeutically effective amount of the polyketal particle comprising a CpG oligodexoynucleotide. Optionally, the polyketal particle can include an imidazoquinoline compound (53).

Biodegradable biocompatible polyketals can be used as drug carriers or preparations for low-invasive surgical procedures. The formulations may be injectable or implantable (54).

Carbohydrate rings of a suitable polysaccharide can be oxidized by glycol-specific reagents, resulting in the cleavage of the carbon-carbon bonds between carbon atoms that are each connected to a hydroxyl group. In Figure 6.18, the oxidative cleavage of a polyfructose is shown.

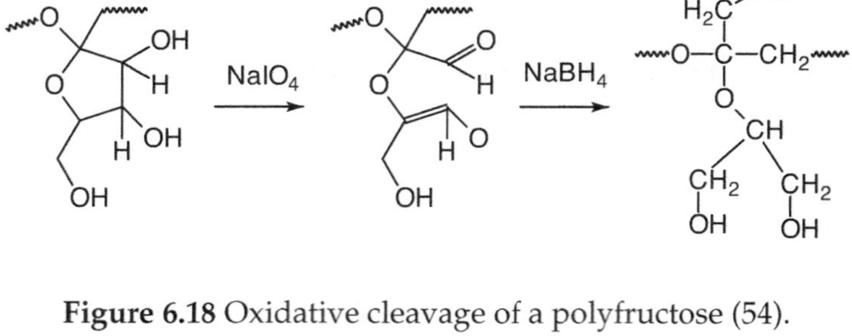

Figure 6.18 Oxidative cleavage of a polyfructose (54).

Particularly preferred polysaccharides are inulin, levans from plants, and bacterial fructans. Suitable glycol-specific oxidizing agents include sodium periodate or lead tetra acetate. Examples of reducing agents include sodium borohydride and sodium cyanoborohydride (54).

Alternatively, a polyketal may be formed by the ring-opening reaction of a monomeric cyclic ketal, as shown in Figure 6.19.

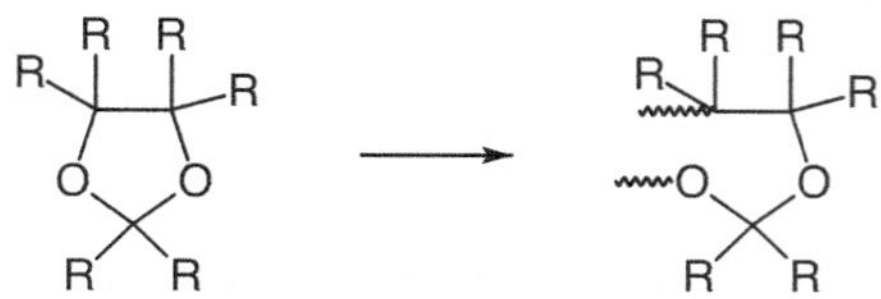

Figure 6.19 Ring-opening of a cyclic ketal (54).

The polyketals can be readily derivatized. For example, aldehyde groups of an intermediate product of a polysaccharide oxidation can be converted into alcohol groups, amines, thioacetals, carboxylic acids, amides, esters, or thioesters.

Examples of pharmaceutically useful groups include hydrophilicity and hydrophobicity modifiers, pharmacokinetic modifiers, antigens, receptor ligands, nucleotides, chemotherapeutic agents, antibacterial agents, antiviral agents, immunomodulators, hormones and their analogs, enzymes, inhibitors, alkaloids, and therapeutic radionuclides.

Suitable chemotherapeutic compounds are topoisomerase inhibitors, anthracyclines, doxorubicin, cisplastin, carboplatin, vincristine, mitromycine, taxol, camptothecin, ribozymes, dactinomycines, antisense oligonucleotides, or therapeutic radionuclides (54).

l-Asparaginase is an amidohydrolase used as a chemotherapeutic agent for the treatment of acute lymphoblastic leukemia (55). The nanoencapsulation of this enzyme is strategic to avoid its immediate immunogenic effects that lead to a decrease in the enzyme half-life.

l-Asparaginase-containing nanoparticles were prepared by double emulsification, through an ultrasonic sonicator or an Ultra-Turrax, using two copolymers of 50:50 (w/w) poly(lactic-*co*-glycolic acid) (PLGA) with different ranges of molecular weight

(24–38 *kDa* and 30–60 *kDa*) and varying the concentration of poly(vinyl alcohol) (PVA) as a stabilizer (0.5, 1.0, 1.5 and 2.0%) as well as the emulsification time (30 *s* and 60 *s*). Using 24–38 *kDa* PLGA and 1.0% PVA, cavitation nanoparticles were obtained with a hydrodynamic diameter of 384 *nm*, a polydispersity index of 0.143 and a zeta potential of −16.4 *mV*, whose *l*-asparaginase encapsulation efficiency was as high as 87±2%.

The encapsulated enzyme showed an activity 22% higher than that of the free enzyme, and no conformational changes were detected by circular dichroism. The enzyme release from nanoparticles entrapped in dialysis bags (500 *kDa* molecular weight cut-off) allowed selecting a controlled system able to release about 60% of the enzyme within 14 *d*. Here, the Korsmeyer-Peppas model showed the best correlation of $R^2 = 0.966$ (55).

6.8 Biorubber

An elastomeric biodegradable polymer is also called biorubber. The elasticity of biorubber recommends it for use in regenerating a variety of tissues. Other uses of biorubbers are summarized in Table 6.2.

Table 6.2 Uses of biorubbers (56).

Field of Application of Biorubbers
Tissue engineering
Nerve regeneration
Mimicking blood vessels
Drug release applications
Coating of metallic stents
Surgical glue
In-vivo sensors and catheters
Diapers
Base for chewing gums

A depolymerized carboxyl alkyl guar with a degree of substitution of up to 3 carboxylic groups per sugar unit can be used as a scale inhibitor in oilfield applications (57). This is a biodegradable scale inhibitor which is effective under severe scaling conditions.

6.9 Collagen

Biodegradable polymers have significant potential in various fields of medicine, such as in (56):

- Tissue engineering,
- Drug delivery, and
- *In-vivo* sensing.

Many biomedical devices are implanted in a mechanically dynamic environment in the body, which requires the implants to sustain against and recover from various deformations without mechanical irritation of the surrounding tissue.

In many cases, the matrices and scaffolds of these implants would ideally be made of biodegradable polymers. These polymers should mimic the functions of the extracellular matrix (ECM). This is a soft, tough and elastomeric proteinaceous network that provides mechanical stability and structural integrity to tissues and organs.

Therefore, an elastomeric biodegradable polymer that readily recovers from relatively large deformations is advantageous for maintaining the proper function of the implant.

Collagen forms fibrous networks that reinforce tissues and provide an extracellular matrix for cells (58). These networks exhibit remarkable strain-stiffening properties that tailor the mechanical functions of tissues and regulate cell behavior.

Recent models can explain this nonlinear behavior as an intrinsic feature of disordered networks of stiff fibers. This theoretical framework was evaluated by measuring the elastic properties of collagen networks over a wide range of self-assembly conditions (58). The model allowed a quantitative relation of both the linear and nonlinear elastic behavior of collagen networks to their underlying architecture.

The network elastic response reveals that the the local coordination number decreases from 3.5 to 3 as the polymerization temperature is raised from 26°C to 37°C, while being weakly dependent on concentration. Furthermore, a Young's modulus of 1.1 *MPa* was found for the collagen fibrils from the linear modulus (58).

Scanning electron microscopy technology could confirm that the local coordination number is between 3 and 4 but is unable to detect the subtle changes in the local coordination number with polymerization conditions. The initial stress-stiffening response of collagen networks was found to be controlled by the negative normal stress that builds up under shear (58).

Collagen and elastin are the major fibrous protein components of ECM. The modulus of collagen is 1.2 *G Pa* and the modulus of elastin is 410 *k Pa*. Collagen provides mechanical strength to ECM, while elastin adds rubber-like elasticity to ECM.

Unfortunately, most biodegradable polymers are not elastomeric. For example, a lactide glycolide copolymer has a modulus of 2 *G Pa* and a maximum elongation of about 2–10%.

Often, biodegradable polymers require surface modification for wettability and cell attachment and are subject to fibrous encapsulation (56). Also, collagen and derived peptides have been described as nutritional supplements for integumental care (59).

6.10 Pyridinium Modified Polymers

Biodegradable polymers can be obtained by introducing pyridinium groups, such as *N*-alkylpyridinium groups or *N*-arylpyridinium groups, into the backbone of a synthetic vinyl polymer and subsequently quaternizing the pyridinium groups with chlorine or bromine ions (60). The copolymer is synthesized from monomers shown in Table 6.3.

Table 6.3 Comonomers for biodegradable copolymers (60).

Monomer A	Monomer B
Methyl methacrylate	4-Vinylpyridine
Acrylonitrile	4-Vinylpyridine
Methyl acrylate 4-Vinylpyridine	

The polymerization is guided as a radical polymerization using dibenzoylperoxide, 2,2′-azobisisobutyronitrile or ammonium persulfate as initiator. The backbone can be modified by the reaction of 4-vinylpyridine with benzyl chloride to get *N*-benzyl-4-vinylpyridine monomers (60).

The biodegradable polymers have high compatibility with microbial cells so that articles molded or otherwise formed from the biodegradable polymers can be easily decomposed by activated sludge or under conditions simulating burial in the ground (60).

6.11 Commercial Biodegradable Polymers

In response to the demand for biopolymers, a number of new biopolymers have been developed which have been shown to biodegrade when discarded in the environment. Commercial biopolymers from various manufacturers are shown in Table 6.4.

Table 6.4 Commercial biodegradable polymers (61–63).

Compounds	Manufacturer	Trade name
Aliphatic-aromatic copolymer	BASF	Ecoflex®
Aliphatic-aromatic copolymer	Eastman Chemical	Eastar Bio®
Poly(ester amide)	Bayer	BAK
Poly(ester amide)	BASF	P1
Modified poly(ethylene terephthalate)	DuPont	Biomax®
Poly(lactic acid)	NatureWorks LLC	Ingeo™
Poly(lactic acid)	Total Corbion	Luminy®
Poly(hydroxyalkanoate)	Monsanto	
Poly(caprolactone)	Union Carbide	Tone®
Poly(caprolactone)	Ingevity	Capa®
Succinate-based polyesters	Showa Highpolymer	Bionolle®

Each of the biopolymers listed in Table 6.4 has unique properties, benefits and weaknesses.

Trade names appearing in the references are shown in Table 6.5.

Table 6.4 (cont) Commercial biodegradable polymers (61–63).

Compounds	Manufacturer	Trade name
Starch-based	DuPont	Bio-PDO™
Starch-based	Novamont	Mater-Bi®
Starch-based	StarchTech	ReNEW®
PHB and poly(hydroxyvalerate)	Goodfellow Corp.	Goodfellow PHB Biopolymer (PHB)
	Rodenburg Biopolymers B.V.	Solanyl®
Aliphatic copolyester	Procter & Gamble	Nodax™
Poly(butylene succinate-*co*-butylene adipate)	Showa Denko K.K.	Bionolle™ 3000
Poly(butylene succinate-*co*-butylene adipate)	SK Chemicals	Skygreen™

Table 6.5 Trade names in references.

Trade name Description	Supplier
Bio-PDO™ 1,3-Propanediol from biodegradable polymer (62)	DuPont

References

1. P.A. Wilbon, F. Chu, and C. Tang, *Macromolecular Rapid Communications*, Vol. 34, p. 8, January 2013.
2. G. Rubulotta and E.A. Quadrelli, Terpenes: A valuable family of compounds for the production of fine chemicals in S. Albonetti, S. Perathoner, and E.A. Quadrelli, eds., *Horizons in Sustainable Industrial Chemistry and Catalysis*, Vol. 178 of *Studies in Surface Science and Catalysis*, chapter 11, pp. 215–229. Elsevier, 2019.
3. K. Yao, J. Wang, W. Zhang, J.S. Lee, C. Wang, F. Chu, X. He, and C. Tang, *Biomacromolecules*, Vol. 12, p. 2171, June 2011.
4. M. Firdaus, L. Montero de Espinosa, and M.A.R. Meier, *Macromolecules*, Vol. 44, p. 7253, September 2011.
5. A. Rehman, A.M.L. Fernández, M.F.M.G. Resul, and A. Harvey, *Journal of CO2 Utilization*, Vol. 29, p. 126, 2019.
6. M. Winnacker, *Angewandte Chemie International Edition*, Vol. 57, p. 14362, 2018.
7. A. Corma, S. Iborra, and A. Velty, *Chemical Reviews*, Vol. 107, p. 2411, June 2007.
8. J. Lu, M. Kamigaito, M. Sawamoto, T. Higashimura, and Y.-X. Deng, *Macromolecules*, Vol. 30, p. 22, January 1997.
9. K. Satoh, H. Sugiyama, and M. Kamigaito, *Green Chemistry*, Vol. 8, p. 878, 2006.
10. P. Yu, A.-L. Li, H. Liang, and J. Lu, *Journal of Polymer Science Part A: Polymer Chemistry*, Vol. 45, p. 3739, August 2007.
11. A.-L. Li, W. Zhang, H. Liang, and J. Lu, *Polymer*, Vol. 45, p. 6533, September 2004.
12. Y. Wang, A.-L. Li, H. Liang, and J. Lu, *European Polymer Journal*, Vol. 42, p. 2695, October 2006.
13. A.-L. Li, X.-Y. Wang, H. Liang, and J. Lu, *Reactive and Functional Polymers*, Vol. 67, p. 481, May 2007.
14. M. Bähr, A. Bitto, and R. Mülhaupt, *Green Chemistry*, Vol. 14, p. 1447, 2012.
15. C.M. Byrne, S.D. Allen, E.B. Lobkovsky, and G.W. Coates, *Journal of the American Chemical Society*, Vol. 126, p. 11404, September 2004.
16. S. Kobayashi, C. Lu, T.R. Hoye, and M.A. Hillmyer, *Journal of the American Chemical Society*, Vol. 131, p. 7960, June 2009.
17. S. Loughmari, A. Hafid, A. Bouazza, A. El Bouadili, P. Zinck, and M. Visseaux, *Journal of Polymer Science Part A: Polymer Chemistry*, Vol. 50, p. 2898, July 2012.
18. A. Matic and H. Schlaad, *Polymer International*, Vol. 67, p. 500, 2018.
19. K. Błażek and J. Datta, *Critical Reviews in Environmental Science and Technology*, Vol. 49, p. 173, 2019.
20. M.A. Sawpan, *Journal of Polymer Research*, Vol. 25, p. 184, 2018.

21. H. Hatakeyama and T. Hatakeyama, *Journal of Renewable Materials*, Vol. 1, p. 113, April 2013.
22. T.J. Nelson, L. Bultema, N. Eidenschink, and D.C. Webster, *Journal of Renewable Materials*, Vol. 1, p. 141, April 2013.
23. A.S. More, B. Gadenne, C. Alfos, and H. Cramail, *Polymer Chemistry*, Vol. 3, p. 1594, 2012.
24. A.R. Mahendran, N. Aust, G. Wuzella, U. Müller, and A. Kandelbauer, *Journal of Polymers and the Environment*, Vol. 20, p. 926, 2012.
25. J.-B. Zeng, Y.-D. Li, W.-D. Li, K.-K. Yang, X.-L. Wang, and Y.-Z. Wang, *Industrial & Engineering Chemistry Research*, Vol. 48, p. 1706, February 2009.
26. M. Desroches, S. Caillol, V. Lapinte, R. Auvergne, and B. Boutevin, *Macromolecules*, Vol. 44, p. 2489, 2011.
27. G. Zhao, X. Fu, and L. Yu, Biodegradable cationic polymers, US Patent 7 700 542, assigned to Nitto Denko Corporation (Osaka, JP), April 20, 2010.
28. H. Brandl, R. Gross, R. Lenz, and R. Fuller, Plastics from bacteria and for bacteria: Poly (β-hydroxyalkanoates) as natural, biocompatible, and biodegradable polyesters in *Microbial Bioproducts*, Vol. 41 of *Advances in Biochemical Engineering/Biotechnology*, pp. 77–93. Springer Berlin Heidelberg, 1990.
29. G.-Q. Chen, *Plastics from Bacteria: Natural Functions and Applications*, Springer, Heidelberg; New York, 2010.
30. K.N. Anith, N.V. Radhakrishnan, and T.P. Manomohandas, *Current Science*, Vol. 83, p. 562, 2002.
31. T. Rawte, M. Padte, and S. Mavinkurve, *World Journal of Microbiology and Biotechnology*, Vol. 18, p. 655, 2002.
32. S.Y. Choi, M.N. Rhie, H.T. Kim, J.C. Joo, I.J. Cho, J. Son, S.Y. Jo, Y.J. Sohn, K.-A. Baritugo, J. Pyo, Y. Lee, S.Y. Lee, and S.J. Park, *Metabolic Engineering*, 2019.
33. M.E. Grigore, R.M. Grigorescu, L. Iancu, R.-M. Ion, C. Zaharia, and E.R. Andrei, *Journal of Biomaterials Science, Polymer Edition*, Vol. 30, p. 695, 2019.
34. A. Shrivastav, S.K. Mishra, I. Pancha, D. Jain, S. Bhattacharya, S. Patel, and S. Mishra, *World Journal of Microbiology and Biotechnology*, Vol. 27, p. 1531, July 2011.
35. K.Y. Sen, M.H. Hussin, and S. Baidurah, *Biocatalysis and Agricultural Biotechnology*, Vol. 17, p. 51 , 2019.
36. Z. Kádár and C. Fonseca, Bio-products from sugar-based fermentation processes in J.R. Bastidas-Oyanedel and J. Schmidt, eds., *Biorefinery*, pp. 281–311. Springer, Cham, 2019.
37. H. Miyagawa, M. Misra, L.T. Drzal, and A.K. Mohanty, *Polymer Engineering & Science*, Vol. 45, p. 487, April 2005.

38. S. Kumar, S.K. Samal, S. Mohanty, and S.K. Nayak, *Polymer-Plastics Technology and Engineering*, Vol. 57, p. 133, 2018.
39. A. Sienkiewicz and P. Czub, *ACS Sustainable Chemistry & Engineering*, Vol. 6, p. 6084, 2018.
40. M. Shen, R. Almallahi, Z. Rizvi, E. Gonzalez-Martinez, G. Yang, and M.L. Robertson, *Polym. Chem.*, Vol. 10, p. 3217, 2019.
41. BASF, Pluronic®, [electronic:] http://worldaccount.basf.com/wa/NAFTA~en_US/Catalog/ChemicalsNAFTA/pi/BASF/Brand/pluronic, 2011.
42. J. Stopek, B. Cuevas, J. Hotter, B. Nentwick, A. Irfan, and S. Tsai, Biomaterial drug delivery and surface modification compositions, US Patent 7 850 982, assigned to Tyco Healthcare Group LP (North Haven, CT), December 14, 2010.
43. K. Panek, Pluronic® F68 block copolymer surfactant, Technical bulletin, BASF Corporation, New Jersey 07932, 2005.
44. N. Kandemir, Y. Xia, P. Duan, W. Yang, and J. Chen, *MRS Advances*, Vol. 3, p. 1719, 2018.
45. K.W. Leong, W. Jie, R.-X. Zhuo, and H.-Q. Mao, Phosphate based biodegradable polymers, US Patent 6 805 876, assigned to Johns Hopkins University (Baltimore, MD), October 19, 2004.
46. K.W. Leong, Biodegradable poly(phosphate esters), US Patent 5 256 765, assigned to The Johns Hopkins University School of Medicine (Baltimore, MD), October 26, 1993.
47. J. Pretula, K. Kaluzynski, and S. Penczek, *Macromolecules*, Vol. 19, p. 1797, July 1986.
48. G. Łapienis, S. Penczek, G.P. Aleksiuk, and V.A. Kropachev, *Journal of Polymer Science Part A: Polymer Chemistry*, Vol. 25, p. 1729, 1987.
49. R.S. Edmundson, Quinquevalent phosphorous acids in S. Tripett, ed., *Organophosphorus Chemistry*, Vol. 8 of *Specialist Periodical Reports*, chapter 4, pp. 102–117. The Chemical Society, Burlington House, London, UK, 1977.
50. R.S. Edmundson, The synthesis of phosphonic and phosphinic acids and their derivatives: Non-functionalized acids in *PATAI'S Chemistry of Functional Groups*. John Wiley & Sons, Ltd, New York, 2009.
51. R.S. Edmundson, O. Johnson, D.W. Jones, and T.J. King, *J. Chem. Soc., Perkin Trans. 2*, Vol. 2, p. 69, 1985.
52. E. Akyol, M. Tatliyuz, F. Demir Duman, M.N. Guven, H.Y. Acar, and D. Avci, *Journal of Biomedical Materials Research Part A*, Vol. 106, p. 1390, 2018.
53. D.M. Klinman and T. Sato, Polyketal particles including a CpG oligodeoxynucleotide for the treatment of lung cancer, US Patent 9 919 058, assigned to The United States of America, as represented by the Secretary, Department of Health and Human Services (Washington, DC), March 20, 2018.

54. M.I. Papisov, Biodegradable polyketal polymers and methods for their formation and use, US Patent 7 951 898, assigned to The General Hospital Corporation (Boston, MA), May 31, 2011.
55. A.E. Medeiros de Brito, A. Pessoa, Jr., A. Converti, C. de Oliveira Rangel-Yagui, J.A. da Silva, and A.C. Apolinário, *Materials Science and Engineering: C*, Vol. 98, p. 524, 2019.
56. Y. Wang, G. Ameer, and R. Langer, Biodegradable polymer, US Patent 7 722 894, assigned to Massachusetts Institute of Technology (Cambridge, MA), May 25, 2010.
57. S. Kesavan, G. Woodward, and F. Decampo, Polysaccharide based scale inhibitor, US Patent 8 215 398, assigned to Rhodia Operations (Aubervilliers, FR), July 10, 2012.
58. K.A. Jansen, A.J. Licup, A. Sharma, R. Rens, F.C. MacKintosh, and G.H. Koenderink, *Biophysical Journal*, Vol. 114, p. 2665, 2018.
59. M. Nimni, *Journal of the American College of Nutrition*, Vol. 37, p. 267, 2018.
60. N. Kawabata, M. Kohsaka, and T. Kurooka, Biodegradable polymers, US Patent 5 344 891, assigned to Mitsubishi Rayon Co., Ltd. (Tokyo, JP), September 6, 1994.
61. K. Khemani, P.J. Andersen, S.K. Hodson, and H. Schmidt, Biodegradable polymer films and sheets suitable for use as laminate coatings as well as wraps and other packaging materials, US Patent 6 573 340, assigned to Biotec Biologische Naturverpackungen GmbH & Co. KG (DE), June 3, 2003.
62. S. Sumanam, Biodegradable polymer composition, US Patent 8 026 301, assigned to BNT Force Biodegradable Polymers Pvt Ltd. (Chennai, IN), September 27, 2011.
63. R. Hatti-Kaul, L.J. Nilsson, B. Zhang, N. Rehnberg, and S. Lundmark, *Trends in Biotechnology*, 2019.

PART II

APPLICATIONS

7

Packaging, Food Applications and Foams

An overview of novel biodegradable polymers for food packaging has been presented (1). The factors influencing degradation and biodegradation of polymers in various environments have been pointed out. Notably, biodegradable polymers are an imperfect alternative to classical polymers.

In addition, there are various approaches in different countries to certify biodegradable polymer materials, which complicates rather than promotes their application. For example, Bangladesh does not allow bags made from poly(ethylene), whereas France only bans non-biodegradable plastic grocery bags. In Denmark and Ireland taxes on bags with a particular wall thickness have been introduced (1).

7.1 Packaging

Attention is already being focused on methods for the recycling of polymers mainly used for packaging by biological methods (2).

7.1.1 *Packaging Materials*

Natural biopolymers provide biodegradable, sustainable solutions for the manufacture of short-life, single-use disposable packaging, consumer goods, and marine plastics. Starch, an anhydroglucose polymer, offers a structural platform to manufacture sustainable, biodegradable packaging (3).

Packaging material and disposable beakers, cups, and cutlery are widely used, which allow food material to be sold or consumed under hygienic conditions. Such disposable materials and objects are highly esteemed by the consumers and the retailers, since they may be simply disposed of after use and do not have to be washed and cleaned like conventional dishes, glasses, or cutlery.

Yet, the widespread and even growing use of such materials result in an increasing amount of litter produced each day. Currently, the plastic waste is either provided to garbage incinerators or accumulates in refuse dumps, with both of the above-mentioned solutions for waste disposal being associated with problems for the environment (4).

The issues concerning the use of wood-derived polymers and those produced from microorganisms, which display remarkable barrier properties, have been reviewed (5). A functional barrier is defined as a layer that prevents substances from migrating from behind the barrier layer into the food.

Wood-based products have received great attention for their air/oxygen resistance. Microorganism-derived biopolymers are comparable to conventional oil-based thermoplastics, but their cost may still be an issue. Both wood- and microorganism-derived biopolymers are challenged when moisture, grease and oxygen resistance are simultaneously required (5).

7.1.2 Lightweight Compostable Packaging

The state-of-the-art in the field of lightweight compostable packaging has been reviewed (6). The review starts with an overview of compostability, biodegradability and presents some definitions. A brief summary of the current use of synthetic commodity polymers used in packaging is followed by an analysis of commercially available, renewable polymers. An overview of the technology of production has been presented, as well as existing and emerging producers (6).

A comparison of the properties of materials is made to indicate a potential replacement of existing materials with emerging commercially available renewable, compostable alternatives. The analysis is limited to products that are commercially available, e.g., starch, cellulose, whole crop polymers, and polyesters such as poly(lactic acid) (PLA), poly(hydroxyalkanoate), and others (6). Commodity plastics used in food packaging are summarized in Table 7.1.

Table 7.1 Commodity plastics used in food packaging (6).

Polymer	Amount/[%]
Poly(vinyl chloride)	6
Poly(styrene)	5
Poly(ethylene terephthalate)	15
Poly(ethylene) (LDPE, LLDPE)	39
Poly(propylene)	12
Oriented Poly(propylene)	5
High density poly(ethylene)	13
Other polyolefines	5

7.1.3 Laminate Coatings

Packaging materials are mostly intended for single-use. Biopolymers such as Biomax, BAK, PHB and PLA tend to be strong but also quite rigid or even brittle. On the other hand, biopolymers such as poly(caprolactone) (PCL), Ecoflex and Eastar Bio are much more flexible. However, they have relatively low melting points.

They even tend to self-adhesion after thermal processing or after heat exposition. To prevent the self-adhesion, it is typically necessary to incorporate silica or other fillers into such films. Another important criteria for sheets and films used in packaging is temperature stability.

In the case of packaged foods being refrigerated, the packaging materials are subjected to fluctuating temperatures. Thus, a biopolymer that is perfectly satisfactory at room temperature may become completely unsuitable when used to wrap heated foods. The situation will become even worse when the food emits significant quantities of hot water vapor.

Improved biodegradable polymers that can be readily formed into sheets and films with strength and flexibility properties that are satisfactory for their use as packaging materials have been developed (7).

A blend of the comparatively stiff Biomax polymer and the relatively flexible Ecoflex polymer has been found to yield blends which have been shown to have strength and elongation properties which are superior to either biopolymer taken alone.

7.1.4 PLA Resins

Films, sheets, and other molded resinous articles, including biodegradable resins such as PLA resins, are known to exhibit a performance equivalent to that of conventional plastics. Among them, PLA resins have very high transparency and are very usable in packaging uses in which transparency is an important factor. In addition, they have water vapor permeability equal to or better than that of conventional oriented poly(propylene) (PP) or oriented poly(styrene) (PS) films and are expected to replace these films (8).

Biodegradable resins have many advantages and are applicable to molded articles such as films and sheets. However, since they have electrical insulating properties typical of resins, they are very susceptible to being electrically charged as in regular resins, and thereby they have various problems due to electrical charges. Such problems include, for example, crawling of ink upon printing, dropping out of contents to be packaged upon packaging, and dust adhesion to products that impair the appearance of the products.

Biodegradable resins also have problems caused by low hydrophilicity typical of plastics. For example, films for food packaging should have transparency so as to see the appearance of food packaged therein, but the surfaces of such films become fogged due to water drops derived from water vapor from the food.

To solve these problems, a kneading process of previously adding a surfactant to a resin has been employed. According to the kneading process, the surfactant bleeds out from the inside to the surface of the resulting molded article to form a surfactant layer, and thereby exhibits performance such as antistatic and antifogging properties. These properties may sustain to some extent even when the surfactant at the surface is wiped off, because the surfactant in the inside of the resin bleeds out again, thereby recovering the properties.

The performance is exhibited as a result of bleedout of the surfactant from the resin according to the kneading process. However, the degree of bleedout is believed to vary significantly depending on the crystallinity of the resin, such as degree of crystallization and degree of orientation of crystals, and the compatibility between the resin and the surfactant. Among such factors, the crystallinity of the resin and the compatibility of the surfactant also significantly affect the appearance of the resin (8).

The structure of a suitable additive to impart antistatic and antifogging properties is shown in Figure 7.1.

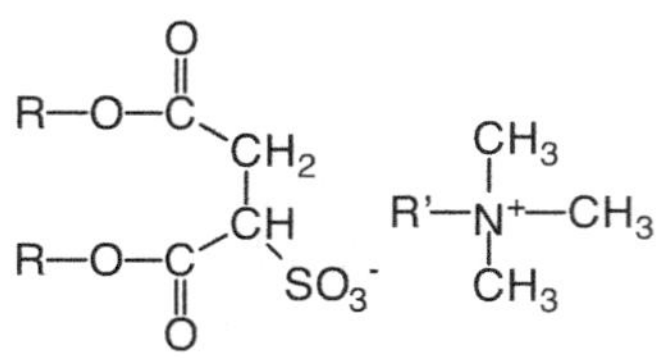

Figure 7.1 Additive for antistatic and antifogging properties (8).

The material shown in Figure 7.1 is basically a sulfosuccinic ester quaternary amine salt.

7.1.5 Protein-Derived Bionanocomposites

A comprehensive overview of the current advances in research and development of protein-derived bionanocomposites used in food packaging applications has been presented (9).

The recent interest in protein-based biomaterials is due to sustainability, renewability, biodegradability and the low carbon footprint. However, the inherent drawbacks of proteins-based materials for food packaging applications are their low mechanical strength, poor thermal and barrier properties, and inferior physicochemical properties.

Nanoreinforced bio-based polymers, also known as bionanocomposites, provide an opportunity to overcome these issues and have an ability to supersede non-biodegradable food packaging plastics produced from petroleum resources (9).

The most studied protein-derived bionanocomposites suitable for food packaging are soy protein isolates and gelatin proteins. Layered silicates are the most promising nanofillers used to increase strength, improve heat resistance and enhance barrier properties of protein-derived materials, while montmorillonite is the most commonly used silicate nanofiller.

The processing strategies used for protein-based biomaterials and their mechanical and moisture barrier properties for food packaging applications have been reviewed (9). Different proteins and nanofillers that have been studied to date in protein-derived food packaging applications have also been discussed in detail (9).

7.1.6 *Lignocellulose*

The suitability of three local lignocellulosic wastes, i.e., almond shell, rice husk and seagrass, as fillers in poly(3-hydroxybutyrate) (PHB)/fiber composites applications has been studied (10). PHB/fiber composites with 10 *phr* and 20 *phr* fiber content were prepared by melt blending.

The influence of the fiber type (size, morphology and origin) and content on the morphological, mechanical and thermal properties of the as obtained composites has been assessed. To evaluate the potential use in food packaging applications, the barrier performance to water, thermoforming ability and disintegration in controlled composting conditions of the composites were also studied. All the fibers could be demonstrated to be apt for their use as fillers in PHB/fiber composites, showing a reinforcing effect without affecting the crystallinity and the disintegration rate of PHB.

The thermal stability and the water barrier performance of the composites were reduced by the presence of the fibers. Nevertheless, the addition of almond shell resulted in the best balance of properties in terms of permeability and mechanical properties, finding an enhancement of the thermoforming ability of PHB when 10 *phr* of almond shell was added (10).

7.1.7 *Tannic Acid*

Natural proteins are good candidates for producing food packaging films due to their excellent functional properties. However, some deficiencies, such as their poor water resistance, need to be addressed before films can be used in daily applications (11).

The use of a low-cost plant-derived phenolic compound, tannic acid, as crosslinking agent for producing plasticized casein films has been investigated (11). Tannic acid is shown in Figure 7.2.

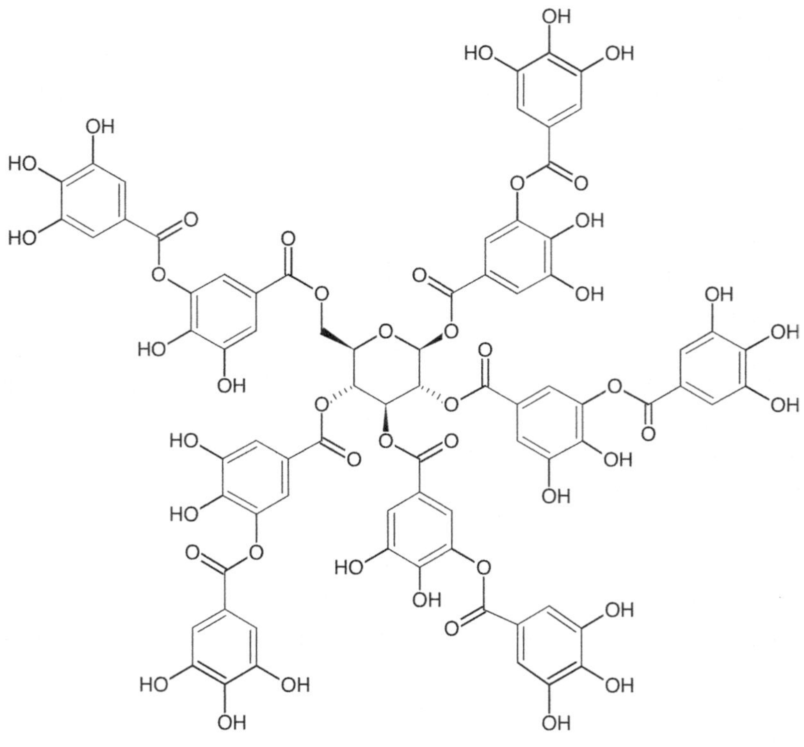

Figure 7.2 Tannic acid.

Fourier transform infrared spectroscopy (FTIR), high-resolution nuclear magnetic resonance spectroscopy, and rheological measurements confirmed the crosslinking reaction between casein amine groups and tannic acid. The influence of the concentration of the crosslinking agent on the mechanical and viscoelastic properties, thermal stability, swelling behavior, water vapor permeability, cytotoxicity and degradation in composting conditions was investigated. The so-obtained results showed that tannic acid is an effective crosslinking agent for the casein protein, so the yielded films, which exhibit improved physicochemical properties, can be used for food packaging applications (11).

7.1.8 Starch Compositions

In recent years, interest in agro-polymers has greatly increased, as these materials are biodegradable and thus eco-friendly (12). Since the fossil fuel-based polymers have created greater environmental concern, these bio-based polymers are addressing the concerns in some important areas, such as food packaging, and contribute significantly to the sustainable development with minimum ecological problems.

Among agro-polymers, starch-based polymers have made major strides towards marketable products in the food packaging field. It is estimated that by the year 2020, the biocomposite materials demand will touch 20% of total plastic utilization.

Non-biodegradable petroleum-based plastics are still the most dominant material used by the food industry for packaging applications (13). Consequently, the widespread usage of these conventional plastics has led to serious negative environmental impacts. Numerous studies have been conducted over the years to substitute these packaging plastics with eco-friendly materials in order to arrest the ongoing plastic waste disposal problems.

Various processing techniques used to produce starch-based polymers and composites with their properties to address the poor properties of starch have been reviewed (12). Low toxicity natural plasticizers with good compatibility are of great interest in the processing of thermoplastic starch. Further emphasis is also placed on essential packaging film properties such as barrier, biodegradation, mechanical, and thermal properties for thermoplastic starch-based materials.

The final properties of thermoplastic starch can be improved by using different fillers, as well as by changing the source of the starch. Since the quest to produce better, cheaper, and eco-friendlier materials never stops, a multidisciplinary approach is required to achieve further improvement in the existing materials and to produce a new class of materials that are eco-friendly, which can extend smarter and efficient services to customers (12).

Owing to their granular structure, natural starches must be destructurized before they can be thermoplastically processed (14). Foaming is not desirable. This generally requires special attention because of the water content of the starch. Foaming can be avoided by degassing or drying the melt prior to melt processing. Alternatively, avoiding the addition of water may be helpful. Urea and sodium, potassium and calcium hydroxides are suitable destructuring agents apart from water (14, 15).

7.1.8.1 *Plasticizers*

Another method to improve the melt processability of starch are plasticizers that reduce the melting point of starch such as dimethyl sulfoxide, polyols, amino, or amide compounds (16). The preferred plasticizer should substantially have a solubility parameter δ of at least 15 $cal^{1/2}cm^{-3/2}$.

Table 7.2 Effect of plasticizers (16).

	Thermal Conversion/[°C]		
Plasticizer[a]	from	center	to
Propylene glycol	150	170	190
Ethylene glycol	140	160	180
Glycerin	145	165	170
Dimethyl sulfoxide	65	80	150
Sorbitol	160	180	190

[a]: 25% Plasticizer, no water

7.1.8.2 *Starch Blends*

Blends of raw starch, poly(vinyl alcohol) (PVA) and talc with glycerol and water can be blown into biodegradable films. Here, preprocessing steps, such as gelatinizing or destructurizing of the starch, are not required. Unprocessed raw starch is directly blended with biodegradable copolymers such as PVA or ethylene vinyl alcohol, a nucleating agent, and a plasticizer (17).

Other compositions contain starch, an ethylene acrylic acid copolymer, stearic acid salt and a glycerol-based lubricant. Increasing the temperature gelatinizes the starch and melts the copolymer. Excess moisture is removed under reduced pressure and the plastic material is extruded (18). An example of a starch polymer composite is given in Table 7.3.

Table 7.3 Starch polymer composite (18).

Ingredient	Amount/[%]
Starch, unmodified pearl	39.6
Calcium Stearate	0.9
Ethyl acrylate copolymer	29.9
Poly(ethylene)	19.9
Other liquids	4.3
Glycerin, 33% in H_2O (pH = 11.0)	5.4

High amylose starch, treated with formamide and dichloroacetic acid and a plasticizer, yield homogeneous flowable quasi-solutions without formation of gels when they are heated, with stirring, to temperatures above 80°C. Amylose is shown in Figure 7.3.

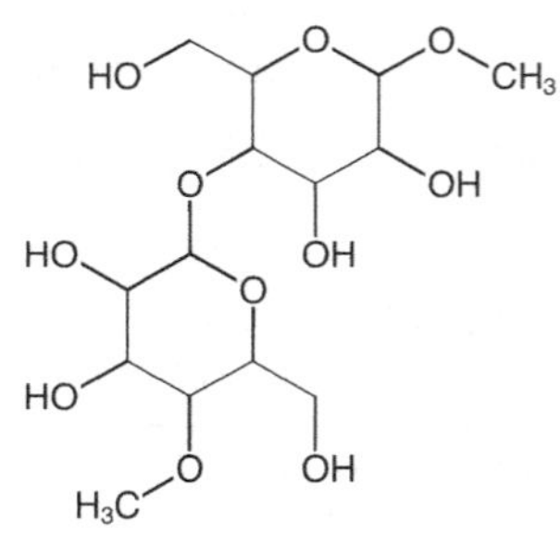

Figure 7.3 Amylose.

Ethylene glycol, triethylene glycol, poly(vinyl alcohol), and glycerol serve as plasticizers (19). The properties of such prepared compositions are shown in Table 7.4.

Other modified starches consist of hydroxypropyl substituted or carboxymethylated starches (14, 20). The use of modified starches increases the elongation at break and burst strength and improved resilience.

An improved composition contains hydroxypropylated amylose, wheat starch, PVA, glycerol, and steric acid (21). All the components are mixed into a free-flowing powder.

Premixing can be carried out in a conventional mixer. The powder is then introduced into a screw extruder and subjected to an elevated temperature by the shearing action of the screw and the

Table 7.4 Properties of high amylose starch (19).

Plasticizer	/[%]	LT[a]/[%]	FT[b]/[μm]	TS[c]/[$kg\,cm^{-2}$]
Ethylene glycol	4.76	89.95	91.6	47.5
Ethylene glycol	9.1	84.93	95.0	45.3
Ethylene glycol	20	89.53	93.3	25.4
Ethylene glycol	33.3	89.10	100.0	6.3
Triethylene glycol	4.76	89.43	71.6	51.3
Triethylene glycol	9.1	87.50	56.0	38.2
Poly(vinyl alcohol)	9.1	62.74	93.3	46.7
Glycerol	9.1	88.96	50.0	57.3
Glycerol	16.6	89.95	76.6	28.6
Glycerol	20	88.10	138.3	19.7

[a] Light transmission
[b] Light film thickness
[c] Tensile stength

application of external heat to the barrel. The temperature is raised to a maximum in the range of 130–160°C. Any liquid components, including additional water, are introduced during this initial phase. The melt that is formed is then propelled toward the die and in moving forward, the temperature is reduced to a value of 85–105°C (21).

7.1.8.3 *Sugar Palm Starch-Based Compositions*

The recent studies on sugar palm starch (SPS)-based films and different modification techniques employed to improve their performance as effective biopackaging material have been reviewed (13).

SPS films were developed using glycerol, sorbitol or their combination as plasticizers at the ratio of 15%, 30% and 45% using a casting technique (22). The addition of plasticizers to SPS film-forming solutions helped to overcome the brittle and fragile nature of unplasticized SPS films.

An increased plasticizer concentration resulted in an increase in the film thickness, moisture content and solubility. In contrast, the density and the water absorption of the plasticized films decreased with increasing plasticizer concentration. Raising the plasticizer content from 15% to 45% showed less effect on the moisture content and water absorption of sorbitol-plasticized films. Films containing glycerol and glycerol-sorbitol plasticizer showed a higher moisture content, solubility and water absorption capacity in comparison to sorbitol-plasticized films (22).

PLA and starch have been extensively studied as potential replacements for nondegradable petrochemical polymers on the basis of their availability, adequate food contact properties and competitive cost (23).

Nevertheless, both polymers exhibit some drawbacks for packaging uses and need to be adapted to the food packaging requirements. In particular, starch is very water sensitive and its film properties are heavily dependent on the moisture content, exhibiting relatively low mechanical resistance. PLA films are very brittle and offer low resistance to oxygen permeation. Their combination as blend or multilayer films could provide properties that are more adequate for packaging purposes on the basis of their complementary characteristics (23).

Environmentally friendly bilayer films from sugar palm starch (SPS) and PLA were investigated (24). The SPS-PLA bilayer films and their individual components were characterized with regard to their physical, mechanical, thermal and water barrier properties. The addition of 50% PLA layer onto 50% SPS layer (SPS 50-PLA 50) increased the tensile strength of neat SPS film from 7.74 to 13.65 *MPa* but reduced its elongation at break from 46.66% to 15.53%.

The incorporation of PLA layer significantly reduced the water vapor permeability as well as the water uptake and solubility of bilayer films, which was attributed to the hydrophobic characteristic of the PLA layer. Furthermore, a scanning electron microscopy (SEM) image of SPS 50-PLA50 revealed the lack of strong interfacial adhesion between the SPS and PLA. In summary, the incorporation of PLA layer onto SPS films enhances the suitability of SPS-based films for food packaging (24).

7.1.8.4 Poly(lactic acid) and Corn Starch

The PLA and PLA-corn starch-based biodegradable films with dibenzoylperoxide and glycidyl methacrylate were developed using a commercial extrusion blown molding method (25).

The addition of corn starch decreased the glass transition temperature from 65.26°C to 59.712°C, and increased the melting temperature from 146.81°C to 150.56°C.

The suitability of capsicum in modified atmospheric packaging was studied using the so-developed biodegradable films. For comparison, polymeric films made from low density poly(ethylene) and linear low density poly(ethylene) were used.

The bio-packages maintained the quality of capsicum up to 12 *d* and 24 *d* at 25°C and 8°C, as compared to unpackaged capsicum with a shelf life of 4 *d* and 9 *d* at 25°C and 8°C, respectively (25).

7.1.8.5 Antimicrobial Packaging

The PHB, starch and poly(lactic-*co*-glycolic acid) (PLGA) materials have unique properties for novel applications in foods, cosmetics, and medicine (26).

In recent years, food packaging has focused on the development of antimicrobial packaging for extended shelf life, quality and safety of food products. Bioplastic materials provide a promising application in the packaging industry as substitutes for environmentally deleterious petrochemical-based plastics. The recent progress in the antimicrobial application of starch, PHB and PLGA as well as their blends and nanocomposites in food packaging research has been summarized (26).

Also, an overview of the antimicrobial application of these materials has been presented, particularly in the food industry and the biomedical industry.

Starch is compatible with a lot of antimicrobial agents and its active films have shown an efficiency against many species of microorganisms. Sweet potato starch films were successfully blended with chitosan and potassium sorbate (27). The active films were observed to effectively suppress the growth of *E. coli* and *S. aureus* and improved the barrier properties of films (28–30)

When various ratios of chitosan and lauric acid were blended with starch, the active films improved the efficacy against *B. subtilis* and *E. coli* (31).

Quinoa starch films blended with gold nanoparticles exhibited a 99% and 98% reduction of *E.coli* and *S. aureus* growth, respectively (32).

Zinc oxide nanorods integrated into sago starch films by sonication demonstrated an excellent inhibition against *S. aureus* growth (33).

It was found that potato starch/cellulose nanocomposite films were found to be UV resistant, and that a higher transparency indicates the nanoscale dispersion (34).

7.1.8.6 *Photo-Degradable Starch*

Starch/TiO_2 bionanocomposites can be produced by photochemical reactions as a biodegradable food packaging material (35). The physical, mechanical, thermal and water-vapor permeability properties were investigated. Also, the photodegradation properties of nanocomposite films were studied. Film-forming solutions were exposed to ultraviolet A (UV-A) for different times.

The results of the study showed that the UV-A irradiation increased the hydrophobicity of the starch films. With increasing UV-A exposure time, tensile strength and Young's modulus of the specimens were decreased.

The elongation at break of the films was increased with increasing UV-A irradiation. The glass transition temperature and melting point of the films were increased by increasing UV-A exposure time.

The results of the study showed that the photodegradation properties of photo-produced starch/TiO_2 nanocomposite were significantly higher than that of virgin starch and virgin starch/TiO_2 films.

A mechanism was developed to describe the photo-production and photodegradation of starch/TiO_2 by UV-A rays. It can be concluded that the modification of a starch-based biopolymer by UV-A and nano-TiO_2 is an easy and accessible process to improve the packaging properties and the photo-degradability of biopolymer-based films (35).

7.1.8.7 *Fiber-Rich Lentil Flour*

Biodegradable composites with edible components, such as starch, glycerol, and lentil flour rich in fiber (0, 0.5% and 1.0%), have been developed (36). The lentil flour obtained from the residue of a commercial lentil protein extraction process, has a micrometer size and, unlike typical lentil flour, has a three times higher concentration of fiber.

Its use leads to an incremental increase in the storage modulus, strength at break, and toughness of the composites and to a decrease in water vapor permeability with respect to the matrix, showing that the additive can act as reinforcement for starch films. Composites with 0.5% flour result in more flexibility due to their homogeneous dispersion in the matrix.

All films are thermally stable up to 240°C and biodegradable in vegetal compost after 4 weeks. These composites are highly promising for use as biodegradable and edible food coatings. They could enrich food nutritional value by the fiber-rich flour addition (36).

7.1.8.8 Poly(vinyl alcohol) with Nanostructured Starch

Ternary films have been synthesized from PVA as polymeric matrix, nanostructured starch as reinforcement phase and hydroxytyrosol (37). Hydroxytyrosol, also known as 2-(3,4-Dihydroxyphenyl)ethanol, is a low-molecular phenolic compound present in olive oil as an antioxidant agent. Hydroxytyrosol is shown in Figure 7.4.

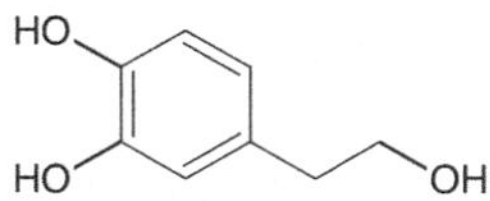

Figure 7.4 Hydroxytyrosol.

Nanostructured starch was considered in the form of starch nanocrystals and nanoparticles obtained by acid hydrolysis and ultrasound irradiation of starch derived from the bread wheat variety Cadenza (WT, with an amylose content of 33%) and a derived high amylose line (HA, amylose content of 75%).

The here developed multifunctional films were characterized with regard to morphological, thermal and optical properties, water absorption capacity, overall and specific migration into a food simulant and antioxidant properties. The studies showed a prolonged release of hydroxytyrosol from all the ternary films. Furthermore, the released hydroxytyrosol retained a strong antioxidant activity. The data, compared to those of PVA/hydroxytyrosol binary films, demonstrated the key role of nanostructured starch in the ternary formulations in promoting a gradual release of hydroxytyrosol. Overall, an PVA film combined with nanoparticles from low amylose starch and hydroxytyrosol resulted in a very promising ternary formulation for food packaging applications (37).

7.1.8.9 *Fiber Reinforced Starch Foam*

Sugarcane bagasse fiber reinforced starch foam composites were prepared with different oregano essential oil contents in the range of 2%–8% to improve the antimicrobial effectiveness of bioactive packaging (38).

FTIR analysis confirmed that the oregano essential oil penetrates into starch foam composite. From the result of antimicrobial activity, the incorporation of 8% oregano essential oil in starch foam composite achieves the minimum inhibitory concentration against *E. coli* and *S. aureus*.

It was found that oregano essential oil is more effective against *S. aureus* (Gram-positive bacteria) than *E. coli* (Gram-negative bacteria). The water absorption capacity and hygroscopicity of the foam composite decrease with increasing content of oregano essential oil.

The lowest monolayer value was observed in a starch foam composite with 8% oregano essential oil. The soil biodegradation rate and flexural strength of samples slightly decrease with increasing oregano essential oil content. So, a starch foam composite containing oregano essential oil can be applied as a bioactive packaging to preserve foods (38).

7.1.8.10 *Starch/Kefiran/ZnO Nanocomposites*

UV-protective biodegradable nanocomposite films composed of starch, kefiran, and ZnO were developed for packaging uses and characterized as a function of the content of nano ZnO (39).

The packaging properties of the nanocomposite films, e.g., moisture sensibility, mechanical properties, and UV protection, were measured. SEM and thermogravimetry coupled with differential scanning calorimetry were used to investigate the microstructures and the thermal properties of the films. Mechanical analysis demonstrated that by increasing zinc content up to 3%, the tensile strength and the Young's modulus of the specimens were increased.

Also, the elongation at break of the nanocomposites was decreased. A UV-Vis spectrum confirmed the UV-protective property of the films. The thermal properties of the samples indicate that T_m increased by increasing zinc content. Also, decomposition temperature was enhanced by nanoparticle application. Furthermore, SEM micrographs confirmed that the aggregation of the zinc particles increased with increasing zinc content (39).

7.1.9 *Heat-Sealable Paperboard*

Heat-sealable paperboard containers composed of various laminate structures are widely used in the container industry. A growing concern regarding the solid waste disposal of containers coated on two sides with nondegradable polymers has arisen, and a solution to the non-readily biodegradable container has been widely sought (40).

A low temperature extrusion coatable grade heat-sealable biodegradable polymer resin is thermoplastic poly(vinyl alcohol) resin or starch-based low density poly(ethylene) polymer resin.

The structure from exterior layer to interior food contact layer is as follows (40):

1. A heat-sealable gloss layer of a degradable polymeric resin, preferably a starch-based poly(ethylene) (PE) resin or a PVA alcohol resin,
2. A substrate layer of high-grade paperboard,
3. An interior adhesive tie layer, preferably an ethylene-based copolymer with grafted functional groups,
4. A barrier layer made from a ethylene vinyl alcohol copolymer, and
5. A heat-sealable layer rendering the finished laminate structure.

The extrusion-coated paperboard is then converted into packages, such as gable-top milk and juice cartons, brick-type aseptic cartons, folding cartons, or paper-based cups or plates.

7.1.10 Packages with Corrosion Inhibitor

Biodegradable plastic films have certain physical and mechanical properties which differ from those of the conventional and well-known polyolefin films such as PE and PP. In general, biodegradable films are hygroscopic and also have a significantly higher gas permeability. Because of these physical properties, biodegradable plastic films have typically offered less protection to metallic articles than is available from the more traditional polyolefin films (41).

However, this disadvantage may be overcome when the biodegradable plastic resin film is combined with a particulate vapor phase corrosion inhibitor dispersed within and through the film or applied as a coating on the surface of the film. The vapor phase corrosion inhibitors used together with the base materials are highly compatible, and offer significant long-term protection to metallic articles in or near the package.

Examples of the selected corrosion inhibiting chemicals suitable for preparing a masterbatch with the biodegradable materials are summarized in Table 7.5.

Table 7.5 Formulations of corrosion inhibitors (41).

No.	Chemical	Parts per Weight Range
1	Disodium sebacate	65 – 70
	Benzotriazole	20 – 25
	Dicyclohexylammonium nitrate	5 – 10
2	Benzotriazole	25 – 30
	Cyclohexylammonium benzoate	60 – 65
	Sodium nitrite	3 – 5
	Dicyclohexylammonium nitrate	3 – 5
3	Cyclohexylammonium benzoate	60 – 70
	Monoethanolammonium benzoate	5 – 10
	Benzotriazole	5 – 10
	Dicyclohexylammonium nitrate	15 – 25
4	Benzotriazole	5 – 10
	Disodium sebacate	55 – 65
	Silica	0 – 5
	Cyclohexylammonium benzoate	15 – 25
	Monoethanolammonium benzoate	5 – 10

The master batches were then blended with a variety of biodegradable materials. As usual, the biodegradable materials are chosen from commercially available brands of PLA, PCL or other suitable polyesters made from adipic acid, succinic acid, butanediol and a small amount of terephthalic acid (41).

Films molded in this way have an unique application for their use as laminated cushioning material, particularly those cushioning products comprising two superposed films with at least one having raised embossments thereon.

With a pair of such films, the raised embossments are preferably in opposed direction so as to form substantially sealed raised cells with entrained air inside. Of course, such cushioning materials are made from commercially available standard thermoplastic films.

In a typical production operation, films to be embossed are draped across the surface of a heated rotary embossing roller in order to create raised embossments of the desired shape and configuration (41).

Vacuum molding techniques are used to create the embossments. With the embossments of the first and second opposed films being in register, one with the other, the surface layers of the film are brought into face-to-face contact under modest pressure, while at a temperature sufficient to provide a fusion or tacky state to create the bond (41). Basically, this technique goes back to the 1960s (42).

7.1.11 Multiwall Package

A method for the fabrication of a multiwall package has been described (43). The package consists of biodegradable paper layers and sheets of adherent films. The film sheet may consist of a corn derivative or PLA. Furthermore, stripes are positioned in such a way that the package is more easily foldable in a lateral direction than in a longitudinal direction.

The layers are fabricated by progressive and controlled extrusion of the adhesive via extruders which are laterally spaced apart. The sheet travels in a longitudinal direction under the extruders. Advantageous properties of the products include (43):

- Scratch resistance,
- Excellent stiffness,

- FDA compliant,
- Good slip and dispensing properties,
- Dimensionally stable,
- Excellent lay-flat capability,
- Easily treated to higher dyne levels, and
- Excellent printability.

The films have been further characterized as sufficiently leak proof to withstand, at least temporarily, any wetness associated with pet food (43).

7.1.12 Cushioning Nuggets

A method and an apparatus for packaging articles has been described where the packages are cushioned with biodegradable, starch-based materials. These materials are initially packed as loose nuggets within the package, and afterwards converted *in situ* to a single, unitary cushion to protect the article (44).

A basic problem associated with the use of cushioning materials is their loose disposition within the package that is holding the article. The materials are subject to movement and shifting due to handling. When the package is opened, the loose fill also creates cleanup and disposal problems.

Another commonly used technique is to use *in-situ* foaming of, for example, poly(urethane) (PU) (45). However, PU is not considered as biodegradable. In addition, such resins also require relatively long cure times before the foam is completely formed and hardened (44).

A method for packaging an article in a container basically consists of the following steps (44):

- Providing a container having a bottom and a sidewall,
- Placing an article for packaging in the container, leaving a space between the article and the container bottom and sidewall,
- Providing a quantity of loose nuggets, said nuggets being defined at least in part by an exposed surface of a starch material,
- Gelatinizing the starch material in at least a portion of the nuggets,

- Packing the nuggets adjacent to each other in the space between the article and the container bottom and sidewall, with nuggets having gelatinized starch surfaces in contact with surfaces of adjacent nuggets, and
- Allowing the gelatinized starch to retrograde, whereby adjacent nuggets adhere together at sites of contact.

An apparatus for carrying out this method for packaging tasks has been described in detail (44).

7.1.13 *Fluid Containers*

7.1.13.1 *Bottled Water*

A particular problem for packaging applications is the growing consumption of plastic containers for bottled water (46). Due to rising health concerns related to the quality of tap water, an increasing number of people have started consuming bottled water. However, the discarded plastic water bottles are harmful to the environment. Not only do disposed of plastic containers fill up our landfills, but they also frequently cause water contamination, as they end up in water streams.

The plastic waste products are subject to photodegradation, which causes the plastic waste products to degrade into small toxic plastic polymers. Over the course of time, the plastic polymers are broken down into smaller and smaller pieces. However, the polymers do not decompose into materials found in nature. These small toxic plastic polymers concentrate in the upper water column and are ultimately ingested by aquatic organisms which reside near the ocean's surface. Plastic waste thus enters the food chain.

Furthermore, many larger aquatic animals, such as orcas, consume the plastic waste during hunting activity, the consumption of which is extremely harmful. The floating particles of plastic waste also frequently resemble zooplankton, which can lead to their being consumed by jellyfish, presenting another venue for entering the ocean food chain. Besides ingestion, other concerns caused by plastic waste ending up in water systems include the frequent entanglement of wildlife.

The degradation of plastic waste also contaminates the air and the soil. As the plastic waste degrades, it can release toxic pollutants such as greenhouse gases. As mentioned above, over time the plastic degrades into smaller and smaller toxic plastic particles, never returning to compounds found in nature.

Corn-based PLA can be used to produce bottles for water products. Such a water bottle can degrade within 75–80 *d* under commercial composting conditions (47).

However, in order to degrade, the bottles need to be placed under specific industrial processing conditions, including being subjected to high heat, particular microorganisms, and high moisture levels (46). Furthermore, elements, such as the cap, are not degradable. These bottles include elements that are not fully biodegradable under standard environmental conditions and they are also expensive. This creates a disincentive for consumers to switch from petroleum-based plastic bottles.

A fluid storage container has been developed that is fully biodegradable, including all auxiliary components. The container has a body, a base, a mouth, and a cap, all of which are biodegradable. The basic design is shown in Figure 7.5.

The body is made of a biodegradable plant fiber-based structural material. Fibers or pulp from sugarcane bagasse can be used. This material is inexpensive and thus decreases production costs. The material provides a solid mechanical support for holding the fluid in the container.

While the mere contact with fluids does not initiate the disintegration of the biodegradable structural material, if directly exposed to fluids for an extended period of time, the structural material will begin to disintegrate. In order to prevent the direct contact between the structural material and the fluid in the container, a thin coating is provided on the inner surface thereof. This coating is made of a biodegradable film or sprayable resin.

This may be formed of cellulose acetate products, i.e., wood pulp. The cellulose acetate products can be heat sealed to offer gas barrier properties and can be tailored to achieve a range of moisture barrier properties. Alternatively, such products can be made from whole grain cereals, or zein, a corn gluten product.

The biodegradable film should have a predetermined shelf life, such as six months, and during its shelf life it is insoluble in fluids and temperature resistant. After its shelf life, it will begin to biodegrade (46).

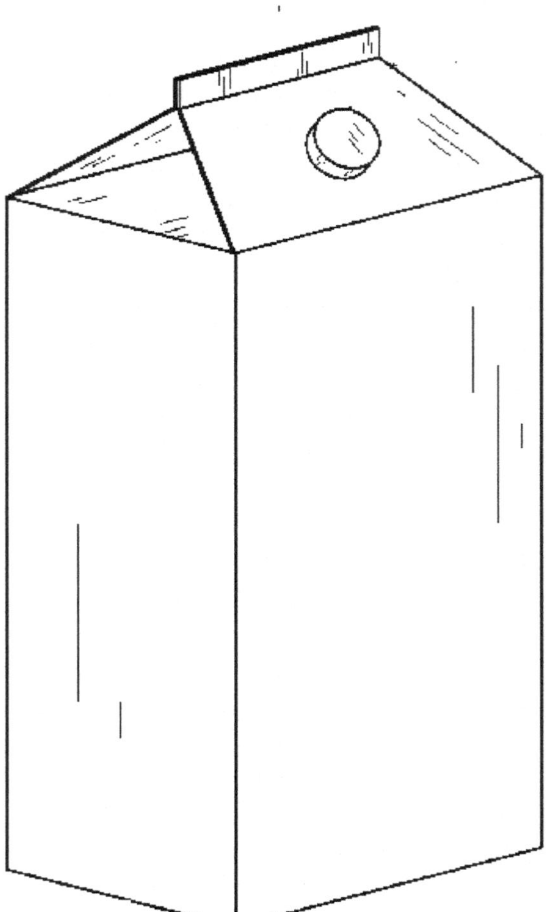

Figure 7.5 Fluid storage container (46).

7.2 Fibers and Nets

The recent developments in the field of renewable fibers and materials for packaging applications have been extensively reviewed (48).

7.2.1 *Multicomponent Fiber*

Disposable absorbent articles generally contain absorbent fibrous webs such as airlaid or bonded carded webs. Such webs are often stabilized with binder fibers during web formation.

These binder fibers are multicomponent fibers with a significant difference in their melt temperature. The fibers are heated at a temperature sufficient to melt the lower melting components, but not the higher melting components (49).

Several binder fibers have been developed that are biodegradable to enhance the disposability of the absorbent article. Most often, biodegradable polymers are formed from aliphatic polyester materials. A multicomponent fiber that contains a high melting aliphatic polyester and a low melting aliphatic polyester has been developed (49).

Here, the low melting component is poly(butylene succinate) (PBSU). The low melting point polyesters may contain a minor portion of other monomer constituents, for example, aromatic monomers such as terephtalic acid, that further improve the strength and tenacity of the fibers. The high melting component is PLA (49).

A wetting agent may be used to improve the hydrophilicity of the fibers. Suitable wetting agents should be compatible with the aliphatic polyesters. Examples of wetting agents are surfactants, ethoxylated alcohols, or acid amide ethoxylates (49).

Typically, the components are extruded in separate extruders, but they may also be spun together. The process of manufacture of the composite fibers has been described in detail (49).

One difficulty encountered in the thermal processing of aliphatic polyester polymers into fibers is the sticky nature of these polymers. Attempts to draw the fibers, either mechanically, or through an air drawing process, will often result in the aggregation of the fibers into a solid mass. Thus, low melting point aliphatic polyester types are selected with a relatively high crystallization temperature, thereby reducing their tackiness.

The multicomponent fibers are substantially biodegradable, yet readily processed into nonwoven structures that exhibit effective fibrous mechanical properties.

7.2.2 *Biodegradable Netting*

The continuous extrusion of plastic netting started in the 1950s (50). Extruded netting is netting in which the strands are extruded from a die, the joints being formed either within the die or immediately outside the die. A variety of configurations are known, such as square, diamond, twill, etc. Common materials used for the preparation are PP, PP, and poly(amide).

Such nets have found a number of uses, including (51):

- Erosion control netting,
- Packaging netting,
- Agricultural netting,
- Filtration, and
- Home furnishings applications.

Netting has also found use in certain composites. In such composites the netting is laminated to one or more fabric overlays. Examples are (51):

- Disposable diapers,
- Incontinent briefs,
- Training pants,
- Bandages,
- Dressings,
- Medical gowns,
- Mattress pads,
- Blankets,
- Sheets,
- Clothing,
- Consumer wipes, and
- Building and construction composites.

Since netting materials are often used in the environment, it has become desirable to provide netting materials which are biodegradable.

In common manufacturing processes, the netting is typically extruded though an annular die and quenched in a water tank. The resulting tubular netting is collapsed in a quench tank, slit, and opened up to a flat sheet before rewinding. To be compatible with this type of web handling system, the polymer material used in the extruded product must be able to withstand being transformed from an annular tube into a collapsed tube.

PLA polymers are well-known biodegradable polymers. However, they are typically high modulus materials. As such, 100% PLA extruded netting, because of its relatively high modulus, does not perform well after being extruded in the subsequent web handling and in the orientation portion of the process, thereby making PLA particularly unsuitable for use as an extruded netting material (51).

However, plasticized PLA can be used for such a purpose (51). The plasticizer is Eastar Bio Ultra® or an Ecoflex® type. These substances are actually polyesters from terephthalic acid, adipic acid, and 1,4-butanediol. In addition, PP may be added in amounts of 7.5%. Other suitable additives include processing aids and antioxidants, as well as slip agents and antiblocking agents.

The oriented netting has been tested for aerobic biodegradation under controlled composting conditions according to ASTM D 5338 (52). The netting materials show complete biodegradation after 90 *d*.

Trays of plants can be encapsulated within a netting tube, which is sealed at each end (53). This approach is relatively difficult to carry out, however, and the presence of netting both beneath and above the tray is associated with a number of disadvantages, including excessive material costs, poor visual presentation of the product to a potential purchaser, and less stable stacking of packages, thereby increasing the likelihood of damage.

7.2.3 Electrospun Nanofibrous Mat

Electrospun chitosan/PVA fibrous mats were successfully fabricated by encapsulating fish-purified antioxidant peptide under the optimal conditions (54).

Here, a flow rate of 0.2 $ml\,h^{-1}$ was used, along with an applied voltage of 15 kV and a tip-to-collector distance of 15 cm. The antioxidant peptide-loaded electrospun fibers displayed uniform and bead-free nanostructures with fiber diameters ranging from 157.9 ± 28.8 *nm* to 195.5 ± 34.4 *nm*. This indicates the miscibility of chitosan and PVA polymers in nanoscale. Atomic force microscopy revealed the dense fibrous nature of the mats, with surface roughness ranging from 180.6 to 279.1 *nm*.

FTIR spectroscopy and X-ray diffraction analysis demonstrated the molecular interaction between components functional groups, which enhanced the mats thermal stability and mechanical properties (e.g., around 3-fold in terms of tensile strength) in contrast to pure PVA. Water vapor permeability of electrospun nanofibers decreased with the inclusion of chitosan and/or peptide into fibrous matrices. Moreover, water contact angle analysis showed that the incorporation of antioxidant peptide made the blend nanofibrous mat more hydrophobic.

Chitosan/PVA/antioxidant peptide ternary fibers showed a high encapsulation efficiency beyond 94%, and the encapsulated peptide retained its antioxidant activity in the electrospun fibers. The loading of the peptide within nanofibers enables a sustained release of the antioxidant peptide. Also, an *in-vitro* cell toxicity experiment displayed a good cytocompatibility of the antioxidant peptide-loaded nanofibers (54).

7.3 Foams

The development of foamable blends containing a bio-based or a biodegradable polymer has proven o be particularly difficult. The techniques of foaming a synthetic resin include the steps of charging styrene beads in a forming mold, adding a vapor forming compound, followed by a decrease in pressure. Alternatively, a synthetic resin is charged in an extruder together with a foaming agent such as an organic solvent to foam the resin due to pressure reduction in the course of extrusion (55).

However, the conventional foaming techniques for foaming a synthetic resin fail to satisfactorily foam a biodegradable resin due to unfavorable relationships between a softening point, melting point of the resin and the foaming temperature of a foaming agent. Thus, there are many known problems which are encountered with techniques of foaming biodegradable resin to a high expansion and forming the foamed resin (55).

7.3.1 *Foamed Articles*

Foam sheet made of a biodegradable resin composition has been reported. The biodegradable resin composition contains a wax or a polyolefin resin. The biodegradable resin is mainly composed of PLA. Natural candelilla wax or paraffin wax is preferable as the wax. PE is preferable as the polyolefin resin. Also disclosed are a foam article and a molded container made of such a foam sheet wherein the degree of crystallinity of the biodegradable resin is 10% or more.

For environmental reasons, foam articles have been developed that are made from biodegradable resins such as PLA, PBSU, PCL, poly(ethylene succinate), and poly(butylene terephthalate-*co*-adipate) (56).

These materials can be degraded by moisture or microbes. Furthermore, the materials can be treated for composting in composters. Among these polymers, PLA in particular has come into use in various applications as an ideal polymer that is a plant-derived raw material produced by polymerizing lactic acid obtained by fermenting materials, such as various starches and sugars, which is eventually reconverted into carbon dioxide gas and water so as to be environmentally recycled on a global scale (56).

However, compositions derived from PLA and aliphatic polyesters exhibit a high gas permeability. This is a drawback for packaging applications for food.

The preparation of a foamed article includes the following steps (56):

A poly(*L*-lactic acid) was fed to a double-screw kneader at 200°C. To this, ethylene glycol dimethacrylate was added as a crosslinking agent and dibutyl peroxide as a radical polymerization initiator. Pellets were formed from this mixture.

These biodegradable resin pellets were dry blended with candelilla wax (Hentriacontane). The mixture thus obtained was fed to a double-screw kneading extruding foam manufacturing machine at 200°C. In this way, a foam sheet was produced at a discharge rate of 100 $kg\,h^{-1}$. Carbon dioxide gas was added as a foaming agent.

The foam sheet was a uniform sheet formed of closed cells with a foaming ratio of 6.0 (56).

7.3.2 *Blends*

Blends of styrenic polymers (PS, high impact poly(styrene)) and biodegradable polymers (PLA) can be extruded and thermoformed to produce very low density food service and consumer foam articles (55, 57). The blends are compatibilized with styrene-based copolymers, a styrene-maleic anhydride copolymer, or a styrene methyl methacrylate copolymer. As blowing agent for foaming the compositions, *i*-pentane is used.

7.3.3 *Starch-Polyester Graft Copolymer*

Thermoplastic polymer foams can be prepared from starch-polyester graft copolymers. Water is used as the plasticizer as well as the blowing agent (3). The foams have improved processability, surface properties, and an extended range of mechanical properties.

An alkylaryl ammonium hectorite clay from the Bentone™ series may also be used as additive. The clay provides excellent mechanical strength, flame retardancy and highly improved gas barrier properties (58).

Still another starch-polyester graft copolymer and a chemically modified starch-polyester graft copolymer composition has been described. The starch-polyester graft copolymers are synthesized by the reaction of starch with biodegradable polyester polymers in the presence of maleic acid and glycerol as plasticizer. Examples of biodegradable polyesters are summarized in Table 7.6.

Maleic acid acts as a transesterification catalyst. The composition can be produced continuously in a twin screw corotating extruder. The starch-polyester graft copolymer can be solvent cast, melt cast and blown into clear transparent film particularly for use in single-use disposable applications and can be biodegradable (58).

7.3.4 *Foamed Gelling Hydrocolloids*

A firming agent and a fibrous material are advantageously added to a foamed gelling hydrocolloid system. When articles prepared

Table 7.6 Biodegradable polyesters (3).

Polyester type
Poly(butylene adipate-*co*-terephthalate)
Poly(caprolactone)
Poly(vinyl acetate-*co*-vinylalcohol)
Poly(lactic acid)
Poly(glycolic acid) or poly(glycolide)
Poly(β-hydroxybutyrate)
Poly(β-hydroxybutyrate-*co*-β-hydroxyvalerate)
Poly(β-hydroxyalkanoate)s
Biodegradable aliphatic polyesters
Aliphatic-aromatic copolyesters

from such compositions are coated with a suitable coating, they are impervious to boiling water and able to withstand microwave heating (59).

Examples for the foamed gelling hydrocolloid system are summarized in Table 7.7.

Table 7.7 Gelling hydrocolloid systems (59).

System	System
Agar	Chitosan/carrageenan
Agarose	Curdlan
Aloe mannans/xanthan	Gellan
Deacetylated aloe mannan	Guar gum/xanthan
Algin/alginates	Hyaluronic acid
Water-insoluble alginates	Konjac
Borate complexes of 1,3-*cis* diols	Deacetylated konjac
Carrageenans	Konjac/xanthan
Cassia gum/xanthan	Locust bean gum/xanthan
Chitosan	Cassia/xanthan
Chitosan/alginate	Pectins

A foaming agent is added for foaming, which should produce a foam that lasts until gelling occurs. The time period will depend on the conditions chosen for foam formation, but will typically be about half an hour or less at room temperature.

Possible polymeric foaming agents include surface active hydrocolloids such as methyl cellulose, hydroxypropyl methyl cellulose, hydroxypropyl cellulose, hydroxyethyl cellulose, and glycol alginates. Suitable firming agents include wheat flour, pea flour, clays, sawdust, starches, and other water-insoluble, or water-absorbing particulates (59).

Examples of surfactants for foaming include sodium dodecyl sulfate, cleaning compositions, lecithins, as well as pertinent derivatives of cellulose and other polysaccharides (59, 60). For physical applications, such as wound-healing dressings, when a surfactant is included in the dried gelled foam, nonionic surfactants, such as the ethoxylates of sorbitan esters, are preferred (60).

When dried gelled foams containing high levels of calcium are soaked in water, the foam structure does not break down due to the high level of crosslinking of the foam. However, the soluble components in the foam, including water-soluble foaming agents such as hydroxypropyl methyl cellulose, will diffuse out of the foam. This loss of foaming agent may be prevented, for example, in a wound-healing application, by using a foaming agent that is not soluble under conditions of use.

Some foaming agents form gels at body temperature. For example, methyl cellulose forms gels above 35°C. When using a foam that comprises methyl cellulose as the foaming agent in an application in which the foam is at body temperature, the methyl cellulose will stay in the gelled state and remain in the foam and contribute to the wet strength of the foam (60).

In the case of alginate-based compositions, the extent of crosslinking can be controlled by either the amount of gelling agent, such as calcium carbonate, or the amount of pH modifier, such as glucono δ-lactone. The extent of crosslinking can also be controlled by the drying time. By initiating and completing the drying before all the calcium is released and reacted with the alginate, the crosslinking will be done to a lesser extent compared to a longer gelling time before the drying process (60).

Several examples for such compositions have been described in great detail (59, 60). Coated articles that are prepared from such compositions have been described that are impervious to boiling water and can withstand microwave heating (59).

The effect of fish gelatine-alginate interactions on the formation and stability of foams was investigated by examining relationships between surface, bulk, and foaming properties of aqueous mixtures of fish gelatine and alginate at 25°C under different values of pH and fish gelatine-to-alginate ratio (61). Replacing a portion of fish gelatine with alginate (fish gelatine-to-alginate ratio of 80:20, 50:50, and 20:80) at pH 5.0 or 7.0 increased the air-liquid surface tension, negative electrophoretic mobility, bulk viscosity, and particle size of fish gelatine-alginate mixtures.

At a pH of 3.5, below the isoelectric point of fish gelatine, the alginate replacement increased the particle size more dramatically; however, it suppressed trends of increasing negative electrophoretic mobility and bulk viscosity, and even reduced the surface tension, due to stronger electrostatic attractions between oppositely charged fish gelatine and alginate molecules and the resulting formation of more charge-neutralized fish gelatine-alginate complexes.

The foaming ability became stronger when the surface tension decreased, the negative electrophoretic mobility approached zero (more charge-neutralized), and the bulk viscosity decreased. However, it was not closely correlated with particle size. Fish gelatine-alginate mixtures had a weaker foaming ability than solutions prepared only with fish gelatine or whey protein concentrate. However, these mixtures exhibited a much higher foam stability during storage at 25°C. Fish gelatine-alginate mixtures prepared at pH 3.5 and a ratio of fish gelatine to alginate of 80:20 showed the best foaming ability and foam stability (61).

A hydrophobic modification of sodium alginate foams via a simple freeze-drying and post-crosslinking induced by zirconium ions was developed (62). All results demonstrated that zirconium ions not only constructed surface microstructure but also lowered surface energy of foams, leading to the hydrophobic character. Hydrophobic and oleophilic foams showed excellent adsorption capacities for different oils and organic solvents (11.2–25.9 $g\,g^{-1}$). Furthermore, a sodium alginate solution can also be coated on porous substrates, such as melamine sponges and Nylon strainers, to give hydrophobic modification by zirconium ion crosslinking. Melamine is shown in Figure 7.6.

These excellent performances made them a promising material for oil adsorption and cleanup (62).

Floatable and porous foam adsorbents were fabricated by encapsulating attapulgite in sodium alginate via a freeze-drying and

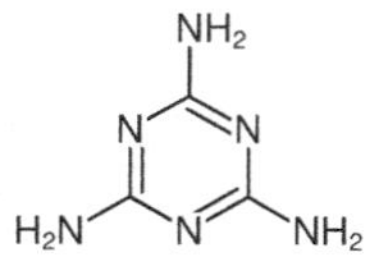

Figure 7.6 Melamine.

post-crosslinking method (63). Both attapulgite and sodium alginate possessed adsorptive sites.

In the adsorption tests, the adsorption capacity was derived from the Langmuir isotherm model, and the maximal adsorption capacity of as-prepared adsorbents was 119.0 mgg^{-1} for Cu(II) and 160.0 mgg^{-1} for Cd(II). Furthermore, these adsorbents presented a strong chemical stability and were readily recyclable because of their floatability in water solution. Thus, the alginate-encapsulated attapulgite foams are potential scalable adsorbents for heavy metal ions removal from polluted water (63).

Alginate bead is a promising strontium adsorbent in seawater. However, highly concentrated Na ions caused overswelling and damaged the hydrogel bead. To improve the mechanical stability of the alginate bead, a flexible foam-type zeolite-alginate composite was synthesized (64).

The Sr adsorption performance was evaluated in seawater. A freeze-dry technique was used to prepare 1–10% zeolite immobilized alginate foams. The immobilization of zeolite into alginate foam converted the macropores into mesopores, which leads to a more compact structure. This resulted in a less swollen composite in seawater medium and exhibited highly improved mechanical stability compared to an alginate bead.

Moreover, the Sr adsorption efficiency and the selectivity were enhanced by the immobilization of zeolite in alginate foam due to the increase of Sr binding sites (zeolite). In particular, Sr selectivity against Na was found to be highly improved. The 10% zeolite-alginate foam exhibited a higher $\log K_d$ of 3.3, while the pure alginate foam exhibited 2.7 in the presence of 0.1 M Na. In a real seawater, the 10% zeolite-alginate foam exhibited 1.5 times higher Sr adsorption efficiency than the pure alginate foam. This result reveals that zeolite-alginate foam composite is an appropriate material for Sr removal in seawater due to its swelling resistance, as well as the improved Sr adsorption performance in complex media (64).

A flexible PU foam was modified with bio-based coatings composed of poly(ether imide) and alginate, fabricated by layer-by-layer self-assembly (65). The layer-by-layer self-assembled coatings modified flexible PU foams were studied for their potential for removing Cu(II) ions from aqueous solutions by using batch adsorption technique as a function of assembled bilayers number, pH value, carrier amount and adsorption time. The maximum adsorption capacity of the modified flexible PU foam was shown to be 54 mgg^{-1} for Cu(II). The results suggested that the layer-by-layer self-assembly could potentially be a promising cost-effective technique for fabricating advanced adsorbent materials for removing pollutants from wastewater (65).

7.4 Biodegradable Adhesive Compositions

Packaging materials that are fully biodegradable also involve the requirement of hot melt adhesive compositions that are biodegradable (66). The fundamental characteristics, general physical and mechanical behavior, and recent developments in the knowledge of hot melt adhesives have been reviewed (67).

Hot melt pressure-sensitive adhesive compositions have been described based on lactic acid polymers (68–70). However, it is claimed that such compositions are not suitable for the fields of food packaging as they have the drawbacks of being rather stiff and having a heat-reactivation temperature that is too high for food packaging applications (66). Hot melt compositions are shown in Table 7.8.

The formulations given in Table 7.8 are representative of hot melt adhesive compositions with open assembly times of greater than 2 s, which makes them suitable for application on rigid supports such as standard cardboards. The hot melt compositions can be prepared by simply mixing their components at a temperature of 100–200°C until a homogeneous mixture is obtained.

Still other ingredients for compositions and suppliers have been described, in particular for breathable biodegradable hot melt adhesives, for their use in personal hygiene articles (71). This issue is detailed in Section 9.1.

Table 7.8 Hot melt compositions.

Formulations		
Chemicals	[%]	[%]
PCL copolymer		35
Polyester from glycol and sebacic acid	35	
Phenolic terpene resin	44.5	44.5
Stearamide	20	
Ethylene bis oleamide		20
Pentaerythritol tetrakis 3-(3,5-di-tert-butyl-4-hydroxyphenyl) propionate	0.5	0.5
Total % by weight	100	100
Properties		
Viscosity 170°C /[*mPa s*]	1560	4800
Ring and ball/[°C]	71	79
Open assembly time/[s]	25	15
Cure time/[s]	6	5

In some cases, hot melt adhesive compositions are tackified, plasticized and reinforced with a variety of resins, oils or waxes that have been originally and initially derived from petroleum feedstocks. These classic compositions are generally very resistant to degradation. However, sucrose benzoate has been investigated and described as a tackifier for water-sensitive or biodegradable hot melt adhesives (72).

7.5 Food Applications

7.5.1 *Edible Packaging*

Edible packaging is a thin layer formed on food surface which can be eaten as an integral part of the food product (73). While an edible coating is formed as thin layer directly on the food surface for improving shelf life of fruits and vegetables, the edible film is formed as thin layer separately and wrapped on food surface later.

Edible films have attracted much interest since they have a potential to overcome the problems associated with plastic packaging. However, their film properties are not as good as the conventional packaging materials such as plastics. On the other hand, the food and beverage industry has much interest in incorporating the benefits of nanotechnology. Nanomaterials have unique characteristics, such as large surface area-to-volume ratio, distinct optical behavior and high mechanical strength, which, when incorporated with the edible films, could improve the film properties of the edible films. Thus, the correct selection and incorporation of nanomaterials can improve the film properties (73).

The recent studies on nanoedible films that have been prepared incorporating the nanofillers, such as nanostarch, nanocellulose, nanochitosan/nanochitin, nanoproteins and nanolipids, have been reviewed (73–75).

7.5.2 *Canola Protein-Based Biodegradable Packaging*

Among various protein sources, canola protein is a novel source for manufacturing polymer films for food packaging (76).

Canola protein can be concentrated or isolated by an aqueous extraction technology followed by protein precipitation. Using this procedure, it was claimed that more than 99% of protein was extracted from the defatted canola meal, and protein recovery was 87.5%. Canola protein exhibits thermoplastic properties when plasticizers are present, including water, glycerol, poly(ethylene glycol), and sorbitol (76). The addition of these plasticizers allows the canola protein to undergo a glass transition and facilitates the deformation and the processability.

Normally, canola protein-based bioplastics showed low mechanical properties which had tensile strength of 1.19 *MPa* to 4.31 *MPa* (76). Various factors were explored to improve it, including blending with synthetic polymers, modifying protein functionality through controlled denaturation, and adding crosslinking agents. Canola protein-based bioplastics were reported to have a glass transition temperature below −50°C, but this strongly depends on the content of plasticizer. Canola protein-based bioplastics have also demonstrated comparable mechanical and moisture barrier properties in comparison to other plant protein-based bioplastics. These materials have a great potential in food packaging applications, including their use as wraps, sacks, sachets, or pouches (76).

7.6 Other Applications

7.6.1 *Chewing Gum*

A chewing gum with two different biodegradable polymers exhibits an improved texture prior to adding softeners. The desired texture properties of the chewing gum can be obtained when combining biodegradable chewing gum polymers (77).

The first polymer is prepared from ε-caprolactone 1,2-propylene glycol using stannous octoate as the catalyst. The second polymer is prepared from dimethyl terephthalate, dimethyl succinate, ethylene glycol and calcium acetate monohydrate.

The fact that biodegradable polymers may actually be configured into a suitable polymer gum base, facilitates the possibility of providing a completely biodegradable chewing gum formulation (77).

7.6.2 *Astaxanthin*

Astaxanthin, a hydrophobic carotenoid found in marine plants and animals, is claimed to exhibit various beneficial biological activities (78). Astaxanthin is shown in Figure 7.7. Its use as a nutraceutical in foods is currently limited by its low water solubility and poor bioavailability.

The goal of a study was to fabricate astaxanthin-loaded colloidal particles to overcome these challenges. Astaxanthin was encapsulated in PLGA nanoparticles coated with chitosan oligosaccharides.

The properties of the loaded nanoparticles were characterized by transmission electron microscopy, SEM, and dynamic light scattering. The influence of PLGA properties on the loading capacity, water solubility, stability, and the release of the astaxanthin were investigated (78).

The nanoparticles were smooth spheres with mean particle diameters around 150 nm and positive surface potentials ($\zeta = +30\ mV$). The encapsulation efficiency (>85%) and loading capacity (>15%) of the astaxanthin in the nanoparticles was relatively high.

X-ray analysis suggested that the encapsulated astaxanthin was in an amorphous form. The nanoparticles had a good dispersibility and stability in aqueous solutions, as well as high cytocompatibility. *In-vitro* studies showed that the astaxanthin was released

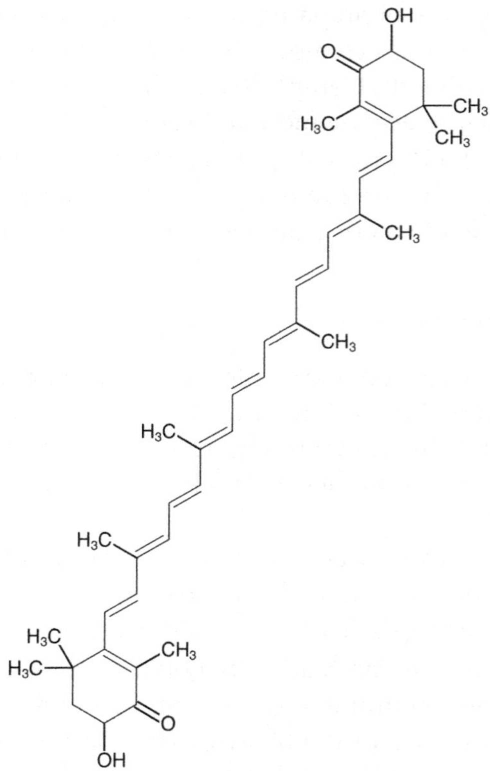

Figure 7.7 Astaxanthin.

from the nanoparticles under simulated gastric and small intestinal conditions. The results of the study suggest that the core-shell nanoparticles developed here may be suitable for encapsulating this important nutraceutical in functional foods and cosmetics (78).

7.6.3 *Edible Films and Coatings*

A comprehensive review of starch-based edible films and coatings described the characteristics of the polymer matrix, methods of film formation, and their properties and potential applications, as well as the influence of the addition of different components into these materials (79). The review also outlined the main current technological advances and future trends in this area.

In a recent study, an attempt was made to standardize the formulation and process protocol for the development of starch-based biodegradable antimicrobial films for the extension of shelf life of food and food products (80). The processing technology of starch edible films was optimized using response surface methodology with different levels of starch (4%, 5%, 6% w/v) incorporated with 0.3% level of sodium alginate and 5% of glycerol at drying temperature 35°C. The starch level was standardized as 6% on the basis of the biochemical characteristics, viz., thickness, penetrability, moisture, water vapor transmission rate, and solubility values.

Different essential oils were selected on the basis of their antimicrobial activity. The standardized formulation of starch was incorporated with 0.5% of essential oils, viz., cinnamaldehyde (3-phenylacrylaldehyde), peppermint oil, clove oil and lemon grass oil. The antimicrobial efficacy was then assessed on the nutrient agar medium against pathogenic microorganisms, viz., *E. coli* and *B. cereus* (80).

Some edible film compositions from starch are summarized in Table 7.9.

Starch can be used to create edible coatings for fresh foods in order to extend their shelf life (82). Wheat starch films with two glycerol contents were formulated to mimic the effects of compounds used to coat fruit. Their structural and functional properties were characterized. It was found that the transfer properties of starch films containing 33% of plasticizer was less effective than film with 50% glycerol.

Table 7.9 Edible film compositions.

Composition	Reference
Corn starch, sodium trimetaphosphate, citric acid	(81)
Wheat starch, glycerol	(82)
Wheat starch, octenyl succinic anhydride	(83)
Rice starch, sucrose ester	(84)
Lemongrass, rosemary oil	(85)
Ulluco	(86)
Aloe vera	(87)

The water diffusivity, oxygen permeability, and water vapor permeability at two different humidity gradients, surface tension, works of surface adhesion and cohesion, and moisture sorption were tested. The glycerol content did not play a significant role in the color or the mechanical properties. Thus, glycerol can strongly affect the functional properties of starch-based coatings and films (82).

A rice starch edible coating blended with sucrose esters was developed for controlling the postharvest physiological activity of Cavendish banana to extend the postharvest quality during ripening at around 20°C (84).

The effectiveness of the coating was assessed against using parameters such as weight loss, titratable acidity, total soluble solids, flesh fruit firmness, ion leakage, color change, respiration, ethylene production, chlorophyll degradation, and starch conversion. Surface morphology studies indicated the binding compatibility of the coating matrix with the fruit peel character and formed a continuous uniform layer over the fruit surface.

The results showed that the coating was effective in delaying the biosynthesis of ethylene and reducing the respiration rate. Other factors were delayed chlorophyll degradation, reduced weight loss and retention of fruit firmness for the first six days.

The shelf life of a coated fruit was prolonged for 12 *d* in comparison with untreated samples. These ripened within 7 *d* (84).

In the case of the essential oils, their inclusion in edible films is limited by their low water solubility (85). Nanoemulsions are thermodynamically stable and transparent systems that can be used as a way of incorporating essential oils into edible film matrices.

Lemongrass and rosemary essential oil nanoemulsions were developed by using a low-energy method.

The effect of the concentration of nanoemulsions incorporated into banana starch edible films was studied (85). It was observed that essential oil nanoemulsions have a plasticizing effect, increasing the water vapor permeability, transparency, and elongation at break of the films. However, the hydrophobic nature of the essential oils led to a decrease in their water solubility (85).

Aloe vera gel is a promising material in food conservation due to its antimicrobial and antioxidant activity (87). However, its application in the formation of edible films and coatings has been small, owing to its low film-forming capability.

The physicochemical properties of film-forming solutions and films prepared using unripe banana starch-chitosan and aloe vera gel at different aloe vera gel concentrations were investigated (87).

The results of the study showed that aloe vera gel considerably affects the rheological and optical properties of the edible coatings, mainly due to increased amounts of solids brought by the aloe vera gel. Also, the film-forming capacity and the physicochemical properties were studied. Most of the film properties were affected by the inclusion of aloe vera gel, which decreased the water vapor permeability, tensile strength and the elongation at break. FTIR studies showed that the inclusion of aloe vera gel disrupts the interaction between starch and chitosan molecules.

It was found that the addition of aloe vera gel creates a crosslinking effect between the phenolic compounds in aloe vera gel and starch molecules, which disrupts the starch-chitosan interaction and greatly affects the properties of both the film-forming solution and edible films (87).

Starch obtained from ulluco (US) was evaluated for its use in edible films prepared using different concentrations of 2.0%, 2.5%, and 3.0% at low temperatures (86). So, the the storage conditions of different foods were simulated. US exhibits a high amylose content (35.3a %), low stability against thermal degradation, and a B-type crystalline structure.

For edible films prepared from US, good barrier properties related to the semicrystalline region were obtained. In addition, good mechanical properties, opacity and stability against thermal degradation were obtained. The extraction and the use of US in the preparation of edible films could be an alternative method for adding value to this crop. Furthermore, the films appear as a potential material for food packaging (86).

The effectiveness of native and heat treated octenyl succinic anhydride modified wheat starch in edible film making was investigated (83). Pasting and rheological properties of the native and octenyl succinic anhydride modified starch as well as antioxidants and sensorial attributes of certain coated and non-coated grapes were evaluated.

The peak and final viscosity of the octenyl succinic anhydride modified wheat starch (1270 *Pa* and 1589 *Pa*) was found to be higher than native starch with 1120 *Pa* and 1490 *Pa*.

The results of the study showed that starch-based coatings have a potential to maintain total phenolic content, total carotenoids and scavenging activities as well as the shelf life and quality of grapes for 13 *d*. The coated grapes were also found microbiologically safe for human consumption. Octenyl succinic anhydride modified starch-based coating was found to be more effective as compared to native wheat starch. Therefore, it was suggested that the edible coating with good water barrier properties can be beneficial for fruit processors to enhance the shelf life of fresh produce (83).

Chemically modified corn starch with sodium trimetaphosphate or citric acid, c.f. Figure 7.8, and grape juice was used to produce edible films (81). Solubility, color, transparency, microstructure and glass transition temperature properties of the films were studied to understand the effects of the grape juice and the modified starch usage in the films.

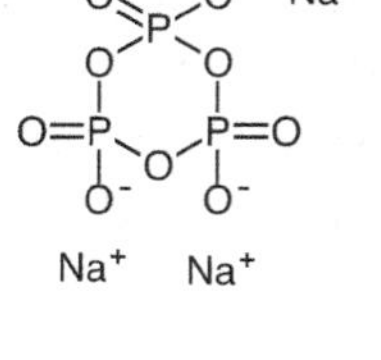

Sodium trimetaphosphate

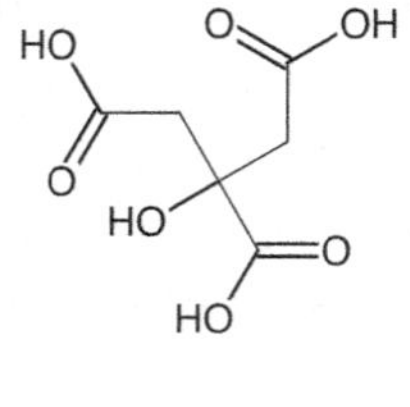

Citric acid

Figure 7.8 Modifiers for corn starch (81).

The usage of starch crosslinked with sodium trimetaphosphate significantly decreased the oxygen permeability, water vapor permeability, solubility, and percent elongation. The chemical reaction between starch and citric acid affected barrier, solubility and elongation properties of the films. The values were higher than the values of sodium trimetaphosphate films (81).

Trade names appearing in the references are shown in Table 7.10.

Table 7.10 Trade names in references.

Trade name Description	Supplier
Biolice® Biodegradable polymer (46)	Limagrain
Biopol® Biodegradable hot melt adhesive (PHV/B) (68, 69)	Zeneca
Cozeen™ 303N Natural coating (46)	Freeman Industries LLC
NatureFlex™ NE 30 Heat-sealable compostable film (46)	Inovia Films Ltd. Comp.

References

1. A. Guzman, N. Gnutek, and H. Janik, *Chemistry & Chemical Technology*, Vol. 5, p. 7, 2011.
2. J.E. Potts, R.A. Clendinning, W.B. Ackart, and W.D. Niegisch, The biodegradability of synthetic polymers in J. Guillet, ed., *Polymers and Ecological Problems*, Vol. 3 of *Polymer Science and Technology*, chapter 4, pp. 61–79. Springer-Verlag, 1973.
3. R. Narayan and Y. Nabar, Thermoplastic and polymer foams and method of preparation thereof, US Patent 7 638 560, assigned to Board of Trustees of Michigan State University (East Lansing, MI), December 29, 2009.
4. F. Scheer, Biodegradable nano-polymer compositions and biodegradable articles made thereof, US Patent 7 927 532, assigned to Cereplast, Inc. (El Segundo, CA), April 19, 2011.
5. K. Helanto, L. Matikainen, R. Talja, and O. Rojas, *BioResources*, Vol. 14, p. 4902, 2019.
6. E. Kosior, R.M. Braganca, and P. Fowler, Lightweight compostable packaging: Literature review, Project: INN003/26, BioComposites Centre, University of Wales, Bangor, Gwynedd, 2006.

7. K. Khemani, P.J. Andersen, S.K. Hodson, and H. Schmidt, Biodegradable polymer films and sheets suitable for use as laminate coatings as well as wraps and other packaging materials, US Patent 6 573 340, assigned to Biotec Biologische Naturverpackungen GmbH & Co. KG (DE), June 3, 2003.
8. M. Goino, Y. Kakida, and S. Obuchi, Biodegradable resin composition, US Patent 7 714 048, assigned to Toho Chemical Industry Co., Ltd. (Tokyo, JP), May 11, 2010.
9. M. Zubair and A. Ullah, *Critical Reviews in Food Science and Nutrition*, Vol. in press, p. 1, 2019.
10. E.L. Sánchez-Safont, A. Aldureid, J.M. Lagarón, J. Gámez-Pérez, and L. Cabedo, *Composites Part B: Engineering*, Vol. 145, p. 215, 2018.
11. M.L. Picchio, Y.G. Linck, G.A. Monti, L.M. Gugliotta, R.J. Minari, and C.I.A. Igarzabal, *Food Hydrocolloids*, Vol. 84, p. 424, 2018.
12. T. Niranjana Prabhu and K. Prashantha, *Polymer Composites*, Vol. 39, p. 2499, 2018.
13. M.L. Sanyang, R.A. Ilyas, S.M. Sapuan, and R. Jumaidin, Sugar palm starch-based composites for packaging applications in M. Jawaid and S.K. Swain, eds., *Bionanocomposites for Packaging Applications*, pp. 125–147. Springer International Publishing, Cham, 2018.
14. F.S. Buehler, J. Casanova, H. Ernst, and H.-J. Schultze, Process and apparatus for producing starch melt and products obtainable by this process, US Patent 5 316 578, assigned to Ems-Inventa AG (CH), May 31, 1994.
15. C. Bastioli, R. Lombi, G. Del Tredici, and I. Guanella, Biodegradable compositions and films or moulded articles obtained therefrom, US Patent 5 569 692, assigned to Novamont S.p.A. (Milan, IT), October 29, 1996.
16. I. Tomka, Thermoplastically processable starch and a method of making it, US Patent 5 362 777, November 8, 1994.
17. J.M. Mayer, M.J. Hepfinger, and E.A. Welsh, Method of producing biodegradable starch-based product from unprocessed raw materials, US Patent 5 322 866, assigned to The United States of America as represented by the Secretary of the Army (Washington, DC), June 21, 1994.
18. D.C. Schiltz, Biodegradable starch-based polymer compositions, US Patent 5 449 708, September 12, 1995.
19. R. Frische, K. Wollmann, R. Gross-Lannert, J. Schneider, and B. Best, Special amyloses and their use for producing biodegradable plastics, US Patent 5 374 304, assigned to Battelle-Institut e.V. (Frankfurt am Main, DE), EMS-Chemie AG (Domat/Ems, CH), December 20, 1994.
20. A.T. Knight, Starch derived shaped articles, US Patent 5 314 754, assigned to Goodman Fielder Wattie Australia Limited (Gladsville, AU), May 24, 1994.

21. Y. Long, S. Coombs, and G.B.Y. Christie, Biodegradable polymer, US Patent 7 495 044, assigned to Plantic Technologies Ltd. (Victoria, AU), February 24, 2009.
22. M.L. Sanyang, S.M. Sapuan, M. Jawaid, M.R. Ishak, and J. Sahari, *Journal of Food Science and Technology*, Vol. 53, p. 326, Jan 2016.
23. J. Muller, C. González-Martínez, and A. Chiralt, *Materials*, Vol. 10, p. 952, Aug 2017.
24. M.L. Sanyang, S.M. Sapuan, M. Jawaid, M.R. Ishak, and J. Sahari, *Carbohydrate Polymers*, Vol. 146, p. 36, 2016.
25. S. Mangaraj, S. Mohanty, S. Swain, and A. Yadav, *Journal of Packaging Technology and Research*, Vol. in press, February 2019.
26. N. Mlalila, A. Hilonga, H. Swai, F. Devlieghere, and P. Ragaert, *Trends in Food Science & Technology*, Vol. 74, p. 1, 2018.
27. X.L. Shen, J.M. Wu, Y. Chen, and G. Zhao, *Food Hydrocolloids*, Vol. 24, p. 285, 2010.
28. C.Y. Basch, R.J. Jagus, and S.K. Flores, *Food Bioprocess Technol.*, Vol. 6, p. 2419, 2013.
29. M.P. Arrieta, M.d.M. Castro-López, E. Rayón, L.F. Barral-Losada, J.M. López-Vilariño, J. López, and M.V. González-Rodríguez, *Journal of Agricultural and Food Chemistry*, Vol. 62, p. 10170, 2014.
30. S. Azlin-Hasim, M.C. Cruz-Romero, M.A. Morris, E. Cummins, and J.P. Kerry, *Food Packaging and Shelf Life*, Vol. 4, p. 26, 2015.
31. E. Salleh, I.I. Muhammad, and Q.A. Pahlawi, *Procedia Chemistry*, Vol. 9, p. 11, 2014. International Conference and Workshop on Chemical Engineering UNPAR 2013 (ICCE UNPAR 2013).
32. C.H. Pagno, T.M. Costa, E.W. de Menezes, E.V. Benvenutti, P.F. Hertz, C.R. Matte, J.V. Tosati, A.R. Monteiro, A.O. Rios, and S.H. Flóres, *Food Chemistry*, Vol. 173, p. 755, 2015.
33. A. Díez-Pascual and A. Díez-Vicente, *International Journal of Molecular Sciences*, Vol. 15, p. 10950, Jun 2014.
34. P. Balakrishnan, S. Gopi, M.S. Sreekala, and S. Thomas, *Starch - Stärke*, Vol. 70, p. 1700139, 2018.
35. V. Goudarzi and I. Shahabi-Ghahfarrokhi, *International Journal of Biological Macromolecules*, Vol. 106, p. 661 , 2018.
36. O. Ochoa-Yepes, C. Medina-Jaramillo, L. Guz, and L. Famá¡, *Starch - Stärke*, Vol. 70, p. 1700222, 2018.
37. F. Luzi, E. Fortunati, A.D. Michele, E. Pannucci, E. Botticella, L. Santi, J.M. Kenny, L. Torre, and R. Bernini, *Carbohydrate Polymers*, Vol. 193, p. 239 , 2018.
38. S. Ketkaew, P. Kasemsiri, S. Hiziroglu, W. Mongkolthanaruk, R. Wannasutta, U. Pongsa, and P. Chindaprasirt, *Journal of Polymers and the Environment*, Vol. 26, p. 311, Jan 2018.
39. A. Babaei-Ghazvini, I. Shahabi-Ghahfarrokhi, and V. Goudarzi, *Food Packaging and Shelf Life*, Vol. 16, p. 103, 2018.

40. C.L. Tanner and C.E. Gibbons, Biodegradable paperboard laminate structure, US Patent 5 213 858, assigned to International Paper (Tuxedo Park, NY), May 25, 1993.
41. B.A. Miksic and R.A. Boyle, Biodegradable corrosion inhibitor packages, US Patent 6 617 415, assigned to Cortec Corporation (St. Paul, MN), September 9, 2003.
42. M.A. Chavannes, Method for making laminated cushioning material, US Patent 3 142 599, assigned to Sealed Air Corp., July 38, 1964.
43. D.F. Brulz, Multi-wall package, US Patent 8 387 797, assigned to Precision Color Graphics, Ltd. (Franklin, WI), March 5, 2013.
44. R.R. Hornstein and G.F. Landrum, Molded biodegradable packaging, US Patent 5 623 815, assigned to Norel (Little Ferry, NJ), April 29, 1997.
45. J.D.E. Remer, T.J. Herringshaw, and J.W. Parent, Method of packaging articles in foam plastic, US Patent 3 204 385, assigned to Continental Aviat. & Eng. Corp., September 7, 1965.
46. C. Tolibas-Spurlock, B. Allred-Forsman, and D. Agnew-Von Moos, Compostable container for storing fluids, US Patent 8 245 848, assigned to Sacred Green, Inc. (Chehalis, WA), August 21, 2012.
47. BIOTA Brands of America, Inc., Telluride, CO, The bottle, 2013.
48. C. Johansson, J. Bras, I. Mondragon, P. Nechita, D. Plackett, P. Šimon, D.G. Svetec, S. Virtanen, M.G. Baschetti, C. Breen, et al., *BioResources*, Vol. 7, p. 2506, 2012.
49. J. Chakravarty, V. Topolkaraev, and G.J. Wideman, Biodegradable multicomponent fibers, US Patent 7 972 692, assigned to Kimberly-Clark Worldwide, Inc. (Neenah, WI), July 5, 2011.
50. M.F. Brian, Production of net-like structures, US Patent 2 919 467, assigned to Plastic Textile Access Ltd., January 5, 1960.
51. H.O. Cederblad, K.E. Misukanis, C.P. Hakanson, and A.J. Thoen, Biodegradable netting, US Patent 7 326 659, assigned to Conwed Plastics LLC (Minneapolis, MN), February 5, 2008.
52. Subcommittee: D20.96, Standard test method for determining aerobic biodegradation of plastic materials under controlled composting conditions, incorporating thermophilic temperatures, ASTM Standard D5338-15, ASTM International, West Conshohocken, PA, 2015. http://www.astm.org/Standards/D5338.htm.
53. R.J.S. Burnet and P.D.M. Parsons, Packaging for plants, fruit or vegetables, US Patent Application 20 190 071 243, assigned to Oceans Of Notions Ltd., March 7, 2019.
54. S.F. Hosseini, Z. Nahvi, and M. Zandi, *Food Hydrocolloids*, Vol. 89, p. 637 , 2019.
55. Y.W. Cheung, D.V. Dobreski, R. Turner, M. Wheeler, and Y.P. Handa, Polymer blends of biodegradable or bio-based and synthetic polymers and foams thereof, US Patent 7 872 056, assigned to Pactiv Corporation (Lake Forest, IL), January 18, 2011.

56. F. Matsuoka, K. Ueda, T. Matsumoto, and Y. Oogi, Biodegradable resin foam sheet, biodegradable resin foam article and biodegradable resin molded container, US Patent 7 972 669, assigned to Unitika Ltd. (Hyogo, JP), July 5, 2011.
57. Y.W. Cheung, D.V. Dobreski, R. Turner, M. Wheeler, and Y.P. Handa, Polymer blends of biodegradable or bio-based and synthetic polymers and foams thereof, US Patent 7 977 397, assigned to Pactiv Corporation (Lake Forest, IL), July 12, 2011.
58. R. Narayan, S. Balakrishnan, Y. Nabar, J.-M. Raquez, and P. Dubois, Starch-polyester biodegradable graft copolymers and a method of preparation thereof, US Patent 7 985 794, assigned to Board of Trustees of Michigan State University (East Lansing, MI), July 26, 2011.
59. D.W. Renn, Biodegradable compositions, articles prepared from biodegradable compositions and manufacturing methods, US Patent 7 618 485, assigned to The Biodegradable Technologies General Partnership (Delta, British Columbia, CA), November 17, 2009.
60. O. Gaserod, P.O. Andersen, and R. Myrvold, Gelled biopolymer based foam, US Patent 7 671 100, assigned to FMC Biopolymer AS (Sandvika, NO), March 2, 2010.
61. N. Phawaphuthanon, D. Yu, P. Ngamnikom, I.-S. Shin, and D. Chung, *Food Hydrocolloids*, Vol. 88, p. 119, 2019.
62. Y. Wang, Y. Feng, and J. Yao, *Journal of Colloid and Interface Science*, Vol. 533, p. 182, 2019.
63. Y. Wang, Y. Feng, X.-F. Zhang, X. Zhang, J. Jiang, and J. Yao, *Journal of Colloid and Interface Science*, Vol. 514, p. 190 , 2018.
64. H.-J. Hong, B.-G. Kim, J. Ryu, I.-S. Park, K.-S. Chung, S.M. Lee, J.-B. Lee, H.S. Jeong, H. Kim, and T. Ryu, *Journal of Environmental Management*, Vol. 205, p. 192, 2018.
65. Y. Pan and H. Zhao, *Polymer-Plastics Technology and Materials*, Vol. 58, p. 384, 2019.
66. C. Robert, Biodegradable hot-melt adhesive composition, US Patent 8 067 492, assigned to Bostik SA (Paris la Defense Cedex, FR), November 29, 2011.
67. W. Li, L. Bouzidi, and S.S. Narine, *Industrial & Engineering Chemistry Research*, Vol. 47, p. 7524, October 2008.
68. G.J. Edgington and C.M. Ryan, Biodegradable/compostable hot melt adhesives comprising polyester of lactic acid, US Patent 5 700 344, assigned to H. B. Fuller Licensing & Financing Inc. (St. Paul, MN), December 23, 1997.
69. G.J. Edgington and C.M. Ryan, Biodegradable/compostable hot melt adhesives comprising polyester of lactic acid, US Patent 5 753 724, assigned to H. B. Fuller Licensing & Financing, Inc. (St. Paul, MN), May 19, 1998.

70. M. Viljanmaa, A. Södergård, and P. Törmälä, *International Journal of Adhesion and Adhesives*, Vol. 22, p. 219, January 2002.
71. S.U. Ahmed and A.W. Emiru, Breathable biodegradable hot melt composition, US Patent 7 910 645, assigned to H.B. Fuller Company (St. Paul, MN), March 22, 2011.
72. M.L. Sharak and C.W. Paul, Sucrose benzoate as a tackifier for water sensitive or biodegradable hot melt adhesives, US Patent 5 574 076, assigned to National Starch and Chemical Investment Holding Corporation (Wilmington, DE), November 12, 1996.
73. J. Jeevahan and M. Chandrasekaran, *Journal of Materials Science*, Vol. in press, p. 1, 2019. in press.
74. J. Jeevahan, M. Chandrasekaran, R.B. Durairaj, G. Mageshwaran, and G.B. Joseph, *Journal of Global Engineering Problems and Solutions*, Vol. 1, p. 9, 2017.
75. M. Zambrano-Zaragoza, R. González-Reza, N. Mendoza-Muñoz, V. Miranda-Linares, T. Bernal-Couoh, S. Mendoza-Elvira, and D. Quintanar-Guerrero, *International Journal of Molecular Sciences*, Vol. 19, p. 705, Mar 2018.
76. Y. Zhang, Q. Liu, and C. Rempel, *Critical Reviews in Food Science and Nutrition*, Vol. 58, p. 475, 2018.
77. L. Andersen and H. Wittorff, Chewing gum comprising at least two different biodegradable polymers, US Patent 7 833 555, assigned to Gumlink A/S (DK), November 16, 2010.
78. C. Liu, S. Zhang, D.J. McClements, D. Wang, and Y. Xu, *Journal of Agricultural and Food Chemistry*, Vol. 67, p. 5113, 2019.
79. F.M. Pelissari, D.C. Ferreira, L.B. Louzada, F. dos Santos, A.C. Corrêa, F.K.V. Moreira, and L.H. Mattoso, Starch-based edible films and coatings: An eco-friendly alternative for food packaging in M.T.P.S. Clerici and M. Schmiele, eds., *Starches for Food Application*, chapter 10, pp. 359–420. Academic Press, 2019.
80. P. Singh, A.S. Chatli, and H.K. Mehndiratta, *International Journal of Development Research*, Vol. 8, p. 23501, 2018.
81. M. Yildirim-Yalçin, M. Şeker, and H. Sadikoğlu, *Food Chemistry*, Vol. 292, p. 6, 2019.
82. E. Basiak, A. Lenart, and F. Debeaufort, *Polymers*, Vol. 10, p. 412, April 2018.
83. S. Punia, K.S. Sandhu, S.B. Dhull, and M. Kaur, *International Journal of Biological Macromolecules*, Vol. 133, p. 110, 2019.
84. R. Thakur, P. Pristijono, M. Bowyer, S.P. Singh, C.J. Scarlett, C.E. Stathopoulos, and Q.V. Vuong, *LWT*, Vol. 100, p. 341, 2019.
85. A.E. Restrepo, J.D. Rojas, O.R. García, L.T. Sánchez, M.I. Pinzón, and C.C. Villa, *Food Science and Technology International*, Vol. 24, p. 705, 2018.
86. A. Galindez, L.D. Daza, A. Homez-Jara, V.S. Eim, and H.A. Váquiro, *Carbohydrate Polymers*, Vol. 215, p. 143, 2019.

87. M.I. Pinzon, O.R. Garcia, and C.C. Villa, *Journal of the Science of Food and Agriculture*, Vol. 98, p. 4042, 2018.

8

Medical Applications

During the past two decades significant advances have been made in the development of biodegradable polymeric materials for biomedical applications. The recent trends, applications and materials have been reviewed (1–5).

Degradable polymeric biomaterials for medical applications are particularly preferred for (1):

- Controlled drug delivery,
- Temporary prostheses, and
- Scaffolds for tissue engineering.

Each of the above-mentioned applications demands materials with tailored properties to provide an efficient therapy.

8.1 Drug Delivery

The development of sustained release technology has proved a number of ways of producing long acting dosage. Of all the methods suggested, the usage of a coating has proven to be one of the most enduring. A historical review of sustained release coating has been given (6).

The issues of drug delivery with polymeric matrices have been viewed also from the opposite view, i.e., drug delivery with non-polymeric matrices (7).

The literature on floating drug delivery systems with special focus on the principal mechanism of floatation to achieve gastric retention has been collected (8). Approaches to design single-unit and multiple-unit floating systems and their classification and formulation aspects have been discussed in detail.

The role of natural polymers for drug delivery has been reviewed (9–11). Natural polymers that are important in drug delivery are summarized in Table 8.1.

Table 8.1 Natural polymers for drug delivery (9).

Natural polymer type	Remark
Acacia	Dried exudate from gum arabic
Agar	Non-branched polysaccharide
Alginate	Polysaccharide
Carrageenan	Marine hydrocolloids
Cellulose	Polysaccharide
Chitosan	Polysaccharide
Dextran	Branched glucan
Dextrin	Low molecular weight carbohydrate
Gellan gum	Anionic polysaccharide
Guar gum	Galactomannan polysaccharide
Inulin	Plant polysaccharide
Karaya gum	Vegetable gum
Konjac glucomannan	From konjac plant
Locust bean gum	Galactomannan vegetable gum
Pectin	Heteropolysaccharide
Psyllium husk	From plant genus *Plantago*
Scleroglucan	Extracellular polysaccharide
Starch	Polysaccharide
Xanthan gum	Polysaccharide

The biodegradation of polymers that are used for controlled drug delivery have been described in a chapter of a monograph (12).

A recently developed nanodelivery strategy involves the use of hydrophilic polymers as carriers of proteins and siRNA (13). By controlling the reaction conditions during the production of the polymer, various degrees of anionic charge, cationic charge, and crosslinking can be added, thereby changing their capabilities as protein and nucleic acid carriers and also promoting effective cell membrane permeation. The recent progress in the ability to modify drug release of hydrophilic polymers nanoparticles has been given in a review (13).

The most common matrix materials for drug delivery are polymers. The field of biodegradable polymers has developed rapidly ever since the synthesis and biodegradability of poly(lactic acid) (PLA) was reported in 1966 (14). Natural polymers used for drug delivery have been reviewed (15).

Examples of other polymers which have been reported as useful as a matrix material for delivery devices include poly(anhydride)s, polyesters such as poly(glycolide)s and poly(lactide-*co*-glycolide)s, poly(amino acid)s such as poly(lysine), polymers and copolymers of poly(ethylene oxide) (PEO), acrylic-terminated PEO, poly(amide)s, poly(urethane)s, poly(orthoester)s, poly(acrylonitrile)s, and poly(phosphazene)s.

Degradable materials of biological origin are well known, including, for example, crosslinked gelatin. Hyaluronic acid has been crosslinked and used as a degradable swelling polymer for biomedical applications.

Biodegradable hydrogels have also been developed for use in controlled drug delivery as carriers of biologically active materials such as hormones, enzymes, antibiotics, antineoplastic agents, and cell suspensions. Temporary preservation of functional properties of a carried species, as well as the controlled release of the species into local tissues or systemic circulation, have been achieved.

Thermo-sensitive hydrogels, characterized by a temperature-dependent sol-gel transition, have been massively used as drug delivery systems for the controlled release of various bioactives (16). Poloxamer 407 is an ABA-type triblock copolymer with a center block of hydrophobic poly(propylene oxide) between two hydrophilic poly(ethylene oxide) lateral chains. Due to its unique thermoreversible gelation properties, poloxamer 407 has been widely investigated as a temperature-responsive material. The gelation phenomenon of poloxamer 407 aqueous solutions is reversible and characterized by a sol-gel transition temperature. The nanoencapsulation of drugs within biocompatible delivery systems dispersed in poloxamer 407 hydrogels is a strategy used to increase the local residence time of various bioactives at the injection site. Examples of biocompatible nanocarriers embedded in P407-based hydrogels are liposomes, niosomes, and ethosomes. The formulations have been detailed in the literature (16).

The state-of-the-art of the most important mixed systems made up of colloidal carriers localized within a poloxamer 407 hydrogel has been reviewed in order to illustrate the possibility of obtaining a controlled release of the entrapped drugs and an increase in their therapeutic efficacy as a function of the biomaterial used (16, 17).

Biocompatible hydrogels with a poorly soluble drug ibuprofen (5%) have been investigated (18). A comprehensive evaluation and comparison of the effect of different bioadhesive polymers on their suitability for application on skin and their physical stability during accelerated and natural aging tests have been presented.

Hydrogels, formulated with xanthan gum 1%, sodium carboxymethyl cellulose 5%, poloxamer 407 16%, and carbomer 1%, were soft pseudoplastic semisolids with thixotropy and biocompatible pH. The type of the polymer significantly affected the apparent viscosity of the hydrogels and miscibility rate with artificial sweat, their physical stability, and shape, size, and aggregation of the drug crystals and degree of crystallization (18).

The drug release in all investigated hydrogels was diffusion-controlled in accordance with the Higuchi model and sustained for 12 *h*, with the drug release rate and the amount of drug released dependent on the polymer (18).

Three dimensional (3D) printing techniques for wood-derived biopolymers for use in biomedical applications and as novel composite materials for electronics and energy devices have been investigated (19). This review highlights recent work on emerging applications of wood-derived biopolymers and their advanced composites with a specific focus on customized pharmaceutical products and advanced functional biomedical devices prepared via 3D printing. Specifically, various biofabrication strategies in which woody biopolymers are used to fabricate customized drug delivery devices, cartilage implants, tissue engineering scaffolds and items for other biomedical applications are discussed in this review (19).

A 3D printing technology was used to precisely construct nanogel discs carrying paclitaxel and rapamycin (20). The 3D-printed nanogel disc rounds (12 *mm* diameter × 1 *mm* thickness) carrying paclitaxel and rapamycin evaded a premature gelation during storage and the initial burst release of the drugs in the dissolution medium. *In-vivo* 3D-printed nanogel discs permitted a successful intraperitoneal delivery of paclitaxel and rapamycin in ES-2-luc ovarian-cancer-bearing xenograft mice. The materials were also shown to be therapeutically effective and capable of preventing postsurgical peritoneal adhesions in the treated xenograft mice (20).

The proper choice of hydrogel macromers can produce membranes with a range of permeability, pore sizes and degradation rates suitable for a variety of applications in surgery, medical diagnosis and treatment. Dispersion systems used for pharmaceutical and cosmetic formulations can be categorized as either suspensions or emulsions.

Natural polymers that can be used as suspending agents and dispersing agents are summarized in Table 8.2.

Table 8.2 Natural polymers for drug delivery (21).

Suspending agents	Dispersing agents
Gum acacia	Heptadecaethylene-oxycetanol
Gum tragacanth	Lecithin
Hydroxypropyl methyl cellulose	Polyoxyethylene sorbitan monoleate
Methyl cellulose	Polyoxyethylene sorbitol monoleate
Sodium alginate	Polyoxyethylene stearate
Sodium carboxymethyl cellulose	

Particularly useful rate-controlling polymers for causing an effective controlled release of administered drug or agent following the administration are summarized in Table 8.3.

Table 8.3 Natural polymers for delivery rate control (22).

Polymer	Polymer
Gum arabic	Agar
Guar gum	Starches
Dextran	Gelatin
Hydroxypropyl cellulose	Hydroxyethyl cellulose
Hydroxypropyl methylcelluose	Sodium carboxymethyl cellulose
Pectin	Carrageenan
Ethyl cellulose	Methyl cellulose
Cellulose acetate	Cellulose acetate butyrate
Cellulose acetate phthalate	Cellulose acetate trimellitate
Hydroxypropyl methyl cellulose phthalate	Hydroxypropyl methyl cellulose acetate succinate

Suspensions are defined as solid particles ranging in size from a few nanometers up to hundreds of microns dispersed in a liquid medium using suspending agents. Solid particles include microspheres, microcapsules, and nanospheres.

Emulsions are defined as dispersions of one liquid in another, stabilized by an interfacial film of emulsifiers such as surfactants and lipids. Emulsion formulations include water-in-oil and oil-in-water emulsions, multiple emulsions, microemulsions, microdroplets, and liposomes. Microdroplets are unilamellar phospholipid vesicles that consist of a spherical lipid layer with an oil phase inside.

Liposomes are phospholipid vesicles prepared by mixing water-insoluble polar lipids with an aqueous solution. The unfavorable entropy caused by mixing the insoluble lipid in the water produces a highly ordered assembly of concentric closed membranes of phospholipid with entrapped aqueous solution.

An implant can be formed *in situ* by dissolving a non-reactive, water-insoluble thermoplastic polymer in a biocompatible, water-soluble solvent to form a liquid, placing the liquid within the body, and allowing the solvent to dissipate to produce a solid implant.

The polymer solution can be placed in the body via a syringe. The implant can assume the shape of its surrounding cavity. In an alternative embodiment, the implant is formed from reactive, liquid oligomeric polymers which contain no solvent and which cure in place to form solids, usually with the addition of a curing catalyst.

Drug-loaded biodegradable microneedles have been described for transdermal delivery technology. The microneedles have been proposed for their use in post-operative pain management through delivery of a local anaesthetic, i.e., lidocaine (23).

8.1.1 Acacia

Gum acacia is mainly used in oral and topical formulations as suspending agent and emulsifier. The synthesis of SNPs from gum arabica with particle size of 35 *nm* has been described (21). The silver nanoparticles show a high salt stability and antimicrobial activity against both gram-positive and gram-negative bacteria. Hence, these gum-stabilized silver nanoparticles can be used for preparing antimicrobial surface layers on solid material surfaces or fabric clothing since these nanoparticles are stable in high salt concentrations and antimicrobial in nature.

In addition, by functionalizing them with target drug we can use them as drug delivery vehicles in nano drug delivery applications.

8.1.1.1 Controlling the Rate of Degradation of PLA

It is possible to control the rate of degradation of lactic acid polymers by homogeneously blending certain additives which are fully miscible with PLA and will not leach out. The blending process is simple and results in stable polymer blends which can be readily thermoformed, such as by injection molding, to form implantable medical devices which will both maintain their physical strength yet biodegrade in a predictable manner (24). Before, the degradation rate was controlled by the use of physical barrier layers to retain the accelerator or by the use of complex polymer systems.

The strength retention of a homogeneous blend of PLA in admixture with an additive in a certain amount can be represented as follows:

$$\frac{\%_{Additive}}{100 M_{Acid}} = \left(\frac{\ln \frac{M_{n,0}}{M_{n,s}} - k_1 t}{k_2 t} \right)^2 - \frac{1}{M_{n,0}} \tag{8.1}$$

Here, $M_{n,0}$ is the initial molecular weight of the polymer, $M_{n,s}$ is the molecular weight of the polymer at which it loses strength, M_{Acid} is the molecular weight of the acid, t is the duration in weeks within which that strength retention is required, and k_1, k_2 are some kinetic constants.

The degradation rates of the additives employed as 2% component in a PLA blend are shown in Table 8.4.

The degradation rate in Table 8.4 is the slope of a graph of $\ln(M_n)$ against the degradation time in weeks. A preferred additive for use is lauric acid. This may be employed as the acid per se or as a derivative, for example, in the form of lauric acid anhydride.

Eventually, an implantable, biodegradable medical device can be formed from a homogeneous polymer blend. The fabrication proceeds in detail as follows (24):

Table 8.4 Degradation rate by acids (24).

Additive	Degradation rate $weeks^{-1}$
Hexanoic acid	−0.0565
Octanoic acid	−0.0448
Decanoic acid	−0.0472
Lauric acid	−0.0326
Myristic acid	−0.0281
Crotonic acid	−0.0489
4-Pentenoic acid	−0.0567
2-Hexenoic acid	−0.0713
Undecylenic acid	−0.07
Petroselenic acid	−0.0542
Oleic acid	−0.0442
Erucic acid	−0.0315
2,4-Hexadienoic acid	−0.0618
Linoleic acid	−0.0488
Linolenic acid	−0.0589
Benzoic acid	−0.0798
Hydrocinnamic acid	−0.0737
4-Isopropylbenzoic acid	−0.0728
Ibuprofen	−0.051
Ricinoleic acid	−0.061
Adipic acid	−0.0373
Suberic acid	−0.0311
Phthalic acid	−0.0855
2-Bromolauric acid	−0.0769
2,4-Hydroxydodecanoic acid	−0.0318
Monobutyrin	−0.0347
2-Hexyldecanoic acid	−0.0339
2-Butyloctanoic acid	−0.0467
2-Ethylhexanoic acid	−0.0473
2-Methylvaleric acid	−0.0411
3-Methylvaleric acid	−0.0587
4-Methylvaleric acid	−0.0553
2-Ethylbutyric acid	−0.053
trans-β-Hydromuconic acid	−0.039

Blends of PLA with lauric acid, in amounts of 2% and 5%, are prepared by first dry blending the solid materials and then solution blending with 90% chloroform by roller mixing.

After complete dissolution of the solids, the solutions are cast onto an open tray, dried in a fume cupboard and then *in vacuo*. The dry films are then compression molded at 195°C to form transparent sheets.

Experiments to elucidate the degradation behavior are done by immersion in a standard phosphate buffer solution at 37°C for 10 weeks. During the test period, the molecular weight of the polymer blend was analyzed in samples to measure the degree of degradation.

Furthermore, the amount of lauric acid in the polymer was measured to determine the leaching of the lauric acid and the amount of lactic acid was measured to estimate the amount of degradation products released into the buffer solution. The lauric acid remaining in the sample was determined by gas chromatography/mass spectroscopy (GC/MS). The decrease in molecular weight over time is shown in Figure 8.1.

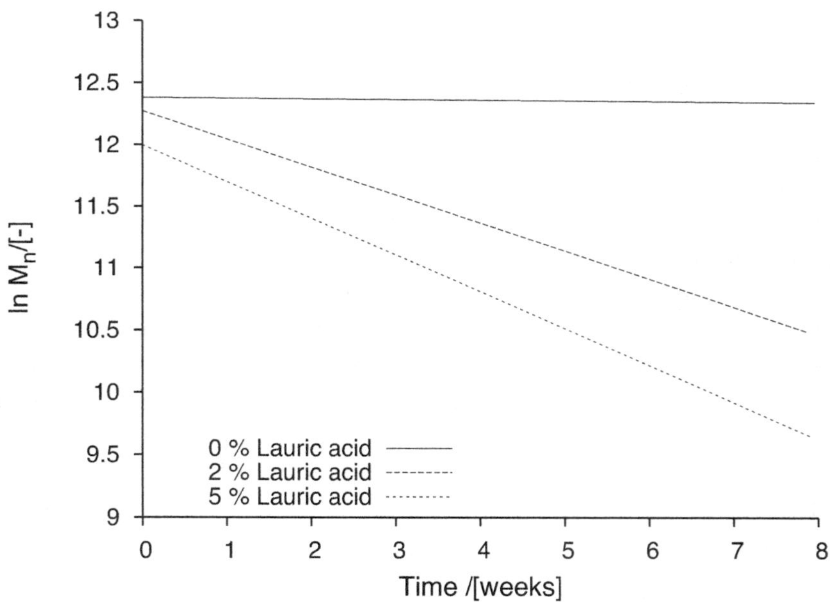

Figure 8.1 Decrease in molecular weight of PLA over time (24).

8.1.2 PLA and PLGA Copolymers

The PLA and poly(lactic-*co*-glycolic acid) (PLGA) copolymers are biodegradable polyesters, which have been widely employed over the last decades for the design of drug delivery systems such as polymeric hydrogels and nanocarriers, e.g., micelles and nanoparticles (25). The coupling with poly(ethylene glycol) (PEG) offers some advantages with respect to PLA and PLGA, including a higher hydrophilicity and a prolonged retention time for nanoparticulate systems, as well as the possibility of preparing thermoresponsive hydrogels.

A large variety of pharmacologically active compounds (small molecules, natural compounds or biomolecules such as proteins, peptides, oligonucleotides) have been formulated and delivered through PEGylated PLA or PLGA copolymers. Due to the high number of papers recently published about the use of these biodegradable copolymers in drug delivery, PEGylated PLA or PLGA copolymers are still considered attractive. Their potential applications have also been broadened by the developing of ligand-functionalized copolymers, enabling an *active drug targeting* for nanoparticulate systems (25).

8.1.2.1 Hepatitis B Virus Drug

To avoid severe exacerbations in the load of hepatitis B virus as a consequence of discontinuous use of anti-hepatitis B virus drugs, entecavir, c.f. Figure 8.2, the first-line anti-hepatitis B virus drug, was primally formulated as extended-release poly(lactic-co-glycolic acid) microspheres in a study (26). Because entecavir is slightly soluble in water and in some other organic solvents used for microsphere preparation, methods for solid-microencapsulation were employed to fabricate the entecavir microspheres. The optimized microspheres were evaluated for their morphology, particle size, drug loading, *in-vitro* drug release, and *in-vivo* pharmacokinetics in rats. The optimized formulation was found to have a mean particle size of 86 μm and drug loading of 13%.

Differential scanning calorimetry (DSC) and powder X-ray diffraction indicated that entecavir existed in crystal, amorphous, and molecular states in the microspheres. *In-vitro* and *in-vivo* re-

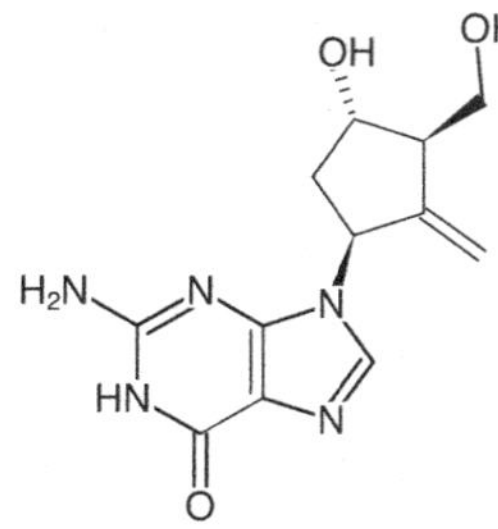

Figure 8.2 Entecavir.

lease revealed that the dissolution of entecavir dominated the release process. The morphology of the microspheres and changes in the morphology during *in-vitro* release were assessed by scanning electron microscopy (SEM). The novel entecavir microspheres described in this study should have great potential for clinical use as an alternative treatment against hepatitis B virus (26).

8.1.3 Poly(γ-glutamic acid)

Biocompatible thermosetting hydrogels are attractive compounds either for the delivery of drugs or as medical devices for tissue regeneration purposes (27). Poly(γ-glutamic acid) and poly(γ-glutamic acid)-based hydrogels were explored to evaluate poly(γ-glutamic acid) and crosslinking agent combinations in order to obtain thermosetting hydrogels suitable for injection.

The attention of the study focused on hydrogels that can form by ionic interaction with compounds bearing amine groups, and/or by interaction with polymers such as chitosan and symmetric triblock copolymer (polyethylene glycol-polypropylene glycol-polyethylene glycol). The characterization parameters were: Rheologic behavior, osmolarity, syringeability and injectability.

The results of the study showed that for poly(γ-glutamic acid) solutions the syringeability was always acceptable, while the injectability test did not meet aspiration by 22G needle. Osmolarity value suitable for injection could be obtained by adding buffering solutions. Only ternary blends made of chitosan, sodium β-glycerophosphate and poly(γ-glutamic acid), or poly(γ-glutamic acid), Pluronic F68, and *l*-lysine, with fixed compositions, resulted in thermosetting gels. The so-obtained thermosetting gels were not reversible (27).

Comblike amphiphilic polymers could be obtained by grafting long paraffinic chains on microbial poly(γ-*d*,*l*-glutamic acid) poly(β-*l*-malic acid) in two steps, i.e., the allylation of the carboxylic side groups followed by UV-initiated thiol-ene click reaction with 1-alkanethiols bearing 8, 12 and 16 carbon atoms (28).

The incubation in water under physiological conditions led to hydrolysis of the lateral ester bonds followed by the scission of the polyamide or the polyester main chain.

As model drugs, theophylline and carbamazepine were efficiently encapsulated in these systems with much better results attained for the latter. Drug release from nanoparticles incubated under physiological conditions occurred with a burst effect and were completely discharged in 24 *h*. The release profiles recorded from drug-loaded films suggested that the drug was delivered in both cases through a diffusion process (28).

8.1.4 Carrageenan

Carrageenan is used as a gelling agent. κ-Carrageenan is shown in Figure 8.3. There are studies in the field of controlled drug release (29–33). The release rate increases both with an increase in the tablet diameter and increase in drug-to-carrageenan ratio in the tablets. The moisture content of carrageenans, storage of tablets at 37°C at 75% relative humidity for 3 months, and incorporation of 10% sodium chloride in the tablets did not show a significant effect on the release rate.

However, the change in ionic strength of a simulated gastric fluid alters the release rate. Carrageenan tablets were found to be relatively insensitive to small changes in formulation parameters and dissolution conditions (29).

The drug release rates of ibuprofen in tablets formulated with carrageenans and cellulose ethers were investigated. Multiple regression analysis was used to predict the time for 50% release as a function of the concentration of the polymers used. The majority of the matrix tablets that contained 10% polymer disintegrated prematurely. Of all the polymer blends that were investigated, the

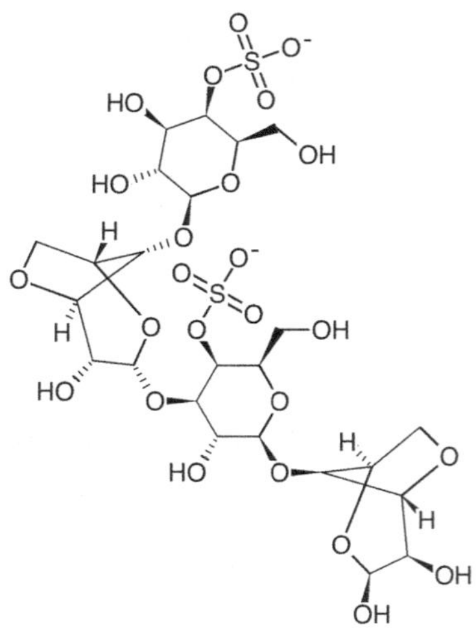

Figure 8.3 κ-Carrageenan.

combination of λ-carrageenan (Viscarin) and hydroxypropyl methyl cellulose gave almost linear release profiles over the entire range of concentration studied (30).

8.1.5 Cellulose

Hydrophilic sustained release matrix tablets are often prepared from cellulose ethers, e.g., hydroxypropyl methyl cellulose. When these materials are exposed to water, the surface of the polymer becomes hydrated. In this way, a gel layer is formed on the glassy core. This changes the barrier properties that are controlling the drug release by a diffusion mechanism.

The exact mechanism that governs the drug release is still somewhat controversial. A swelling-controlled mechanism is often suggested based on a simple data fitting to a mathematical model (34, 35).

A drug release mechanism from compressed cellulose ether matrices using dimensionless analysis has been discussed (36). A non-Fickian mechanism could be dismissed when calculating the Deborah and the swelling interface numbers from relaxation, penetrant diffusion, swelling and drug diffusivity data (36).

The Deborah number is the ratio of the characteristic polymer relaxation time and a characteristic diffusion time (37, 38). It is often used in rheology to characterize the fluidity of materials under specific conditions of flow.

Also, the effect of pH on the transport mechanisms was investigated. A Fickian diffusion is suggested for pure water at pH 6 and Case II transport at pH 2 (36).

8.1.6 *Chitosan*

Chitosan is a derivative of chitin. Chitin exhibits an antimicrobial activity. The main natural sources of chitin are insects and yeast (15). Chitosan is comparatively reactive and can be produced in various forms such as powder, paste, film, or fiber (39, 40). Chitosan is shown in Figure 8.4.

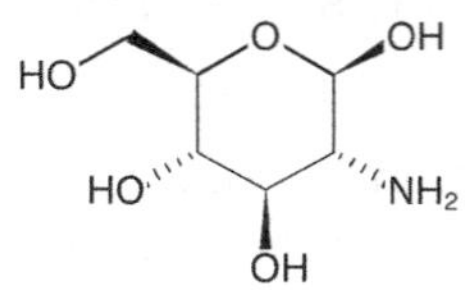

Figure 8.4 Chitosan.

It was suggested that chitosan-based tablets are safe for human use (41). Chitosan-based micro- or nanoparticulate drug delivery systems have been reviewed (42).

8.1.7 *Gellan Gum*

Gellan gum is an exocellular polysaccharide that is secreted by *Pseudomonas elodea*. This gum had been investigated as tablet disintegrant, binder, gelling agent, and for controlled release. The release kinetics of amoxicillin from gellan beads follows the diffusion model for an inert porous matrix (43). Amoxicillin is shown in Figure 8.5.

Gellan gum microparticles coated with colon-specific films based on retrograded starch and pectin were developed for enhancing the oral release of insulin (44). The developed system promoted an impressive protection of insulin (80%) after 120 *min* of incubation with trypsin and α-chymotrypsin, while only 3% of the free insulin

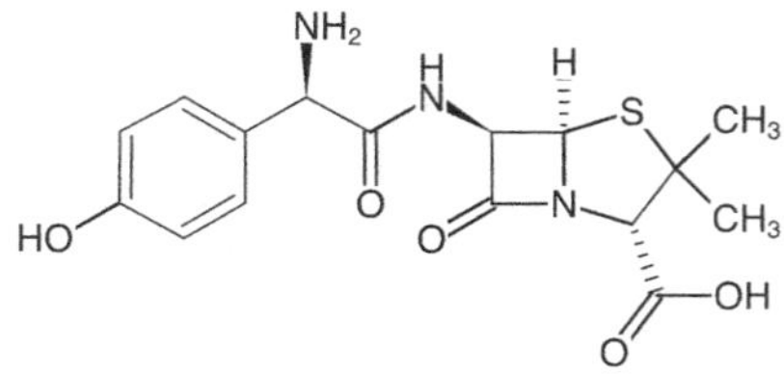

Figure 8.5 Amoxicillin.

remained intact after the same time, possibility due to the calcium chelating activity of the polymers in inhibiting the proteolytic activity.

In-vitro insulin release in media simulating the gastrointestinal portions revealed a pH-dependent behavior, as well as the significance of the coating in lowering the release rates in relation to their counterparts. The permeability of insulin on Caco-2 cell monolayers and excised rat intestine were significantly improved, mainly due to the influence of the anionic polymers on tight junctions opening, along with the excellent mucoadhesive properties of the gellan gum. All these features together contributed greatly to the hypoglycemic effect observed after the oral administration of the insulin-loaded microparticles in diabetic rats, with a reduction of up to 51% of the blood glucose levels. These findings could contribute to the advances on the search alternatives to the oral administration of insulin (44).

Gellan gum was used to fabricate spherical porous beads that are suitable as sustained drug delivery systems for oral administration (45). The gellan gum was crosslinked with calcium ions to prepare polymeric beads.

Rheological studies and preliminary experiments of beads preparation allowed identifying the gellan gum and the $CaCl_2$ concentrations suitable for obtaining stable and spherical particles. Gellan gum beads were formed through ionotropic gelation technique, with and without the presence of the synthetic clay laponite.

The resultant beads were analyzed for dimensions (before and after freeze-drying), morphological aspects and ability to swell in different media, miming biological fluids, namely simulated gastric fluid (HCl 0.1 *M*) and simulated intestinal fluid (simulated intestinal fluid, phosphate buffer, 0.044 *M*, pH 7.4).

The swelling degree was lower in simulated gastric fluid than in simulated intestinal fluid and further reduced in the presence of laponite. The gellan gum and gellan gum-layered silicate composite beads were loaded with two model drugs having different molecular weight, namely theophylline and cyanocobalamin (vitamin B12), c.f. Figure 8.6, and subjected to *in-vitro* release studies in simulated gastric fluid and simulated intestinal fluid. The presence of laponite in the bead formulation increased the drug entrapment efficiency and slowed down the release kinetics of both drugs in the gastric environment. A moving-boundary swelling model with diffuse glassy-rubbery interface was proposed in order to describe the swelling behavior of porous freeze-dried beads. Consistent with the swelling model adopted, two moving-boundary drug release models were developed to interpret release data from highly porous beads of different drugs: drug molecules, e.g., theophylline, that exhibit a typical Fickian behavior of release curves and drugs, such as vitamin B12, whose release curves are affected by the physical/-chemical interaction of the drug with the polymer/clay complex. The theoretical results support the experimental observations, thus confirming that laponite may be an effective additive for fabricating sustained drug delivery systems (45).

8.1.8 *Guar Gum*

The use of guar gum as carrier in the formulation of sustained release systems has been assessed. Tablets with diltiazem as model drug were prepared. Diltiazem is shown in Figure 8.7. Guar gum is an economical alternative (46).

Guar gum-based three-layer matrix tablets for oral controlled delivery of highly soluble metoprolol tartrate as a model drug have been evaluated (47).

8.1.9 *Hyaluronic Acid Derivatives*

Hyaluronan or hyaluronic acid is a linear mucopolysaccharide constituted by *N*-acetyl-*D*-glucosamine and *D*-glucuronic acid repeating units. The structure of hyaluronic acid is shown in Figure 8.8.

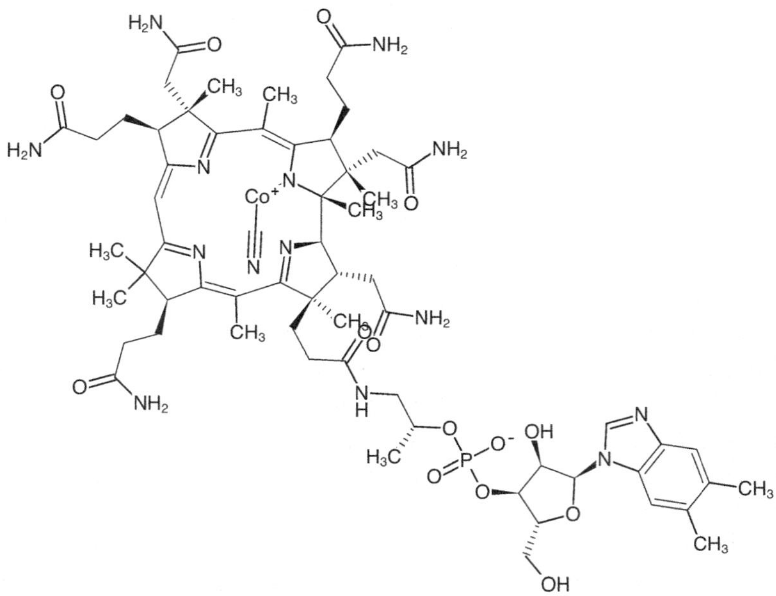

Figure 8.6 Cyanocobalamin.

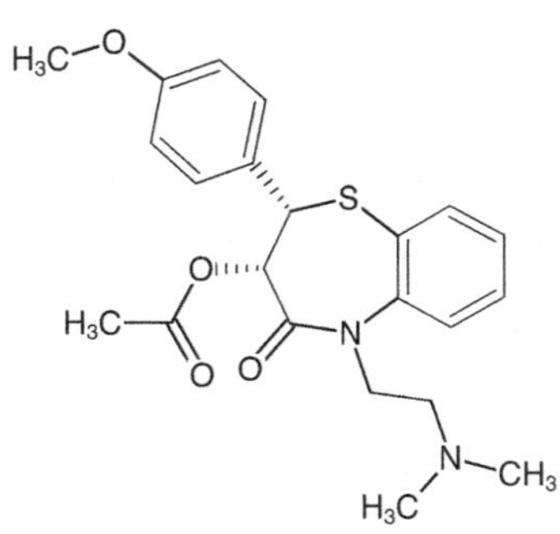

Figure 8.7 Diltiazem.

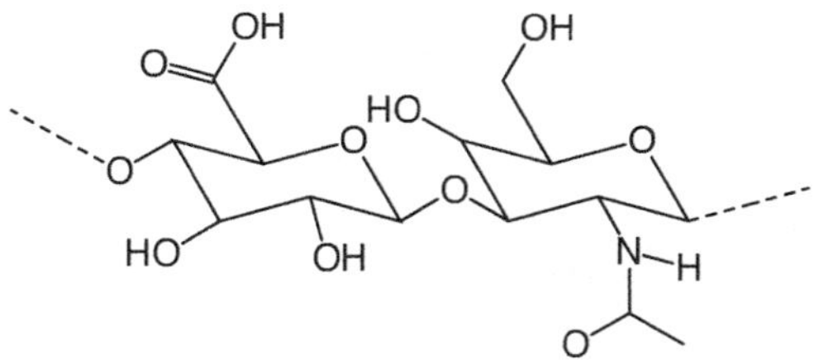

Figure 8.8 Hyaluronic acid.

Hyaluronic acid was first found in the vitreous body of cattle eye by Meyer and Palmer in 1934 (48), and then was found in other tissues such as extracellular matrix and synovial fluid of the joints.

Hyaluronic acid is a viscoelastic fluid filled in the space between cells and collagenous fibers and coated on some epidermal tissues. Hyaluronic acid plays an important role in the biological organism as a mechanical support of the cells of many tissues such as the skin, the tendons, the muscles and cartilage. Hyaluronic acid also performs other functions in biological processes such as moistening of tissues, lubrication, and cellular migration (49).

Hyaluronic acid and its salts have been used in pharmaceutical, surgical and cosmetic fields. However, since hyaluronic acid is expensive and is quite easily biodegraded, its application has been rather limited. Several methods have been developed to modify hyaluronic acid in order to increase its resistance to biodegradation. A series of publications deal with salts and ester derivatives, including crosslinked substances (50, 51).

Hyaluronic acid has also been modified with a long-chain alkyl group or a biodegradable prepolymer (49). The prepolymer may be hydrophobic, hydrophilic, or amphiphilic. Isocyanates may be used to support the grafting reaction. The pendent hydroxyl groups in hyaluronic acid may be used for the addition of isocyanate groups in forming urethane moieties.

The synthesis of an octyl urethane derivative of hyaluronic acid is subsequently illustrated:

Preparation 8–1: A quaternary ammonium salt of hyaluronic acid is dissolved in dimethyl sulfoxide. Octyl isocyanate as reactant and di-*n*-butyltin dilaurate as the catalyst are added. The reaction is done at 65°C for 8 *h*. Afterwards, dibutylamine is added to stop the reaction. Eventually, the reaction mixture is dialyzed in saturated sodium chloride aqueous solution, purified, exchanged from the quaternary ammonium salt to a sodium salt, and finally freeze-dried.

A pharmaceutically active or bioactive molecule can be entrapped in the hyaluronic acid derivative micelles to form a pharmaceutically active or a bioactive composition with a stable controlled released effect (49).

8.1.10 Khaya Gum

Karaya gum is a hydrophilic polysaccharide obtained from the *Khaya grandifiola* tree. Khaya gum is used as a binding agent in tablets and for controlled release. Khaya gum provides a controlled release in paracetamol (4-acetamidophenol) tablets for 5 *h*. Paracetamol is shown in Figure 8.9.

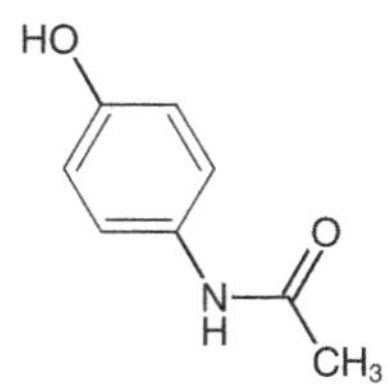

Figure 8.9 Paracetamol.

A combination of khaya gum and hydroxypropyl methyl cellulose showed a zero-order release kinetics. Therefore, tablet matrices containing khaya gum can be used for sustained release (52, 53).

8.1.11 Locust Bean Gum

Locust bean gum has been investigated for its use as superdisintegrant additive. An orodispersible tablet material of nimesulide with locust bean gum has been fabricated. Nimesulide is shown in Figure 8.10. It was found that a tablet formulation with 10% locust bean gum disintegrates in only 13 *s* (54).

A mixture of locust bean gum and chitosan was used as a mucoadhesive component in buccal tablets (55). Propranolol hydrochloride

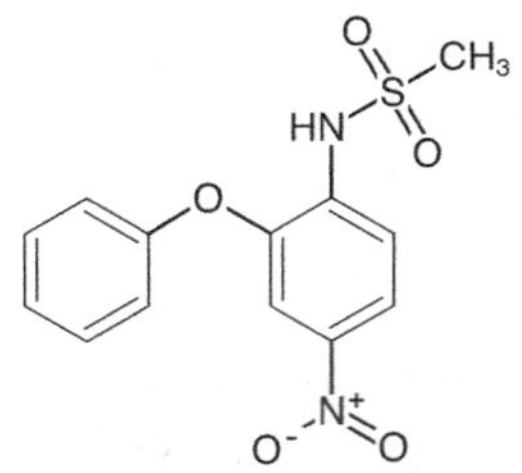

Figure 8.10 Nimesulide.

buccal tablets with various weight ratios of locust bean gum and chitosan were prepared and coated with 5% w/v ethyl cellulose on one face, and oral tablets containing 10 *mg* propranolol hydrochloride alone, were formulated using a direct compression technique.

The strength of mucoadhesion of the tablets was quantified based on the tensile force required to break the adhesive bond between a model membrane, i.e., porcine buccal mucosa and the test polymer. The study revealed that locust bean gum and chitosan in a weight ratio of 2:3 not only releases the drug unidirectionally from the dosage form, but also results in buccal tablets which are sufficiently mucoadhesive for clinical applications (55).

8.1.12 Pectin

The major sources of pectin are plant cell walls. These include sunflower, orange, lemon, carrot, mango, guava, and papaya. Pectin is the methoxyester of pectic acid. Pectin exhibits a stability in acidic media.

Pectin-hydroxypropyl methyl cellulose has been used as compression coating material (56). This envelope was found to be a promising drug delivery system for those drugs to be delivered to the colon. The kinetics of release of 5-aminosalicylic acid is shown in Figure 8.11.

Colonic drug delivery systems using a pectin matrix system have been reported (57, 58).

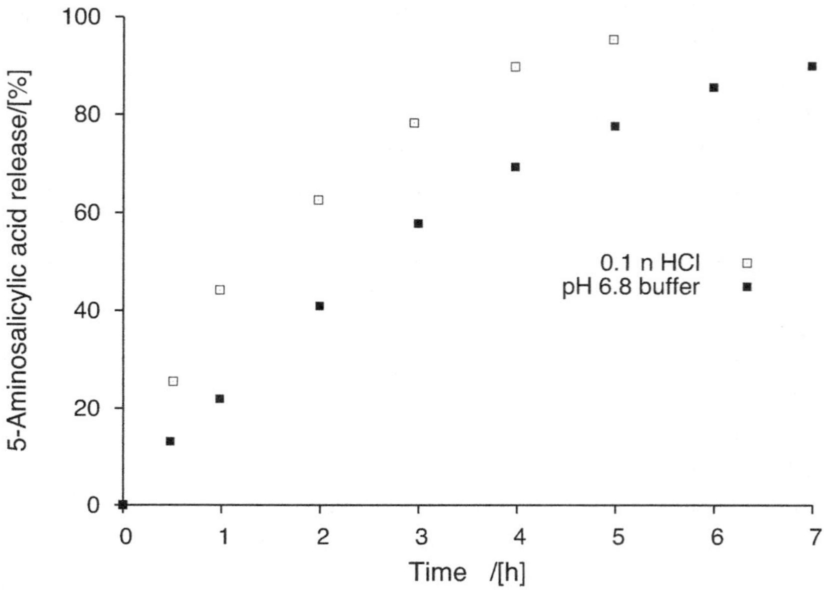

Figure 8.11 Release of 5-aminosalicylic acid over time (56).

8.1.13 *Xanthan Gum*

Xanthan gum is a hydrophilic, anionic bacterial heteropolysaccharide. Xanthan gum can be obtained from the fermentation of the gram-negative bacterium *Xanthomonas campestris* (15).

Sustained release tablets for metoprolol tartrate have been prepared using xanthan gum and tragacanth (59).

8.1.14 *Tragacanth Gum*

Recent progress in the industrial and biomedical applications of tragacanth gum has been reviewed (60).

Tragacanth gum is an anionic polysaccharide. As a biomaterial, tragacanth gum has been used in industrial settings, such as food packaging and water treatment, as well as in the biomedical field as drug carrier and for wound-healing purposes.

The structure, properties, cytotoxicity, and degradability as well as the recent advances in industrial and biomedical applications of tragacanth gum have been detailed (60).

Tragacanth gum-based silver nanocomposite hydrogels were prepared using acrylamide as monomer with a redox polymerization method (61). Tragacanth gum acts as stabilizer to produce uniform and large silver nanoparticles in the presence of *Terminalia chebula* leaf extract in the hydrogel network.

Fourier transform infrared spectroscopy (FTIR) demonstrated the structural units of functional groups of polymeric backbone of the hydrogels. The formation of the silver nanoparticles in the tragacanth gum-based hydrogels could be confirmed by UV-Vis spectra. The morphology of homogeneously dispersed silver nanoparticles throughout hydrogel networks was confirmed by SEM. Transmission electron microscopy (TEM) indicated that silver nanoparticles with average diameters of around 5 *nm* were formed within the hydrogel networks.

The antibacterial activity of the silver nanoparticles against gram-negative *Escherichia coli* and gram-positive *Bacillus subtilis* bacteria was evaluated (61). The so-obtained results proved that these hydrogels have great potential for their use in wound-healing as well as for water purification applications.

A graft polymerization of itaconic acid on tragacanth gum was carried out using potassium persulfate as initiator to develop smart hydrogels for drug delivery systems (62). In the study, the effect of the grafting parameters on the degree of grafting was investigated. The grafting was significantly influenced by the reaction medium, the reaction temperature and the monomer concentration. The monomer dependency of the system was found to be 1.52.

The hydrogels were characterized using FTIR spectroscopy, TG, DSC, and XRD. The grafted tragacanth gum had an amorphous nature and exhibited excellent swelling in water and strong dependence on the pH of the medium. The grafted tragacanth gum showed a pH-dependent release of ciprofloxacin. Ciprofloxacin is shown in Figure 8.12.

This behavior offers the possibility of using these materials in colon-specific drug targeting for human healthcare systems (62).

Electrospun curcumin-loaded poly(ε-caprolactone)/tragacanth gum for nanofibers has been used for wound-healing in diabetic rats (63). Curcumin is shown in Figure 8.13.

These scaffolds with antibacterial property against methicillin-resistant *Staphylococcus aureus* as gram-positive bacteria and extended

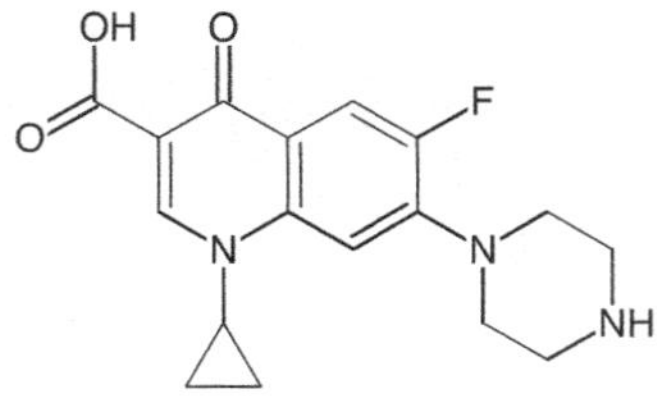

Figure 8.12 Ciprofloxacin.

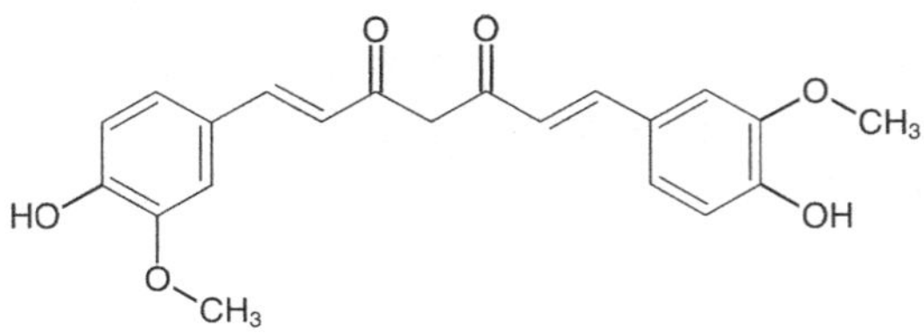

Figure 8.13 Curcumin.

spectrum β-lactamase as gram-negative bacteria were applied in two forms of acellular and cell-seeded grafts for assessing their capability in healing full-thickness wound on the dorsum of rats.

After 15 *d*, a pathological study showed that the application of these nanofibers caused markedly fast wound closure with well-formed granulation tissue dominated by fibroblast proliferation, collagen deposition, complete early regenerated epithelial layer and formation of sweat glands and hair follicles (63).

8.1.15 *Electrospinning*

Electrospinning was already described in the early 19th century (64, 65). Details of modern electrospinning techniques have been described (66, 67).

The process of electrospinning is an intermediate of dry spinning and electrospraying. Typically, very fine drawn fibers can be fabricated in the range of micrometers and nanometers.

An image using light microscopy of electrospun fibers is shown in Figure 8.14.

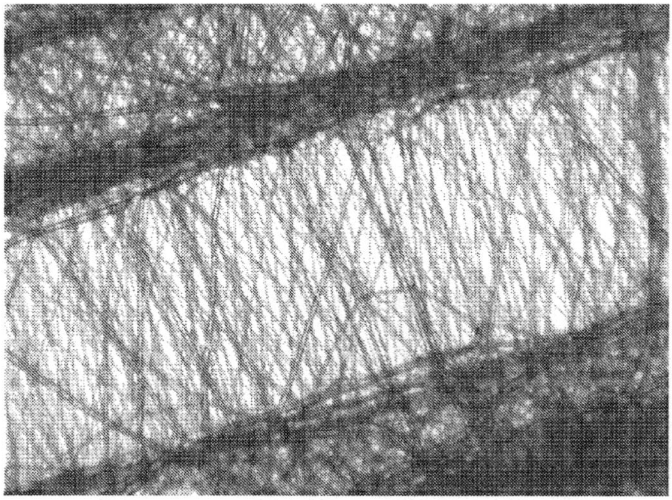

Figure 8.14 Light microscopy image of electrospun fibers (67).

Electrospinning techniques are used to form particles and fibers as small as one nanometer in a principal direction. The phenomenon of electrospray involves the formation of a droplet of polymer melt at an end of a needle, the electric charging of that droplet, and an expulsion of parts of the droplet because of the repulsive electric force due to the electric charges. In electrospraying, a solvent present in the parts of the droplet evaporates and small particles are formed, but not fibers. The electrospinning technique is similar to the electrospray technique. However, in electrospinning and during the expulsion, fibers are formed from the liquid as the parts are expelled (66).

Biological materials and biocompatible materials that can be electroprocessed have been described (68–70).

8.1.15.1 Drug Release from Electrospun Fibers

Biocompatibility is a desirable attribute in compositions designed for drug delivery (68). With surgical and subdermal implants, for example, the substance to be delivered is often contained in a matrix composed of synthetic polymers. Where natural products are used in making bandages, the products typically comprise wood products such as cellulose or other materials that are not readily absorbed by the body of the recipient. Accordingly, such bandages must eventually be removed. Implants compressed from natural materials that may be absorbed by the body are a viable way to improve the biocompatibility.

The cell death-inducing drug diclofenac was encapsulated in an electrospun PLA scaffold (71). The scaffold offers a system for a sustained and controlled delivery of the cytotoxic diclofenac over time, making it clinically favorable by achieving a prolonged therapeutic effect. Diclofenac is shown in Figure 8.15.

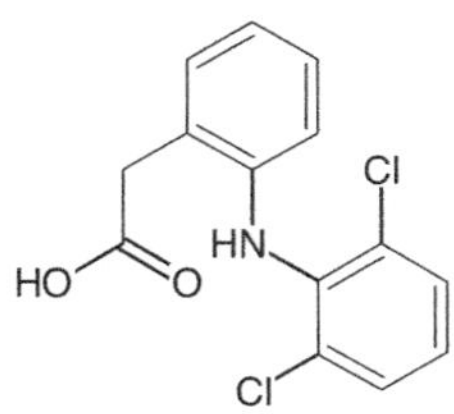

Figure 8.15 Diclofenac.

Human dermal fibroblasts were exposed to the drug-eluting scaffold and multiphoton microscopy and fluorescence lifetime imaging microscopy were used for analysis. These methods were suitable for noninvasive and marker-independent assessment of the cytotoxic effects. Released diclofenac induced changes in cell morphology and glycolytic activity. Furthermore, it was shown that drug release can be influenced by adding dimethyl sulfoxide as a cosolvent for electrospinning. Interestingly, without affecting the drug diffusion mechanism, the resulting PLA scaffolds showed an altered fiber morphology and enhanced initial diclofenac burst release (71).

Coaxial electrospinning is capable of simultaneously using two different solvent systems (72). However, the coaxial electrospinning process is extremely complex and numerous parameters might impact the incorporation as well as release kinetics of the drugs.

To address this gap, coaxial fibers for controlled co-delivery of two different drugs were fabricated: bovine serum albumin as a model protein and acyclovir as a small molecule (72). Acyclovir is shown in Figure 8.16.

The controlled release of both embedded drugs was achieved, and drug release profiles could be modulated through modifying

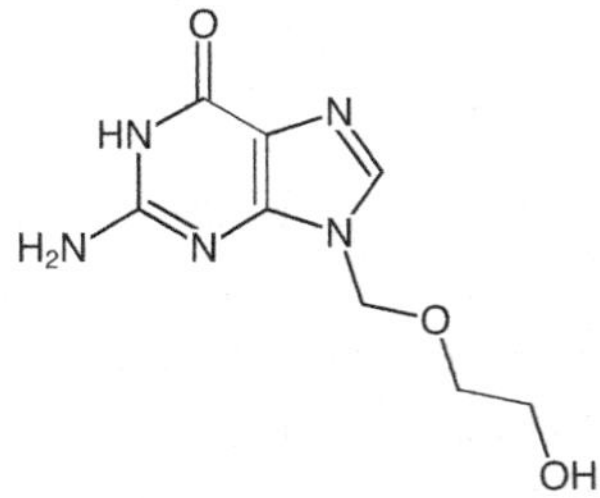

Figure 8.16 Acyclovir.

the spinning techniques as well as the compositions and the process parameters of the core fluid.

The interactions of the co-incorporated drugs were found to be of key significance for affecting the drug distribution within the fibers, and thus ultimately impacting drug release kinetics. Based on these findings, key factors for influencing drug encapsulation and release kinetics from coaxial fibers could be identified, providing a valuable guidance for the rational development of coaxial fiber-based drug carriers (72).

Tetracyclines. Tetracyclines are a family of broad-spectrum antibiotic compounds. The release profiles of tetracycline hydrochloride from electrospun fibers and the cast films are shown in Figure 8.17.

The materials are obtained by electrospinning solutions containing tetracycline along with PLA, ethylene vinyl acetate (EVA) or a combination of PLA and EVA. The release of tetracycline hydrochloride was determined by spectroscopic measurements.

Fibrin. Fibrin is an excellent extracellular matrix because it is eventually absorbed by the body of the recipient (70, 73). During and after the time in which fibrin serves its healing and rebuilding purpose, the body will naturally degrade it. This is part of the natural healing mechanism. There would be no lingering foreign matter, only the cells constituting the replacement cells, tissues or organs.

Electroprocessed fibrin may be mixed with other fibrillar materials to obtain different rates of degradation. Native fibrin de-

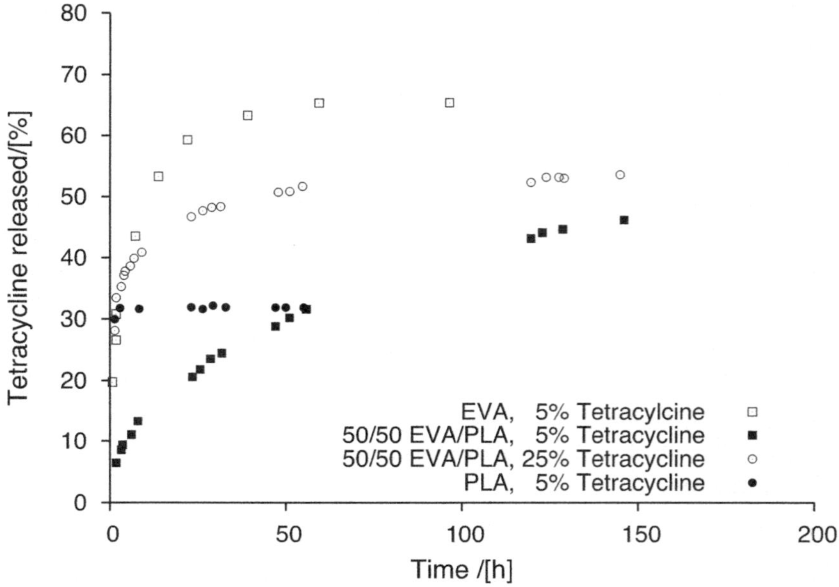

Figure 8.17 Release profiles of tetracycline hydrochloride (68).

grades rapidly, collagen more slowly, poly(glycolic acid) and PLA still more slowly. Crosslinking the fibrin and other materials to various degrees will stabilize the materials to different extents. Electroprocessed fibrin may be slightly crosslinked for a small increase in half life, or highly crosslinked to provide extended duration of use (70).

Electroprocessed fibrin may be employed as a vehicle for drug delivery. The material can be used as a biodegradable depot system, placing high concentrations of a substance such as a drug at a specific location. Electroprocessed fibrin forms a three-dimensional, space-filling network, and can be aligned during formation by flow dynamics. Once formed, the fibrin matrix can be aligned by mechanical traction, magnetic force or other means. Drug release can be tailored by altering the structure of the electroprocessed fibrin or the rate at which it dissolves. Any drug or other substance may be combined with the electroprocessed fibrin (70).

Some preferred drugs or substances include estrogen, androgen, cortisone, cyclosporin, peptide growth factors, nerve growth factors, and bone growth factors. Topical antibiotics include, but are not limited to, penicillin, gentimycin, and tetracycline. Antimycotics may also be employed (70).

Electroprocessed fibrin can be electrodeposited inside a specifically shaped mold. For instance, when a particular type of organ or tissue needs to be replaced with a specific shape, such as a skin patch to fit a biopsy site or a large scalp area following a wide area removed after discovering a malignant melanoma, the desired shape is then reproduced and created inside a mold designed to mimic that shape. This mold could be filled by electrodepositing the fibrin into the mold. In this way, the fibrin matrix exactly mimics the mold shape. Creating an extracellular fibrin matrix that has a specific shape can be very important in creating a new organ. The shape of the matrix can induce cells seeded into the fibrin matrix to differentiate in a specific manner.

Growth factors or other substances may be incorporated as discussed elsewhere herein. This can result in a more effective, more natural-like organ or tissue being created. Hollow and solid organs can be made. Mixing cells with the material during electrospraying forms cells within the matrix so that they do not have to migrate into the gel (70).

The main electrospinning parameters, i.e., polymer concentration in the injectable solution, solvents used and their proportion, flow rate, voltage and distance to collector, have been systematically modified to analyze their particular influence on the fiber diameter of electrospun membranes of PLA, poly(caprolactone) (PCL) and a mixture of these polymers (74).

As a result of this analysis, procedures to fabricate membranes of these polymers and blend with under- and above-micron-sized fibers were established, in which the solvents' ratio (chloroform/methanol and dichloromethane/dimethylformamide) and voltage were found to play a major role. Moreover, the plausible differential effect of these fibers' diameters (0.8 μm and 1.8 μm) in the controlled release of a molecule of interest was explored using bovine serum albumin, proving that the most effective configuration for the release of bovine serum albumin among those studied was the PLA-PCL combination in membranes of above-micron-sized fiber diameter (74).

Dexamethasone. Electrospinning technique has been explored to produce nanofibers incorporated with drugs as alternative drug delivery systems for therapeutic purposes in various organs and tissues. Before such systems could potentially be used, their biocompatibility must be evaluated (75).

Dexamethasone acetate-loaded poly(ε-caprolactone) nanofibers were developed for targeted delivery in the vitreous cavity in the treatment of retinal diseases (75). Dexamethasone is shown in Figure 8.18.

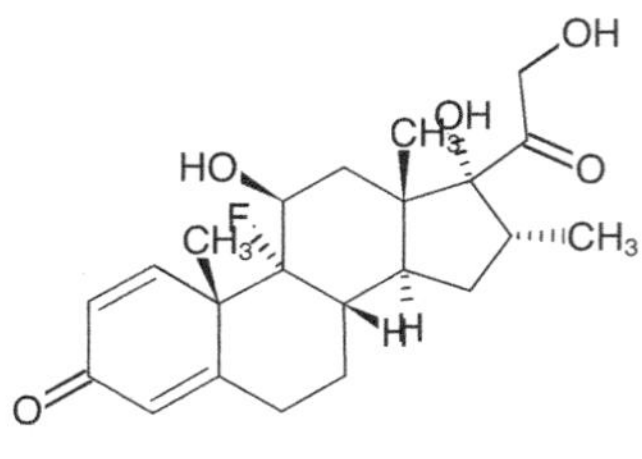

Figure 8.18 Dexamethasone.

The ocular biocompatibility was tested *in vitro* and *in vivo*. The nanofibers were characterized by SEM and FTIR and the *in-vitro* drug release from nanofibers was evaluated. The *in-vitro* biocompatibility of the nanofibers was tested on both ARPE-19 and MIO-M1 cells using the cytotoxicity (MTT) test by morphological studies based on staining of the actin fibers in ARPE-19 cells and GFAP in MIO-M1 cells.

The *in-vivo* biocompatibility of the nanofibers was investigated after intravitreous injection in a rat eye, using spectral-domain optical coherence tomography (OCT) imaging of the retina. The SEM measurements indicated that nanometric fibers were interconnected in a complex network, and that they were composed of polymer. The FTIR studies showed that polymer and drug did not chemically interact after the application of the electrospinning technique.

The PCL nanofibers provided a controlled dexamethasone release for 10 *d* (75). The dexamethasone release rate is shown in Figure 8.19.

The nanofibers were found to be noncytotoxic to the ocular cells, allowing for the preservation of actin fibers and the glial fibrillary

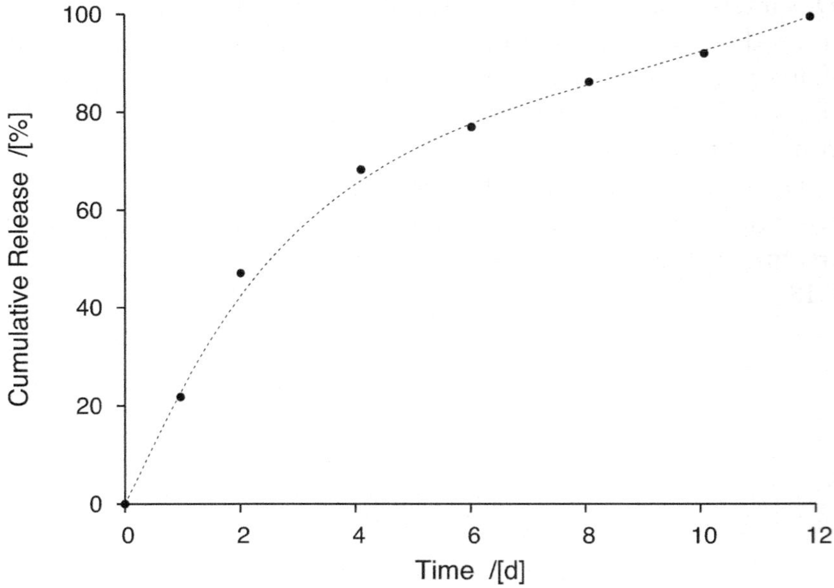

Figure 8.19 Dexamethasone release (75).

acidic protein (GFAP (76)) in the cytoplasm of ARPE-19 and MIO-M1 cells, respectively, which are biomarkers of these ocular cell populations.

These nanofibers did not affect the retinal and choroidal structures, and they did not induce abnormalities, hemorrhages, or retinal detachment, suggesting that the nanofibers were well tolerated. In eyes receiving the nanofibers, spectral-domain optical coherence tomography (SD-OCT) images were corroborated with a histological analysis of neuroretina and choroid, which are ocular tissues that are extremely sensitive to toxic agents.

The preservation of cone and rod photoreceptors indicated the light sensitivity of the animals. So, these nanofibers exhibited ocular biocompatibility and safety in the rodent eye and allow the release of dexamethasone (75).

8.1.16 Mucoadhesive Drug Delivery

Mucodhesion refers to bonds formed between two biological surfaces or a bond connecting a biological and a synthetic polymer surface (77).

The literature concerning mucoadhesive drug delivery has been reviewed with a special focus on the role of natural polymers on mucoadhesive drug delivery systems. Polymers used in the formulation in such systems are listed in Table 8.5.

Table 8.5 Polymers used for the formulation (77).

Gum Type	Formulation and examples for usage
Karya gum	Developed and evaluated gastric retentive gel for rodents *in situ*, using a mixture of sodium alginate and karaya gum
Xanthan gum	Sustained release and swelling characteristics of xanthan gum/ethylcellulose-based injection molded matrix tablets
Guar gum	Sustained release of diltiazem
Tragacanth	Common natural ingredients used in food, drugs and cosmetics
Pectin	Gelling behavior of nopal (*Opuntia ficus indica*) low metholoxyl pectin
Chitosan	Propranolol hydrochloride, buccal film, buccal patches containing cetylpyridinium chloride, curcumin, or propranolol hydrochloride
Gum arabic	Encapsulation of endoglucanase
Locust bean gum	Ibuprofen compression-coated tablets
Grewia gum	Suspending agent on ibuprofen pediatric formulation
Bhara gum	Controlled release of famotidin
Mango gum	Evaluation of disintegrating properties of *Mangifera indica*
Gelatin	Mucoadhesive bilayered patches for administration of sumatriptan succinate
Fenugreek gum	*Trigonella foenumgraecum* L. seed mucilage as suspending agent
Tamarind gum	Buccal delivery of drugs

Under this drug delivery, buccal mucosa is the preferred site for both systemic and local drug action because the mucosa has a rich blood supply and it is relatively permeable. Different bioadhesive dosage forms are available on the market such as chewing gum, tablets, patches, hydrogel, and thiolated tablets (77).

8.2 Tissue Engineering

Tissue engineering is a new field that has developed with the progress of science (78). Tissue engineering involves concepts and techniques from various fields of science such as life science, engineering, medical science and the like.

Tissue engineering aims to understand the relationship between the structure and function of a body tissue and producing a substitute for a damaged body tissue or organ for transplantation purposes so as to maintain, improve or restore the function of a human body.

Biomaterials for tissue engineering applications and the very methods of use have been detailed (79). One typical tissue engineering technique consists of the following steps (78):

- Removing a required tissue from a patient body,
- Isolating a cell from the removed tissue,
- Proliferating the isolated cell,
- Seeding the cell in the biodegradable porous polymer scaffolds,
- Culturing the cell in vitro for a predetermined period, and
- Transplanting the obtained hybrid-type cell/polymer structure into the human body.

After the transplantation is achieved, oxygen and nutrients are provided to the transplanted cells in biodegradable porous polymer due to the diffusion of bodily fluids until a blood vessel is newly formed.

When the blood vessel is formed, the cells are cultivated and divided in order to form a new tissue and organ. During the formation of a new tissue and organ, the polymer scaffolds become degraded and eventually disappear (78).

8.2.1 Scaffolds for Tissue Engineering

The interest in modified poly(γ-glutamic acid) as a functionalized biomaterial for scaffold and drug delivery has considerably increased in the last few years (80).

8.2.1.1 Dual-Pore Scaffolds

A simple process has been presented for the fabrication of biodegradable dual-pore scaffolds for tissue engineering, which have a high surface area and porosity without toxic substance secretion. Dual pores may be formed by (78):

1. Forming a large pore using an effervescent mixture, while forming a small pore using a solution containing nonsolvent, or
2. Forming two types (in size) of pores by controlling the particle size of the effervescent mixture.

In detail, dual-pore polymer scaffolds are prepared as follows (78):

Preparation 8–2: A PLGA copolymer with lactic acid and glycolic acid in the weight ratio of 50:50 is dissolved in a mixture of dioxane and water to provide a 5% PLGA solution. Then, an effervescent mixture of sodium hydrogen carbonate and citric acid is added and uniformly mixed therein.

The mixed solution is poured into a frame made of silicone material which is placed in liquid nitrogen. Then, the polymer sample is added to a mixture of water and ethanol and subjected to foaming for 24 *h*. After foaming, the sample is freeze-dried for 20 *h* to provide biodegradable dual-pore polymer scaffolds. The overall porosity was about 98%, which was analyzed by mercury porosimeter.

An SEM image of the external surface of the dual-pore scaffolds is shown in Figure 8.20.

Using effervescent materials, the pore size of the polymer scaffolds can be easily controlled, and biodegradable polymer scaffolds can be more simply prepared compared to conventional methods such as salt leaching technique, phase separation technique, or gas foaming technique.

Since the biodegradable polymer scaffolds have a high surface area and porosity without the formation of a skin layer, they can be advantageously used for regenerating almost all living tissues and organs in tissue engineering (78).

8.2.1.2 Hard Tissue Regeneration

Porous hydroxyapatite scaffolds are commonly used for hard tissue regeneration because of their biocompatibility and osteoconduction properties, but they are limited with respect to bioactivity

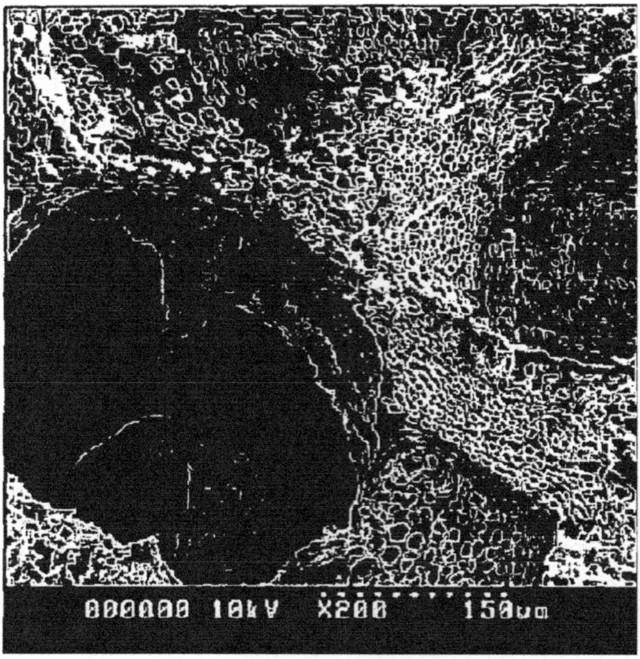

Figure 8.20 External surface of dual-pore scaffolds (78).

and osteoinduction (81). The fabrication of porous hydroxyapatite scaffolds has been investigated that are coated with PLGA/bioactive glass composite microfibers using the sponge replica method and an electrospinning process for improved bioactivity and osteoinductivity during osteogenesis.

The characterization of these scaffolds was carried out by examining morphology and ion release. The biological evaluation of scaffolds was carried out by assessing cytotoxicity, cell proliferation, and cell differentiation using MC3T3-E1 preosteoblasts (81). The results showed that all these scaffolds exhibited a high porosity of 89.2% and had porous structures coated with a layer of composite containing bioactive glass ions (Si, Ca, Na, P). These scaffolds enabled the controlled release of the Si, Ca, Na, and P ions for up to 28 *d*. There was no significant difference in cytotoxicity between the scaffolds. Cell proliferation on the scaffolds was increased from day 1 to 3.

Cell differentiation, as shown by alkaline phosphatase activity and western blot analyses, indicated that the scaffolds with 20 *min* and 30 *min* of coating induced higher levels of osteogenesis-related markers compared to other scaffolds.

Immunocytochemistry experiments indicated osteopontin expression. Alizarin red staining indicated that the scaffolds with 20 *min* of coating were more effective than the porous hydroxyapatite scaffolds in terms of mineralization.

In conclusion, porous hydroxyapatite scaffolds coated with PLGA/bioactive glass composite microfibers for 20 *min* are promising materials for osteogenic activity and may be a potential bone substitute for tissue engineering (81).

8.2.2 3D Bioprinting

The aim of 3D bioprinting is to create custom scaffolds that are biologically active and accommodate the desired size and geometry (82). A thermoplastic backbone can provide mechanical stability similar to native tissue while biologic agents offer compositional cues to progenitor cells, leading to their migration, proliferation, and differentiation to reconstitute the original tissues/organs.

Unfortunately, many 3D printing compatible, bioresorbable polymers, such as PLA, are printed at temperatures of 210 °C or higher. These high temperatures are detrimental to biologics. On the other hand, PCL, a different type of polyester, is a bioresorbable, 3D printable material that has a gentler printing temperature of only 65°C.

Therefore, it was thought that a decellularized extracellular matrix contained within a thermally protective PLA barrier could be printed within a PCL filament and remain in its functional conformation.

The production of a PCL filament with embedded poly(*l*-lactide) (PLLA) microspheres which contain decellularized matrices for 3D printing of structural tissue engineering constructs has been described (82).

In the study, an osteochondral repair was the application for which the hypothesis was tested. A porcine cartilage was decellularized and encapsulated in PLLA microspheres, which were then extruded with PCL into a filament to produce 3D constructs via fused deposition modeling (82).

8.2.3 *Periodontal Tissue Engineering*

The periodontium consists of gingiva, periodontal ligament, cementum, and alveolar bone. It is a hierarchically organized tissue, whose primary role is to provide physical and mechanical support to the teeth (83).

Periodontitis is a chronic disease that reduces the integrity of the periodontal system and causes damage to the periodontal tissues, eventually leading to a loss of teeth (84).

Various tissue engineering approaches were investigated for their ability to control the critical temporo-spatial wound-healing events that are essential for achieving periodontal regeneration (83).

The recent progress in the field of periodontal tissue engineering has been reviewed with a focus on advanced 3D multiphasic tissue engineering constructs (83). Also, the future of periodontal tissue engineering was addressed, including scaffold customization for individual periodontal defects, functionalization strategies for imparting enhanced bioactivity, periodontal ligament fiber guidance, and the utilization of chair-side regenerative solutions that can facilitate clinical translation (83).

Nanoparticles of chitosan, PLGA, and silver were produced and the optimal composite ratio of these three materials for periodontal tissue regeneration was investigated (85).

The single nanoparticles were found to have no cytotoxicity and were able to promote the cell mineralization. Chitosan and Ag in low concentrations showed antibacterial activity; however, Ag inhibited the cell proliferation (85).

A major drawback of current periodontal membranes is the lack of tissue regeneration properties (86). A multifunctional periodontal membrane was fabricated for the guided tissue regeneration of lost periodontal tissues. Nanofibrous membranes based on PCL with tunable biochemical and biophysical properties were developed for periodontal tissue regeneration (86).

The engineered membranes were surface coated using biomimetic poly(dopamine) to promote the adhesion of therapeutic proteins and cells. The cellular localization on the surface of the engineered membrane by morphological patterning was shown. Poly(dopamine) accelerates the osteogenic differentiation of dental-derived stem cells by promoting hydroxyapatite mineralization.

Such multiscale designs can mimic the complex extracellular environment of periodontal tissue and serve as functional tissue constructs for periodontal regeneration.

In a periodontal defect model in rats, the engineered periodontal membrane successfully promoted the regeneration of periodontal tissue and bone repair (86).

8.2.4 *Cell Carriers*

Nanofibrous PLLA microspheres are extensively studied for use as cell carriers in the field of tissue engineering because their unique structure can promote cell proliferation and migration (87). But as injectable scaffold materials, PLLA microspheres easily run off to the soft tissue space because of the lack of cohesive force. This will affect the treatment efficiency and even cause additional inflammatory response.

In order to overcome this disadvantage, superparamagnetic γ-Fe_2O_3 nanoparticles assisted with oxidative polymerization of dopamine were used for the surface modification of PLLA microspheres (87).

The results of this study showed that this surface modification had no obvious cytotoxicity, and the modified microspheres possessed the ability to carry seed cells to controllably move to the defect sites with the guidance of magnetic field, which may be able to increase the repair efficiency. Moreover, the characteristic nanofibrous structure was not destroyed after the modification procedure, which was able to promote biological activity of cells (87).

8.3 Tissue Markers

For tissue markers, biodegradable polymers and copolymers can be combined with or without ionic and nonionic contrast agents, depending upon the application. For example, no contrast agent is needed in an application where there is no need to view a site at a later date.

Formulations that include contrast agents are designed to allow the radiopacity of the markers to last for an extended period of time so that they can be seen under X-ray or other imaging means for further diagnostic or intervention procedures at a date that may be weeks or months after implanting of the marker.

The plug is treated so that it is visible under ultrasound, magnetic resonance imaging, and other imaging techniques if an application requires such visibility. Accordingly, it may contain or be impregnated with a contrast solution containing radium, iodine, beryllium, or other contrasting agent.

Alternatively, the plug includes a means of transmission adapted to transmit signals in the electromagnetic spectrum that are detectable by receivers adapted to receive signals in the electromagnetic spectrum. Polymers used there are based upon well-known polymers such as poly(lactide)s, poly(glycolide)s, PCL, poly(dioxanone), collagen, and gelatin.

Some of these polymers and copolymers do not possess the mechanical properties that are required for certain applications. For example, as implants for soft tissue, poly(lactide)s, poly(glycolide)s, PCL, poly(dioxanone) and their copolymers may require increased flexibility and a modulus of elasticity that is closer to soft tissue.

Some hydrogels, due to their water content, provide a more flexible structure that is similar to soft tissue. Therefore, by combining the polymers that lack certain required mechanical properties such as a suitable modulus of rigidity or modulus of elasticity with hydrogels, suitable degradation and drug delivery properties are obtainable and an exterior is achieved that has mechanical properties similar to the mechanical properties of soft tissue.

The above-mentioned polymers are designed to provide a sustainable and gradual degradation and therefore a slow release of drugs in the course of the degradation of the polymeric substrate.

Hydrogels are not suitable for prolonged drug delivery without using additional bonding of the drug to the hydrogels, especially with small-molecule pharmaceutical agents (88).

Copolymers from glycolide, *D,L*-lactide, and ε-caprolactone have been synthesized. The properties are shown in Table 8.6.

Furthermore, starting materials include contrast agents such as sodium diatrizoate and 5-(*N*-2,3-dihydroxypropylacetamido)-2,4,6-triiodo-*N,N*′-bis(2,3-dihydroxypropyl) isophthalamide (iohexol). The contrast agent is added in amounts of 5–40%. These contrast agents are shown in Figure 8.21.

Table 8.6 Properties of copolymers (88).

Composition/[%]			
Glycolide	*D,L*-Lactide	ε-Caprolactone	Melting point /[°C]
	20	80	46
	75	25	—
19.4		90.6	55

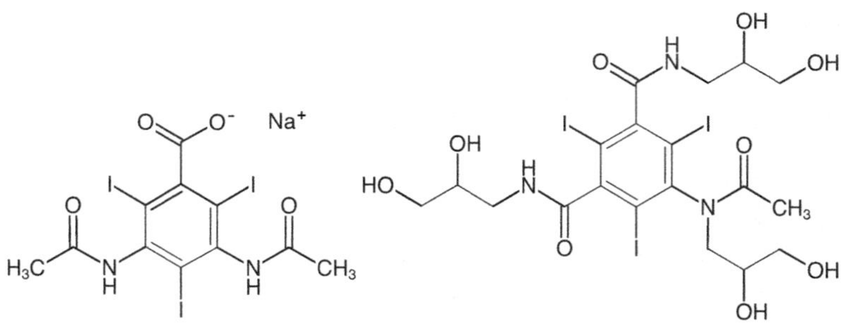

Sodium diatrizoate

Iohexol

Figure 8.21 Contrast agents.

The biomarkers may be fabricated by extrusion, injection molding, or compression molding. Multiple sizes and shapes may be fabricated. Some applications are listed in Table 8.7.

Table 8.7 Applications of biomarkers (88).

Use
Diagnosis of hepatic carcinoma
Heart biopsy
Prostate cancer biopsy
Tubular urethral stent
Treatment of malignant brain tumors

8.4 Hydrogels

Hydrogels are crosslinked, macromolecular polymeric materials arranged in a three-dimensional network, which can absorb and retain large amounts of water. Hydrogels are commonly used in clinical practice and experimental medicine for a wide range of applications, including drug delivery, tissue engineering and regenerative medicine. The design, synthesis and application in drug delivery and regenerative medicine of hydrogels has been described in a monograph (89).

In-situ polymerizable hydrogels are useful for several tasks in medicine, including (90):

- Tissue augmentation,
- Coating of medical devices,
- Protection of sensitive tissue,
- Prevention of postsurgical adhesions,
- Microencapsulation of tissue,
- Embolization of biopsy needle tracks,
- Local delivery of a therapeutic substance,
- Entrapment of enzymes, and
- Coating of tissues and organs.

Macromers for use in forming hydrogels in situ generally consist of a water-soluble crosslinking agent and a water-soluble crosslinkable polymer (90). Examples of both kinds of compounds are shown in Table 8.8.

Table 8.8 Macromers for hydrogels (90).

Water-soluble crosslinker
Activated poly(ethylene glycol) Acrylated PEG and PLA
Water-soluble polymer
Poly(*L*-lysine)

8.5 Microporous Materials

Porous polymeric biomaterials and implants composed of biodegradable polymers have been presented in numerous papers and patents. A number of techniques that have been proposed to build such porous polymer structures are collected in Table 8.9.

Table 8.9 Methods for manufacture of porous biomaterials.

Technique	Reference
Melt blending of two immiscible polymers	(91)
Using solid porogens	(92)
Freeze-drying	(93)
Gel casting	(94)
Leaching with solvent	(95, 96)
Lyophilization process	(97)
Thermally induced phase separation	(98)
Gas forming and particulate leaching	(99)

A SEM micrograph of a foam prepared from deaminotyrosyl-tyrosine ethyl ester is shown in Figure 8.22; and deaminotyrosyl-tyrosine ethyl ester is shown in Figure 8.23.

Biocompatible foam structures are formed by solidifying a solution of a solvent and a bioabsorbable polymer to form a solid, then subliming the solvent out of the solid.

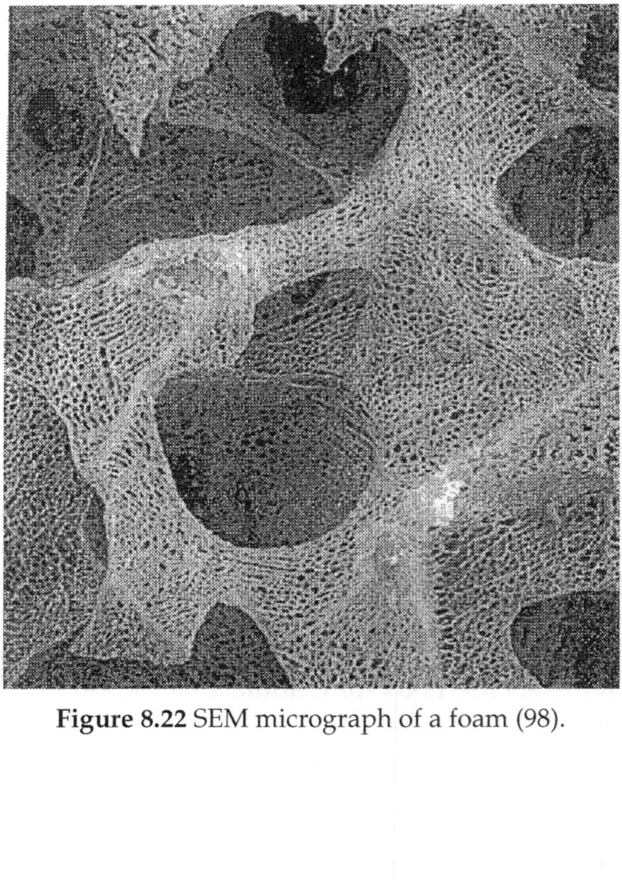

Figure 8.22 SEM micrograph of a foam (98).

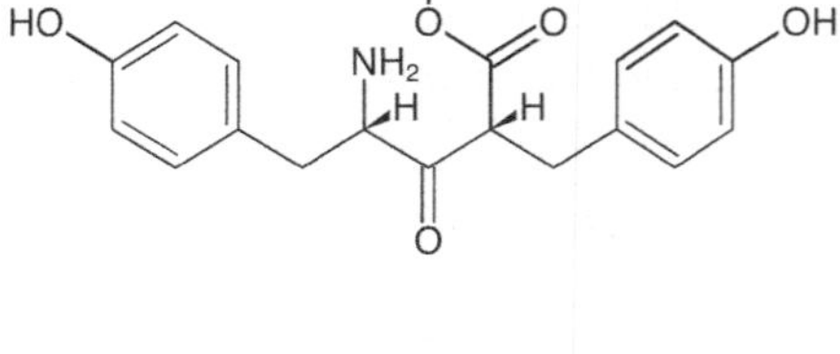

Figure 8.23 Deaminotyrosyl-tyrosine ethyl ester.

Porous polymeric biomaterials have been promoted as carriers or scaffolds for repairing tissues, delivering drugs and bioactive compounds, encapsulating cells or bioengineering artificial tissues. The repair or regeneration of bone, cartilage, skin, liver, etc., has attracted interest in such biodegradable porous polymer materials (96).

A microporous biodegradable polymeric material is prepared from a biodegradable polymer (A) and another polymer (B), no matter whether biodegradable or not. However, the biodegradable polymer should be at least partially immiscible with (A). Furthermore, a polymeric compatibilizer (C) for the polymers (A) and (B) is needed (96).

The components are then melt blended. Thereby a compatibilized polymer blend is prepared , wherein the polymers (A) and (B) have an essentially continuous morphology. Upon cooling the blend to room temperature, its morphology should be retained.

Finally, the polymer (B) and the compatibilizer (C) is extracted at least partially from the polymer blend by dissolving (B) and (C) in a solvent that is a nonsolvent of polymer (A). In this way, the article becomes essentially continuously porous. A void volume of 10–90% can be obtained. The pore diameters show a unimodal pore size distribution.

In particular, a microporous PLLA composition can be prepared using poly(styrene) (PS) and as compatibilizer a copolymer from a lactide and styrene. Binary blends and compatibilized ternary blends are prepared by melt mixing the polymers and copolymer in a Brabender internal mixer with roller blades under a constant high flow of dry nitrogen. Dry nitrogen is required to avoid a dramatic melt degradation of the PLLA. Prior to blending, PLLA and PS are dried for 48 *h* in a vacuum oven at 70°C.

The PLLA and the PS are added simultaneously in the mixing chamber, whereas the compatibilizer is introduced immediately afterwards. The mixing blades were maintained at 50 *rpm* at a temperature of 200°C. Cyclohexane extraction of the PS is done at 50–60°C in a Soxhlet extraction apparatus. After the extraction time, the specimens are dried in a vacuum oven for 10 *h* at 50°C (96).

8.6 Implants

Bioabsorbable implants are typically made from polymeric materials such as lactone-based polyesters. These bulk-eroding materials break down over time due to chemical hydrolysis to produce water-soluble, low molecular weight fragments (100).

PLA is now the most commonly used biodegradable polymer in clinical applications (101). Examples range from drug delivery systems, tissue engineering, temporary and long-term implantable devices.

This is owed greatly to the polymer's favorable biocompatibility and to its safe degradation products. Once coming in contact with biological media, the polymer begins breaking down, usually by hydrolysis, into lactic acid or into carbon dioxide and water.

These products are metabolized intracellularly or excreted in the urine and breath. Bacterial infection and foreign-body inflammation enhance the breakdown of PLA through the secretion of enzymes that degrade the polymeric matrix. The biodegradation occurs both on the surface of the polymeric device and inside the polymer body, by diffusion of water between the polymer chains. The median half-life of the polymer is 30 weeks. However, this can be lengthened or shortened to address the clinical needs (101).

The kinetics of degradation can be tuned by determining the molecular composition and the physical architecture of the device. For example, using L- or D-chirality of the lactic acid will greatly lengthen or shorten the degradation rates, respectively (101).

Implantable medical devices provide an effective therapeutic approach for neurological and cardiovascular diseases (102). With the development of transient electronics, a new power source with biocompatibility, controllability, and bioabsorbability becomes an urgent demand for medical sciences. Various fully bioabsorbable natural-materials-based triboelectric nanogenerators, *in vivo*, have been developed.

The modification of silk fibroin encapsulation film makes the operation time of the bioabsorbable natural-materials-based triboelectric nanogenerator tunable from days to weeks. After completing its function, the bioabsorbable natural-materials-based triboelectric nanogenerator can be fully degraded and resorbed in

Sprague-Dawley rats, which avoids a second operation and other side effects (102).

The bioabsorption process of implants is typically associated with a gradual decline, over months to years, in structural strength and integrity and may also be associated with cellular responses, such as phagocytosis, that may confound interpretation of efficacy and safety endpoints (103). Additionally, as these implants bioabsorb, they become increasingly difficult to isolate histologically and thus imaging modalities, such as microCT, become very valuable to determine the original location of the implants and to assess the remodeling response in tandem with histopathology.

The various types of bioabsorbable implants and commonly used bioabsorbable materials have been reviewed. Some of the most common challenges and pitfalls that are confronting histologists and pathologists in collecting, handling, imaging, and preparing tissues were addressed (103).

Bioabsorbable steroid-releasing implants (mometasone furoate, 370 μg) are effective for improving postsurgical outcomes in the frontal sinus ostia (104). The effect of these implants on frontal outcomes were evaluated in various patient subgroups with chronic rhinosinusitis using pooled data from two randomized, controlled trials.

A total of 160 subjects were enrolled in two randomized, controlled trials. After surgery, subjects were randomized to receive an implant in one frontal sinus ostia with the contralateral side as control. Data through day 90 from the two studies were pooled and subgroup analysis was performed.

It was found that bioabsorbable steroid-releasing sinus implants improve outcomes of frontal sinus surgery through 90 d, irrespective of asthma status, previous endoscopic sinus surgery, extent of surgery, extent of polyps, or Lund-Mackay computed tomography stage in the frontal sinus ostia (104).

8.6.1 *Inflammatory Problems with Implants*

Bioabsorbable implants are typically made from polymeric materials such as lactone-based polyesters. These bulk eroding materials break down over time due to chemical hydrolysis to produce water-soluble, low molecular weight fragments.

These fragments can be attacked by enzymes to produce lower molecular weight metabolites. Acid fragments that are produced during degradation of the polymer backbone have been shown to cause local tissue inflammation (100).

The inflammation has been observed in vascular systems, the extent of which depends on the pH of the acid that in turn is dependent on the type and amount of acid produced during degradation.

Such an inflammation is not typically observed in polymers that degrade by surface erosion, such as poly(orthoester)s and poly(anhydride)s, as the amount of acid released at a given time is too small to cause tissue inflammation (100).

An *in-situ* release of nitric oxide (NO) from medical devices can enhance their performance via NO's potent anti-thrombotic, bactericidal, anti-inflammatory, and angiogenic activity (105). However, NO-releasing semicrystalline thermoplastic systems are limited and the relationship between polymer crystallinity and the NO release profile is not known.

The functionalization of poly(ether-block-amide), Nylon 12, and poly(urethane) (PU) tubes, as examples of semicrystalline polymers, with the NO donor S-nitroso-*N*-acetylpenicillamine within, could be demonstrated via a polymer swelling method (105). S-Nitroso-*N*-acetylpenicillamine is shown in Figure 8.24.

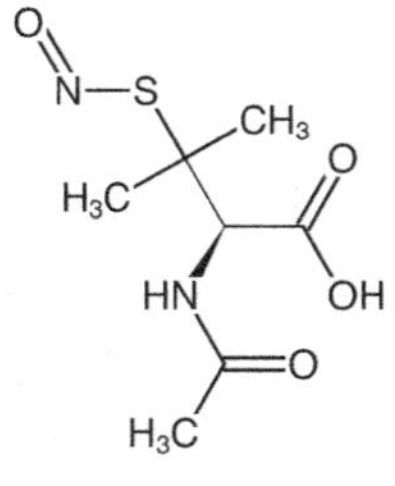

Figure 8.24 S-Nitroso-*N*-acetylpenicillamine.

The degree of crystallinity of the polymer plays a crucial role in both S-nitroso-*N*-acetylpenicillamine impregnation and NO release. Nylon 12, which has a relatively high degree of crystallinity, exhibits an unprecedented NO release duration of over 5 months at a low

NO level, while poly(ether-block-amide) tubing exhibits NO release over days to weeks.

As a new biomedical application of NO, the NO-releasing poly(ether-block-amide) tubing was examined as a cannula for continuous subcutaneous insulin infusion. The released NO is shown to enhance insulin absorption into the bloodstream probably by suppressing the tissue inflammatory response, and thereby could benefit insulin pump therapy for diabetes management (105).

Renal tissue regeneration using functional scaffolds with biomaterials has attracted a great deal of attention due to limited donor organ availability (106). A bioinspired scaffold was developed that can efficiently induce renal tissue regeneration. The bioinspired scaffold was designed using PLGA, magnesium hydroxide ($Mg(OH)_2$), and decellularized renal extracellular matrix. The $Mg(OH)_2$ inhibited materials-induced inflammatory reactions by neutralizing the acidic microenvironment formed by degradation products of PLGA, and the acellular renal extracellular matrix helped to restore the biological function of kidney tissues.

When the PLGA/renal extracellular matrix/$Mg(OH)_2$ scaffold was implanted in a partially nephrectomized mouse model, it led to the regeneration of renal glomerular tissue with a low inflammatory response (106).

The PLGA/renal extracellular matrix/$Mg(OH)_2$ scaffold was able to restore the renal function more effectively than the control groups. These results suggest that the bioinspired scaffold can be used as an advanced scaffold platform for renal disease treatment (106).

8.6.1.1 Self-Regulating System for Controlling Acidic Effects

A biodegradable material can be combined with a buffering agent (100). There, a second (another) biodegradable material encapsulates the buffering agent.

When the encapsulating biodegradable material is degrading, the buffering agent is set free. This causes a reduction in the pH level at the site in the patient's body resulting from the hydrolysis of the first biodegradable material.

This device is configured so that upon exposure to an acidic environment, an amount of the second biodegradable material degrades and releases the encapsulated buffering agent in an amount that provides a stoichiometric balance between the amount of buffering agent released and the total amount of acid released from the device, such that the pH level is raised to a neutral pH level (100).

Suitable buffering agents for encapsulation are summarized in Table 8.10.

Table 8.10 Buffering agents for encapsulation (100).

Compound	Compound
Calcium hydroxyapatite	Carbonated apatite
Tricalcium phosphate	Calcium carbonate
Sodium bicarbonate	Calcium phosphate
Carbonated calcium phosphates	Magnesium hydroxide
Amine monomers	Lactate dehydrogenase

Acidic byproducts resulting from degradation can generate vigorous inflammatory reactions, often leading to a clinical failure (107). An approach has been presented to prevent acid-induced inflammatory responses associated with biodegradable polymers, such as poly(lactide-*co*-glycolide), by using oligo(lactide)-grafted magnesium hydroxide ($Mg(OH)_2$) nanoparticles. These neutralize the acidic environment.

In particular, it was demonstrated that incorporating the modified $Mg(OH)_2$ nanoparticles within degradable coatings on drug-eluting arterial stents efficiently attenuates the inflammatory response and in-stent intimal thickening by more than 97% and 60%, respectively, in the porcine coronary artery, compared with that of drug-eluting stent control. It was also observed that decreased inflammation allows better reconstruction of mouse renal glomeruli in a kidney tissue regeneration model. Such modified $Mg(OH)_2$ nanoparticles may be useful to extend the applicability and improve the clinical success of biodegradable devices used in various biomedical fields (107).

8.6.2 *Eye Implants*

Implants for the sustained release of drugs for treating the eyes have been tested (108). The delivery of more bioactive agents can be localized at a desired treatment site, particularly the choroid and the retina.

The polymer used is mostly PCL. It has been found that PCL is well tolerated by the retinal tissue and can deliver bioactive agents without eliciting inflammatory response or complications. PCL can elute a steroid for a period of at least 4 weeks without eliciting inflammatory response or complications (108).

An implant may be rod-shaped and may include the corticosteroid triamcinolone acetonide in a PCL polymer matrix. The implant may be fabricated by (108):

- Solubilizing PCL in chloroform,
- Adding a bioactive agent to the solution,
- Evaporating the solvent,
- Transferring the solid loaded polymer to an extrusion device, and
- Shaping the implant thereby.

A clinical examination of the test with rabbits has been described in much detail (108).

Achieving local ocular administration of a drug may require direct injection or application, but in order to achieve long-lasting drug delivery, specialized devices may be required (109). For example, a drug-eluting implant sized and shaped to rest in the irido-corneal angle of the eye without abrading or inflaming adjacent tissue can be is utilized for localized and/or long-term treatment of the eye. Use of a drug-eluting implant allows the targeted delivery of a drug to a specific ocular tissue, such as, for example, the macula, the retina, the ciliary body, the optic nerve, or the vascular supply to certain regions of the eye. Use of an anterior chamber drug-eluting implant particularly allows for the targeted delivery of a drug to the iris and cornea and other tissues located in the anterior chamber (109).

Implants for ocular drug delivery have been described that contain an elongated outer shell having a proximal and distal end, and are shaped to define an interior lumen and at least one therapeutic agent positioned within the lumen, wherein the outer shell has at least a first thickness, the outer shell comprises one or more regions of drug release, and the implant is dimensioned for implantation within the anterior chamber of the eye (109). Examples of drugs are collected in Table 8.11. Some drugs, i.e., mydriatic-cycloplegic agents, are shown in Figure 8.25.

Table 8.11 Drugs for ocular implants (109).

Drug	Drug	Drug
	Anti-angiogenesis agents	
Angiostatin	Anecortave acetate	Thrombospondin
	Anti-VEGF drugs	
Ranibizumab (Lucentis®)	Bevacizumab (Avastin®)	
Pegaptanib (Macugen®)	Sunitinib	
Sorafenib		
	Beta-blocker agents	
Atenolol & propranolol	Metipranolol	Betaxolol
Carteolol	Levobetaxolol	Levobunolol
Timolol		
	Adrenergic agonists or sympathomimetic agents	
Epinephrine	Dipivefrin	Clonidine
Aparclonidine	Brimonidine	
	Parasympathomimetics or cholingeric agonists	
Pilocarpine	Carbachol	Phospholine iodine
Physostigmine	Salicylate	Acetylcholine chloride
Eserine	Diisopropyl fluorophosphate	Demecarium bromide
	Carbonic anhydrase inhibitor agents	
Acetozolamide	Brinzolamide	Dorzolamide
Methazolamide	Ethoxzolamide	Diamox
Dichlorphenamide		
	Mydriatic-cycloplegic agents	
Atropine	Cyclopentolate	Succinylcholine
Homatropine	Phenylephrine	Scopolamine
Tropicamide		
	Prostaglandin analog agents	
Bimatoprost	Latanoprost	Travoprost
Unoprostone		

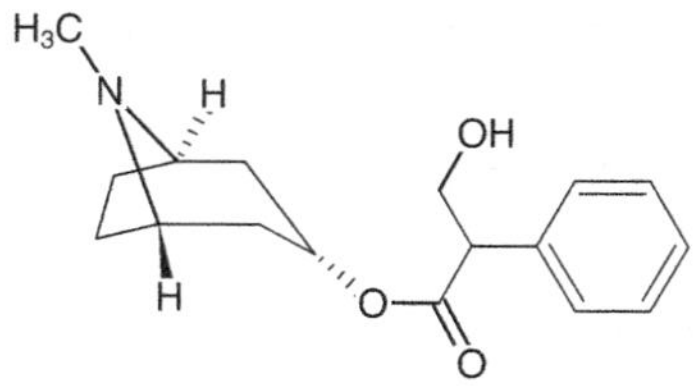

Atropine

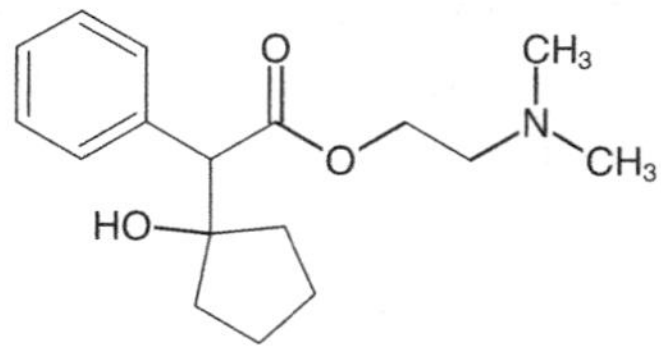

Cyclopentolate

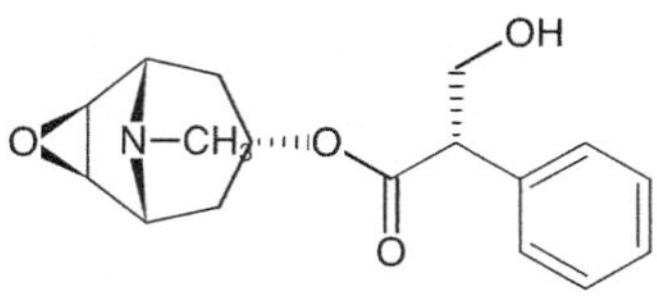

Scopolamine

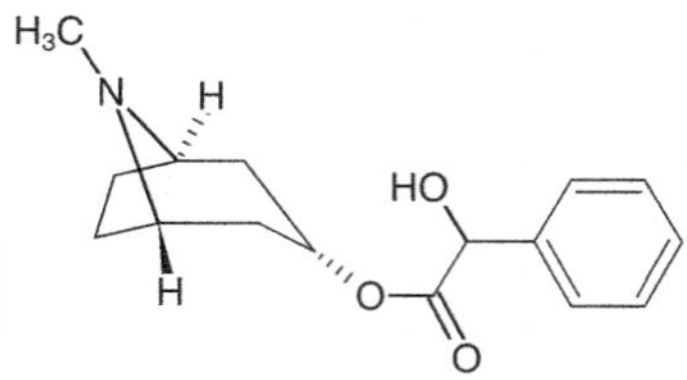

Homatropine

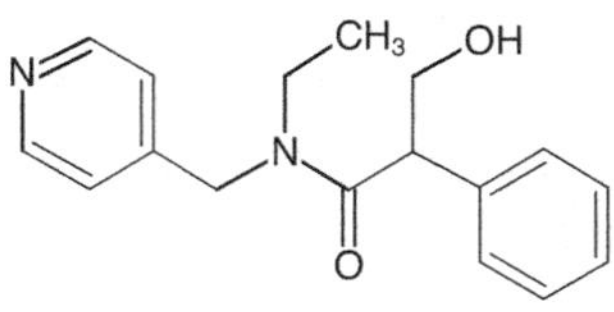

Tropicamide

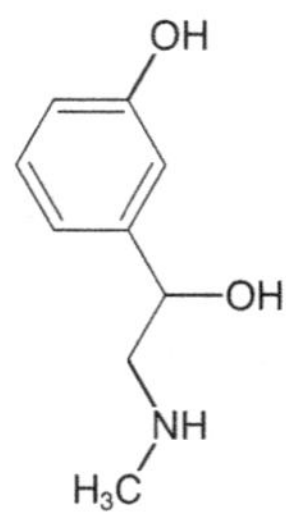

Phenylephrine

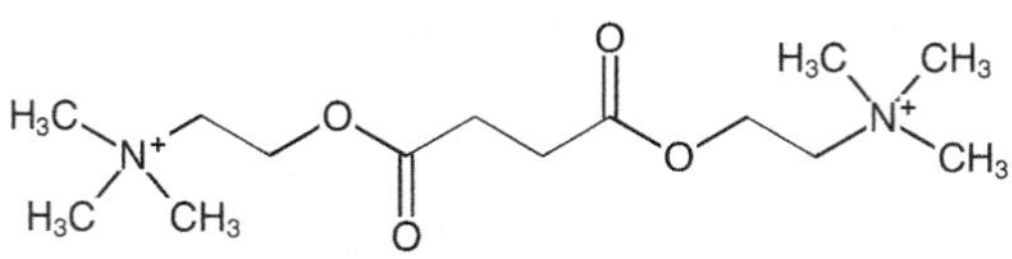

Succinylcholine

Figure 8.25 Mydriatic-cycloplegic agents.

8.6.3 *Thermosetting Implants*

Polymeric implants are useful as delivery systems or as mechanical barriers. Implants, such as preformed membranes or films, are common. However, many of these implants have limited properties and produce inferior results either as mechanical barriers or delivery systems (110).

Several implant techniques have been used in medical and dental applications. One application of interest is the use of implants in the treatment of periodontal disease.

Surgery alone does not result in restoration of lost periodontium. Successful periodontal restoration is known to occur if periodontal ligament cells are allowed to colonize root surfaces preferentially over gingival epithelial cells, gingival fibroblasts or osteoblasts.

Microporous membranes have been used for periodontal tissue regeneration. Typically, the periodontal flap is cut, and the membrane is surgically inserted to cover the surface of the tooth root and to physically occlude epithelial cells from apically migrating along the root surface.

Those membranes, however, have a drawback as a second surgical entry is needed to remove the membrane after tissue regeneration has been achieved if the membranes are not biodegradable. In order to preclude the surgical removal of an implant after doing its task, membranes made of bioabsorbable material, such as microfibrillar collagen, PLA, and poly(galactin), have been used.

However, the therapeutic effect of these membranes has been unpredictable. The degradation time of membranes composed of collagen is variable. Therefore, the risk of adverse immunological reaction to this foreign protein material in the body is a major concern.

Therefore, a composition that does not suffer from these drawbacks is desirable. Thus, the development of an implant that will eliminate the need for its surgical removal after its purpose has been achieved is highly useful (110).

Furthermore, a composition may be administered to an implant site in liquid form, which subsequently solidifies to form the actual implant. It is also desirable that the implant is capable of drug delivery.

8.6.3.1 *Base Polymer System*

The composition of a base polymer system must be a liquid formulation of a thermosetting oligomeric substance that is capable of crosslinking or hardening to provide a microporous gelatinous or solid matrix suitable for use as an implant.

Both the thermosetting prepolymers and the resulting crosslinked polymers must be biocompatible, and biodegradable or bioerodible. A preferred polymer matrix and implant is prepared from poly(*D*,*L*-lactide-*co*-caprolactone).

Lactide and γ-caprolactone are copolymerized in the presence of a multifunctional polyol. For example, a diol, such as ethylene glycol, may be included to produce a bifunctional hydroxyl-terminated polymer; or a triol, such as trimethylolpropane, may be used to produce a trifunctional polymer hydroxyl-terminated polymer.

This polyol-terminated prepolymer is then converted to an acrylic ester-terminated prepolymer, for example, by reaction acryloyl chloride, i.e., by a Schotten-Baumann reaction.

A curing agent, such as an organic peroxide, may be added to the acrylic prepolymer mixture to enhance the crosslinking of the prepolymers and thus the subsequent solidification of the resulting polymer to form a matrix.

After the addition of the curing agent, the prepolymer mixture preferably remains in liquid form for a period of time effective to allow administration of the composition to the implant site.

Here, the crosslinking reaction continues until a solid or gelatinous polymer matrix is produced. So, the prepolymer mixture cures in situ to form a polymer matrix which is capable of biodegradation or bioabsorption over time.

8.6.3.2 *Other Additives*

To the basic system, other additives may be added, such as a (110):

- Pore-forming agent,
- Delivery system, and
- Biocompatible solvents.

A pore-forming agent may be included to generate additional pores in the polymer matrix. The pore-forming agent is soluble in body fluids and water as well as in the organic solvents. The pore-forming agent is further capable of diffusing or dispersing out of the coagulating polymer matrix and into the adjacent fluids, whereupon pores are generated in the polymer matrix.

Suitable pore-forming agents include, for example, sugars such as sucrose and dextrose, salts such as sodium chloride and sodium carbonate, and polymers such as hydroxylpropylcellulose, carboxymethyl cellulose, PEG, and poly(*N*-vinyl-2-pyrrolidone).

To promote tissue growth, a biologically active agent can be added. This may be either a hard or soft tissue promoting substance or combinations thereof. Suitable tissue growth-promoting agents are shown in Table 8.12.

Table 8.12 Tissue growth promoters (110).

Agent	Abbreviation
Fibronectin	FN
Endothelial cell growth factor	ECGF
Cementum attachment extracts	CAE
Human growth hormone	HGH
Fibroblast growth factor	FGF
Platelet-derived growth factor	PDGF
Epidermal growth factor	EGF
Interleukin-1	IL-1
Transforming growth factor	TGFβ-2
Insulin-like growth factor II	ILGF-II
Human alpha thrombin	HAT
Osteoinductive factor	OIF
Bone morphogenetic protein	BMP

The delivery system can contain a large number of biologically active agents. Examples of these biologically active agents are shown in Table 8.13.

Advantageously, the polymer compositions may be administered to an implant site, as, for example, by injection with a syringe, whereupon the composition will solidify in situ. The resulting implant may thus be administered to an implant site without the need for surgical incision (110). The methods described here have been reported, but seem not to have been exemplified.

Table 8.13 Biologically active agents (110).

Agent	Exemplary Compounds
Anti-inflammatory	Hydrocortisone, prednisone, fludrotisone, triamcinolone, dexamethasone, betamethasone
Antibacterial	Penicillins, cephalosporins, vancomycin, bacitracin, polymycins, tetracyclines, chloramphenicol, erythromycin, streptomycin
Antiparasitic	Quinacrine, chloroquine, quinine
Antifungal	Nystatin, gentamicin, miconazole, tolnaftate, undecyclic acid and its salts
Antiviral	Vidarabine, acyclovir, ribarivin, amantadine hydrochloride, iododeoxyuridine, dideoxyuridine, interferons
Antineoplastic	Methotrexate, 5-fluorouracil, bleomycin, tumor necrosis factor, tumor-specific antibodies conjugated to toxins
Analgesic	Salicylic acid, salicylate esters and salts, acetaminophen, ibuprofen, morphine, phenylbutazone, indomethacin, sulindac, tolmetin, zomepirac
Local anaesthetics	Cocaine, benzocaine, novocaine, lidocaine
Vaccines	Hepatitis, influenza, measles, mumps, rubella, hemophilus, diphtheria, tetanus, rabies, polio
Growth factors	Colony stimulating factor, epidermal growth factor, erythropoietin, fibroblast growth factor, neural growth factor, human growth hormone, platelet-derived growth factor, insulin-like growth factor
Hormones	Progesterone, estrogen, testosterone, follicle stimulating hormone, chorionic gonadotrophin, insulin, endorphins, somatotropins
Antihistamines	Diphenhydramine, chlorpheneramine, chlorcyclizine, promethazine, cimetidine, terfenadine
Cardiovascular	Verapamil hydrochloride, digitalis, streptokinase, nitroglycerine papaverine, disopyramide phosphate, isosorbide dinitrate
Anti-ulcer	Cimetidine hydrochloride, isopropamide iodide, propantheline bromide
Bronchodilators	Metaproternal sulfate, aminophylline, albuterol
Vasodilators	Theophylline, niacin, nicotinate esters, amylnitrate, minoxidil, diazoxide, nifedipine

Similar thermosetting materials have been proposed for displacing tissues in order to improve radiological results (111). As in the previous literature, it is not clear whether these methods and materials are in fact used as described.

8.6.4 *Neurotoxin Implants*

Biodegradable neurotoxin implants include a neurotoxin, a biodegradable polymer component, and an acidity regulating component. The biodegradable polymer component is effective in controlling the release of the neurotoxin from the implant when the implant is located in the body of a patient (112).

A biodegradable implant with a stabilized botulinum toxin is prepared by combining a botulinum toxin, a biodegradable polymeric material, and biodegradable monomers and oligomers. PLGA is used as biodegradable polymer. The monomers and oligomers of PLGA are prepared by hydrolyzing the PLGA.

The implants may be administered to a patient in which a therapeutic effect is desired for prolonged periods of time (112).

Botulinum neurotoxin can be used for the treatment of disorders associated with melanocyte hyperactivity (113).

The methods of administration permit the neurotoxin to be administered locally to a selected target tissue. The methods of administration include injection of a solution or composition containing the neurotoxin, and the implantation of a controlled release system that controllably releases the neurotoxin to the target tissue. Such controlled release systems reduce the need for repeated injections. Diffusion of biological activity of a botulinum toxin within a tissue appears to be a function of dose and can be graduated. For example, the neurotoxin can be administered so that it primarily effects dermal cells believed to be involved in the generation of a melanin-related affliction (113).

8.6.5 *Water-Soluble Glass Fibers*

Silicon dioxide is used in common silicate glasses. If instead of silicon dioxide, phosphorous pentoxide is used, a kind of phosphate glass is obtained. Some phosphate glasses are water-soluble as well as biocompatible and can be used in biodegradable tissue formulations.

A biodegradable composite material for tissue repair has been described, formulated from water-soluble glass fibers impregnated with a biodegradable polymer, PCL (114). Such a biodegradable composite is particularly useful for the repair of nerve or bone tissue, especially the bones of the skull.

8.6.6 Bone Repair

8.6.6.1 Bioactive Glass

Bioactive glass is a biomaterial that is well suited for bone and tissue regeneration (115). Bioactive glass is capable of generating strong chemical interactions with the surrounding bone tissue and, by forming a calcium phosphate layer when exposed to physiological fluids, can promote osteointegration.

Also, the ionic release from bioactive glass may stimulate gene expression and, thus, osteoinduction. Still, its use in biomedicine is limited by its poor mechanical performance. The combination with biodegradable polymers to form composites that explore the flexibility of these materials with the stiffness and bioactive character of the bioactive glass has opened a window to many more uses (115).

Both synthetic and natural origin polymers have been combined with bioactive glass nanoparticles/nanofibers with successful outcomes. Bioactive glass-polymer nanocomposites more closely mimic the structure of natural bone.

The most promising bioactive glass-polymer combinations and the advantages introduced in bone regeneration by these composites have been reviewed and some of the most commonly used synthesis techniques applied to the production of bioactive glass-polymer composites were detailed (115).

8.6.6.2 Porous Scaffolds

Three-dimensional composite porous scaffolds containing hydroxyapatite and a polymer matrix showed wide applications in bone repair because of their improved biocompatibility, bioactivity, and mechanical properties (116). Mesoporous hydroxyapatite surface-modified by different amounts of poly(γ-benzyl-l-glutamate), from

11% to 50%, was synthesized by the *in-situ* ring-opening polymerization of γ-benzyl-l-glutamate *N*-carboxy anhydride. Then the composite films were prepared to check the biological performance of the composites. Porous scaffolds of these composites were fabricated through modified solvent casting/particulate leaching method to demonstrate the ability of *in-vivo* bone defect repair.

In-vitro cytological assay indicated that enhanced cell expansion on these composites with 11% poly(γ-benzyl-l-glutamate) amounts and improved osteogenic differentiation on the composites with 33% and 50% poly(γ-benzyl-l-glutamate) amounts were achieved. The porous scaffolds exhibited a high porosity and interconnected pores.

Results of the *in-vivo* rabbit radius defect repair indicated that rapid mineralization and new bone formation could be observed on the composites with 22% and 33% poly(γ-benzyl-l-glutamate). So, this study revealed that these composites might have potential applications in clinical bone repair (116).

8.6.6.3 PLGA Membrane

A PLGA membrane containing fluvastatin was evaluated for bone regeneration in bone defects in rat calvaria and tibia for its possible use as a guided bone regeneration membrane (117).

PLGA and fluvastatin-containing PLGA membranes were prepared and mechanical properties were evaluated (117). Fluvastatin is shown in Figure 8.26. In addition, standardized bony defects were created in rat calvaria and the right tibia, and covered with a PLGA or PLGA-fluvastatin membrane.

The effect of bone regeneration was evaluated using image analysis based on histologic examination. At 4 and 8 weeks after membrane implantation, the PLGA-fluvastatin group displayed enhanced new bone formation around the edge of the defect compared with the PLGA membrane group in the calvarial model. Thick bone regeneration was observed in tibia-defect sites in the PLGA-fluvastatin membrane group.

These results suggest that the PLGA-containing fluvastatin membrane prepared in this study may potentially be used as a guided bone regeneration membrane (117).

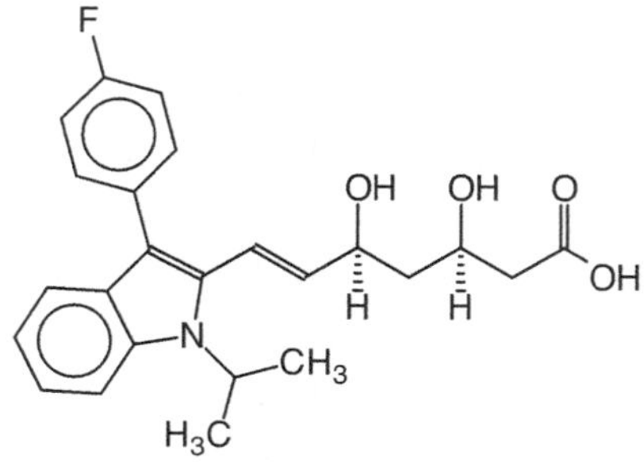

Figure 8.26 Fluvastatin.

8.6.6.4 Barrier Membrane

A barrier membrane can be used in guided bone regeneration applications to cover and aid the healing of bone defects (118). Lauric acid and nanoapatite were incorporated into PLGA matrices to form triple-layered composite membranes for potential use in guided bone regeneration applications. Lauric acid and nanoapatite were added to introduce antimicrobial and bioactive properties, respectively, to the composite membrane.

The membranes were fabricated using the combined techniques of solvent casting, thermally induced phase separation, and solvent leaching, in a single step (118). The *in-vitro* degradation behavior of this composite membrane system was studied for 24 weeks in phosphate buffer saline at 37°C at a pH of 7.4 to match the bone healing period in the guided bone regeneration applications.

An immersion of membrane samples was carried out at predetermined time intervals of 1, 2, 4, 8, 12, 16, 20 and 24 weeks. Physical changes, such as weight loss and water uptake, were measured after each time period and relative pH changes were monitored in post-immersed phosphate buffer saline solutions. SEM with energy dispersive X-ray spectroscopy was used to assess changes in morphology and the presence of nanoapatite particles on the membrane surfaces after withdrawing them from the phosphate buffer saline. The entire weight loss for the membranes was only up to 22% over 24 weeks of incubation, which suggests its structural integrity and potential use as a physical barrier in guided bone regeneration applications (118).

8.7 Shape Memory Polymers

Shape memory is the ability of a material to remember its original shape, either after mechanical deformation, which is a one-way effect, or by cooling and heating, which is a two-way effect. This phenomenon is based on a structural phase transformation.

Potential uses of shape memory polymers in medicine are to prepare sutures that have a rigid enough composition to provide for ease of insertion, but upon attaining body temperature, soften and form a second shape that is more comfortable for the patient while still allowing healing (119, 120). Another potential use is in the area of catheters. A catheter can be rigid below body temperature for its ease of insertion, but, after warming to body temperature, can soften to provide comfort to the patient (120).

The first materials known to exhibit shape memory were shape memory metal alloys. Shape memory polymers are being developed to replace the use of shape memory alloys, in part because the polymers are lightweight, high in shape recovery ability, easy to manipulate, and economical in comparison to shape memory alloys.

Shape memory polymers have been reviewed (121, 122). Shape memory polymers are characterized as phase-segregated linear block copolymers having a hard segment and a soft segment. The hard segment is typically crystalline, with a defined melting point. The soft segment is typically amorphous, with a defined glass transition temperature (120). The glass transition temperature of the soft segment is substantially less than the glass transition temperature of the hard segment.

When the shape memory polymer is heated above the glass transition temperature of the hard segment, the material can be shaped. This original shape can be memorized by cooling the shape memory polymer below the glass transition temperature of the hard segment.

When the shaped shape memory polymer is cooled below the melting point of the soft segment while the shape is deformed, a new temporary shape is fixed. The original shape is recovered by heating the material above the melting point of the soft segment but below the glass transition temperature of the hard segment. In another method for setting a temporary shape, the material is deformed at a temperature lower than the melting point of the soft segment, resulting in stress and strain being absorbed by the soft segment.

When the material is heated above the melting point of the soft segment, but below glass transition temperature of the hard segment, the stresses and strains are relieved and the material returns to its original shape. The recovery of the original shape, which is induced by an increase in temperature, is called the thermal shape memory effect. Properties that describe the shape memory capabilities of a material are the shape recovery of the original shape and the shape fixity of the temporary shape (120).

Conventional shape memory polymers are segmented PUs and have hard segments that include aromatic moieties (123, 124). But a series of other thermoplastic block copolymers with thermally induced shape memory effects have been summarized (122).

Biodegradable shape memory polymers are multiblock copolymers that are prepared from α,ω-dihydroxy [oligo(ethylene glycol glycolate) ethylene oligo (ethylene glycol glycolate)] as the hard segment (120).

As the soft segment, either poly(ε-caprolactone) or α,ω-dihydroxy[oligo(*L*-lactate-*co*-glycolate) ethylene oligo (*L*-lactate-*co*-glycolate)] are used. The polymers are chain extended using trimethylhexane-1,6-diisocyanate (120).

The hydrolytic degradation behavior of the materials was tested in buffer solution pH 7 at 37°C. It was shown that the polymers are completely degradable and their degradation rate can be adjusted by the concentration of easily hydrolyzable ester bonds.

The current standard for embolization treatment utilizes platinum coils to occlude the aneurysm and promote thrombogensis (125). Issues with platinum coils include material cost and a significant incidence of recanalization and retreatments due to coil packing over time. Hydrogel coated platinum wires were developed to reduce the aneurysm recurrence rates and are the intermediary device between the platinum coil embolism treatments and shape memory polymer coated treatments (125).

A process has been described for attaching a shape memory polymer foam to a backbone wire by synthesizing the foam directly over the nitinol wire (nickel titanium wire) or the polymer suture backbone, while holding the current properties of the foam constant in the immediate vicinity of the wire. The specimen is then cut to the desired diameter using an Excimer Laser, resistive wire cutter, or biopsy punch. This eliminates the need to optimize the process of crimping the foam to the wire and alleviates multiple issues within the device design and fabrication, including foam slippage and detachment from the coil, and the use of an epoxy-like adhesive (125).

8.7.1 *Shape Memory Polyesters*

Dihydrocarvone, c.f. Figure 8.27, is a compound that can be found in caraway oil.

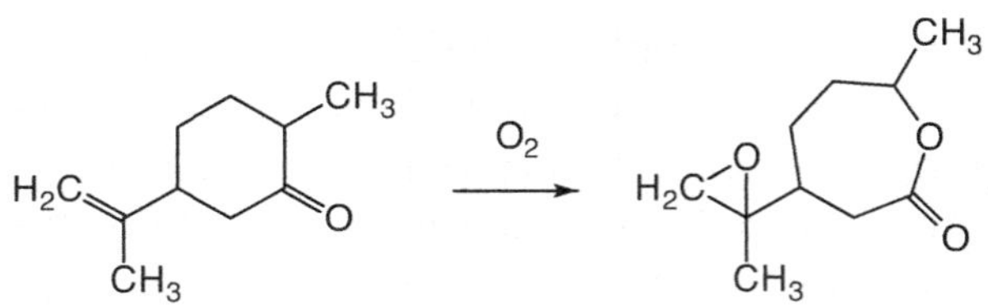

Figure 8.27 Epoxydation of dihydrocarvone.

Dihydrocarvone can be epoxidized to give the corresponding epoxylactone (126). This compound can be used as a multifunctional monomer and as a crosslinking agent in ring-opening metathesis polymerization.

The homopolymerization of the epoxylactone using diethylzinc and tin(II) 2-ethylhexanoate as catalysts yields only low molecular weight oligomers with a number average molecular weight of less than 2.5 $kg\,mol^{-1}$. However, the copolymerization with ε-caprolactone results in flexible crosslinked materials in a one-step synthesis. These copolymers exhibit a near perfect shape memory property, even after repeated bending (126).

8.8 Stents

A variety of medical situations requires the use of a mechanism to expand and support a constricted vessel and to maintain an open passageway through the vessel. In these situations, devices, commonly known as stents, are useful to prevent the stenosis of a dilated vessel, or to eliminate the danger of occlusion caused by flaps resulting from intimal tears that may be associated with angioplasty, or to hold two ends of a vessel in place (127).

Stents should also optimize flow through a vein or artery. Therefore, there is a need for stents which deliver agents or drugs to blood passing through the vein or artery that are generally beneficial to the recipient. Also desired are stents which can deliver drugs or biologically active agents at a controlled rate to blood passing through the vessel lumen as well as to the vessel wall (127).

In the treatment of vascular disease, such as arteriosclerosis, intracoronary stent placement is a common adjunct to balloon angioplasty (128). Stents, however, can also cause undesirable side effects. For example, the continued exposure of a stent to blood can lead to thrombus formation and the presence of a stent in a blood vessel can cause the blood vessel wall to weaken over time, creating the potential for arterial rupture or formation of aneurisms. A stent can also become overgrown by tissue after its implantation such that its utility is diminished or eliminated, while its continued presence may lead to a variety of complications such as the foregoing. To ameliorate the above situation, stents can be fabricated from materials that are biodegradable and, if desired, bioabsorbable (128).

In addition, the materials must not only be biocompatible and biodegradable, they must exhibit sufficient mechanical properties for this particular application (128).

PLA, poly(lactic acid-*co*-glycolic acid) and PCL are biodegradable. However, these materials typically biodegrade by bulk erosion. This may result in large particles breaking away from the degrading stent. Such particles, when released into the bloodstream, may cause emboli or other complications. Therefore, materials are desirable that degrade by surface erosion rather than bulk erosion.

In order to solve the thrombosis and restenosis problem in cardiovascular stent implantation for cardiovascular artery disease, chondroitin 6-sulfate, c.f. Figure 8.28, with heparin has been used as drug carrier layers and alternatively covalently bonded onto gold (Au)-dimercaptosuccinic acid-thiolized cardiovascular metallic (SUS316 L stainless steel) stents (129).

Sirolimus, a model drug, was encapsulated in the chondroitin 6-sulfate-heparin alternative layers. Sirolimus is shown in Figure 8.29.

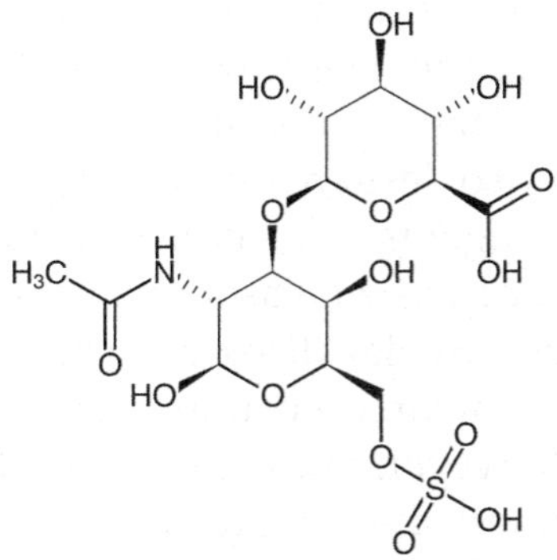

Figure 8.28 Chondroitin 6-sulfate.

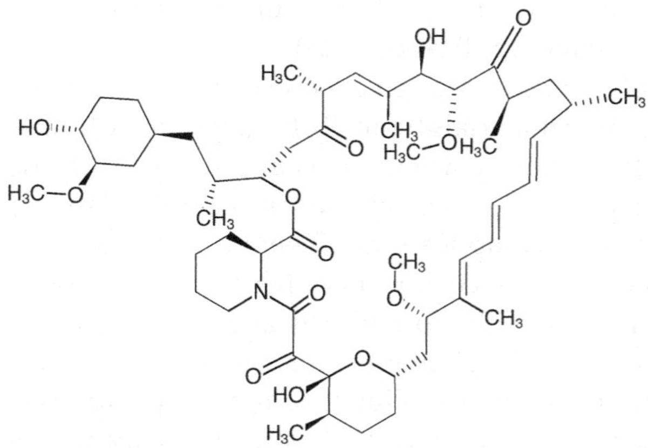

Figure 8.29 Sirolimus.

The behavior of the drug in releasing and suppressing the growth of smooth-muscle cells was evaluated with 5-layer chondroitin 6-sulfate-heparin coating on the stents. Moreover, hemocompatibility of blood clotting time and platelet adhesion was performed. The results showed that the 5-layer chondroitin 6-sulfate-heparin-modified stents displayed the greatest hemocompatibility, showing prolonged blood clotting time of the activated partial thrombin time (> 500 *s*) and less platelet adhesion to reduce thrombosis. Furthermore, sirolimus can be released continuously for more than 40 *d* with the 5-layer chondroitin 6-sulfate-heparin coating and is beneficial for inhibiting the growth of smooth-muscle cells. However, it does not affect the proliferation of endothelial cells, which can avoid restenosis formation. Therefore, the multilayers of chondroitin 6-sulfate-heparin grafted onto the Au-DMSA-cardiovascular stents provide a high potential for their use as drug-eluting stents (129).

8.8.1 Surface Erosion

Degradation by surface erosion is more likely with hydrophobic polymers. The overall hydrophobic nature of the polymer precludes or at least inhibits the incursion of water into the interior of the polymer, while water-labile linkages exposed on the surface of the polymer hydrolyze, resulting in the degradation of the polymer from the outermost surface of the bulk polymer. Therefore, hydrophobic polymers are more suitable for degradable medical implants than hydrophilic polymers (128).

Biodegradable hydrophobic polymers useful for the fabrication or coating of implantable medical devices are summarized in Table 8.14.

Therapeutic agents may be optionally added to the polymers to create a composition useful for localized sustained delivery of the agents to a patient at the site of implantation of a medical device. The therapeutic agent may be incorporated during the polymerization process or it may be blended with the polymer after it is formed (128).

Some therapeutic agents can be chemically incorporated into the backbone of the polymer, or can be attached as a pendant group. Therapeutic agents include salicylic acid, nitric oxide, PEG, heparin, low molecular weight heparins, heparinoids and hyaluronic acid.

Table 8.14 Hydrophobic polymers useful for stents (128).

Polymer
Poly(maleic acid-*co*-sebacic acid)
Poly(*L*-lactic acid-*co*-*L*-aspartic acid)
Poly(*D,L*-lactic acid-*co*-*L*-aspartic acid)
Poly(*L*-lactic acid)
Poly(*D,L*-lactic acid)
Poly(*L*-lactic acid-*co*-ethylene glycol)
Poly(*D,L*-lactic acid-*co*-ethylene glycol)
Poly(ethylene glycol-*co*-butylene terephthalate)
Poly(4-hydroxy-L-proline ester)
Poly(1,10-decanediol-*co*-*L*-lactic acid)
Poly(1,10-decanodiol-*co*-*D,L*-lactic acid)
Poly(1,2,6-hexanetriol-*co*-trimethylorthoacetate)
Poly(hydroxybutyrate)
Poly(hydroxyvalerate)
Poly(hydroxy-butyrate-valerate)

The polymers are useful for the fabrication or coating of implantable medical devices, in particular stents. They not only have the requisite mechanical properties but they also biodegrade by surface erosion rather than by bulk erosion. They thus would be expected to drastically reduce the formation and the release of large particles during biodegradation (128).

8.8.2 *Tubular Main Body*

The tubular main body of the stent includes a substantially biodegradable matrix having collagen IV and laminin that enclose voids within the matrix. The tubular main body also contains PLA as a biodegradable strengthening material.

The rate of drug release can either accelerate or slow down the rate of degradation of the biodegradable material. Thus, the rate of release of a drug acts as a control quantity for the rate of degradation (127).

For instance, a coiled stent device could include a coiled main body made from a strip having four layers. A first layer, contacting the vessel wall, could incorporate the drug fibronectin. A second layer contacting the first layer, could incorporate the drug fibronectin at a lower concentration than the first layer. A third layer contacting the second layer could incorporate fibronectin at a lower concentration than either the first or the second layer. A fourth layer could contact the fluid passing through the lumen and could incorporate the drug heparin.

The drug fibronectin accelerates growth of cells surrounding the stent. The accelerated growth of cells accelerates the resorption reactions of the first layer of the stent. A reduced fibronectin concentration in the second and third layers slows down the resorption reactions so that the degradation of the first three layers will proceed at a cumulative rate that is compatible with the degradation of the fourth layer.

Drugs are incorporated into the biodegradable stent by (127):

- Mixing or solubilizing with polymer solutions,
- Dispersing into the biodegradable polymer during the extrusion of melt spinning process, or
- Coating onto an already formed film or fiber.

In particular, hollow fibers, which contain anti-thrombogenic drugs, can be arranged in a parallel concentric configuration with solid fibers for added support for use on the outer surface of the main body of the stent (127).

8.8.3 *Multilayer Stents*

Stents have been made from a biodegradable fiber having an inner core and an outer layer (130). The inner core is a copolymer of about 75 *mol%* glycolide and about 25 *mol%* caprolactone. The outer layer is a blend of two polymer components; this is a lactide/glycolide (90/10) copolymer and a lactide/glycolide (85/15) copolymer. The molecular weights of these copolymers are different. The two copolymers are mixed into a 60/40 weight ratio before being finally melt blended and pelletized using a twin screw extruder.

The inner core has a first degradation rate, and the outer layer has a second degradation rate. The second degradation rate is slower than the first degradation rate. The fiber softens in vivo such that the stent is readily passed from the lumen as a softened fragment or filament after a predetermined period of time through normal flow of body fluids passing through the lumen (130).

Endovascular treatment using bare metal stents or drug-eluting stents is routinely performed to restore the blood flow in the coronary and peripheral arteries (131). However, intimal hyperplasia and late stent thrombosis are still major issues leading to a failure of the stent and also often to the need for reintervention.

Covered stents have emerged as an alternative approach to avoid intimal hyperplasia. They are characterized by the presence of a membrane that spans the struts of the stent and acts as a physical barrier to block the ingrowth of the neointimal tissue into the lumen.

Membrane materials that are currently used include mainly synthetic polymers. Their main limitation of these polymers is their lack of hemocompatibility, which can induce thrombosis and ultimately reocclusion, thus impairing the long-term performance of these devices. To overcome this issue, the use of biologically derived materials has been proposed with the aim of enhancing the biocompatibility and capability of this type of stent to support endothelialization (131).

A comprehensive overview is given of the applications of bio-based covered stents as well as the different strategies followed for their development from the perspective of the material selection, fabrication approaches, and validation of the different concepts in animal and clinical trials (131).

8.9 Thermogelling Materials

Materials that gel in situ have recently gained attention as promising implantable drug delivery systems as well as injectable matrices for tissue engineering (132).

In-situ gelation is the basis of injectable systems that eliminates the need for surgical procedures and offers the advantage of the ability to form any desired implant shape.

A thermogelling biodegradable aqueous polymer solution has been developed that consists of a PEG block and a biodegradable polyester block, i.e., PLGA (132). The graft copolymer PEG-*g*-PLGA can be synthesized by a one-step ring-opening polymerization of

D,L-lactide, glycolide, and an epoxy-terminated PEG using stannous octoate as a catalyst (132). At high concentrations, the PEG-*g*-PLGA aqueous solution undergoes a sol-to-gel transition with increasing temperature.

The material can be applied for a short-term delivery of bioactive agents such as pharmaceutical drugs, e.g., proteins, anticancer drugs, as well as a carrier or delivery system for bioactive agents used in tissue engineering (132).

8.10 Cancer Therapy

8.10.1 Anticancer Peptide

Amphiphilic PLGA nanoparticles with improved physicochemical properties were formulated for the delivery of the novel peptide (CK-10) to be used for targeting of the cancerous/tumor tissue (133). This was achieved by blending of various amphiphilic polymers with PLGA, especially by using a novel microfluidic technique which can overcome several problems of the conventional techniques like the double emulsion technique, e.g., low peptide loading efficiencies, large sizes, and high polydispersity index.

The loading efficiency was measured by a modified Lowry assay. The size and the zeta potential were characterized by dynamic light scattering and tuneable pore resistive sensing techniques. The images were scanned by SEM and TEM. The stability and the interaction were confirmed by high performance liquid chromatography mass spectrometry (HPLC/MS), FTIR, DSC and capillary zone electrophoresis.

PLGA/Poloxamer nanoparticles exhibited a higher peptide loading than the other types of PLGA nanoparticles (56.13 % *m/m* for the novel microfluidic technique). PLGA/Poloxamer prepared by the microfluidics technique had the smallest size of 208.90 *nm* with the lowest polydispersity index of 0.11, which is a vital parameter for targeting.

The successful development of better physicochemical properties for the CK-10 loaded PLGA nanoparticles can improve the RAN blocking by CK-10 (133).

8.10.2 *Synergistic Cancer Therapy*

Near-infrared light-induced photothermal therapy has been of great interest as an alternative to conventional treatments for cancer therapy because of its minimal invasiveness and high selectivity (134, 135)

A fibrous system for synergistic cancer therapy has been developed (136). The system consists of PLGA fibers with a core encapsulating an anticancer drug and a shell entrapping gold nanorods as a photothermal agent.

Upon exposure to near-infrared light, the photothermal agent generates heat to raise the local temperature of the fibers. If the temperature is above a glass transition of the polymer, the polymer chains will be mobile, increasing the free volume size within the shell. As a result, a rapid release of the drug can be achieved (136).

When the near-infrared light is turned off, the release will stop with inactivity of the photothermal agent, followed by freezing the segmental motion of the polymer chains. The on–off switching of near-infrared light in a time-controllable manner allows a repeated and accurate release of the drug, leading to the significant enhancement of anticancer activity in combination with the hyperthermia effect arising from the photothermal agent (136).

8.11 Wound Dressings

A polymer blend solution is prepared by electrospinning with a potential difference of 20–40 *kV* at a gap distance of 15–40 *cm* (137). Microfibers containing poly(lactide-*co*-glycolide) and poly(*N*-vinyl pyrrolidone) are used in a ratio of 70:30. The microfibers are prepared by an electrohydrodynamic method, i.e., electrospinning (138). To obtain a totally biodegradable, hydrophilic unwoven absorbent, the electrohydrodynamic method for solution spinning can be applied. The method involves spraying the solution of a polymer blend through a capillary nozzle onto a substrate (138). When the material is used for wound and burn healing acceleration, it may contain living human cells like keratinocytes or fibroblasts previously grown on the material as on the solid porous scaffold.

To provide a prolonged and controlled drug release to the surface of wounds or burns, the material may contain two or more microfiber layers. Different layers may have different compositions.

Each layer includes the biodegradable polymer with or without poly(*N*-vinyl)lactam. The layers may also have different biodegradable polymers (138).

8.12 Bioceramics

Sol-gel-derived ceramic materials have many applications in various fields. Bioceramics is one of the most promising and interesting fields that still needs much developmental work for optimizing the properties of the material in a biological environment.

The sol-gel process starting from a liquid phase enables an easy control of the pore structure of the material and an introduction of other components in different kinds of composites, especially in the case of silica-based materials. Processing of sol-gel-derived silica fibers is well established.

The main parameters controlling the process are the functionality of the silica precursors and the degree of branching of the silica clusters. The latter critically affects the spinnability and has commonly been characterized by rheological measurements.

The biodegradation of silica fibers can be controlled by controlling the viscosity of the spinning solution and, thus, the biodegradation of the silica fibers can be varied even when the same recipe is used (139).

Factors affecting the viscosity are the stage of spinnability, the temperature of the silica sol and the amount of solvent in the spinning sol. The silica sol remains spinnable within a certain period of time. The viscosity of the silica sol increases continuously during that period of time.

In the earlier stage of spinnability the silica polymers are somewhat smaller and they are packed easier, forming denser structures than the larger silica polymers of the later stage of spinnability. Higher viscosity inhibits the orientation of the silica polymers, leaving the structure more open. Therefore, the fibers spun in the early stage of the spinnability period degrade more slowly than fibers spun in the later stage of the spinnability. Another parameter that controls the spinnability and the viscosity is the temperature.

Fibers that are spun from the silica sols at a lower temperature degrade faster than the corresponding fibers spun at higher temperatures. Of course, the viscosity becomes higher at lower temperatures. The reason for this effect is obviously the same as varying the stage of time when the fibers are spun.

The introduction of biologically active agents into the porous structure of the fiber opens alternatives for the design of biomedical applications. Biodegradable and nontoxic materials that are able to work directly and locally in humans or animals are beneficial as implants used as drug delivery devices or temporary implants in bone repairs. Suitable biologically active agent are usually in such applications and have been listed in detail (139).

An effective amount of a biologically active agent can be added to the reaction mixture at any stage of the process. For example, it can be mixed with the starting materials. It can also be added to the reaction mixture at the sol stage before condensation reactions take place or during the condensation reactions, or even afterwards.

The silica sols are prepared from tetraethyl orthosilicate, deionized water and ethanol. HNO_3 or NH_3 are used as catalysts for successful performance of the sol-gel method (139). The biodegradation of the samples was studied in vitro using a simulated body fluid as described in the literature (140).

8.13 Conjugates

Biologically active agents can be covalently linked to a polymer. The polymer is biodegradable, such as is poly(hydroxymethylethylene hydroxymethylacetal) (141). The polymer can be crosslinked with epibromohydrin or epichlorohydrin to form a gel. Other suitable polymers are listed in Table 8.15.

A conjugate of a biologically active agent and a polymer refers to a covalent bond or an ionic bond between the two species. The biologically active agent is preferably a protein such as an extracellular enzyme.

Conjugates with a biological agent and a biodegradable polymer can further be conjugated to one or more other moieties. For example, they may be linked to a label, e.g., to permit detection of the conjugates in vivo. The conjugates may also be linked to a targeting agent that will target the conjugate to the appropriate site within a

Table 8.15 Polymers for conjugates (141).

Polymer or Copolymer
Poly(lactic acid) polymers
Poly(glycolide-*co*-caprolactone)
Hyaluronic acid derivatives
Carboxymethyl cellulose
Poly(β-hydroxybutyrate)
Poly(*p*-dioxanone)
Polyesters of oxalic acid
Poly(dihydropyrane)s
Poly(β-malic acid)
Poly(ethylene adipate)
Glycolide *L*-lactide copolymers
Lactide trimethylene carbonate copolymers
Lactide tetramethylglycolide copolymers
Lactide α-valerolactone copolymers
Methacrylate *N*-vinyl-pyrrolidone copolymers
Lactic acid ethylene oxide copolymers

subject. A targeting agent may be, for example, an antibody. For example, when the biological agent is an apyrase, it may be desirable to target a conjugate comprising the apyrase to endothelial cells. This can be accomplished, for example, by linking to the apyrase or to the polymer an antibody binding specifically to membrane antigens of endoplasmic cells. Other targeting agents are known in the art for targeting various sites. Targeting agents can be linked to either the biological agent or the polymer (141).

The conjugates can be used as therapeutics. For example, a conjugate containing an apyrase can be used for treating and preventing thrombosis, atherosclerotic plaque complications and vascular disorders (141).

Trade names appearing in the references are shown in Table 8.16.

Table 8.16 Trade names in references.

Trade name Description	Supplier
Teflon® Tetrafluoro polymer (21)	DuPont

References

1. L.S. Nair and C.T. Laurencin, *Progress in Polymer Science*, Vol. 32, p. 762, August 2007.
2. S. Kalia and L. Avérous, eds., *Biopolymers: Biomedical and Environmental Applications*, Wiley-Scrivener, Salem, MA, 2011.
3. S. Ahmed, S. Ikram, S. Kanchi, and K. Bisetty, eds., *Biocomposites: Biomedical and Environmental Applications*, CRC Press, 2018.
4. S. Misra, A.K. Srivastava, S. Raghuwanshi, V. Sharma, and P.S. Bisen, Medical grade biodegradable polymers: A perspective from gram-positive bacteria in S. Thomas, P. Balakrishnan, and M. Sreekala, eds., *Fundamental Biomaterials: Polymers*, Woodhead Publishing Series in Biomaterials, chapter 12, pp. 267–286. Woodhead Publishing, 2018.
5. J. Wróblewska-Krepsztul, T. Rydzkowski, I. Michalska-Pozoga, and V.K. Thakur, *Nanomaterials*, Vol. 9, p. 404, March 2019.
6. G.V. Savage and C.T. Rhodes, *Drug Development and Industrial Pharmacy*, Vol. 21, p. 93, January 1995.
7. J.W. Gibson, S.S. Miller, J.C. Middleton, and A.J. Tipton, High viscosity liquid controlled delivery system and medical or surgical device, US Patent 7 833 543, assigned to Durect Corporation (Cupertino, CA), November 16, 2010.
8. S. Arora, J. Ali, A. Ahuja, R.K. Khar, and S. Baboota, *AAPS PharmSciTech*, Vol. 6, p. E372, September 2005.
9. N. Rajesh, N. Uma, and R. Valluru, Natural polymers – A boon for drug delivery in V. Mittal, ed., *Renewable Polymers: Synthesis, Processing, and Technology*, chapter 10, pp. 429–472. Wiley Blackwell (John Wiley & Sons), October 2011.
10. T. Ayyappan, S. Shanmugam, and T. Vetrichelvan, *Trends in Drug Delivery*, Vol. 3, p. 1, 2016.
11. L.K. Garg, B. Singh, S. Pal, A. Saini, and S. Thakur, *Journal of Pharmacy Research*, Vol. 12, p. 794, 2018.
12. C.G. Pitt and A. Schindler, Biodegradation of polymers in S.D. Bruck, ed., *Controlled Drug Delivery. Basic Concepts*, Vol. 1 of *CRC Revivals*, chapter 3. CRC Press, Boca Raton, FL; Cambridge, England, 2019.
13. F. Abedini, M. Ebrahimi, A.H. Roozbehani, A.J. Domb, and H. Hosseinkhani, *Polymers for Advanced Technologies*, Vol. 29, p. 2564, 2018.
14. R.K. Kulkarni, K.C. Pani, C. Neuman, and F. Leonard, *Archives of Surgery*, Vol. 93, p. 839, 1966.
15. S. Kumar and S.K. Gupta, *Polimery w Medycynie*, Vol. 42, p. 191, 2012.
16. E. Giuliano, D. Paolino, M. Fresta, and D. Cosco, *Medicines*, Vol. 6, 2018.

17. P. Zarrintaj, Z. Ahmadi, M.R. Saeb, and M. Mozafari, *Materials Today: Proceedings*, Vol. 5, p. 15516, 2018. INN International Conference/Workshop on Nanotechnology and Nanomedicine (NTNM2017), 2-3 May 2017.
18. L. Djekic, M. Martinović, V. Dobrić, B. Čalija, D. Medarević, and M. Primorac, *Journal of Pharmaceutical Sciences*, Vol. 108, p. 1326, 2019.
19. W. Xu, X. Wang, N. Sandler, S. Willför, and C. Xu, *ACS Sustainable Chemistry & Engineering*, Vol. 6, p. 5663, 2018.
20. H. Cho, U. Jammalamadaka, K. Tappa, C. Egbulefu, J. Prior, R. Tang, and S. Achilefu, *Molecular Pharmaceutics*, Vol. 16, p. 552, 2019.
21. J. Kunzler and J. Salamone, Drug delivery systems, US Patent 7 544 371, assigned to Bausch + Lomb Incorporated (Rochester, NY), June 9, 2009.
22. J. Swanson, R.A. Jain, R. Hontz, J.G. Devane, K.I. Cumming, M.J.A. Clancy, J.E. Codd, and G.G. Liversidge, Controlled release nanoparticulate compositions, US Patent Application 20 130 011 447, assigned to Alkermes Pharma Ireland Limited, January 10, 2013.
23. A. Nayak and D.B. Das, *Biotechnology Letters*, Vol. 35, May 2013.
24. J. Rose and S. Hardwick, Biodegradable polymer systems, US Patent 7 524 891, assigned to Smith & Nephew PLC (GB), April 28, 2009.
25. D.R. Perinelli, M. Cespi, G. Bonacucina, and G.F. Palmieri, *Journal of Pharmaceutical Investigation*, Vol. in press, p. 1, 2019.
26. C. Zhang, A. Wang, H. Wang, M. Yan, R. Liang, X. He, F. Fu, H. Mu, and K. Sun, *International Journal of Pharmaceutics*, Vol. 560, p. 27, 2019.
27. E. Chiesa, I. Genta, R. Dorati, T. Modena, and B. Conti, *Reactive and Functional Polymers*, Vol. 140, p. 93, 2019.
28. A. Lanz-Landázuri, A.M. de Ilarduya, M. García-Alvarez, and S. Muñoz-Guerra, *Reactive and Functional Polymers*, Vol. 106, p. 143, 2016.
29. V. Gupta, M. Hariharan, T.A. Wheatley, and J.C. Price, *European Journal of Pharmaceutics and Biopharmaceutics*, Vol. 51, p. 241, May 2001.
30. J. Nerurkar, H.W. Jun, J.C. Price, and M.O. Park, *European Journal of Pharmaceutics and Biopharmaceutics*, Vol. 61, p. 56, September 2005.
31. A. Kelemen, G. Buchholcz, T. Sovány, and K. Pintye-Hódi, *Journal of Pharmaceutical and Biomedical Analysis*, Vol. 112, p. 85, 2015.
32. S. Ghayas, M.H. Shoaib, F. Qazi, R. Bushra, F.R. Ali, M. Maboos, and F. Khalid, *Polymer Bulletin*, Vol. in press, p. 1, May 2019.
33. M. Ghanbarzadeh, A. Golmoradizadeh, and A. Homaei, *Phytochemistry Reviews*, Vol. 17, p. 535, Jun 2018.
34. M.D. Chavanpatil, P. Jain, S. Chaudhari, R. Shear, and P.R. Vavia, *International Journal of Pharmaceutics*, Vol. 316, p. 86, June 2006.
35. N.L. Prasanthi, S.S. Manikiran, and N.R. Rao, *Int. J. Pharm. Tech. Res.*, Vol. 2, p. 2506, 2010.

36. C. Ferrero, D. Massuelle, D. Jeannerat, and E. Doelker, *Journal of Controlled Release*, Vol. 170, p. 175, September 2013.
37. M. Reiner, *Physics Today*, Vol. 17, p. 62, 1964.
38. C.S. Brazel and N.A. Peppas, *Biomaterials*, Vol. 20, p. 721, 1999.
39. R. Muzzarelli, *Chitin in Nature and Technology*, Plenum Press, New York, 1986.
40. G. Braek, *Chitin and Chitosan: Sources, Chemistry, Biochemistry, Physical Properties, and Applications*, Elsevier Applied Science, London, New York, 1989.
41. N.S. Tapola, M.L. Lyyra, R.M. Kolehmainen, E.S. Sarkkinen, and A.G. Schauss, *Journal of the American College of Nutrition*, Vol. 27, p. 22, 2008.
42. S.A. Agnihotri, N.N. Mallikarjuna, and T.M. Aminabhavi, *Journal of Controlled Release*, Vol. 100, p. 5, November 2004.
43. R.J. Babu, S. Sathigari, M.T. Kumar, and J.K. Pandit, *Current Drug Delivery*, Vol. 7, p. 36, 2010.
44. A.B. Meneguin, E. Beyssac, G. Garrait, H. Hsein, and B.S.F. Cury, *European Journal of Pharmaceutics and Biopharmaceutics*, Vol. 123, p. 84, 2018.
45. A. Adrover, P. Paolicelli, S. Petralito, L. Di Muzio, J. Trilli, S. Cesa, I. Tho, and M.A. Casadei, *Pharmaceutics*, Vol. 11, p. 187, April 2019.
46. S.A. Altaf, K. Yu, J. Parasrampuria, and D.R. Friend, *Pharmaceutical Research*, Vol. 15, p. 1196, 1998.
47. S. Al-Saidan, Y. Krishnaiah, V. Satyanarayana, P. Bhaskar, and R. Karthikeyan, *European Journal of Pharmaceutics and Biopharmaceutics*, Vol. 58, p. 697, November 2004.
48. K. Meyer and J.W. Palmer, *Journal of Biological Chemistry*, Vol. 107, p. 629, 1934.
49. J.-H. Chen, B.-H. Tsai, H.-T. Chang, M.-L. Chen, Y.-H. Chen, S.-H. Jan, and M.-J. Liu, Biodegradable hyaluronic acid derivative and biodegradable polymeric micelle composition, US Patent 7 780 982, assigned to Industrial Technology Research Institute (Chutung, Hsinchu, TW), August 24, 2010.
50. F. della Valle and A. Romeo, Total or partial esters of hyaluronic acid, US Patent 5 336 767, assigned to Fidia S.p.A. (Abano Terme, IT), August 9, 1994.
51. F. della Valle, A. Romeo, and S. Lorenzi, Salts and mixtures of hyaluronic acid with pharmaceutically active substances, pharmaceutical compositions containing the same and methods for administration of such compositions, US Patent 5 442 053, assigned to Fidia S.p.A. (Abano Terme, IT), August 15, 1995.
52. O.A. Odeku and O.A. Itiola, *Drug Development and Industrial Pharmacy*, Vol. 28, p. 329, January 2002.

53. O.A. Odeku and J.T. Fell, *Journal of Pharmacy and Pharmacology*, Vol. 56, p. 1365, November 2004.
54. K. Malik, G. Arora, and I. Singh, *Polimery w Medycynie*, Vol. 41, 2011.
55. C. Vijayaraghavan, S. Vasanthakumar, and A. Ramakrishnan, *Die Pharmazie - An International Journal of Pharmaceutical Sciences*, Vol. 63, p. 342, 2008.
56. M. Turkoglu and T. Ugurlu, *European Journal of Pharmaceutics and Biopharmaceutics*, Vol. 53, p. 65, January 2002.
57. P. Mura, F. Maestrelli, M. Cirri, M.L. González Rodríguez, and A.M. Rabasco Alvarez, *Journal of Drug Targeting*, Vol. 11, p. 365, July 2003.
58. K.P. Rao, B. Prabhashankar, A. Kumar, A. Khan, S.S. Biradar, S.P. Srishail, and B. Satyanath, *The Yale Journal of Biology and Medicine*, Vol. 76, p. 149, 2003.
59. A. Rasul, M. Iqbal, G. Murtaza, M.K. Waqas, M. Hanif, S.A. Khan, and N.S. Bhatti, *Acta Pol. Pharm.*, Vol. 67, p. 517, 2010.
60. E.N. Zare, P. Makvandi, and F.R. Tay, *Carbohydrate Polymers*, Vol. 212, p. 450, 2019.
61. K.M. Rao, A. Kumar, K.S.V. Krishna Rao, A. Haider, and S.S. Han, *Journal of Polymers and the Environment*, Vol. 26, p. 778, Feb 2018.
62. C. Verma, P. Negi, D. Pathania, V. Sethi, and B. Gupta, *Polymer International*, Vol. 68, p. 344, 2019.
63. M. Ranjbar-Mohammadi, S. Rabbani, S.H. Bahrami, M. Joghataei, and F. Moayer, *Materials Science and Engineering: C*, Vol. 69, p. 1183, 2016.
64. J.F. Cooley, Apparatus for electrically dispersing fluids, US Patent 692 631, assigned to Eastman, (Boston, MA), February 4, 1902.
65. W.J. Morton, Method of dispersing fluids, US Patent 705 691, July 29, 1902.
66. A.L. Andrady and D.S. Ensor, Electrospray/electrospinning apparatus and method, US Patent 7 762 801, assigned to Research Triangle Institute (Research Triangle Park, NC), July 27, 2010.
67. S.B. Mitchell and J.E. Sanders, Method for controlled electrospinning, US Patent 7 981 353, assigned to University of Washington (Seattle, WA), July 19, 2011.
68. G.L. Bowlin, G.E. Wnek, and D.G. Simpson, Electroprocessed material made by simultaneously electroprocessing a natural protein polymer and two synthetic polymers, US Patent 7 374 774, assigned to Virginia Commonwealth University Intellectual Property Foundation (Richmond, VA), May 20, 2008.
69. D.G. Simpson, G.L. Bowlin, G.E. Wnek, P.J. Stevens, M.E. Carr, J.A. Matthews, and S. Rajendran, Electroprocessed collagen and tissue engineering, US Patent 7 615 373, assigned to Virginia Commonwealth University Intellectual Property Foundation

(Richmond, VA); Organogenesis, Inc. (Canton, MA), November 10, 2009.

70. G.L. Bowlin, G.E. Wnek, D.G. Simpson, P. Lam, and M.E. Carr, Jr., Electroprocessed fibrin-based matrices and tissues, US Patent 7 759 082, assigned to Virginia Commonwealth University Intellectual Property Foundation (Richmond, VA), July 20, 2010.
71. G. Piccirillo, D.A. Carvajal Berrio, A. Laurita, A. Pepe, B. Bochicchio, K. Schenke-Layland, and S. Hinderer, *Scientific Reports*, Vol. 9, p. 3446, 2019.
72. J. Wang and M. Windbergs, *International Journal of Pharmaceutics*, Vol. 556, p. 363, 2019.
73. G.E. Wnek, M.E. Carr, D.G. Simpson, and G.L. Bowlin, *Nano Letters*, Vol. 3, p. 213, 2003.
74. M. Herrero-Herrero, J.A. Gómez-Tejedor, and A. Vallés-Lluch, *European Polymer Journal*, Vol. 99, p. 445, 2018.
75. G.R. Da Silva, T.H. Lima, G.M. Fernandes-Cunha, R.L. Oréfice, A.D. Silva-Cunha, M. Zhao, and F. Behar-Cohen, *European Journal of Pharmaceutics and Biopharmaceutics*, Vol. in press, 2019.
76. Wikipedia contributors, Glial fibrillary acidic protein — Wikipedia, the free encyclopedia, https://en.wikipedia.org/w/index.php?title=Glial_fibrillary_acidic_protein&oldid=862463090, 2018. [Online; accessed 9-June-2019].
77. A. Garg, S. Garg, M. Kumar, S. Kumar, A.K. Shukla, S. Prakash, and C. Kaushik, *Advance Pharmaceutical Journal*, Vol. 3, p. 38, 2018.
78. D.K. Han, K.-D. Ahn, H.J. Jung, and K. Park, Method of preparing biodegradable dual pore polymer scaffolds for tissue engineering, US Patent 7 842 305, assigned to Korea Institute of Science and Technology (Seoul, KR), November 30, 2010.
79. J.A. Burdick and R.L. Mauck, eds., *Biomaterials for Tissue Engineering Applications*, Springer-Verlag Wien, 2011.
80. S. Muñoz-Guerra, M. García-Alvarez, and J.A. Portilla-Arias, *Journal of Renewable Materials*, Vol. 1, p. 42, January 2013.
81. J.-H. Ryu, J.-S. Kwon, K.-M. Kim, H.J. Hong, W.-G. Koh, J. Lee, H.-J. Lee, H.-J. Choi, S. Yi, H. Shin, and M.-H. Hong, *ACS Omega*, Vol. 4, p. 2302, 2019.
82. S.M.S. Gruber, P. Ghosh, K.W. Mueller, P.W. Whitlock, and C.-Y. Lin, *Journal of Visualized Experiments*, Vol. 143, p. e58720, 2019.
83. C. Vaquette, S.P. Pilipchuk, P.M. Bartold, D.W. Hutmacher, W.V. Giannobile, and S. Ivanovski, *Advanced Healthcare Materials*, Vol. 7, p. 1800457, 2018.
84. J. Mota, N. Yu, S.G. Caridade, G.M. Luz, M.E. Gomes, R.L. Reis, J.A. Jansen, X.F. Walboomers, and J.F. Mano, *Acta Biomaterialia*, Vol. 8, p. 4173, 2012.

85. Y. Xue, X. Hong, J. Gao, R. Shen, and Z. Ye, *International Journal of Nanomedicine*, Vol. 14, p. 483, January 2019.
86. M. Hasani-Sadrabadi, P. Sarrion, N. Nakatsuka, T.D. Young, N. Taghdiri, S. Ansari, T. Aghaloo, S. Li, A. Khademhosseini, P.S. Weiss, and A. Moshaverinia, *ACS Nano*, Vol. 13, p. 3830, 2019.
87. M. Qu, W. Xiao, J. Tian, S. Wang, H. Li, X. Liu, X. Yang, B. Li, and X. Liao, *Journal of Biomedical Materials Research Part B: Applied Biomaterials*, Vol. 107, p. 511, 2019.
88. J.S. Fisher, Biodegradable polymer for marking tissue and sealing tracts, US Patent 7 972 619, assigned to Biopsy Sciences, LLC (Clearwater, FL), July 5, 2011.
89. T.R.R. Singh, H.L. McMillan, R. Sheshala, I.A. Tekko, F. Alshammari, and P. Kesharwani, In situ forming phase–inversion injectable hydrogels for controlled drug release in T.R.R. Singh, G. Laverty, and R. Donelli, eds., *Hydrogels: Design, Synthesis and Application in Drug Delivery & Regenerative Medicine*, chapter 13, pp. 228–249. CRC Press, Boca Raton, FL, 2018.
90. A.S. Sawhney, In situ polymerizable hydrogels, US Patent 6 818 018, assigned to Incept LLC (Lexington, MA), November 16, 2004.
91. N.R. Washburn, C.G. Simon, Jr, A. Tona, H.M. Elgendy, A. Karim, and E.J. Amis, *Journal of Biomedical Materials Research*, Vol. 60, p. 20, 2002.
92. A.G. Mikos, Biodegradable bone templates, US Patent 5 522 895, assigned to Rice University (Houston, TX), June 4, 1996.
93. L. Shapiro, R. Glicklis, and S. Cohen, Method of forming polysaccharide sponges for cell culture and transplantation, US Patent 6 334 968, assigned to Ben Gurion University of the Negev (Be'er-Sheva, IL), January 1, 2002.
94. B.D. Boyan, C.M. Agrawal, and J.D. Heckman, Biodegradable implant for fracture nonunions, US Patent 5 492 697, assigned to Board of Regents, Univ. of Texas System (Austin, TX), February 20, 1996.
95. T.H. Barrows, M.T. Truong, and P.R. Suszko, Biocompatible porous matrix of bioabsorbable material, US Patent 5 856 367, assigned to Minnesota Mining and Manufacturing Company (St. Paul, MN), January 5, 1999.
96. B.D. Favis, P. Sarazin, J. Li, and Z. Yuan, Microporous articles comprising biodegradable medical polymers, method of preparation thereof and method of use thereof, US Patent 8 007 823, assigned to Corporation de l'Ecole Polytechnique de Montreal (Montreal, CA), August 30, 2011.
97. M.N. Vyakarnam, M.C. Zimmerman, A.G. Scopelianos, M.B. Roller, and D.V. Gorky, Porous tissue scaffoldings for the repair or regeneration of tissue, US Patent 6 365 149, assigned to Ethicon, Inc. (Somerville, NJ), April 2, 2002.

98. H.B. Levene, C.M. Lhommeau, and J.B. Kohn, Porous polymer scaffolds for tissue engineering, US Patent 6 337 198, assigned to Rutgers, The State University (New Brunswick, NJ), January 8, 2002.
99. L. Harris, D.J. Mooney, and L. Shea, Open pore biodegradable matrices, US Patent 6 281 256, assigned to The Regents of the University of Michigan (Ann Arbor, MI), August 28, 2001.
100. V.B. Dave, G. Landau, and P. Patel, Biodegradable medical implant with encapsulated buffering agent, US Patent 7 785 615, assigned to Cordis Corporation (Miami Lakes, FL), August 31, 2010.
101. D. da Silva, M. Kaduri, M. Poley, O. Adir, N. Krinsky, J. Shainsky-Roitman, and A. Schroeder, *Chemical Engineering Journal*, Vol. 340, p. 9, 2018. Smart nanomaterials and nanostructures for diagnostic and therapy.
102. W. Jiang, H. Li, Z. Liu, Z. Li, J. Tian, B. Shi, Y. Zou, H. Ouyang, C. Zhao, L. Zhao, R. Sun, H. Zheng, Y. Fan, Z.L. Wang, and Z. Li, *Advanced Materials*, Vol. 30, p. 1801895, 2018.
103. S.D. Rousselle, Y. Ramot, A. Nyska, and N.D. Jackson, *Toxicologic Pathology*, Vol. 47, p. 358, 2019.
104. A. Singh, A.U. Luong, K.J. Fong, R.A. Ow, J.K. Han, R. Gerencer, J.P. Stolovitzky, J.W. Stambaugh, and A. Raman, *International Forum of Allergy & Rhinology*, Vol. 9, p. 131, 2019.
105. X. Wang, A. Jolliffe, B. Carr, Q. Zhang, M. Bilger, Y. Cui, J. Wu, X. Wang, M. Mahoney, A. Rojas-Pena, M.J. Hoenerhoff, J. Douglas, R.H. Bartlett, C. Xi, J.L. Bull, and M.E. Meyerhoff, *Biomater. Sci.*, Vol. 6, p. 3189, 2018.
106. E. Lih, W. Park, K.W. Park, S.Y. Chun, H. Kim, Y.K. Joung, T.G. Kwon, J.A. Hubbell, and D.K. Han, *ACS Central Science*, Vol. 5, p. 458, 2019.
107. E. Lih, C.H. Kum, W. Park, S.Y. Chun, Y. Cho, Y.K. Joung, K.-S. Park, Y.J. Hong, D.J. Ahn, B.-S. Kim, T.G. Kwon, M.H. Jeong, J.A. Hubbell, and D.K. Han, *ACS Nano*, Vol. 12, p. 6917, 2018.
108. S.E. Varner, N.R.F. Beeley, and E. de Juan, Sustained release implants and methods for subretinal delivery of bioactive agents to treat or prevent retinal disease, US Patent 8 003 124, assigned to Surmodics, Inc. (Eden Prairie, MN), August 23, 2011.
109. H.A. Heitzmann, D.S. Haffner, K.M. Curry, and T.W. Burns, Bioresorbable ocular drug delivery device, US Patent Application 20 190 125 581, assigned to DOSE Medical Corporation, San Clemente, CA, May 2, 2019.
110. R.L. Dunn, A.J. Tipton, G.L. Southard, and J.A. Rogers, Biodegradable polymer composition, US Patent 5 599 552, assigned to Atrix Laboratories, Inc. (Fort Collins, CO), February 4, 1997.
111. W.R. Noyes, Fillers and methods for displacing tissues to improve radiological outcomes, US Patent 7 744 913, assigned to Incept, LLC (Lexington, MA), June 29, 2010.

112. P.M. Hughes and O. Olejnik, Stabilized biodegradable neurotoxin implants, US Patent 8 007 828, assigned to Allergan, Inc. (Irvine, CA), August 30, 2011.
113. B.P.S. Jacky, S.Z. Malik, J. Wang, Y. Liu, A. Brideau-Andersen, L.E. Steward, L.Q. Le, and E.C. Hsai, Botulinum neurotoxin for treatment of disorders associated with melanocyte hyperactivity and/or excess melanin, US Patent Application 20 180 344 820, assigned to Allergan Inc., December 6, 2018.
114. D.M. Healy and T. Gilchrist, Biodegradable composite material for tissue repair, US Patent 7 531 005, assigned to Tyco Healthcare Group LP (Mansfield, MA), May 12, 2009.
115. H.P. Felgueiras and M.T.P. Amorim, Production of polymer-bioactive glass nanocomposites for bone repair and substitution in A.-M. Holban and A.M. Grumezescu, eds., *Materials for Biomedical Engineering*, chapter 12, pp. 373–396. Elsevier, 2019.
116. L. Li, X. Shi, Z. Wang, M. Guo, Y. Wang, Z. Jiao, and P. Zhang, *ACS Biomaterials Science & Engineering*, Vol. 5, p. 2466, 2019.
117. H. Zhang, Y. Moriyama, Y. Ayukawa, Y.D. Rakhmatia, Y. Tomita, N. Yasunami, and K. Koyano, *Odontology*, Vol. 107, p. 37, 2019.
118. J.-T.K. Thevar, N.A.N.N. Malek, and M.R.A. Kadir, *Materials Chemistry and Physics*, Vol. 221, p. 501, 2019.
119. S.W. Shalaby, ed., *Biomedical Polymers: Designed-to-Degrade Systems*, Hanser Publishers, Munich, 1994.
120. R.S. Langer, A. Lendlein, A. Schmidt, and H. Grablowitz, Biodegradable shape memory polymers, US Patent 6 160 084, assigned to Massachusetts Institute of Technology (Cambridge, MA), December 12, 2000.
121. A. Lendlein and R. Langer, *Science*, Vol. 296, p. 1673, May 2002.
122. A. Lendlein and S. Kelch, *Angewandte Chemie International Edition*, Vol. 41, p. 2034, June 2002.
123. S. Hayashi, Shape memory polyurethane elastomer molded article, US Patent 5 145 935, assigned to Mitsubishi Jukogyo Kabushiki Kaisha (Tokyo, JP), September 8, 1992.
124. S.P. Bitler, R.F. Stewart, D.A. Kamp, R.G. Freelin, and V.Y. Yoon, Thermoplastic elastomers, US Patent 5 665 822, assigned to Landec Corporation (Menlo Park, CA), September 9, 1997.
125. R. Muschalek, K. Hearon, L.D. Nash, and D.J. Maitland, Medical devices coated with shape memory polymer foams, US Patent Application 2 018 021 461, assigned to Texas A&M University System, August 2, 2018.
126. J.R. Lowe, W.B. Tolman, and M.A. Hillmyer, *Biomacromolecules*, Vol. 10, p. 2003, July 2009.

127. P.J. Buscemi, E.A. Stejskal, D.F. Palme, II, and L. Wang, Biodegradable drug delivery vascular stent, US Patent 7 806 925, assigned to Boston Scientific Scimed, Inc. (Maple Grove, MN), October 5, 2010.
128. S.F.A. Hossainy and D. Dutta, Biodegradable polymers for use with implantable medical devices, US Patent 7 875 283, assigned to Advanced Cardiovascular Systems, Inc. (Santa Clara, CA), January 25, 2011.
129. L.-Y. Huang, M.-C. Yang, H.-M. Tsou, and T.-Y. Liu, *Colloids and Surfaces B: Biointerfaces*, Vol. 173, p. 470, 2019.
130. A. Datta, S.T. Huxel, D.D. Jamiolkowski, and Y. Li, Biodegradable stent, US Patent 7 166 134, assigned to Ethicon, Inc. (Somerville, NJ), January 23, 2007.
131. S. Ichihashi, A. Fernández-Colino, F. Wolf, D.M. Rojas-González, K. Kichikawa, S. Jockenhoevel, T. Schmitz-Rode, and P. Mela, *Tissue Engineering Part B: Reviews*, Vol. 25, p. 135, 2019.
132. B.M. Jeong and A. Gutowska, Thermogelling biodegradable aqueous polymer solution, US Patent 6 841 617, assigned to Battelle Memorial Institute (Richland, WA), January 11, 2005.
133. A.M. Faheem, A.A. Elkordy, M. ElTanani, and S.M. Girgis, *British Journal of Pharmacy*, Vol. 4, 2019.
134. D.V. Peralta, Z. Heidari, S. Dash, and M.A. Tarr, *ACS Applied Materials & Interfaces*, Vol. 7, p. 7101, 2015.
135. W.-H. Jian, T.-W. Yu, C.-J. Chen, W.-C. Huang, H.-C. Chiu, and W.-H. Chiang, *Langmuir*, Vol. 31, p. 6202, 2015.
136. J.H. Choi, H. Seo, J.H. Park, J.H. Son, D.I. Kim, J. Kim, G.D. Moon, and D.C. Hyun, *Colloids and Surfaces B: Biointerfaces*, Vol. 173, p. 258, 2019.
137. B.G. Belenkaya, V.I. Sakharova, and V.N. Polevov, Biodegradable absorbents and methods of preparation, US Patent 7 309 498, December 18, 2007.
138. B. Belenkaya, V. Polevov, and V. Sakharova, Biodegradable absorbents and methods of preparation, US Patent 7 858 837, December 28, 2010.
139. M. Jokinen, T. Peltola, S. Veittola, M. Ahola, and P. Kortesuo, Biodegradable ceramic fibres from silica sols, US Patent 7 897 166, assigned to DelSiTech Oy (Turku, FI), March 1, 2011.
140. C. Ohtsuki, T. Kokubo, and T. Yamamuro, *Journal of Non-Crystalline Solids*, Vol. 143, p. 84, 1992.
141. D.R. Elmaleh, S.C. Robson, and M.L. Papisov, Conjugates comprising a biodegradable polymer and uses therefor, US Patent 7 785 618, August 31, 2010.

9

Personal Care and Sanitary Goods

9.1 Breathable Biodegradable Composition

Films that are used in disposable applications, including, for example, personal hygiene, should be highly permeable to water vapor and should exhibit a good softness. The permeability to water vapor is sometimes referred to as breathability.

Compositions have been developed that include a biodegradable aliphatic-aromatic copolyester, i.e., poly(tetramethylene adipate-*co*-terephthalate) and a plasticizer. The composition exhibits a moisture vapor transmission rate of at least 400 $g\,m^{-2}d^{-1}$ (1).

The plasticizer is selected from esters of citric acid or benzoic acid. Moreover, a tackifying resin can also be added. The tackifier may be a rosin derivative, a terpene derivative, poly(lactic acid), or poly(hydroxyvalerate butyrate) (1).

9.2 Personal Hygiene Applications

A personal hygiene product is an inseparable part of urban society (2). It has given comfort, reliability, and flexibility to sick men, women, and children.

Hygiene items containing superabsorbent polymer (hydrogels) for absorbing large amounts of body fluids are the attractive inventions of modern science. The hydrogels swell and imbibe body fluids in the presence of hydrophilic functional groups in the polymeric backbone.

The current trend of using acrylate-based superabsorbent materials in hygiene products is creating a significant portion of urban garbage. This effects not only shrinking land sites but also causes great harm to the environment due to the nondegradability of superabsorbent materials existing in the core of the hygiene product. In spite of the high water-holding capacity of petrochemical-based superabsorbent polymer, its hidden curse is its nondegradable nature, which poses a health risk.

Cellulose is the most abundant biocompatible matter on Earth which basically originates from plants. It is also a naturally occurring long-chain polymer that plays a vital role in the food cycle of animals. Cellulose and its derivatives have a large application in various fields. As cellulose and its etherified and esterified derivatives have attractive physicochemical and mechanical properties, hydrogels synthesized from cellulose and its derivatives can be an alternative to synthetic superabsorbent polymer (2).

Cellulose-based hydrogels have found application in various fields like agriculture, biomedicine, tissue engineering, wound dressing, pharmaceuticals, etc. Among various applications, some products are available on the market and some are in the research stage. Due to fast swelling and other extraordinary properties, i.e., biocompatibility and biodegradability, cellulosic materials, such as cellulose-originated hydrogels, can be applied in personal hygiene products so that superabsorbents from nonrenewable materials are partially or completely replaced. The history of using superabsorbent polymers in hygiene products and hydrogel synthesis, and the health and environmental risks related to non-cellulosic absorbent materials have been discussed (2).

9.3 Sanitary Goods

A material used in sanitary hygiene goods should fulfill the following conditions (3):

1. No water solubility to cause absorption of moisture in air during storage to become sticky.

2. Capability of being used just the same as usual without disintegration when exposed to body fluid when in use, and having water disintegratability in which the shape disintegrates upon contact with a large amount of water when discharged in a toilet bowl.
3. Biodegradability, which allows decomposition by microorganisms and the like to take place after being discharged and treated in a toilet bowl, thus minimizing the effect on the natural environment.
4. Moldability to maintain good sense of touch such as texture or good appearance.

A polymer material with both an imine moiety and a particular biodegradable unit together does not have water solubility, such as stickiness or shape disintegration under ordinary conditions, and facile disintegration of its shape is only initiated upon contact with a large amount of water, and its shape disintegrates faster when in contact with acidic water. Monomers for a polymer with an imine unit are listed in Table 9.1.

Table 9.1 Monomers for polymer intermediate with amine unit (3).

Aldehyde component	Amine component	Molar ratio
Terephthal aldehyde	2-Ethanolamine	1:2
Terephthal aldehyde	2-(Aminoethoxy)ethanol	1:2
Hydroxyacetone	2-Ethanolamine	1:1
2-Hydroxyethyl pyruvate	2-Ethanolamine	1:1

The condensation of the components in Table 9.1 yield *N,N´*-(1,4-phenylenedimethylidene) bis(ethanolamine), *N,N´*-(1,4-phenylenedimethylidene)bis(2-(2-aminoethoxy)ethanol), 2-(2-hydroxyethylimino)propane-1-ol, and 2-hydroxyethyl-2-(2-hydroxyethylimino)propanoate, respectively. The products are shown in Figure 9.1.

As counterparts to the imine containing intermediates, various ester oligomers from 1,4-butanediol and succinic acid were produced having acid end groups. This means that succinic acid is used in excess in comparison to 1,4-butanediol. These oligomers are con-

(1)

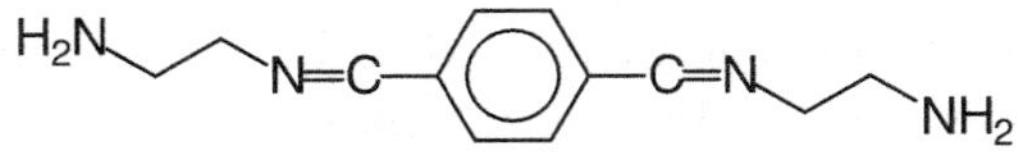

(2)

(3)

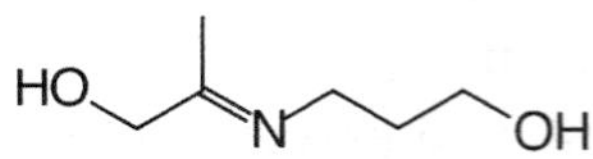

(4)

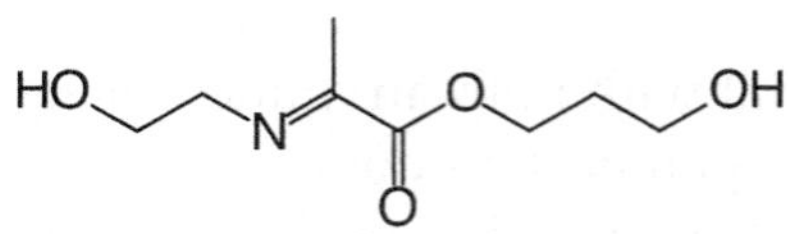

Figure 9.1 (1) *N,N´*-(1,4-Phenylenedimethylidene) bis(ethanolamine), (2) *N,N´*-(1,4-Phenylenedimethylidene)bis(2-(2-aminoethoxy)ethanol), (3) 2-(2-Hydroxyethylimino)propane-1-ol, and (4) 2-Hydroxyethyl-2-(2-hydroxyethylimino)propanoate.

densed with the imine containing intermediates. A typical structure is shown in Figure 9.2.

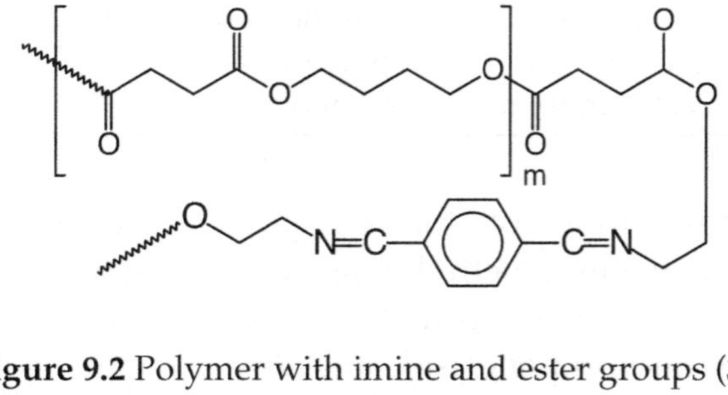

Figure 9.2 Polymer with imine and ester groups (3).

Nonwoven fabrics have been produced from these polymers. The materials have a good appearance, good sense of touch, flexibility, and sufficient strength. They also demonstrate excellent water disintegratability (3).

Life-cycle analysis of nonwovens has shown that they can bring significant benefits in a range of applications and can increase the sustainability of many products due to the lower impacts of production (4). The various factors that are involved when establishing and improving the sustainability of nonwoven materials have been dealt with in a chapter of a monograph (4). The rise of developing economies and an aging population will place extra strain on resources, and there is a need to develop materials that have a much lower environmental impact and are sustainable. The use of biodegradable polymers can further reduce the long-term impacts of nonwovens, which can be especially important for single-use disposables. The use of recycled products and biopolymers can also reduce the carbon and water footprint of producing nonwoven materials (4).

9.4 Superabsorbent Materials

The issues of biodegradable superabsorbent polymers have been reviewed in a chapter of a monograph (5). Superabsorbents are tridimensional crosslinked polymeric materials which are able to absorb water or physical fluids and biological fluids up to multiple times of their own weight in a short time, but are not dissolved when brought into contact with water.

Superabsorbent materials are frequently used in disposable absorbent articles to help improve the absorbent properties of such articles. Superabsorbent materials are generally polymer-based and are available in many forms such as powders, granules, microparticles, films and fibers. Upon contact with fluids, superabsorbent materials swell by absorbing fluid into their structure. The superabsorbent materials quickly absorb fluids insulted into the articles and retain such fluids to prevent leakage and provide a dry feel even after fluid insult (6).

Conventional superabsorbent materials are water insoluble, and although they can absorb many times their weight of fluids, they are relatively resistant to biodegradation. For this reason, a class of superabsorbent materials has been developed which focuses on biodegradation. This class is known as *biodegradable superabsorbents*. Unlike conventional polyacrylate superabsorbents, biodegradable superabsorbents become biodegradable over time under conditions that promote microbial growth.

Polymers for biodegradable superabsorbents are listed in Table 9.2.

Table 9.2 Polymers for biodegradable superabsorbents (7).

Polymer	Polymer	Polymer
Starch	Bamboo	Carrageenan
Proteins	Collagen	Guar gum and succinic anhydride
Gelatin	Cellulose	Chitin and succinic anhydride

Certain biodegradable superabsorbent materials, such as those composed of crosslinked polysaccharides, tend to exhibit lower absorbency properties, particularly absorbency under load and gel bed permeability, in comparison to the conventional non-biodegradable polyacrylate superabsorbents (6).

So, it is desirable to improve the absorbency of biodegradable superabsorbents. However, increasing these properties typically suffers from drawbacks in other areas. For example, increasing the absorbing capacity in biodegradable superabsorbents can result in sloppy gels with a significantly reduced permeability. Similarly, an improvement of the permeability can result in a reduction in the absorbency under load (6).

An absorbent composition is made from a biodegradable absorbent polymer and a surface-modifying agent. The superabsorbent material is a carboxylated polysaccharide and carboxymethyl cellulose. The latter is a self-crosslinking hydrophilic soft cationic polymer.

The cationic polymer is selected from salts of poly(vinyl amine) or copolymers from 2-(acryloyloxy)ethyl-trimethylammonium chloride and 3-(trimethoxysilyl)propyl methacrylate (6). These compounds are shown in Figure 9.3.

Some bacteria, such as *Acetobacter xylinum*, can synthesize bacterial cellulose. Cellulose-based hydrogels are superabsorbent materials which make 3D networks. Chemical bonds or other cohesive forces, such as hydrogen bonding or ionic interactions, connect the cellulose chains together. Hydrogels can swell and absorb water and

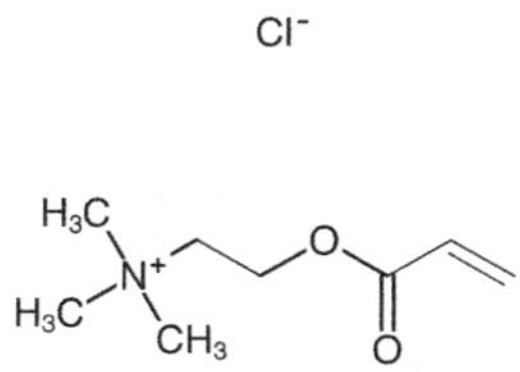

2-(Acryloyloxy)ethyl-trimethylammonium chloride

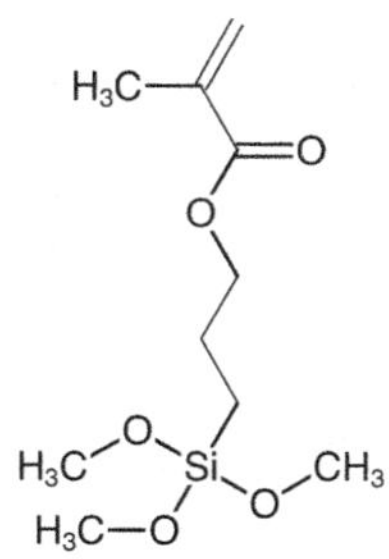

3-(Trimethoxysilyl)propyl methacrylate

Figure 9.3 Acrylic monomers.

other aqueous fluids in their 3D networks, but they are insoluble in them.

The excellent biocompatibility of cellulose has prompted the large use of cellulose-based personal care products. Cellulose hydrogel is used for these products as a thickener and stabilizing agent or as moisturizing agent to improve the skin feel of the product. Hygienic cellulosic absorbent products, such as diapers, panty liners, tampons, paper towels, and tissue papers, are used as personal care products. These products are available in different absorbency ratings from junior/light to ultra-absorbency. Superabsorbent products are made using a cellulose-based hydrogel. Applications of cellulose-based hydrogels in personal care products were reviewed (7). The desired properties of personal care products are listed in Table 9.3.

Table 9.3 Properties of an ideal superabsorbent for personal care products (7).

Property	Property
The highest absorption capacity	Photostability
The highest absorbency under load	The lowest price
Nontoxicity	Odorless
Desired rate of absorption	Colorless
The lowest soluble content	pH-neutrality after swelling
The highest biodegradability	Lowest rewetting
The highest durability in the swelling medium	

In general, the synthesis of polysaccharide-based superabsorbent can be performed by two main methods (7):

1. Graft copolymerization of suitable vinyl monomer on polysaccharide in the presence of a crosslinker, and
2. Direct crosslinking of the polysaccharide.

In the first method, a polysaccharide enters a reaction with an initiator. Therefore, functional groups of polysaccharide and initiator interact to form redox pair-based complexes. Then, by homogeneous cleavage of the saccharide C–C bonds, carbon radicals on the polysaccharide substrate are produced. These free radicals initiate the graft polymerization of the vinyl monomers and crosslinking agent on the substrate.

In the second method, direct crosslinking of polysaccharides, polyvinylic compounds like divinyl sulfone, or polyfunctional compounds like glycerol, epichlorohydrin, and glyoxal is often performed (8).

References

1. S.U. Ahmed and A.W. Emiru, Breathable biodegradable hot melt composition, US Patent 7 910 645, assigned to H.B. Fuller Company (St. Paul, MN), March 22, 2011.
2. M.O. Haque and M.I.H. Mondal, Cellulose-based hydrogel for personal hygiene applications in M.I.H. Mondal, ed., *Cellulose-Based Superabsorbent Hydrogels*, pp. 1339–1359. Springer International Publishing, Cham, 2019.
3. M. Shimoda, T. Kawabata, T. Nobori, and M. Sukegawa, Biodegradable polymer, production method thereof, molded product thereof, and use thereof, US Patent 7 928 180, assigned to Mitsui Chemicals, Inc. (Minato-ku, Tokyo, JP), April 19, 2011.
4. P. Goswami and T. O'Haire, Developments in the use of green (biodegradable), recycled and biopolymer materials in technical nonwovens in G. Kellie, ed., *Advances in Technical Nonwovens*, Woodhead Publishing Series in Textiles, chapter 3, pp. 97–114. Woodhead Publishing, 2016.
5. S. Sinha, Biodegradable superabsorbents: Methods of preparation and application – A review in S. Thomas, P. Balakrishnan, and M. Sreekala, eds., *Fundamental Biomaterials: Polymers*, Woodhead Publishing Series in Biomaterials, chapter 14, pp. 307–322. Woodhead Publishing, 2018.
6. D.A. Soerens, Biodegradable absorbent material, US Patent 7 777 095, assigned to Kimberly-Clark Worldwide, Inc. (Neenah, WI), August 17, 2010.
7. A. Bashari, A. Rouhani Shirvan, and M. Shakeri, *Polymers for Advanced Technologies*, Vol. 29, p. 2853, 2018.
8. M.J. Zohuriaan-Mehr and K. Kabiri, *Iranian Polymer Journal*, Vol. 17, p. 451, June 2008.

10

Miscellaneous Applications

10.1 Flooring Materials

Most flooring materials are conventionally prepared from poly(vinyl chloride) (PVC). This material usually contains more than 30% of phthalate plasticizers to produce sufficient flexibility and impact strength.

Alternative sheet flooring materials prepared from the poly(lactide)-based composites have been described (1). The polymeric material can include 30–50% of PVC, poly(ethylene glycol) (PEG), poly(glycolide), ethylene vinyl acetate, poly(carbonate), poly(caprolactone) (PCL), poly(β-hydroxyalkanoate)s, or ionomer modified polyolefins.

A plasticizer and a compatibilizer are used. As a plasticizer, an epoxidized vegetable oil or esterified and epoxidized vegetable oil is used in an amount of 10–50%. Acetyl tributyl citrate, shown in Figure 10.1, is very effective.

As a compatibilizer, a polyolefin modified with polar functional groups, in an amount of between 5–10%, may be added.

Other additives that can be included in the polymeric material include dyes, inks, and antioxidants. An antistatic characteristic is also important for some applications. Many antistatic additives are compounds with hydrophilic and hydrophobic segments. A common material of this type is a mono ester of a polyol, such as glycerol, with a long-chain fatty acid such as stearic acid. In Table 10.1 some common inorganic fillers are listed.

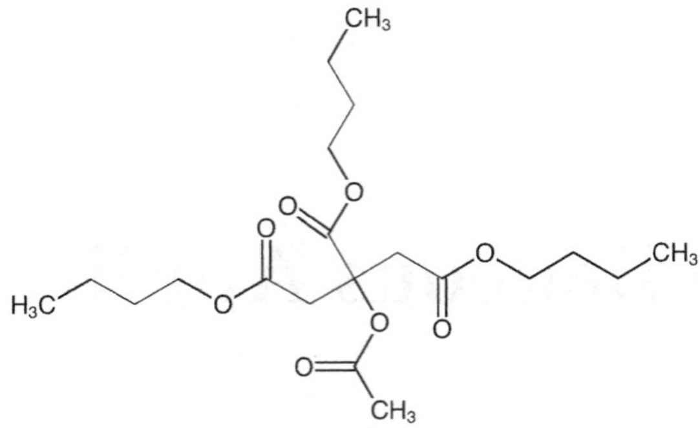

Figure 10.1 Acetyl tributyl citrate.

Table 10.1 Inorganic fillers (1).

Compound	Density/[$g\ ml^{-1}$]
Calcium carbonate	2.7
Talc	2.9
Mica	2.6
Glass fibers	2.6
Silica	2.5
Wollastonite	2.9
Aluminium trihydrate	2.4
Magnesium hydroxide	2.3
Titanium dioxide	4.2
Exfoliated nanoclay	2.6

Whiting filler can be used to increase the opacity. The optical properties of titanium dioxide make it a particularly good pigment for obtaining a white color with good opacity. Such a color is desirable in the layer upon which a printed design is placed. The design layer is located below a transparent wear layer or in the bulk of a through color homogeneous wear layer. Lower levels of titanium dioxide can be employed if a white filler, such as calcium carbonate, is used at moderate levels in this layer.

Calcium carbonate is of particular utility. Hardness, stiffness, heat deflection temperature, slip resistance, stress crack resistance, weldability, printability, and antiblocking characteristics are improved. Thermal shrinkage and elongation, as well as water vapor and oxygen permeability, are decreased.

Talc is another filler well suited to enhance the polymer formulations for floor covering. It has a lamellar structure in contrast to the low aspect particulate structure of calcium carbonate. This lamellar form allows talc to be more effective than calcium carbonate with regard to increasing stiffness, heat deflection temperature and dimensional stability. The disadvantage of talc relative to the calcium carbonate centers on the reduced impact strength, matt surface, and lower thermo-oxidative stability. Mica also has a lamellar structure and has similar advantages and disadvantages.

High aspect ratio fillers, such as wollastonite and glass fibers can have an even stronger effect than talc and mica on increasing the modulus of elasticity, tensile strength, and heat-distortion temperature of poly(lactic acid) (PLA)-based systems.

The improvements provided by high aspect ratio inorganic additives would be of particular assistance in these floor covering systems made using a permanent plasticizer or processing aid, such as liquid paraffin. In these cases, the stiffening action of such additives would compensate for the loss of stiffness produced by the liquid paraffin.

Silica in its fumed or precipitated forms can be useful at low levels of 0.1–1.5% in PLA formulations when antiblocking and printability is of importance. In the floor covering system these would be in the wear layer and in the layer upon which the printed design is applied.

Alumina trihydrate and magnesium hydroxide in the correct particle sizes, which for most systems is less than 40 microns in diameter, can provide the same type of property enhancement provided by calcium carbonate. In addition, they can provide useful fire resistance and smoke control characteristics.

Suitable reinforcing fibers are kenaf, jute, hemp, sisal, corn stalk, grass fibers, and wood fibers.

Suitable thermal and light stabilizers include hindered phenol types and hindered amine light stabilizer (HALS). Good thermal stability can be obtained using a high molecular weight hindered phenol, such as Irganox™ 1010 from Ciba-Geigy, together with secondary antioxidants such as thioethers and phosphorus compounds. Distearylthiodipropionate and Ultranox® 626 from General Electric are examples of these types of antioxidants. A Polymeric HALS, such as Luchemy® HA-B18 from Atochem, is particularly effective as it has the additional advantage of showing no antagonism towards a lot of other additives.

The flammability and smoke generation of the floor coverings, including the polymeric sheets, can be improved using aluminum trihydrate and magnesium hydroxide. These compounds release water at elevated temperatures. Also, phosphorous compounds, borates, and zinc oxide improve the fire characteristics of PLA-based systems.

Methods for preparation include melt calendering or extrusion, which are detailed subsequently. These melt calendering operations can all be done in a continuous way using a series of calendering rolls, or they may be done in a segmented fashion with a single layer being applied, followed by a wind-up operation with additional layers being added in separate operations. Likewise, a combination of continuous and discontinuous calendering operations can be employed.

For example, a formulation can be applied to a glass fiber web followed by a foamable layer on top and a base layer beneath. These three operations are carried out in a consecutive way as the material passes through three different sets of calender rolls before wind-up. Additional processing steps can be placed between and among calendering operations. A heat treatment step can follow the application of the wear layer, either in a continuous or discontinuous fashion. The heat treatment can expand the various layers through the formulation of a chemical foam.

The floor covering structure can also be prepared by melt extrusion. In such a process, one or more polymer layers can be applied to a continuous web or substrate in a single extrusion operation. When co-extrusion is used to provide multiple layers in a single pass, a separate extruder is used to feed each melt to the sheet die block.

Extrusion operations can be intermixed with other processing steps in preparing the final structure. A thermal treatment can follow the application of a wear layer in either a continuous or discontinuous fashion. This treatment can enhance the final product by expansion of layers containing chemical blowing agents or crosslinking of layers containing crosslinking systems.

A significant advantage of this polymeric composite is that it is readily biodegradable and in at least some embodiments is naturally occurring, and it can also be used together with a biodegradable plasticizer.

In particular, the use of a biodegradable polymer and a biodegradable plasticizer can significantly reduce environmental problems caused by the migration of phthalate plasticizers out of the material after disposal. One advantage of using the compatibilizer is that it tends to coat the material and keep water away, which prohibits significant decomposition of the materials in use, but allows the decomposition when the material is removed and placed in a landfill or other suitable location where the material is subject to environmental conditions, or, optionally, in the presence of suitable enzymes to break down the polymers.

The compositions can be used in decorative surface coverings such as a floor coverings. The floor covering may also include one or more additional layers such as wear layers, foamed or foamable layers, top coat layers and design layers (1).

10.2 Abrasives and Polishing Compositions

10.2.1 Cleansers

A wide variety of general purpose cleansers are available with abrasive materials to facilitate the cleansing process. Conventional cleansers are petroleum-based and use pumice, sand, metal meshes, foams, rigid and textured plastics, and various composites as abrasive materials. Although these materials are effective, they are often too harsh for use in sensitive or soft areas. Furthermore, many of the abrasives and their associated adhesives are not biodegradable, so they remain in the plumbing and in the water system long after their use (2).

Similar concerns arise with single-use products such as wet-wipes, scouring pads, or disposable cleaning heads. Thus, a definitive need exists for abrasive products that are more readily disposable.

Gently abrasive natural coatings and films have been developed that incorporate agricultural fibers and grains.

These fibers include mainly wheat straw and rice straw. Grains used in abrasive products include corn, wheat, sorghum, rice, oats, barley, soy meal, and puffed starch beads. The films and coatings so manufactured are generally flushable in conventional toilets.

Starch from various surplus crops is used as well as agricultural fibers with only little or no economic value. This is particularly important because there are limitations on burning or otherwise disposing of these fibers.

Polymeric materials and binders are combined with water so that the materials are suspended in the resulting mixture. The mixture is then heated to 70–80°C to achieve complete dissolution, and the plant-based abrasive and calcium carbonate are added to the formulation. The formulation is then mixed and blended to the desired degree of homogeneity and allowed to cool to room temperature (2).

The components of a typical composition are shown in Table 10.2.

Table 10.2 Components of an abrasive composition (2).

Polymeric materials	Fibers	Binders	Cereal grains
Starch	Rice straw	Alginates	Ground corn
Modified starch	Wheat straw	Guar gum	Cracked wheat
Poly(vinyl alcohol)	Corn fiber	Gum arabic	Ground soy
Soy meal	Rice husk	Oats	
Puffed starch beads	Pulp paper	Barley	
Poly(isoprene)		Sorghum	
Chitosan			
Pectin			

Skin cleansing compositions routinely include abrasive particles for a variety of reasons such as cleansing the skin, e.g., removal of dirt, sebum, or oil, or improving the appearance of the skin.

Abrasive particles were commonly made from petroleum-based synthetic polymers like poly(styrene) (PS), poly(urethane) (PU), polyolefins, or waxes. Unfortunately, these petroleum-based synthetic polymers may pose long-term disposal problems and could have adverse effects on the environment. A skin cleansing composition having biodegradable abrasive particles has been developed (3). An example of such a composition is shown in Table 10.3. Cocoamidopropyl betaine is shown in Figure 10.2.

Table 10.3 Water-based body wash (3).

Compound	Amount/[%]
Sodium laureth sulfate, 3 ethoxylate units per molecule	6.5
Cocoamidopropyl betaine	1.0
Sodium laureth sulfate, 1 ethoxylate unit per molecule	2.5
Trihydroxystearin	0.5
Sodium benzoate	0.2
Methylchloroisothiazolinone/Methylisothi-axolinone	0.03
Fragrance	1.0
Disodium EDTA	0.1
Abrasive particles (poly(3-hydroxybutyrate-*co*-4-hydroxybutyrate) (PHBV))	0.2
Citric acid	0.2
NaCl	2.0
Water	QS

QS: *Quantum satis*

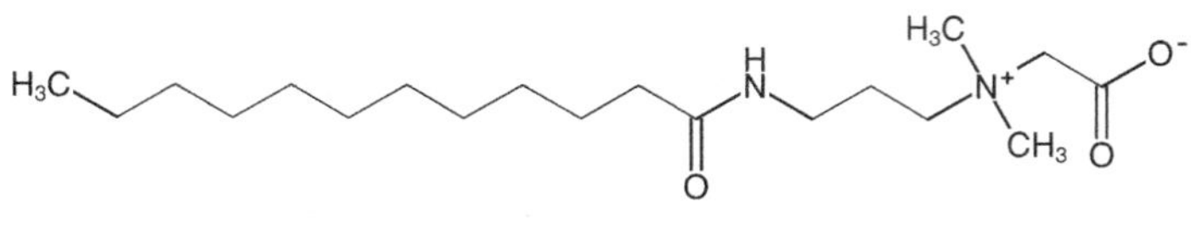

Figure 10.2 Cocoamidopropyl betaine.

The PHBV biodegradable abrasive particles can have a circularity of about 0.3 to about 0.5, an area-equivalent diameter particle size from 850 μm to 1400 μm and a solidity of about 0.7 to about 0.8 (3).

10.2.2 Polishing Pads

Chemical-mechanical polishing processes are used in the manufacturing of microelectronic devices to form flat surfaces on semiconducting wafers, field emission displays, and many other microelectronic workpieces. For example, the manufacture of semiconductor devices generally involves the formation of various process layers, selective removal or patterning of portions of those layers, and deposition of yet additional process layers above the surface of a semiconducting workpiece to form a semiconducting wafer (4).

Typically, polishing pads are prepared from non-biodegradable PUs. While many of these polishing pads are suitable for their intended use, such polishing pads typically are discarded after use and end up in the solid waste stream, headed for rapidly vanishing and increasingly expensive landfill space. While some efforts at recycling have been made, the nature of polymers and the way they are produced and converted to products limits the number of possible recycling applications. A polishing pad based on a biodegradable polymer, PLA, has been reported (4).

The introduction of crosslinks into the polymer can modify the physical properties of the polymer. For example, crosslinking can enhance the impact and tensile strength, the ductility, and elasticity of the polymer. In addition, crosslinking can reduce the water solubility of the polymer. Accordingly, the degree of crosslinking imparted to the biodegradable polymer will depend on the physical properties that are desired for a given polishing application.

Crosslinking can be carried out in situ or *ex situ*. Suitable polymer crosslinking agents for *in-situ* crosslinking include those selected from the group consisting of 5,5′-bis(oxepane-2-one), spiro-bis-dimethylene carbonate (5) (c.f. Figure 10.3), trifunctional isocyanates, epoxides, maleic anhydride, dicumyl peroxide, adipic acid, and 1,4-dibutanediol.

Ex-situ crosslinking can be carried out by blending the PLA with acrylic acid derivatives, which can form crosslinks with the polymer upon exposure to ultraviolet irradiation, or with silane-functionalized compounds, which can form crosslinks with the polymer upon exposure to elevated temperatures in the presence of water (4).

Thus, the crosslinked biodegradable material acts as a matrix for

Spiro-bis-dimethylene carbonate 5,5´-Bis(oxepane-2-one)

Figure 10.3 Crosslinking agents.

embedded abrasive particles. Abrasive particles are listed in Table 10.4.

Table 10.4 Abrasive particles (4).

Oxides		
Titania	Zirconia	Alumina
Silica	Magnesia	Germania
Ceria	Chromia	Iron oxide
Nitrides		
Boron nitride	Silicon nitride	

Other abrasive materials include silicon carbide particles, diamond particles, and ceramic particles.

Micropellets of PLA can be dispersed in a substantially non-biodegradable polymer having a lower melting temperature and extruded. Micropellets also can be joined together through sintering to produce porous materials having closed or open, interconnected pores. Sometimes it is desirable to combine the biodegradable micropellets with a binder to produce a porous pad material.

If a water-soluble PLA is used, the resulting sheet can be washed with water to remove the PLA micropellets, yielding a microporous, non-biodegradable polymer sheet. Similarly, water-soluble PLA micropellets can be extruded with another water-soluble polymer to produce a solid polymer sheet. By crosslinking the PLA to render it water insoluble, the secondary water-soluble polymer can be extracted to produce a microporous PLA-based polishing pad (4).

Moreover, PLA can be spun into fibers and thus used to form

a nonwoven web. Nonwoven webs can be used in soft polishing applications, for example, for polishing rigid disks, magnetic heads, or in second step copper polishing. Preferably, such a polishing pad contains a polishing layer and a subpad layer.

10.3 Lubricants

Nearly 50% of the lubricants leak into the environment. Therefore, there is a need for the development of biodegradable lubricants (6).

Vegetable oil-based biolubricants and their applications in the power distribution industry have been assessed in a monograph (7). Vegetable oil-based lubricants can offer significant advantages over petroleum-based lubricants, including biodegradability, cost-effectiveness, renewability, and lower environmental effects. The physical and chemical properties of vegetable oil have been discussed, and also their effects on its applications in tribology.

Fatty acid esters would be suitable lubricants; however, their oxidation resistance and their tribological properties are not satisfactory and need to be improved (6). Here, the epoxidation of the fatty acids with an eventual alcoholysis of the epoxide groups is helpful in these aspects.

For example, the epoxidation of fatty acid methyl esters from sunflower oil can be conducted by an environmentally benign route in the presence of acidic solid catalysts and *tert*-butylhydroperoxide at high conversion and selectivity.

Monoricinolein and diricinolein are renewable and biodegradable lubricants (8). These compounds are shown in Figure 10.4.

Monoricinolein and diricinolein were prepared by the enzymatic glycerolysis of castor oil in a solvent-free system. Three commercial immobilized lipases (Novozym 435, Lipozyme RM IM and Lipozyme TL IM) were screened and compared. Novozym 435 showed the best catalytic performance among the tested lipases. The effects of reaction variables (reaction pressure, reaction time, temperature, enzyme load, substrate ratio and water content) on the enzymatic glycerolysis of castor oil were investigated.

A high conversion of the castor oil (92.48±1.02%) and the maximum biolubricant yields (54.47±0.58% of monoricinolein and

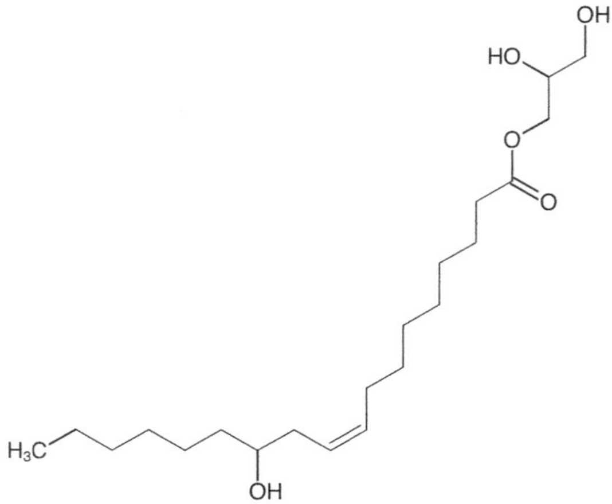

Figure 10.4 Monoricinolein.

38.01±0.50% of diricinolein) were achieved under the following conditions: Enzyme load 5% (w/w, based on the total mass), castor oil/glycerol ratio at 1:6, 90°C, and 3 *h*.

The activation energies (E_a) for castor oil conversion and the formation of monoricinolein and diricinolein were calculated as 33.75 $kJ\,mol^{-1}$, 22.26 $kJ\,mol^{-1}$ and 24.37 $kJ\,mol^{-1}$, respectively. The process showed a high selectivity toward the formation of monoricinolein (8).

10.4 Renewable Cards

A wide variety of cards, cardstock, and card products are manufactured using materials of petrochemical origin. Examples of such cards include credit cards, debit cards, loyalty cards, gift cards, telephone cards, prepaid purchase cards, cellular communication cards, membership cards, student cards, identification cards and transit cards. These cards contain conventional PVC, PS, etc. (9).

In particular, the parts that include information in a machine readable component are currently manufactured using plastic and other similar materials which are incapable of breaking down under an

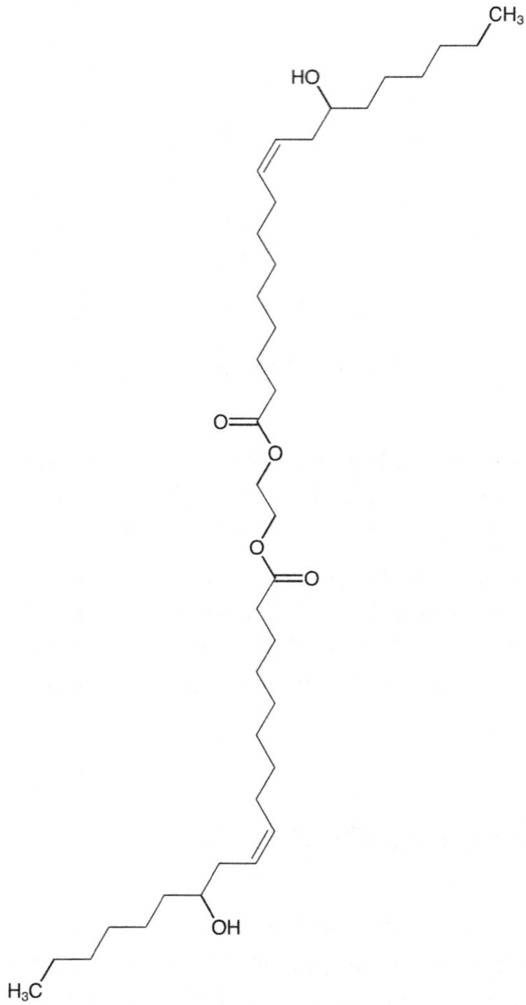

Figure 10.4 (cont) Diricinolein.

industry standard, such as International Standards for Compostable Materials, and thus are not environmentally friendly (10).

The replacement of the petrochemically derived polymers with more environmentally friendly materials may decrease the environmental impact of cards, cardstock, and card products (9).

Certain designs using renewable materials for an information card contain a substantially planar base layer fabricated from an opaque poly(lactide), and they further include an annotation region configured to retain information for electronically accessing data associated with the information card. The annotation region may be printed on the first surface of the base layer and may include an ink made from a vegetable-based oil (9).

Suitable magnetic strips or bar codes can be both writeable and readable by currently available equipment. In a degradable design, the card is desirably coated on at least one side with a film laminate that preferably consists of a bio-based film, a biopolymer material, a cellulose film, or an organic-based film (10). Also, a pressure-sensitive adhesive may be used that affixes the protective layer to the base layer in a manner that protects the annotation region from degradation (9).

10.5 Biodegradable Irrigation Pipe

Typically, a network of irrigation pipes used to irrigate a field crop is laid down in a field in which the crop is grown at the beginning of the crop growing season and removed from the field at the end of the season. The same irrigation pipes are generally used repeatedly for a number of growing seasons until damage and wear renders the pipes inefficient for use, at which time they are discarded.

Irrigation pipes are conventionally made from non-biodegradable petroleum-based plastics such as poly(ethylene) or poly(propylene) (PP). As a result, conventional irrigation pipes that are discarded are generally not readily recycled into the environment and add to the stress of environmental pollution. In addition, because petroleum is a nonrenewable raw material, it is expected that raw material costs for producing irrigation pipes from petroleum-based plastics will increase (11).

However, recently a biodegradable irrigation pipe formed from a renewable material has been described (11). The biodegradable irrigation pipe is formed having a pipe wall with a plurality of layers. The layers may be fabricated from a core layer sandwiched between inner and outer protective layers.

The core layer provides body and mechanical strength to the irrigation pipe and is formed from a biodegradable material, which degrades relatively quickly when exposed to degrading elements such as water, wet soil, or heat. The core layer can be formed from PLA, starch polymers polymerized seed oils, or polyesters. The protective layers are thin layers of a suitable inorganic material such as titania or silica (11).

10.6 Thermosensitive Material

A biodegradable copolymer has been described that consists of an amphiphilic diblock copolymer with a hydrophilic segment and a hydrophobic segment (12).

The hydrophilic segment is an end-capped PEG derivative. The hydrophilic segment is a random polymer polymerized from lactide or related compounds and other monomers that are shown in Table 10.5 and in Figure 10.5. Also, the reaction is shown in Figure 10.6.

Table 10.5 Comonomers for biodegradable materials (12).

Monomer	Abbreviation for the polymer
γ-Propiolactone	mPEG-PPLA
δ-Valerolactone	mPEG-PVLA
ε-Caprolactone	mPEG-PCLA
γ-Thiobutyrolactone	mPEG-PSLA
2-Iminothiolane hydrochloride	mPEG-PITLA
N-Methyl-2-Pyrrolidone	mPEG-PNLA
γ-Butyrolactam	mPEG-PPDLA
2-Oxo-4-thiazolidinecarboxylic acid	mPEG-POTLA
2-Thioxo-4-thiazolidinecarboxylic acid	mPEG-PTTLA
D,L-N-Acetylhomocysteine thiolactone	mPEG-PACTLA
5-Thiazolecarboxaldehyde	mPEG-PTCALA

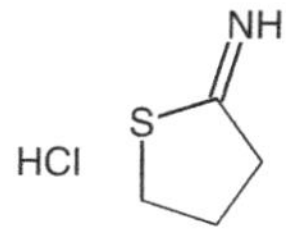

2-Iminothiolane hydrochloride

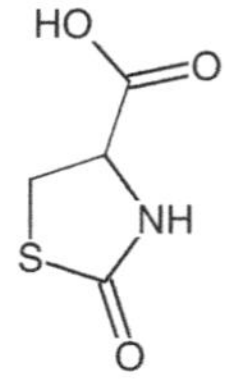

2-Oxo-4-thiazolidinecarboxylic acid

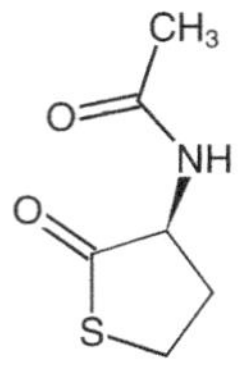

D,L-N-Acetylhomocysteine thiolactone

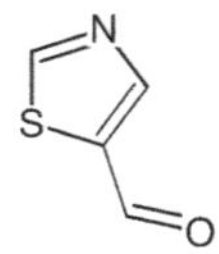

5-Thiazolecarboxaldehyde

Figure 10.5 Monomers.

There is no coupling agent between the hydrophilic and hydrophobic segments. The polymer is formed by a one-pot ring-opening polymerization technique (12). The polymer can be dissolved in water to form a thermosensitive material having a phase-transfer temperature of 25–50°C. It can be used for biological activity factor delivery, tissue engineering, cell culture and as a biological glue.

Table 10.6 Properties of biodegradable materials (12).

Monomer	Adhesion strength for the polymer/[$gf\ mm^{-2}$]
γ-Propiolactone	52
δ-Valerolactone	58
ε-Caprolactone	100

Recently, a lot of applications of thermosensitive biodegradable materials in the medical field have been reported (13–20).

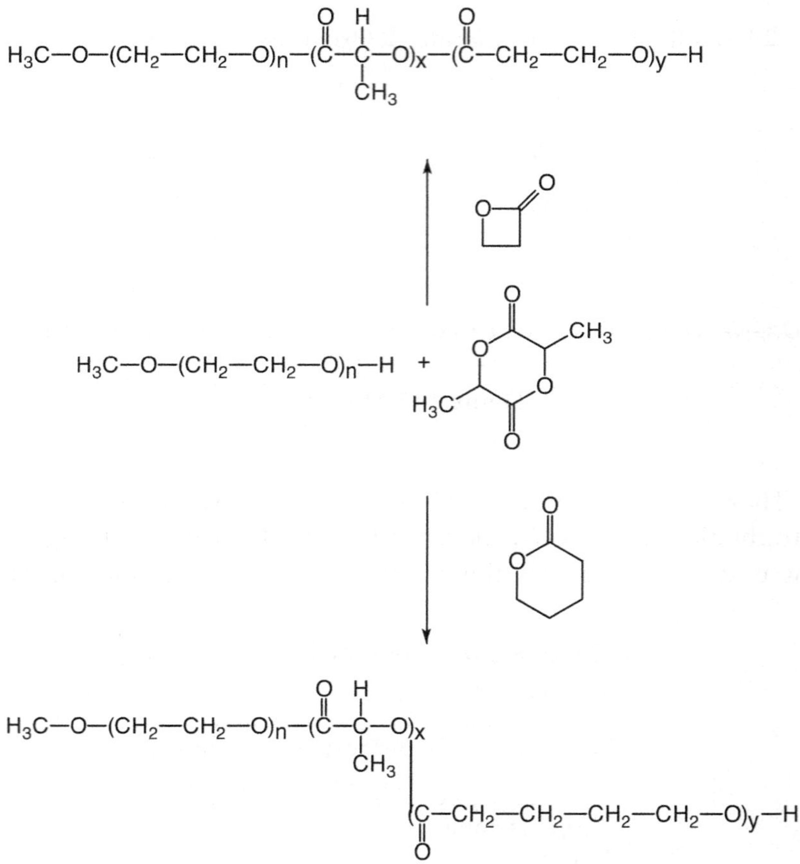

Figure 10.6 Polymers from γ-propiolactone and δ-valerolactone (12).

10.7 Biodegradable Scale Inhibitors

10.7.1 Phosphorus-Containing Polymer

Polymers having phosphorus-containing end-caps, especially polymers comprising acrylic acid, are widely used as scale inhibitors, particularly in oilfield applications (21).

The biodegradability of poly(acrylic acid) (PAA), which is generally poorly biodegraded, can be greatly improved by reducing the weight average molecular weight below 700 *D*. The degradation behavior drops off sharply for a molecular weight of PAA above this limit, as measured by the balance of carbon dioxide in the course of degradation (22).

To get scale inhibitors with improved biodegradability, during the radical polymerization, chain transfer agents have been used. Conventional chain transfer agents are halogen compounds and thiols.

However, the use of chain transfer agents containing at least one P–H bond allows the production of polymers which combine excellent activity as oilfield scale inhibitors with good biodegradability, often in excess of that required by the OECD 306B standard (23).

Polymers which contain phosphorus end-capping groups can have a desirably high biodegradability as well as being effective as oilfield scale inhibitors (21).

The reagents for a recipe ready for polymerization are shown in Table 10.7.

Table 10.7 Reagents for scale inhibitor (21).

Reagent	Amount/[*mol*]
Acrylic acid (99%)	1.98
Sodium hydroxide (46–48%)	0.71
Trisodium 1-phosphono-2-phosphino ethane (17.2%)	0.20
Sodium persulfate	0.10

Trisodium 1-phosphono-2-phosphino ethane is used as an end-capper. This compound is also known as a telomeric adduct from vinyl phosphonic acid and sodium hypophosphite (24).

In order to improve the scale inhibition efficiency of existing poly(epoxysuccinic acid) and study the impact of its molecular structure on scale inhibition efficiency, a series of poly(epoxysuccinic acid) types with linear and hyperbranched structure have been synthesized through a copolymerization reaction with glycidyl and epoxy succinate (25).

The scale inhibition behavior of poly(epoxysuccinic acid) with linear and hyperbranched structure against $CaCO_3$ and $CaSO_4$ scales was evaluated using a static scale inhibition method (25). Also, their ability to retard the deposition of $CaCO_3$ was examined. The experimental results showed that, for $CaCO_3$ and $CaSO_4$, the poly(epoxysuccinic acid) with the hyperbranched structure provides a scale inhibiting efficiency as high as 95.9% and 94.3%, respectively, at an inhibitor concentration of 15 $mg\,l^{-1}$.

In addition, the processes of crystal nucleation, growth and crystal morphology were analyzed. The experimental results show that the poly(epoxysuccinic acid) with hyperbranched structure not only prolongs the induction period of $CaCO_3$ crystal nucleation, but also reduces the number of crystal nuclei and changes the size and morphology of the $CaCO_3$ crystal. Fourier transform infrared spectroscopy, scanning electron microscopy (SEM) and X-ray diffraction (XRD) studies showed that the poly(epoxysuccinic acid) with hyperbranched structure can induce the irregularity of growing $CaCO_3$ crystal, destroy the formation of crystals and change the polymorphs of calcium scale crystal. This indicates that the prepared poly(epoxysuccinic acid) with hyperbranched structure has great potential for application in the treatment of industrial water (25).

10.8 Nanocomposites

An injection molding formulation has been reported and is summarized in Table 10.8. 2,5-Dimethyl-2,5-di(*tert*-butyl peroxy) hexane is shown in Figure 10.7.

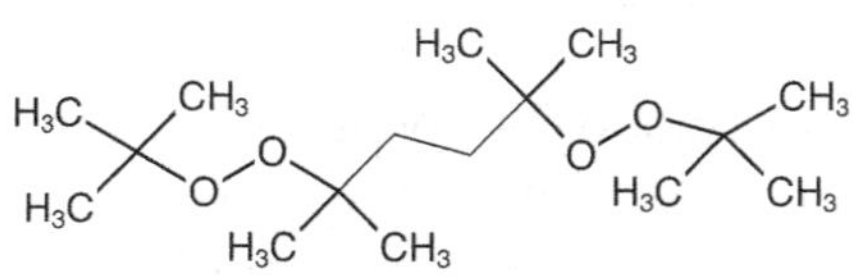

Figure 10.7 2,5-Dimethyl-2,5-di(*tert*-butyl peroxy) hexane.

Table 10.8 Injection molding formulation (26).

Component	Amount/[%]
Poly(lactic acid)	74.5
Co-polyester with adipic acid	5
Magnesium silicate (talc)	15
Finely ground 99.99% pure silica	6
2,5-Dimethyl-2,5-di(*tert*-butyl peroxy) hexane	0.5

It is crucial that the mineral nanoparticles are added without creating aggregates. The nanoparticles should not be introduced directly into the barrel of the extruder but through a tower, letting the nanoparticles fall and mix smoothly with the molten material.

10.9 Molded Articles from Fruit Residues

Biodegradable molded articles can be produced from residues obtained from lees, i.e., residual yeast of Sake (*Sake kasu*), c.f. Figure 10.8, and also squeezed and strained lees of fruits (27).

This method enables the residues to be effectively reutilized by producing therefrom molded articles which can be used in a variety of fields. Being biodegradable, the molded article also permits easy disposal without creating an environmental problem. Some formulations are shown in Table 10.9.

The mixtures from Table 10.9 were poured into a mold which had been heated to 130–150°C, and molded under the pressure of 50 $kp\,cm^{-2}$ for 30–60 s. Then, the product was taken out from the mold, and then cooled off at room temperature, thereby obtaining the molded article (27).

10.10 Fluorescent Biodegradable Particles

Fluorescent particles are used for a variety of applications. One application is their use in aerosol studies, such as large-scale air dispersal to track particulate migration over vast areas, or for urban particle dispersion studies. Conventionally, researchers perform such studies to investigate the movement and flow of aerosols in urban environments (29).

Figure 10.8 *Sake kasu* (28). Reproduced from Wikipedia under the terms of the GNU Free Documentation License.

Table 10.9 Formulations from fruit residues (27).

No.	Ingredient	Amount/[kg]
1	Squeezed and strained residues of orange	100
	Egg white	10
	Water amount 5–50%	
2	Residues of bean curd	100
	Glycerin	20
3	Residues of sake	120
	Glycerin	10
	Unrefined soybean oil	6
4	Squeezed and strained residues of orange	100
	Glycerin	20
	Unrefined soybean oil	3
	Starch	5
	Sorbitol	30

Newly developed responsive particles consist of a non-biological but biodegradable carrier and natural fluorophores encapsulated in the carrier. These particles would provide safety benefits. Furthermore, these particles could be designed with the appropriate density and shape of a microorganism to mimic its aerodynamic movement. Also, such a particle can be used as a simulant for mimicking the fluorescence properties of microorganisms.

However, the particle need not mimic the fluorescent characteristics of a microorganism, but rather it can be incorporated into one or more natural fluorophores as a means of fluorescence detection.

A single fluorophore or combinations of fluorophores are encapsulated to produce a desired fluorescent effect such as particles that fluoresce at 370 nm maxima. The particles can therefore be tuned to the excitation wavelength of a fluorescence detector.

Most of the fluorescence in biological organisms comes from natural fluorophores such as the aromatic amino acids; tryptophan, phenylalanine, and tyrosine; NADH, picolinic acid, and flavins. These microspheres combine the fluorescence characteristics of the microorganisms, the handling characteristics of the PS latex microspheres, and the safety characteristics of the biodegradable polymer microspheres.

If these natural fluorophores are doped into biodegradable polymer microspheres in the correct ratios, then those spheres will act as fluorescent analogs to microorganisms. By varying the concentrations of the various fluorophores, different agents can be simulated (29).

10.11 Test Cylinder Mold for Testing Concrete

Various tests have been developed to ensure that concrete used in particular applications satisfies specifications. One of these is a compression test for testing concrete cylinders. The concrete cylinder is typically formed using a test cylinder mold. The mold is a hollow cylindrical container including a bottom wall used to test specimens of concrete at a construction site. Such molds are generally disposable (30).

Initially, such molds were made of paperboard with metal bottoms. Subsequently, the molds are injection molded from PP which can be recycled. Furthermore, a biodegradable test cylinder mold for testing concrete specimens has been developed. The cylindrical container is formed exclusively from corn starch or from a blend of 40% corn starch and 60% PP (30).

10.12 Flexographic Inks

Flexographic printing is a printing process that uses a flexible relief plate. The basic principles of this method can be found elsewhere (31, 32). The first patent for a flexographic-type printing press dates from 1909, even though such machines were fabricated earlier (33).

Today, flexographic presses and letterpress presses for newspaper printing use plates made of photosensitive polymeric materials that can be formed directly from a light image. The polymeric cylinder, when properly developed, presents raised areas that reproduce the image to be printed.

This plate is mounted on a cylindrical roll that rotates against an ink cylinder so as to become coated in its raised areas. These areas rotate against a substrate (generally paper) which is backed by a hardened rubber roller (34).

Letterpress and flexographic systems can be used with either solvent-based or aqueous-based ink systems. In either case, current practice consists of mixing a solvent or water in the presence of a binder resin such as, for example, nitrocellulose, maleic anhydride, an acrylic copolymer, or various starch derivatives. The pigment is applied over the paper by the printing roller and the binder serves to hold the pigment particles in place.

In some instances, the pigment is carbon black directly suspended in a high boiling point aliphatic hydrocarbon solution. This ink has no binder but instead is fixed to the paper by diffusion of the oil into the paper, thus leaving a highly blackened layer of carbon black and oil upon the surface of the paper.

These oil-based systems suffer from a number of disadvantages, including the fact that since the print never chemically bonds to the paper it can easily smudge when handled. Furthermore, there is a tendency for the inks to contaminate the various guide rollers in the system and to produce a background (34).

Water-based flexographic inks usually contain between 5–20% of organic solvents. Although they are harmful, they cannot easily be eradicated because they help to reduce the drying time of the ink once it is applied on a surface (paper, cardboard, plastic, etc.).

Only a few other concrete efforts have been devoted to improving the environmental aspect of flexographic inks. Therefore, they are still composed of poorly biodegradable or non-biodegradable pigments and polymer resins that generally accumulate in post-treated sludge resulting from the de-inking processes, or are buried in soils when the final product is disposed therein.

Biodegradable pigments already exist, but they do not provide better properties than synthetic pigments. Conventional flexographic ink compositions are normally made of acrylic, methacrylic, epoxy or styrene-based polymer resins. Actually, acrylic and epoxy resins are present in 90% of the flexographic ink compositions.

A flexographic ink composition that uses a biodegradable polymer, a binder, and a solvent has been described. The polymer may be selected from the group consisting of PLA, poly(lactic-*co*-glycolic acid), PCL, poly(vinyl alcohol), a polymer derived from adipic acid or aminocaproic acid, poly(butylene succinate), and still others. The binder is a triblock compound that has both hydrophilic and hydrophobic properties, e.g., a poly(ethylene glycol) distearate. The concentration of the binder is between 1–10% (34).

Pigments are the second major component in the flexographic water-based ink composition. The pigment is applied over the paper by the printing roller and the binder serves to hold the pigment particles in place. In some instances, the pigment is carbon black directly suspended in a high boiling point aliphatic hydrocarbon solution.

This ink has no binder but instead is fixed to the paper by diffusion of the oil into the paper, thus leaving a highly blackened layer of carbon black and oil upon the surface of the paper. These oil-based systems suffer from a number of disadvantages, including the fact that since the print never chemically bonds to the paper it can easily smudge when handled. Furthermore, there is a tendency for the inks to contaminate the various guide rollers in the system and to produce a background (34).

10.13 Audio Systems

Loud speaker housings use PLA as the base material. In addition, the composition contains aluminum hydroxide as inorganic filler and a hydrolysis inhibitor (35).

The inorganic filler allows adjustment of the desired acoustical and mechanical properties. The hydrolysis inhibitor comprises at least one member selected from a carbodiimide compound, an isocyanate compound, and an oxazoline compound.

The hydrolysis inhibitor is added to prevent the acoustic apparatus housing from easily decomposing due to moisture in the air during the period of use. Preferred hydrolysis inhibitors are carbodiimides. A carbodiimide compound can be melt-kneaded with the biodegradable polymer compound. Examples are *N,N′*-dicyclohexylcarbodiimide or *N,N′*-diisopropylcarbodiimide (35). In order to definitely obtain the required mechanical strength, a rubber component is added.

10.14 Automotive Uses

The number of light vehicle registrations dramatically increases the environmental impact of automobiles worldwide (36). The automotive industry must implement fundamental issues such as fuel economy and reduced emissions. Many design parameters, such as rolling resistance, aerodynamics, drivetrain design, friction, and vehicle weight, can influence the environmental properties of the vehicles.

Desirable vehicles combine an ultralight design with a likely hybrid or fuel cell engine technology. Advanced lightweight materials, such as plastics or composites, are highly appreciated and end-of-life recycling and recovery is of critical importance. The trend will be towards fewer materials and parts in the design of a vehicle.

A compostable interior panel for use in a vehicle has been described (37). A typical interior panel is shown in Figure 10.9.

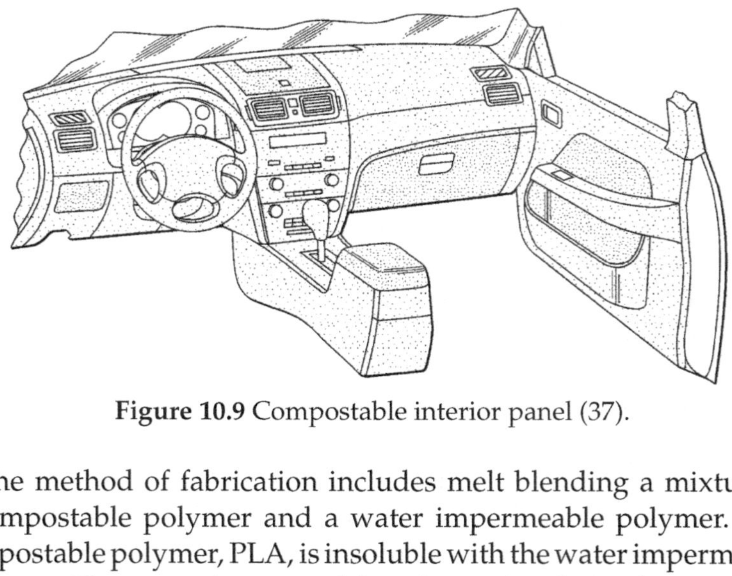

Figure 10.9 Compostable interior panel (37).

The method of fabrication includes melt blending a mixture of a compostable polymer and a water impermeable polymer. The compostable polymer, PLA, is insoluble with the water impermeable polymer. The water impermeable polymer is admixed in amounts of 0.1–5%.

The method also includes fountain-flowing the mixture into a closed mold. The layer is substantially impenetrable to water and is a biaxially oriented thermoplastic polymer such as a polyolefin.

In addition, a permeation resistance-improving additive can be introduced, which is a polymer-clay nanocomposite.

The interior panel is formed by solidifying the mixture, and has a core and an encapsulating layer. The interior panel includes the core comprising the compostable polymer. The layer encapsulating the core comprises the water impermeable polymer. The interior panel is then removed from the mold.

Metal-plated plastics are increasingly employed in the automotive industry. The recycling of metal-plated plastics is an important subject for both resource recycling and environmental protection. An attempt to recycle metal-plated plastics has been discussed (38).

A hammer crusher was designed to liberate the coating from the plastic substrate. The size distribution of particles was analyzed and could be described by the Rosin-Rammler function model (39–41).

The optimum retention time of materials in the crusher is 3 *min*. There, 87.3% of the materials can be liberated. A two-step magnetic separation was adopted to avoid excessive crushing and to guarantee the quality of products. With regards to both the separation efficiency and grade of products, the optimum rotational speed of the magnetic separator is 50–70 *rpm* (38).

10.15 Oil Well Environment

Two hydrolyzable polymers, poly(methylene oxide) and poly(glycolic acid), were tested for association with mechanical systems to promote diversion of formation acid treatments in oil wells (42).

Samples were confined to chemical products commonly employed in oil well completion processes, HCl 15% and NaCl 23% aqueous solutions, in separate metallic cylindrical cells, under conditions observed in southeastern Brazilian offshore fields (50°C and 75°C and a pressure of 41.4 *MPa*). Pristine samples were characterized by 13C-*NMR*. Hydrolytic influences were evaluated by differential scanning calorimetry, SEM, wide-angle XRD, and computerized microtomography.

The intense effects observed showed a specific correspondence with each tested polymer and were evidenced by post-exposure analysis. The results indicated that both polymers can be used in association with mechanical diverting systems to promote the chemical treatment distribution and, therefore, improve the oil well productivity (42).

10.16 Green Hot Melt Adhesives

The use of agricultural products for green hot melt adhesives has been proposed. In particular, it has been proposed to replace paraffin wax with soybean wax (43).

In comparison to paraffin wax, soybean wax shows a natural difference in molecular and crystalline structures, but it exhibits some similar or even unique characteristics in its melt temperature and crystallization temperature, onset time of crystallization, hardness, and microstructure. For these reasons, soybean wax is a promising candidate for the formulation of a green hot melt adhesive with an acceptable functionality (43).

10.17 Mechanistic Studies

10.17.1 Olefin Isomerization

Acyclic diene metathesis (ADMET) is a special type of olefin metathesis that can be used to transform terminal dienes into polyenes with olefin ejection, as shown in Figure 10.10.

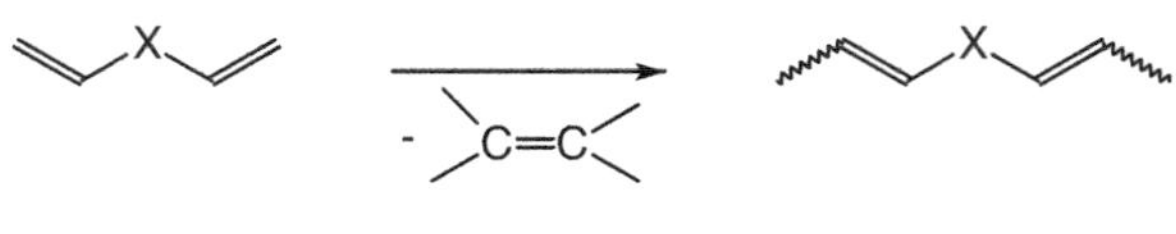

Figure 10.10 ADMET polymerization.

ADMET polymerization is a versatile technique for the preparation of a wide variety of linear polymers (44, 45). However, olefin isomerization side reactions may result in polymers with irregular repeating units (46).

Olefin isomerization side reactions that occur during an ADMET polymerization can be studied by preparing polyesters via ADMET and subsequently degrading these polyesters by a transesterification with methanol (46). The resulting diesters, representing the repeating units of the previously prepared polyesters, are analyzed by gas chromatography-mass spectroscopy.

This method allows the quantification of the olefin isomerization that may occur in the course of ADMET polymerization using second generation ruthenium metathesis catalysts.

Furthermore, it was demonstrated that the addition of benzoquinone to the polymerization mixture prevents the olefin isomerization. Therefore, second generation ruthenium metathesis catalysts can be used for the preparation of well-defined polymers via an ADMET technique, causing little isomerization (46).

Trade names appearing in the references are shown in Table 10.10.

Table 10.10 Trade names in references.

Trade name Description	Supplier
Citroflex® A-4 Acetyltri-*n*-butyl citrate (1)	Morflex Inc.
Eastar® Bio Compostable copolyester (1)	Eastman Chemical Company
Irganox® 1010 Pentaerythritol tetrakis(3-(3,5-di-*tert*-butyl-4-hydroxyphenyl)propionate), phenolic antioxidant (1)	Ciba-Geigy
Kraton® Styrenic block copolymer (1)	Kraton Polymers LLC
L9000® Poly(lactic acid) (1)	Biomer
Luchem® HA-B18 Polymeric HALS (1)	Atochem Inc., M&T Chemicals, Pennwalt Corp.
Luperox® P *tert*-Butyl peroxybenzoate (1)	Arkema Inc.
Perkadox® 14-40B Di-(2-*tert*-butylperoxyisopropyl) benzene on inert filler (1)	Akzo Nobel Chemicals Inc.
Polysar® EPDM rubber (1)	Bayer
Surlyn® Ionomer resin (1)	DuPont
Teflon® Tetrafluoro polymer (1)	DuPont
Triganox® 101-45B 2,5-Dimethyl-2,5-di-(*tert*-butylperoxy)hexane on inert filler (1)	Akzo Nobel Chemicals Inc.
Triganox® 17-40B *n*-Butyl 4,4-di-(*tert*-butylperoxy)valerate on inert filler (1)	Akzo Nobel Chemicals Inc.
Vikoflex® (Series) Epoxidized esters, plasticizers for PVC (1)	Viking Chemical Co.

References

1. A.K. Mohanty, L.T. Drzal, B.P. Rook, and M. Misra, Floor covering made from an environmentally friendly polylactide-based composite formulation, US Patent 7 354 656, assigned to Michigan State University, Board of Trustees (East Lansing, MI), April 8, 2008.
2. S.H. Imam and G.M. Glenn, Biodegradable abrasive compositions, assigned to The United States of America, as represented by the Secretary of Agriculture (Washington, DC).
3. G.J. Guskey, D.A. Gonzales, and K.S. Wei, Skin cleansing compositions comprising biodegradable abrasive particles, US Patent 9 717 674, assigned to The Procter & Gamble Company (Cincinnati, OH), August 1, 2017.
4. A. Prasad, Polishing pad comprising biodegradable polymer, US Patent 7 264 641, assigned to Cabot Microelectronics Corporation (Aurora, IL), September 4, 2007.
5. D.W. Grijpma, E. Kroeze, A.J. Nijenhuis, and A.J. Pennings, *Polymer*, Vol. 34, p. 1496, 1993.
6. P. Gallezot, *Catalysis Today*, Vol. 121, p. 76, March 2007.
7. D. Gnanasekaran and V.P. Chavidi, *Vegetable Oil Based Bio-lubricants and Transformer Fluids: Applications in Power Plants*, Materials Forming, Machining and Tribology book series, Springer, Singapore, 2018.
8. S. Sun, G. Wang, and P. Wang, *Journal of Cleaner Production*, Vol. 188, p. 530 , 2018.
9. D. Engel and W.J. Garland, Renewable cards, US Patent Application 20 080 277 481, November 13, 2008.
10. J.F. Starkey, Environmentally friendly degradable paper products, US Patent 8 354 174, assigned to Pres-On Tape and Gasket Corporation (Addison, IL), January 15, 2013.
11. T. Yankovitz and A. Schweitzer, Biodegradable irrigation pipe, US Patent 7 862 873, assigned to Netafim, Ltd. (Tel Aviv, IL), January 4, 2011.
12. Y.-J. Yu, C.-F. Chen, T.-Y. Lin, S.-J. Yeh, S.-Y. Wang, and P.-L. Lai, Biodegradable copolymer and thermosensitive material, US Patent 7 884 142, assigned to Industrial Technology Research Institute (TW), February 8, 2011.
13. S. Xu, X. Du, G. Feng, Y. Zhang, J. Li, B. Lin, L. Yang, S. Fu, and J. Wu, *Oncotarget*, Vol. 9, p. 282, December 2017.
14. S. Shi, X. Wen, T. Li, M. Pagel, and C. Li, *Journal of Nuclear Medicine*, Vol. 59, p. 192, 2018.
15. C. Casadidio, M.E. Butini, A. Trampuz, M.D. Luca, R. Censi, and P.D. Martino, *European Journal of Pharmaceutics and Biopharmaceutics*, Vol. 130, p. 260, 2018.

16. P.S. Chan, J.W. Xian, Q. Li, C.W. Chan, S.S.Y. Leung, and K.K.W. To, *The AAPS Journal*, Vol. 21, p. 59, Apr 2019.
17. H. Liu, X. Shi, D. Wu, F. Kahsay Khshen, L. Deng, A. Dong, W. Wang, and J. Zhang, *ACS Applied Materials & Interfaces*, Vol. 11, p. 19700, 2019.
18. L. Vojtova, L. Michlovska, K. Valova, M. Zboncak, M. Trunec, K. Castkova, M. Krticka, V. Pavlinakova, P. Polacek, M. Dzurov, V. Lukasova, M. Rampichova, T. Suchy, R. Sedlacek, M.-P. Ginebra, and E.B. Montufar, *International Journal of Molecular Sciences*, Vol. 20, p. 391, Jan 2019.
19. K. Shi, B. Xue, Y. Jia, L. Yuan, R. Han, F. Yang, J. Peng, and Z. Qian, *Nano Research*, Vol. 12, p. 1389, Jun 2019.
20. M. Zou, R. Jin, Y. Hu, Y. Zhang, H. Wang, G. Liu, Y. Nie, and Y. Wang, *J. Mater. Chem. B*, p. in press, 2019.
21. G. Woodward, G.P. Otter, K.P. Davis, and K. Carmichael, Biodegradable polymers, US Patent 7 879 964, assigned to Rhodia Consumer Specialties Limited (Hertfordshire, GB), February 1, 2011.
22. R.J. Larson, E.A. Bookland, R.T. Williams, K.M. Yocom, D.A. Saucy, M.B. Freeman, and G. Swift, *Journal of Polymers and the Environment*, Vol. 5, p. 41, 1997.
23. OECD, Biodegradability in seawater, OECD Guidelines for the Testing of Chemicals 306, OECD Publishing, Paris, France, 1992.
24. K.P. Davis, D.R.E. Walker, G. Woodward, and A.C. Smith, Novel phosphino derivatives, EP Patent 0 861 846, assigned to Albright & Wilson UK Ltd., September 02, 1998.
25. H. Huang, Q. Yao, Q. Jiao, B. Liu, and H. Chen, *Journal of Saudi Chemical Society*, Vol. 23, p. 61 , 2019.
26. B.D. Favis, P. Sarazin, J. Li, and Z. Yuan, Microporous articles comprising biodegradable medical polymers, method of preparation thereof and method of use thereof, US Patent 8 007 823, assigned to Corporation de l'Ecole Polytechnique de Montreal (Montreal, CA), August 30, 2011.
27. S. Andou, Biodegradable molded articles, US Patent 5 470 382, assigned to Nissei Kabushiki Kaisha (Osaka, JP), November 28, 1995.
28. Wikipedia, Sake kasu — Wikipedia, the free encyclopedia, 2013.
29. S.I. Martin, D.P. Fergenson, A. Srivastava, M.J. Bogan, V.J. Riot, and M. Frank, Safe biodegradable fluorescent particles, US Patent 7 781 224, assigned to Lawrence Livermore National Security, LLC (Livermore, CA), August 24, 2010.
30. G. Workman, Biodegradable test cylinder mold, US Patent 7 694 580, assigned to Deslauriers, Inc. (La Grange Park, IL), April 13, 2010.
31. H. Kipphan, ed., *Handbook of Print Media: Technologies and Production Methods*, Springer, Berlin, Heidelberg, 2001.
32. K.-H. Meyer, ed., *Technik des Flexodrucks*, Rek & Thomas Medien, St. Gallen, Switzerland, 5th edition, 2006.

33. C. Holweg, An improved method of printing paper bags, GB Patent 16 517, April 29, 1909.
34. L. Masaro, P. Lapointe, and J.-C.J. Gayet, Biodegradable flexographic ink, US Patent 7 588 632, assigned to BioMatera Inc. (CA), September 15, 2009.
35. S. Yamada, Y. Fujihira, H. Mori, T. Noguchi, K. Tokura, and M. Uryu, Biodegradable material for audio systems, US Patent 7 576 154, assigned to Sony Corporation (Tokyo, JP), August 18, 2009.
36. J.W. McAuley, *Environmental Science & Technology*, Vol. 37, p. 5414, December 2003.
37. D.F. Mielewski, A.M. Harris, and E.C.-c. Lee, Compostable interior panel for use in a vehicle and method of manufacture, US Patent 8 197 924, assigned to Ford Global Technologies, LLC (Dearborn, MI), June 12, 2012.
38. M. Xue, J. Li, and Z. Xu, *Environmental Science & Technology*, Vol. 46, p. 2661, March 2012.
39. P. Rosin and E. Rammler, *Glückauf*, Vol. 68, p. 529, 1932.
40. P. Rosin and E. Rammler, *J. Inst. Fuel*, Vol. 7, p. 29, 1933.
41. P. Rosin and E. Rammler, *Kolloid-Zeitschrift*, Vol. 67, p. 16, April 1934.
42. A.Z.I. Pereira, M.C. Delpech, F.G. da Costa, F.B. da Cruz, and P.R. Tavares, *Journal of Applied Polymer Science*, Vol. 136, p. 47362, 2019.
43. W. Li, X.H. Kong, E. Kharraz, L. Bouzidi, and S.S. Narine, *The Journal of Adhesion*, Vol. 87, p. 95, February 2011.
44. P.M. O'Donnell, K. Brzezinska, D. Powell, and K.B. Wagener, *Macromolecules*, Vol. 34, p. 6845, September 2001.
45. T.E. Hopkins and K.B. Wagener, *Macromolecules*, Vol. 37, p. 1180, February 2004.
46. P.A. Fokou and M.A.R. Meier, *Macromolecular Rapid Communications*, Vol. 31, p. 368, February 2010.

11

Biofuels

Increasing energy demands have directed attention to renewable resources for the fabrication of biofuels. The exploitation of feedstocks containing polysaccharides other than cellulose is of interest for biofuel production. The aspects of the production of biofuels, including biodiesel, bioethanol, methane, and hydrogen, are the subject of a detailed monograph (1)

Biomass feedstocks available decentrally will be more commodious for a localized biorefinery approach than the exhaustive large-scale and centralized plants driven by cost-intensive technology (2). A simplified scheme of biorefinery operations is given in Figure 11.1.

11.1 Xenobiotics

The biodegradation processes of both xenobiotics such as aromatic compounds, plastics such as poly(vinyl alcohol) (PVA), polyesters, poly(ethylene), and nylon, as well as polymers of biological origin and blends of both types have been reviewed (3). Synthetic materials with groups susceptible to hydrolytic microbial attack have also been dealt with.

The biodegradation process can be classified into an aerobic and anaerobic degradation (4–6). The basic reactions are given in Eq. 11.1.

$$
\begin{aligned}
\text{Polymer} + O_2 &\rightarrow CO_2 + \text{Biomass} + \text{Residue} \\
\text{Polymer} &\rightarrow H_2O + CH_4 + \text{Biomass} + \text{Residue}
\end{aligned}
\qquad (11.1)
$$

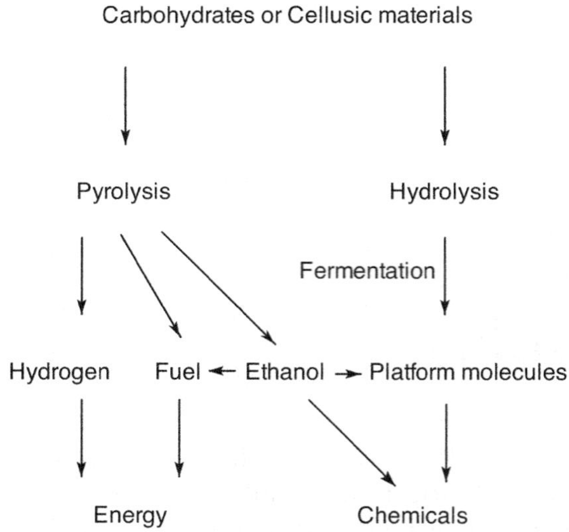

Figure 11.1 Biorefinery operations (7).

If oxygen is present the aerobic biodegradation occurs and carbon dioxide is produced. In contrast, without oxygen, an anaerobic degradation occurs and methane is produced instead of carbon dioxide.

11.2 Biopolymers

11.2.1 Poly(l-lactide)

In general, the biodegradability by microorganisms is studied in a real environment that contains a natural mixture of fungi and bacteria. The purely fungal degradation of poly(*l*-lactide) has been studied in order to diversify the role of fungi in the real process of biodegradation (8). Respirometric tests were done in soil at 30°C, and further in compost at 30°C and 58°C.

The percentage of degradation is determined by monitoring the formation of carbon dioxide and comparing the amount of carbon dioxide with the stoichiometric amount:

$$\%CO_2 = \frac{m\,CO_2 - c\,CO_2}{t\,CO_2} \times 100 \tag{11.2}$$

Here, $m\,CO_2$ is the amount of carbon dioxide delivered from the sample, $c\,CO_2$ is the amount of carbon dioxide delivered from a control system, a blank beaker, and $t\,CO_2$ is the theoretical stoichiometric amount of carbon dioxide delivered. The degradation progress at 30°C is shown in Figure 11.2.

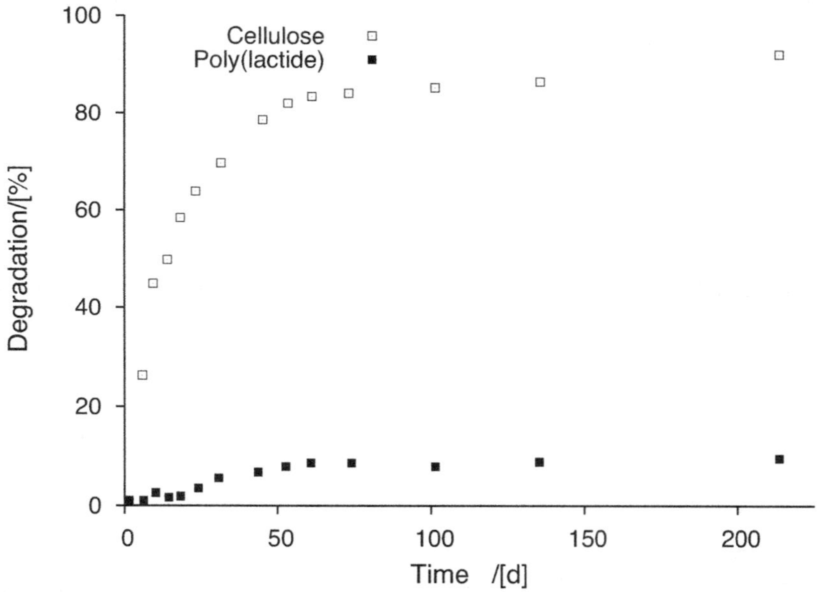

Figure 11.2 Degradation progress in sterilized and inoculated soil by fungi at 30°C (8).

Whereas the degradation process for cellulose reaches completeness, the situation for poly(*l*-lactide) is different, as only minimal degradation takes place. However, at elevated temperature the situation changes, as can be seen in Figure 11.3.

The results suggest that the temperature is the predominant parameter that governs the fungal degradation of poly(*l*-lactide). In real compost, the kinetics of biodegradation of the poly(*l*-lactide) is synergetic between bacteria and fungi.

A semiempirical model used to predict the biodegradation profile has been presented (8). This is based on a mathematical model from Hill (9–11). The Hill equation was originally developed for the

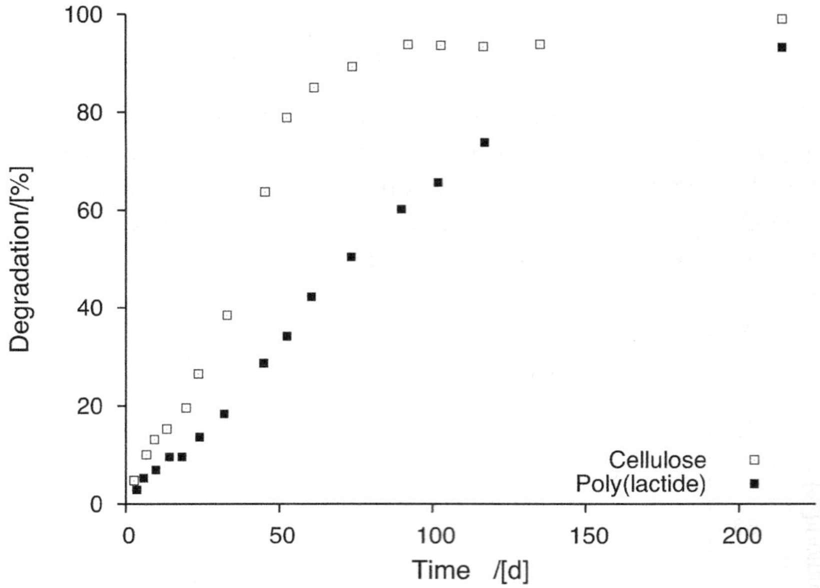

Figure 11.3 Degradation progress in sterilized and inoculated soil by fungi at 58°C (8).

binding of a ligand to a macromolecule. For fitting the points in Figure 11.2 and Figure 11.3 into curves, the equation looks like:

$$y = y_{max}\frac{t^n}{k^n + t^n} \quad . \tag{11.3}$$

Matrix-assisted laser desorption ionization time of flight mass spectrometry was used to characterize the oligomers formed in the course of degradation. In the case of poly(*l*-lactide) the degradation happens at the chain ends. The oligomers formed show two kinds of chain ends (8):

1. Oligomers with alcohol and carboxylic acid at the chain ends, and
2. Oligomers with alcohol and carboxylate at the chain ends.

11.3 Bioethanol

Bioethanol is a versatile transportation fuel and fuel additive that offers excellent performance and reduced air pollution compared to conventional fuels (12). Its production and use adds little, if any, net release of carbon dioxide into the atmosphere, thus dramatically reducing the potential for global climate change. The state-of-the-art of bioethanol production has been detailed (13). Also, the production of bioethanol from food crops has been detailed in a book (14).

The recently listed materials for bioethanol production are summarized in Table 11.1. Some of these methods are detailed below.

Table 11.1 Materials for bioethanol production.

Material	References
Bamboo	(15)
Cardoon	(16)
Cellulose	(17)
Corn	(18)
Corn and corn stover	(19)
Lichenan	(2)
Lignocellulose	(20)
Microalgae feedstock	(21)
Moso bamboo	(22)
Olive tree biomass	(23)
Palm wood	(24)
Pineapple waste	(25)
Recycled paper sludge	(26)
Rice	(27)
Starch	(28)
Sugar beet roots	(29)
Sugarcane	(30)
Sweet potato	(31)
Sweet sorghum	(32)

The total energy from biomass contributes about 9–13% of the global energy supply (33). The production of bioethanol has been reviewed (34). A lot of attempts are being made to develop a technology for converting biomass for commercial production of bioethanol as a renewable energy source.

Ethanol derived from biomass, one of the modern forms of biomass energy, has the potential to be a sustainable transportation fuel, as well as a fuel oxygenate that can replace gasoline (35). In particular, ethanol used as a liquid transportation fuel could reduce domestic consumption of fossil fuels, particularly petroleum.

Worldwide, ethanol is mostly produced by fermentation, accounting for up to 90% of ethanol produced globally. In Brazil, most bioethanol is produced from sugarcane or molasses and in the United States from corn. However, other starch materials, such as wheat, barley and rye, are also suitable as raw material. The starch-containing grains have to be converted into sugar. Thus, in Europe, starch grains, e.g., wheat and sugar beet, are the major source of bioethanol production (36).

In Brazil, the best bioethanol yield from sugarcane is 7.5 $m^3 ha^{-1}$ (37). However, bioethanol has only 64% of the energy content of biodiesel. This would require planting sugarcane over a very large area, which may not even be available (30).

An estimation was made of how much bioethanol can potentially be produced from starch, sugar crops, and agricultural residues. These crops include corn, barley, oat, rice, wheat, sorghum, and sugarcane (33). The potential yield of bioethanol from various biopolymers, i.e., crops, is listed in Table 11.2.

Table 11.2 Yield of bioethanol from dry crops (33).

Crop Material	Yield/[$l\,kg^{-1}$]	Crop Material	Yield/[$l\,kg^{-1}$]
Oat straw	0.26	Sorghum straw	0.27
Rice straw	0.28	Bagasse	0.28
Corn stover	0.29	Wheat straw	0.29
Barley straw	0.31	Wheat	0.40
Barley	0.41	Oat	0.41
Sorghum	0.44	Corn	0.46
Rice	0.48	Sugarcane	0.50

The global potential bioethanol production from wasted crops and in particular from the various natural polymers has been detailed (33).

The technical and economic results of alternative approaches taken to integrate various unit operations into processes for converting biomass to ethanol have been reviewed (38).

11.3.1 Pretreatment Methods

Raw biomass is extremely persistent to enzymatic degradation. Therefore, methods of thermochemical pretreatment have been developed to improve the digestibility of biomass (39). For example, ammonia explosion, aqueous ammonia recycle, controlled pH, dilute acid, flow through, and lime treatment were applied to prepare corn stover, i.e., the leaves and stalks of maize, for a subsequent conversion to sugars.

In general, the digestibility of cellulose is improved by the removal of lignin and hemicellulose (17). This can be achieved by a treatment with an excess of calcium hydroxide. Alkaline treatments degrade the lignin but leave both hemicellulose and cellulose intact. For example, of the initial lignin, 87.5% could be removed (40). This pretreatment has no inhibitory effect on the ethanol fermentation.

On the other hand, an acidic pretreatment will hydrolyze the hemicellulose but the cellulose and lignin remain intact (41). An acidic pretreatment may also result in high concentrations of furfural compounds in the liquid phase (34). Such compounds may act as fermentation inhibitors.

A problem that is typically encountered in the bioethanol fermentation process is the presence of degradation products arising from the pretreatment of lignocellulosic feedstocks (42). These degradation products often act as fermentation inhibitors. In high temperature pretreatments, sugar degradation products, such as furfural and hydroxymethylfurfural, can be formed in high temperature processes. Furthermore, acetic acid is ubiquitous in lignocellulose pretreatments, since hemicellulose and lignin are thereby acetylated. Formic acid is also often formed, as there are a variety of monomeric phenolic compounds derived from lignin.

Another problem is the bacterial contamination of fermentation mixtures. Bacterial contamination can hardly be avoided under non-sterile conditions. Lactic acid bacteria, in particular *Lactobacillus* species, are the primary bacterial contaminants of fuel ethanol fermentations. A bacterial contamination reduces the yield of ethanol.

It was established that a range of concentrations exists in which fermentation inhibitors derived from pretreatments of lignocellulosic feedstocks do not affect the yeast, but will inhibit further growth of lactic acid bacteria. By an optimization of the levels of fermentation inhibitors the fermentation of biomass can be conducted under non-sterile conditions. Here, the yield of ethanol is comparable to that achieved under sterile conditions (42). Optimized inhibitor levels can be achieved by controlling the ratio of water to biomass of a lignocellulosic biomass during and after the pretreatment.

An environmentally friendly and inexpensive process that requires no detoxification or delignification by chemically acid hydrolyzing biomass is sonication, optionally in combination with French pressing (36). During the sonication process bubbles are generated in a medium containing the cellulose, which collapse implosively after growing. While the bubbles are collapsed, the local temperature inside the bubbles is raised to about 5100 *K* (43). In addition, the pressure may increase to 1000 *at*. Due to the high temperature and pressure, the cellulose material is disrupted (36).

11.3.2 Cellulases and Hemicellulases

In sugar chemistry, in general, the affix *ase* refers to an enzyme that can degrade the corresponding *ose*. This originates from a standard about enzyme nomenclature by the International Union of Biochemistry (44).

Hemicellulose belongs to the group of matrix polysaccharides. It is present in plant cell walls together with cellulose. Cellulose is crystalline, and highly resistant to hydrolysis. It has a degree of polymerization of 7,000–15,000. In contrast, hemicellulose exhibits a random, amorphous structure with a much smaller degree of polymerization of 500–3,000. Hemicelluloses are composed from mostly *D*-pentose monomers.

Cellulases are enzymes that hydrolyze cellulose, i.e., the β-1,4-glucan or β-*D*-glucosidic linkages. This results in the formation of glucose, cellobiose, or cellooligosaccharides. Glucose and cellobiose are shown in Figure 11.4.

Cellulases have been traditionally divided into the following major classes (45):

1. Endoglucanases (EG),
2. Exoglucanases,
3. Cellobiohydrolases (CBH), and
4. β-Glucosidases (BG).

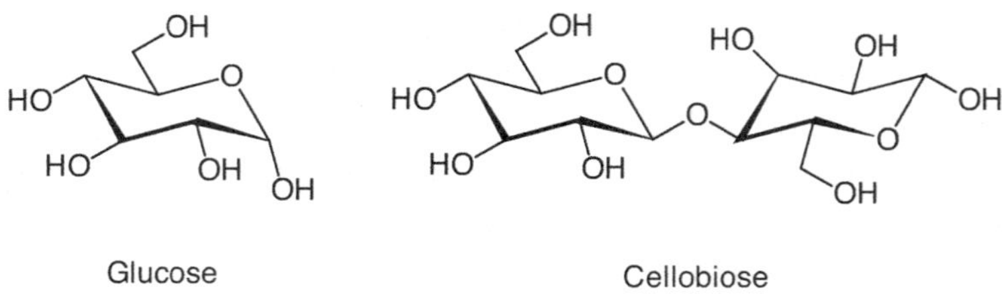

Figure 11.4 Glucose and cellobiose.

Endoglucanases act mainly on the amorphous parts of the cellulose fiber, whereas cellobiohydrolases are also able to degrade crystalline cellulose.

Cellulases are known to be produced by a large number of bacteria, yeast and fungi. Certain fungi produce a complete cellulase system capable of degrading crystalline forms of cellulose, such that the cellulases are readily produced in large quantities via fermentation. Filamentous fungi play a special role since many yeasts, such as *Saccharomyces cerevisiae*, lack the ability to hydrolyze cellulose (45–47).

The economical production of viable biofuels requires metabolic engineering of microbial platforms to convert cellulosic biomass into biofuels with high titers and yields. By the integration of traditional and novel engineering strategies with advanced engineering toolboxes, the development of more robust microbial platforms was possible. Recent trends concerning the metabolic engineering of microbial platforms, such as the industrial yeasts *Saccharomyces cerevisiae* and *Yarrowia lipolytica*, for the production of renewable fuels have been reviewed (20).

Hemicellulases are a highly diverse group of enzymes that hydrolyze hemicelluloses. These enzymes have a lot of biotechnological applications. The catalytic and non-catalytic domains of hemicellulases have been elucidated, which reveal the principles of catalysis and specificity for these enzymes (48). Also, the production of cellulase in microorganisms has been thoroughly investigated (49). For example, the filamentous fungus *Trichoderma reesei* produces and secretes superabundant amounts of enzymes that act synergistically to degrade cellulase and related biomass components.

Of importance is to improve performance of cellulase and to decrease the amount of enzyme needed for the fermentation process.

Most attractive are cellobiohydrolases, as they constitute some 60–80% of natural cellulases.

Site-directed mutagenesis was used to generate novel *Hypocrea jecorina* cellobiohydrolase variants. Several mutants that were expressed in *T. reesei* showed improved thermostability and reversibility (45).

A series of mutants obtained by genetic and protein engineering of *Thermoanaerobacterium saccharolyticum* glucose isomerase have been tested for their ability to produce bioethanol. Glucose isomerase mutants have been found that exhibit an improved catalytic activity in the conversion of *D*-xylose to *D*-xylulose under high temperature conditions (50). The glucose isomerase mutants can be used in the degradation of hemicellulose and the production of bioethanol.

11.3.3 Production from Starch

The production of ethanol from sugar derived from starch and sucrose has been commercially dominated by the yeast *S. cerevisiae* (28). However, the sugar obtained from biomass is a mixture of hexoses and pentoses. Now, most wild-type strains of *S. cerevisiae* do not metabolize xylose (34).

Strengthening methods have been undertaken to increase the yields of fermentation (34):

1. To add additional pentose and to use genetic engineering for other ethanologens, and
2. To improve the yield of ethanol by genetic engineering of microorganisms.

Even when these methods are successful, they are not ripe for commercial use. Moreover, in contrast to the clean sugar streams derived from starch and sucrose, hydrolysates derived from biomass tend to have fermentation inhibitors such as acetic acid or furfural. These compounds must be either removed when their concentrations are high. Or else, robust strains must be developed that are resistant to the above inhibitors.

The improvements in starch hydrolysis coupled with price subsidies have enabled bioethanol, largely from corn wet milling and dry milling operations, to add several billion liters to the total ethanol capacity and flip the once dominant *ethanol from ethylene* process (38).

11.3.4 *Production from Lignocellulose*

Cellulosic fuels are expected to have great potential in industrial applications, but they still face technical challenges to become cost-competitive fuels, thus presenting many opportunities for improvement (20).

The lignin content has been an obstacle to efficient transformation of lignocellulosic biomass to bioethanol, feeds, organic precursors and other useful products (51). The enzymatic conversion of lignocellulosic polysaccharides to fermentable sugars typically requires high enzyme loadings to overcome comparatively low efficiencies of enzymatic catalysis.

The comparatively inefficient hydrolysis of cellulosic substrates in pretreated lignocellulosic biomass is partly due to obstruction by lignin of enzyme access to catalytically productive cellulose binding.

The compactness and complexity of lignocellulose is much more difficult to enzymatically degrade to fermentable sugars than by using starch. The lignin component of the biomass material poses challenges for enzymatic degradation because of its nonproductive binding and inactivation of cellulases. The conversion process consists of three steps (34):

1. Thermochemical pretreatment,
2. Enzymatic saccharification, and
3. Fermentation of the released sugars by specialized organisms.

The activity of cellulases for hydrolysis of the lignocellulose can be improved when weak lignin-binding enzymes are used. Certain naturally occurring cellulases differ significantly in their affinities for lignin (52).

The location and the structure of lignin affects the enzymatic hydrolysis much more than the absolute quantities of lignin. The modification of the lignin surface by oxidative treatments with laccase and a delignification treatment results in an increased hydrolysis of the lignocellulose (53).

An approach to reduce the deleterious effects of lignin is delignification of pretreated lignocellulosic biomass. Another approach is the introduction of additives to hydrolysis mixtures, including surfactants, proteins and other lignin-binding polymers. In particular, poly(ethylene glycol) has been found to be effective (51).

Hydrolysis mixtures prepared using high dry matter content, in the presence of PEG and surfactant, can be readily used in simultaneous saccharification and fermentation (SSF) and other processes involving the fermentation in the presence of lignin.

The optimization of the concentration of PEG in cellulose hydrolysis of steam-pretreated wheat straw and in SSF at high dry matter contents of 25% insoluble dry matter is shown in Figure 11.5.

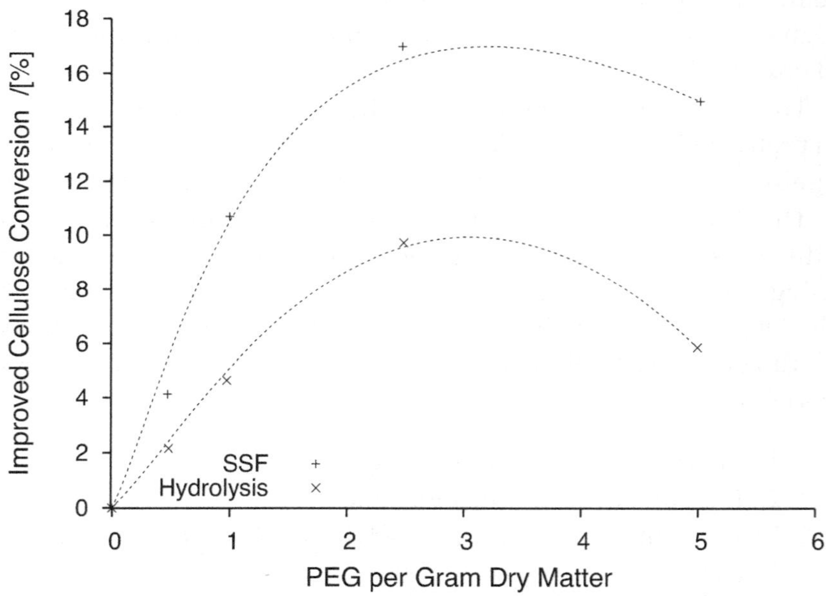

Figure 11.5 Optimization of the concentration of PEG (51).

11.3.5 Production from Lichenan

Bioethanol can be fabricated from polymeric lichenan with lichenase and a thermotolerant yeast (2). Lichen is present in a wide range of habitats. *Parmelia* is a large genus of a lichenized fungus with an essential global distribution. It occurs even in the Arctic and Antarctic regions (54, 55), but mostly in temperate regions.

Lichen consists mainly of the polysaccharide lichenan, with some 62% of the total carbohydrate content. Lichenan is a linear polysaccharide from β-D-glucopyranose that is linked by both 1,3 and 1,4 linkages (56).

A hydrolysis grade of 73–76% can be obtained in 24 h for lichenan from *Cetraria islandica*, *Usnea barbata* and *Parmelia* sp. by the use of lichenase from an alkalothermophilic *Thermomonospora* sp. With the commercial enzyme, Accellerase™ 1000, a hydrolysis of 100% can be reached.

β-Glucosidase plays a synergistic role in the hydrolysis of lichenan. A yield of ethanol of 0.45–0.48 g/g can be reached. This corresponds to a theoretical conversion efficiency of 93–96%.

Thermotolerant yeast can be isolated from rotten strawberries. The yeast cells are immobilized on calcium alginate and can be reused eight times at 40°C with a 100% fermentation efficiency (2).

The degree of hydrolysis over time using Accellerase™ 1000 at 400 $U\,g^{-1}$ with a substrate loading of 100 $g\,l^{-1}$ is shown in Figure 11.6.

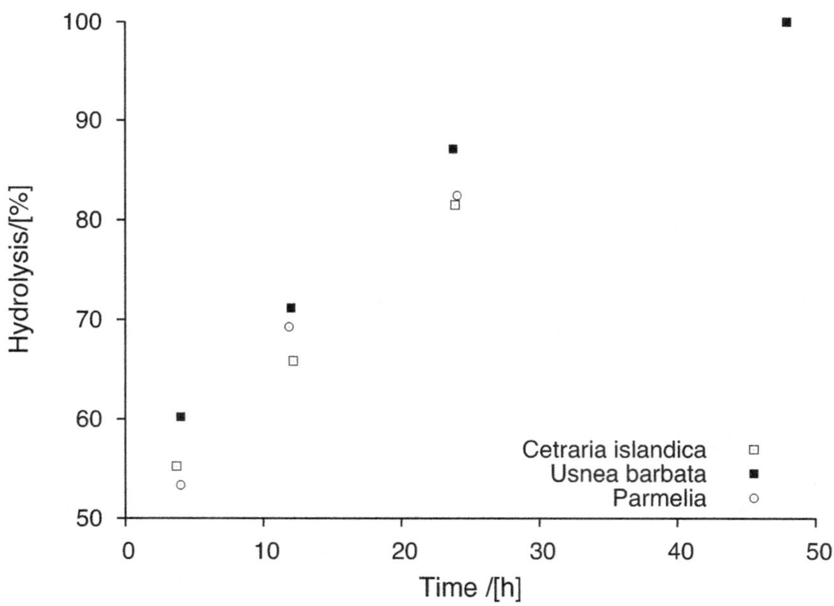

Figure 11.6 Saccharification of lichenan (2).

11.3.6 Production from Other Sources

11.3.6.1 Olive Tree Biomass

Olive tree biomass was pretreated with sulfuric acid under previously optimized conditions according to two different criteria optimizations (23). Then, two schemes of saccharification and fermentation were tested to evaluate the potential of bioethanol production from olive tree biomass pretreated under both optimal conditions.

The best results of ethanol production were achieved from olive tree biomass pretreated at 15% solids, 164°C and 0.89% H_2SO_4 (23). However, when olive tree biomass was pretreated at 35% solids, 160°C, and 1.72% H_2SO_4, more concentrated sugar solutions were obtained, but the ethanologenic microorganism was not capable of assimilating all sugars in the medium. Therefore, the highest ethanol yield of 0.45 *g* ethanol per *g* sugar (88% of the theoretical ethanol yield), corresponding to 14.4 *g* ethanol per 100 *g* raw olive tree biomass, was achieved when the olive tree biomass was pretreated at 15% solids using a new bioconversion scheme for this feedstock (23).

This consisted of the enzymatic hydrolysis of the whole slurry from the pretreatment followed by the co-fermentation of cellulosic and hemicellulosic sugars by a recombinant ethanologenic *Escherichia coli* MM160 (23).

11.3.6.2 Pineapple Waste

Pineapple waste, an abundant agro-industrial residue, was studied as a low-cost material for the generation of different value-added products (25). The work succeeded in obtaining bioethanol and proteolytic enzymes from these residues and, accordingly, integrated approaches for pineapple waste valorization combining the production of bioethanol and bromelain in a unique process are suggested. Bromelain is shown in Figure 11.7.

Bromelain is mainly used in the food and pharmaceutical industries, and it also has applications in cosmetics, textiles, leather and detergents (25). Proposals of the study were based on the optimization of bioethanol production through different fermentation and saccharification processes: Direct fermentation of the liquor, consecutive saccharification and fermentation of the solid waste and simultaneous saccharification and fermentation of the solid waste.

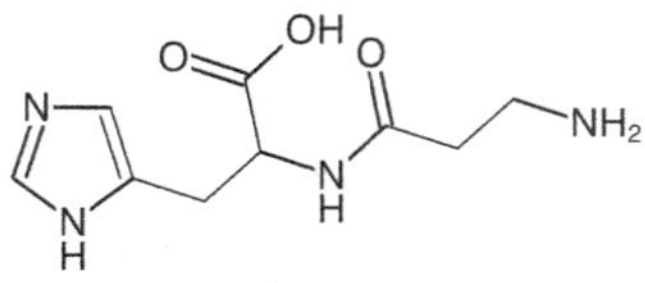

Figure 11.7 Bromelain.

Simultaneous saccharification and fermentation increased the ethanol production (5.4±0.1% v/v) as compared to direct fermentation (4.7±0.3%) and saccharification and fermentation of the solid waste (4.9±0.4% v/v) processes.

On the other hand, the separation of bromelain was accomplished using membrane separation techniques, i.e., microfiltration and ultrafiltration, and further stabilization of the concentrated stream by freeze-drying. An increased protein concentration after downstream processes was confirmed by the Lowry analytical method (11.5±1.2 to 21.0±1.3 $mg\,ml^{-1}$ in the retentate), and the proteolytic activity of the lyophilized powder was estimated to be 340–805 GDU (gelatin digestion units). The resulting permeate successfully underwent fermentation for bioethanol production (25).

11.3.6.3 Moso Bamboo

Moso bamboo was pretreated with industrially derived crude glycerol obtained from different sources at 150°C–160°C for 3 *h* (22). This bamboo, pretreated with base biodiesel glycerol with the pressure filtration removal method, showed a high glucose yield of 94.95% and an ethanol yield of 73.10% of the theoretical value.

The major glycerol content could be removed by pressure filtration, leaving a small amount of fatty acid soap in the pretreated sample, which formed an emulsion that reduced lignin redisposition onto the biomass surface and effectively blocked lignin absorption of cellulase, allowing greater enzymatic hydrolysis and fermentation system function.

The surface was more hydrophilic and a higher lignin removal was achieved: 39.24% with base biodiesel glycerol pretreatment compared to 26.08% with sodium hydroxide glycerol pretreatment (22).

11.3.6.4 Microalgae Feedstock

The production of bioethanol and biodiesel from a microalgae feedstock was investigated (21). Various pretreatment methods were used to determine the maximum recovery of sugars from *Scenedesmus* sp. The total sugar yield of 93% was obtained when the biomass was pretreated by acid hydrolysis.

The hydrolysate produced 86% ethanol (theoretical yield) after the fermentation using *Saccharomyces cerevisiae*. Enzyme-catalyzed direct transesterification of the biomass was performed using dimethyl carbonate as a solvent. A maximum yield of 92% methyl ester, 1.86% glycerol carbonate and 4.93% glycerol dicarbonate was achieved. The integrated process of bioethanol and biodiesel production was optimally achieved when a direct transesterification was done first, followed by ethanol fermentation yielding 92% and 93% of methyl ester and ethanol, respectively (21).

11.3.6.5 Sweet Potato

The production of sweet potato bioethanol was conducted at a small industrial plant in Brazil and integrated into the biodiesel production through ethanolysis. The specific objectives were (31):

1. Monitoring the bioethanol production at a small industrial plant, considering production yield, and
2. Studying the ethanolysis of sunflower oil using sweet potato bioethanol with different grades (98, 99 and 99.4°GL: degrees Gay Lussac of alcohol), with the focus on biodiesel yield and purity.

Bioethanol was produced under the following conditions (31): Sweet potato dilution in water (1:1 in *wt*%); enzymatic hydrolysis with alpha-amylase (90°C) and glucoamylase (60°C) enzymes (700 $ml\,t^{-1}$); and, fermentation (30°C) with *Saccharomyces cerevisiae* (3.3 $kg\,t^{-1}$).

The ethanolysis was performed as follows: 6:1 ethanol:oil molar ratio, 1.0% NaOH, 45°C, 1 *h*. The average yield of bioethanol was 161.4 $l\,t^{-1}$, corresponding to 10,598 $l\,ha^{-1}$. The biodiesel synthesis using bioethanol of different grades showed that both the yield and purity of the product increased with the grade of the ethanol used, with values ranging from 63% to 83%, and 50% to 94% for yield and purity, respectively. These results showed that a high alcohol grade is required to make a viable integrated production of biodiesel and bioethanol (31).

11.3.6.6 Rice

Rice is a major food crop containing large amounts of lignocellulosic residues which are usable for biofuels (27).

Transgenic rice plants were collected that overproduced *Trichoderma reesei* β-1,4-d-glucosidase (BGL I) into the cell walls of the mature straws. Without any pretreatment, the transgenic rice straws showed a consistently higher biomass enzymatic saccharification than the wild-type cultivar, in particular when 1% Tween-80 or 0.5% PEG-4000 was co-supplied into the enzymatic hydrolysis. Under a mild alkali pretreatment (1% NaOH at 5°C for 2 *h*), the desirable transgenic line exhibited a complete biomass enzymatic hydrolysis, resulting in the highest bioethanol yield of 21% (% dry matter) when compared with the rice and other bioenergy crops subjected to stronger pretreatment conditions.

Despite the relatively low hexose yields obtained under 1% H_2SO_4 pretreatment, the transgenic rice straw showed high bioethanol production at 18% due to an almost complete sugar-ethanol conversion rate. Chemical analysis indicated that the transgenic rice straw had significantly increased biomass porosity and reduced cellulose features, which contributed to the largely enhanced biomass enzymatic hydrolysis. In addition, the raised arabinose level in hemicellulose and the lignin H-monomer proportion may also positively affect the biomass enzymatic saccharification in the transgenic rice straw. Thus, the study demonstrated a cost-effective and green lignocellulose conversion technology for high bioethanol production in transgenic rice straw. It also provided a strategy for the potential genetic modification of plant cell walls in bioenergy crops (27).

11.4 Biobutanol

The recent trends in biobutanol production have been reviewed (57). Among various liquid biofuels, biobutanol is considered as a suitable and sustainable alternative to gasoline. Low butanol yield and its toxicity are the major bottlenecks to its use. The use of metabolic engineering and integrated fermentation and product recovery techniques has the potential to overcome these challenges. The usage of various nanocatalysts to overcome the existing challenges is gaining much interest.

Conventional processes for the synthesis of butanediol use petrochemical feedstocks for their starting materials. For example, acetylene is reacted with 2 molecules of formaldehyde in the Reppe reaction, followed by a catalytic hydrogenation resulting in 1,4-butanediol (58). It has been estimated that 90% of the acetylene produced in the U.S. is consumed for butanediol production. It has been proposed to use succinic acid as the key biologically produced intermediate for the manufacture of the butanediol. However, succinic acid is costly to isolate and purify and requires high temperatures and pressures for catalytic reduction to butanediol.

The industrial production of biobutanol started in 1916 as a fermentation process using the bacteria *Clostridium acetobutylicum* (59). This type of fermentation is termed acetone-butanol-ethanol (ABE) fermentation and is used with molasses and cereal grains. The materials used for ABE fermentation are detailed in Table 11.3.

ABE fermentation with clostridia has interesting features for the bioconversion of wastes. Unlike most ethanol-producing microorganisms, clostridia are able to produce the solvents from pentoses and starch as effective as hexoses. The different aspects of biobutanol production have been reviewed (60). The potential biobutanol production from agricultural residues using ABE fermentation has been presented in a chapter of a book (61).

The acidogenic and solventogenic metabolic pathways in clostridia have been detailed (62). Clostridial bacteria can also ferment other cheaper alternative biomass, such as lignocellulosic materials, due to their saccharolytic ability (63). There, an acidic or enzymatic hydrolysis of lignocellulosic materials is necessary to convert them into monosaccharides before they can be used as substrates in the ABE fermentation.

Algae are considered as a superior feedstock for biofuel production because of their diversity (64). Some of the major benefits of algae are their extremely fast growth rate and the ability of sequestration of carbon dioxide with high oil and carbohydrate contents

Table 11.3 Materials suitable for ABE fermentation (65).

Raw material	Composition
Barley straw	42% Cellulose, 28% hemicellulose, 7% lignin, 11% ash
Wheat straw	38% Cellulose, 29% hemicellulose, 24% lignin, 6% ash
Corn fiber	20% Starch, 50-60% non-starch polysaccharides
Corn stover	38% Cellulose, 26% hemicellulose, 23% lignin, 6% ash
Switchgrass	37% Cellulose, 29% hemicellulose, 19% lignin
Domestic organic waste	59% Sugars, 13% lignin, 17% ash
Sago	86% Starch, small amounts of mineral and nitrogenous matter
Degermed corn	73% Starch, 3% ash, 13% proteins
Extruded corn	61% Starch, 3.8% corn oil, 8.0% protein, 11.2% fiber
Liquefied corn starch	39% Starch, 45% moisture
Cassava	70% Starch, 2.7% protein, 2.4% fiber, 0.2% ash

that can be easily transformed into biodiesel or other gasoline components such as butanol.

Butanol has become an attractive biofuel as a byproduct of algal biomass processing after lipid extraction for biodiesel, due to its higher energy content, lower vapor pressure, and less hygroscopicity than ethanol. The current status of microalgae for biodiesel and butanol production as eco-friendly alternatives for liquid fossil fuels has been reviewed (64).

Macroalgal biomass types, which contain significant quantities of carbohydrates, with low amounts of lignin and fermentation inhibitors, are interesting alternative substrates for fermentation (66).

Various modes of fermentation processes for the production of biobutanol have been examined, e.g., batch, fed-batch, and continuous processes (65).

Continuous processes seem to be more economical than batch and fed-batch processes. Namely, continuous processes offer some advantages such as reduction in sterilization and re-inoculation time, and superior productivity (67). Recently, a non-naturally occurring microorganism with one or more gene disruptions has been developed for the production of 1,4-butanediol (58).

Biobutanol is now regarded as a biofuel of the second generation (68). Analysis has determined that butanol performs equally as well as ethanol in a direct injection spark-ignition engine from the viewpoint of regulated emissions and combustion, however, with the advantage of a decreased fuel consumption (69). In addition, butanol has several advantages over ethanol such as higher energy content, lower water absorption, lower volatility, better blending ability, and usability in conventional combustion engines without modification (70, 71).

On the other hand, the energy yield of *n*-butanol is about half that of ethanol from corn or switchgrass by the current ABE technology (72). This is a serious disadvantage for *n*-butanol since the feedstock costs represent a significant portion of the fuel price. However, it seems that how this issue should be handled is controversial. Other authors state that with the use of new strains, inexpensive substrates, and superior reactor designs, the economic potential of ABE fermentation has been found to be highly attractive (73, 74). It seems that new fermentation strategies are responsible, as the production of biobutanol is now considered as promising and economically feasible. For example, separate hydrolysis and fermentation (SHF) or a combination of SHF with simultaneous saccharification and fermentation processes have been developed (75).

The combustion properties of butanol have been investigated and a detailed chemical kinetic model has been presented (76). In addition, the flame chemistry during the combustion of butanol isomers has been investigated by molecular beam mass spectrometry (77).

A noticeable reduction of CO and NO_x emissions was observed in diesel engine combustion when an *i*-butanol blended fuel was used (78).

The technologies and markets for chemicals from biobutanol have been reviewed (73, 79). Apart from fuel applications, butanol can be used as (65):

- Solvent in printing inks,
- Butyl esters,
- Chromatographic applications,
- Extractant in the production of drugs and natural substrates,

- Additive in polishes cleaning agents, and
- Anti-icing additive.

The production of hydrogen through steam reforming of a biobutanol using supported cobalt-based catalysts has been described (80, 81). A Co–Ir/ZnO catalyst is a good choice to optimize the hydrogen production by steam reforming.

A comparative analysis of different pretreatment strategies was performed on hydrolysis and fermentation of brewers' spent grain for biobutanol production by *Clostridium beijerinckii* (82). Ozone (2.7%, 30 *min*, 40–60% humidity), sodium hydroxide (1–5% NaOH, 30 *min*, 120°C, 5–15% brewers' spent grain) and hydrogen peroxide (5% H_2O_2, pH 11.5, 60–180 *min*, 50°C, 5–15% brewers' spent grain) were the analyzed pretreatment methods.

Ozonolysis was not very effective for either the degradation of lignin or the recovery of fermentable sugars in the enzymatic process under the tested operation conditions. Maximum butanol and ABE concentrations, 5.3±0.2 *g* butanol per liter and 6.8±0.2 $g\,l^{-1}$ ABE, were observed when a low moisture content of 40% was used. The peroxide alkaline pretreatment was found to be the most effective for brewers' spent grain when it was conducted at 5% brewers' spent grain for 60 *min* (82).

Under these conditions, sugar enzymatic yields of 62.8% for glucose and 28.1% for xylose and arabinose, referred to untreated brewers' spent grain, and product concentrations of 11.0±0.2 *g* butanol per liter and 13.7±0.2 $g\,l^{-1}$ ABE, were achieved. The alkaline pretreatment at 15% brewers' spent grain with 1% NaOH also provided high butanol and ABE concentrations (7.3±0.1 $g\,l^{-1}$ butanol and 8.9±0.1 $g\,l^{-1}$ ABE). Both the sodium hydroxide alkaline and peroxide alkaline methods were highly successful as pretreatment methods of brewers' spent grain for ABE production, with overall yields of 44.4 *g* butanol per *kg* brewers' spent grain and 54.0 $g\,kg^{-1}$ ABE brewers' spent grain for NaOH pretreatment (15% brewers' spent grain, 1% NaOH) and 45.1 $g\,kg^{-1}$ butanol brewers' spent grain and 56.1 $g\,kg^{-1}$ ABE brewers' spent grain for H_2O_2 pretreatment (5% brewers' spent grain, 60 *min*) (82).

In addition to the study presented above, a microwave-assisted hydrothermal pretreatment technology has been developed to recover fermentable sugars from brewer's spent grain (83).

Microwave hydrothermal pretreatment is considered to be a greener pretreatment, as no acid or alkali are used as catalysts.

The optimal conditions were 192.7°C and 5.4 *min*, resulting in a hemicellulosic sugar recovery of 64%, a 70% glucose recovery in enzymatic hydrolysate and 2.4 $g\,l^{-1}$ total inhibitors.

The liquid fraction obtained under optimal conditions was fermented with *Clostridium beijerinckii*, reaching a butanol concentration as low as 1 $g\,l^{-1}$. The butanol concentration could be improved by operating at higher solid loadings in pretreatment, which would increase the fermentable sugar concentration. Enzymatic hydrolysis of pretreated grains yielded a sugar solution, which was also fermented, resulting in a butanol concentration and overall yield of 8.3 $g\,l^{-1}$ and 46 $kg\,t^{-1}$ grain, respectively (83).

For the production of bioenergy from lignocellulosic biomass, most biomass pretreatment processes need to use some chemical reagent as the catalyst to overcome the biomass recalcitrance barrier (84). Such reagents are usually severe inhibitors for the subsequent microbial fermentation process. Therefore, in many cases, the liquid prehydrolysates fraction is discarded after the pretreatment, which is a tremendous waste of materials and leads to additional pollution.

Biobutanol produced from the acetone, butanol and ethanol (ABE) fermentation process has been of great interest recently due to its high value as a bioenergy source or biochemical. During ABE fermentation, acetic acid is produced and then reassimilated as a carbon source. Thus, acetic acid is a substrate rather than an inhibitor for biobutanol production (84).

Acetic acid was employed as the chemical catalyst for the pretreatment of switchgrass, which then was used for ABE production through simultaneous saccharification and fermentation with hyper-butanol producing *Clostridium saccharoperbutylacetonicum* N1-4.

Due to a systematic investigation of the pretreatment conditions and fermentation, it was concluded that the optimized condition for switchgrass pretreatment was with 3 $g\,l^{-1}$ acetic acid at 170°C for 20 *min*. Both prehydrolysates fraction and solid cellulosic fraction (SCF) of the pretreated biomass were highly fermentable. In the fermentation with the SCF/prehydrolysates fraction mixture as the substrate, 8.6 $g\,l^{-1}$ butanol corresponding to a yield of 0.16 $g\,g^{-1}$ was obtained (84).

As the most abundant crop residue in the United States, corn stover is a readily available feedstock for biorefineries (85). In order to estimate the amount of collectible corn stover special to biobutanol production, a quantitate model was presented. In this model, the residue potential quantity, job creation, butanol blends potential quantity, greenhouse gas emissions and social costs, forecasting future trends, economic evaluation, and risk analysis were considered.

It was estimated that a total quantity of 84 *Mt* of stover for the production of about 10.48 *Mt* gigaliters of biobutanol can potentially be collected. Around 9.5 *Gl* of gasoline can be saved and about 64.6 *Gl* of butanol-gasoline blend (Bu16) can potentially be produced from corn residues in the United States, which is equivalent to 11.8% of total domestic gasoline consumption (85).

A hydrotropic reagent, i.e., sodium xylene sulfonate, was used to treat wheat straw for efficient butanol production (86). The effects of temperature, time, and sodium xylene sulfonate concentration on the pretreatment were evaluated. A modified sodium xylene sulfonate pretreatment method with pH adjusted to 3.5 by formic acid was also conducted for efficient wheat straw conversion by removing the hemicellulose.

Composition analysis, structure characterization, enzymatic hydrolysis and fermentation tests were conducted. The results showed that modified sodium xylene sulfonate pretreatment can be considered as an efficient method for improving wheat straw conversion efficiency for ABE production. The hexoses and pentoses in the enzymatic hydrolysates can be used by *Clostridium acetobutylicum* for butanol production with 12.41 $kg\,l^{-1}$ ABE produced, and an overall ABE yield of 100 $g\,kg^{-1}$ of wheat straw was obtained (86).

A simultaneous saccharification and fermentation with delayed yeast extract feeding was conducted in a 2-liter bioreactor equipped with *in-situ* recovery using a gas stripping in order to enhance biobutanol production from lignocellulosic biomass of oil palm empty fruit bunch (87).

This study showed that 2.88 $kg\,l^{-1}$ of biobutanol has been produced from simultaneous saccharification and fermentation with a similar yield of 0.23 $kg\,g^{-1}$ as compared to separate hydrolysis and fermentation. An increase of 42% of the biobutanol concentration was observed when delayed yeast extract feeding was introduced in the simultaneous saccharification and fermentation at 39 *h* of fermentation operation. The biobutanol production was further enhanced up to 11% with a total improvement of 72% when *in-situ* recovery using a gas stripping was implemented to reduce the solvents inhibition in the bioreactor (87).

11.5 Biodiesel

The interest in biodiesel has been growing globally due to the need for liquid fuels and its potential as a biodegradable nontoxic substitute for petroleum diesel (64).

The production of ester-based fuels, such as biodiesel or jet fuel, from renewable starting materials, such as lignocellulosic material or algae, has been described (88). The pulping and saccharification of the renewable starting materials produces carboxylic acids, such as fatty acids or rosin acids, which are esterified via a gas sparged, slurry form of heterogeneous reactive distillation to yield ester-based fuels.

The process for the production of ester-based fuel from renewable starting material consists of (88):

1. The comminution of renewable starting material,
2. The isolation of cellulose and other soluble and insoluble sugars, including isolation by hydrolysis or saccharification of the comminution product,
3. The isolation of fatty acids or rosin acids,
4. The addition of a C_1-C_8 alcohol to the acid and esterification, and
5. Refining of the resulting ester to produce an ester-based fuel.

Materials destined for cellulosic ethanol production have been evaluated, and they were found to contain low relative concentrations of fatty acids. Relative to the amount of ethanol produced, the amount of fatty acid byproduct is actually quite significant. Assuming a typical yield of 20% ethanol and 2% fatty acid means that a minimum of 10% of an ethanol producer's high value products could be in the form of fatty acids (88). It has been claimed that microalgal biodiesel is a better alternative than bioethanol from sugarcane (34).

11.5.1 Transesterification Methods

The performance of biodiesel production from various feedstocks has been summarized (89). The existing literature survey deals with the potential and important feedstocks, like edible oil, non-edible oil, animal fat and algae oil, for the production of biodiesel. The various sources have different yields due to process variables. The yield of biodiesel differs with the feedstocks due to physicochemical properties of sources and the process variables. Recent research has been focusing on cheap, abundant feedstocks, novel production and purification technologies for biodiesel (89).

11.5.1.1 Transesterification by Enzymes

Enzymatic transesterification is advantageous from the viewpoint of conversion, yield and reusability compared to the low yield of chemical catalyst catalyzed transesterification reactions in recent years (89). Lipase mediated transesterification has increased the rate of reaction followed by high conversion. However, the activity of free enzyme is reduced as stability is low.

In order to overcome this drawback, the immobilized lipase mediated transesterification methodology has been introduced in recent studies. Nanobiocatalyst focuses exclusively on the transesterification of oils using methanol to produce fatty acid methyl esters. Importantly, the lipase binds onto magnetic particles with various size ranges, confirming stability and giving more reactive centers. Analytical methods, such as Fourier transform infrared spectroscopy and transmission electron microscopy, are used to characterize the structure of nanoparticles, which exhibit better resistance to temperature and pH, stirring speed, enzyme loading, viscosity of oil and alcohol/oil molar ratio and free fatty acid (89).

Also, the optimum conditions of transesterification, e.g., methanol-to-oil ratio, amount of catalyst, reaction temperature and reaction time, have been assessed in a review (90).

11.5.1.2 In-Situ Transesterification

The current research in *in-situ* transesterification targeting biodiesel using wet microalgae has been reviewed (91). In *in-situ* transesterification, the variable conditions are highly dependent on the water contained in raw microalgae. Nevertheless, the water content in most *in-situ* transesterification was overlooked only as one of the process variables. So, research into wet *in-situ* transesterification technologies using microalgae was accessed. The here developed technologies were classified based on the reaction temperature applied; the increment in the temperature of wet *in-situ* transesterification compensates for the hindrance of water, therefore wet *in-situ* transesterification tends to have distinct features depending on the reaction temperature.

The detailed variables or reaction conditions are characterized by the three different reaction temperature zones of low, under 100°C, middle, 101°C to 200°C, and high, above 200°C.

The future prospects of wet *in-situ* transesterification were suggested based on the techno-economic analysis of existing works that used wet *Nannochloropsis gaditana* (91).

11.5.1.3 Enzymatic Transesterification

The various aspects of the enzymatic transesterification method to convert used cooking oil to biodiesel have been assessed (92).

A thorough overview has been given of general biodiesel production processes, reaction conditions, challenges, and solutions for higher biodiesel production yield through introducing various microorganisms that are capable of producing the enzymes required to convert used cooking oil into biodiesel. The characteristics, composition, and advantages of the used cooking oil as feedstock for biodiesel were also discussed (92).

In addition, the existing transesterification methods including homogeneous alkali-catalyzed, homogeneous acid-catalyzed, non-catalytic reaction under supercritical conditions, and enzyme-catalyzed reactions were explained. Also, the advantages of the enzymatic method over other methods, and the enzymes which are the key elements of such reactions, are discussed. Lipases are the most promising enzymes currently known for biodiesel conversion. The physiological and physical properties of microbial lipases, the catalytic mechanisms of the enzymes, various methods of enzyme immobilization, such as adsorption, covalent and affinity binding, entrapment, and whole cell immobilization, were also reviewed (92).

11.5.2 *Production from Microalgae Beats*

Biodiesel production from microalgae has been recognized as a promising route for sustainable energy supply (93). Many microalgae can be converted to biodiesel (30).

It has been concluded that microalgal biodiesel is the only renewable biodiesel that has the potential to completely displace liquid transport fuels derived from petroleum. An existing demand for liquid transport fuels could be met sustainably with biodiesel from microalgae, but not with bioethanol from sugarcane (30).

The extraction process of oil from microalgae usually generates a large amount of lipid-extracted algal biomass, which has not been economically utilized until now (93). In order to explore the economic potential of the residual biomass, lipid-extracted algal from extraction of microalgae strain, *Nannochloropsis salina*, was employed as the filler in biocomposite fabrication with PVA. It was noted that the increase of the lipid-extracted algal loading reduced the mechanical properties, but enhanced the thermal stability of biocomposite materials compared to neat PVA.

When poly(diallyldimethylammonium chloride) was incorporated, mechanical properties of PVA/lipid-extracted algal/poly(diallyldimethylammonium chloride) composites were significantly improved (93). Poly(diallyldimethylammonium chloride) was compatible with negatively charged lipid-extracted algal and PVA matrix as poly(diallyldimethylammonium chloride) possesses positive charges. PVA was compounded with 20% lipid-extracted algal and 12% poly(diallyldimethylammonium chloride) to synthesize the biocomposite. In contrast to neat PVA, this biocomposite recorded a similar mechanical property but an enhanced thermal property.

This type of biocomposite film can be applied in commercial industries as specialty materials such as 3D printing material. This helps to improve the economic feasibility of microalgae-based biofuel production (93).

11.5.3 *Two-Step Catalytic Conversion*

Intensive efforts have been made to enhance biodiesel yield through a two-step reaction in the production process (94).

These processes have been performed under various conditions, including the type of feedstock oil, alcohol, catalyst, heating mode and reaction time.

The chemical property of the feedstock oil is the most important factor related to the selection of the best suitable two-step reaction technology. Two double-step reaction technologies have been discussed (94).

The purpose of the two-step reaction is to optimize the operating cost of the production process such as by lowering alcohol ratio, reducing reaction time, and minimizing yield loss. If free fatty acid content is more than 2%, then the first step esterification will be followed by the second step transesterification. The target of the first step esterification is to decrease the free fatty acid content in feedstock oils as much as possible in order to reduce the soap formation in the second step transesterification. Soap formation prevents the separation of biodiesel and glycerol phase and also needs a higher amount of water for washing. It is the main reason for loss of biodiesel due to ester turning to soap. The T1+T2 technology should be applied to low free fatty acid content feedstock oils to reduce the production cost (94).

11.5.4 Improvement of Diesel Fuel Properties by Terpenes

A simple gas method has been proposed for the improvement of the properties of diesel fuel (95).

There is no need for the addition of surfactants, chemicals or other pollutants. A natural product from *Pinus halepensis* trees can be used to remove water residues from diesel fuels.

The method relies on the property of the resin to adsorb water, especially when the water amount is dissolved into hydrocarbon fluids. The removal of water improves the physicochemical properties of the diesel fuel up to 633 $J\,g^{-1}$ for the heat of combustion, 69% for humidity and up to 74% for conductivity (95).

Trade names appearing in the references are shown in Table 11.4.

Table 11.4 Trade names in references.

Trade name Description	Supplier
Accellerase™ 1000	DuPont
Enzyme for unlocking C5 and C6 sugars (51)	
Agrimul™ NRE 1205	Diamond Shamrock Corp.
Nonionic surfactant (51)	
Avicel®	FMC BioPolymer
Microcrystalline cellulose (51)	
Berol® 08	AkzoNobel
Fatty alcohol ethoxylate (51)	
Brij® (Series)	ICI Surfactants
Ethoxylated fatty alcohols (51)	
Celluclast®	Novozymes NS
Cellulase (51)	
Celluzyme®	Novozymes A/S
Detergent enzymes (51)	
Cereflo®	Novozymes A/S
Cellulase (51)	
Emulgen® (Series)	Kao Chemicals
Polyoxyethylene ether (51)	
Laminex®	Genencor International
Enzyme preparations for use in the brewing industry, wine and fruit juice industry and dairy industry (51)	
Lutensol® AT (Series)	BASF
Alkylpolyethylene glycol ethers (nonionic surfactants) (51)	
Novozyme® 188	Novozymes NS
Cellobiase from *Aspergillus niger* (42)	
Retsch™ SM 2000	Retsch GmbH
Cutting mill (51)	
Rohament® 7069 W	Röhm
Cellulase (51)	
Softanol™ (Series)	Nippon Shokubai Co., Ltd.
Ethylene oxide polymer surfactants (51)	
Spezyme™ CP	Danisco US Inc.
Cellulase (51)	
Tergitol® 15-S (Series)	Union Carbide Corp.
Ethoxylated C11-15-secondary alcohols, surfactant (51)	

Table 11.4 (cont) Trade names in references.

Trade name Description	Supplier
Thermosacc™ Stress tolerant yeast (42)	Lallemand Specialties Inc.
Tween® (Series) Ethoxylated fatty acid ester surfactants (51)	Uniqema
Ultraflo® Cellulase (51)	Novozymes A/S

References

1. A. Vertes, N. Qureshi, H. Yukawa, and H. Blaschek, eds., *Biomass to Biofuels: Strategies for Global Industries*, Wiley, Hoboken, N.J, 2010.
2. V. Menon, R. Divate, and M. Rao, *Fuel Processing Technology*, Vol. 92, p. 401, March 2011.
3. K. Leja and G. Lewandowicz, *Polish J. Environ. Stud.*, Vol. 19, p. 255, 2010.
4. G. Swift, *Polymer Degradation and Stability*, Vol. 59, p. 19, January 1998.
5. S. Grima, V. Bellon-Maurel, P. Feuilloley, and F. Silvestre, *Journal of Polymers and the Environment*, Vol. 8, p. 183, 2000.
6. I. Kyrikou and D. Briassoulis, *Journal of Polymers and the Environment*, Vol. 15, p. 125, April 2007.
7. P. Gallezot, *Catalysis Today*, Vol. 121, p. 76, March 2007.
8. Z. Saadi, A. Rasmont, G. Cesar, H. Bewa, and L. Benguigui, *Journal of Polymers and the Environment*, Vol. 20, p. 273, June 2012.
9. A.V. Hill, *The Journal of Physiology*, Vol. 40, p. iii, 1910.
10. A. Calmon, F. Silvestre, V. Bellon-Maurel, J.-M. Roger, and P. Feuilloley, *Journal of Environmental Polymer Degradation*, Vol. 7, p. 135, 1999.
11. S. Domenek, P. Feuilloley, J. Gratraud, M.-H. Morel, and S. Guilbert, *Chemosphere*, Vol. 54, p. 551 , 2004.
12. C. Wyman, ed., *Handbook on Bioethanol — Production and Utilization*, Routledge, 2018.
13. R.G. Hemansi, G. Yadav, G. Kumar, A. Yadav, J.K. Saini, and R.C. Kuhad, Second generation bioethanol production: The state of art in N. Srivastava, M. Srivastava, P.K. Mishra, S.N. Upadhyay, P.W. Ramteke, and V.K. Gupta, eds., *Sustainable Approaches for Biofuels Production Technologies: From Current Status to Practical Implementation*, Vol. 7 of *Biofuel and Biorefinery Technologies*, chapter 8, pp. 121–146. Springer, 2018.
14. R. Ray, ed., *Bioethanol Production from Food Crops: Sustainable Sources, Interventions, and Challenges*, Academic Press, an imprint of Elsevier, London, United Kingdom, 2019.
15. H. Yang, Z. Shi, G. Xu, Y. Qin, J. Deng, and J. Yang, *Bioresource Technology*, Vol. 274, p. 261, 2019.
16. M.C. Fernandes, M.D. Ferro, A.F.C. Paulino, H.T. Chaves, D.V. Evtuguin, and A.M.R.B. Xavier, *Industrial Crops and Products*, Vol. 111, p. 633, 2018.
17. S. Kim and M.T. Holtzapple, *Bioresource Technology*, Vol. 97, p. 583, March 2006.
18. S.K. Mohanty and M.R. Swain, Bioethanol production from corn and wheat: Food, fuel, and future in R.C. Ray and S. Ramachandran, eds., *Bioethanol Production from Food Crops*, chapter 3, pp. 45–59. Academic Press, 2019.

19. S. Chen, Z. Xu, X. Li, J. Yu, M. Cai, and M. Jin, *Bioresource Technology*, Vol. 258, p. 18, 2018.
20. J.K. Ko and S.-M. Lee, *Current Opinion in Biotechnology*, Vol. 50, p. 72, 2018.
21. R. Sivaramakrishnan and A. Incharoensakdi, *Fuel*, Vol. 217, p. 458, 2018.
22. L. Ji, F. Lei, W. Zhang, X. Song, J. Jiang, and K. Wang, *Bioresource Technology*, Vol. 276, p. 300, 2019.
23. J.C. Martínez-Patiño, E. Ruiz, C. Cara, I. Romero, and E. Castro, *Biochemical Engineering Journal*, Vol. 137, p. 172, 2018.
24. E.R. Sathendra, G. Baskar, R. Praveenkumar, and E. Gnansounou, *Bioresource Technology*, Vol. 271, p. 345, 2019.
25. L.S. Gil and P.F. Maupoey, *Journal of Cleaner Production*, Vol. 172, p. 1224, 2018.
26. D.G. Gomes, S. Serna-Loaiza, C.A. Cardona, M. Gama, and L. Domingues, *Bioresource Technology*, Vol. 267, p. 347 , 2018.
27. Y. Li, P. Liu, J. Huang, R. Zhang, Z. Hu, S. Feng, Y. Wang, L. Wang, T. Xia, and L. Peng, *Green Chem.*, Vol. 20, p. 2047, 2018.
28. Y. Lin and S. Tanaka, *Applied Microbiology and Biotechnology*, Vol. 69, p. 627, February 2006.
29. C. Marzo, A.B. Díaz, I. Caro, and A. Blandino, Status and perspectives in bioethanol production from sugar beet in R.C. Ray and S. Ramachandran, eds., *Bioethanol Production from Food Crops*, chapter 4, pp. 61–79. Academic Press, 2019.
30. Y. Chisti, *Trends in Biotechnology*, Vol. 26, p. 126, March 2008.
31. J.O. Virgínio e Silva, M.F. Almeida, M. da Conceição Alvim-Ferraz, and J.M. Dias, *Renewable Energy*, Vol. 124, p. 114, 2018.
32. D. Jiang, M. Hao, J. Fu, K. Liu, and X. Yan, *Journal of Cleaner Production*, Vol. 220, p. 225, 2019.
33. S. Kim and B.E. Dale, *Biomass and Bioenergy*, Vol. 26, p. 361, April 2004.
34. K.A. Gray, L. Zhao, and M. Emptage, *Current Opinion in Chemical Biology*, Vol. 10, p. 141, April 2006.
35. S. Kim and B.E. Dale, *The International Journal of Life Cycle Assessment*, Vol. 7, p. 237, July 2002.
36. K.P. Yoon, Method for pretreating biomass to produce bioethanol, US Patent 8 278 080, assigned to Industry Academic Cooperation Foundation Keimyung University (Daegu, KR), October 2, 2012.
37. J. J. K. Bourne, *Natl. Geogr. Mag.*, pp. 41–59, October 2007.
38. F.A. Keller, Integrated bioprocess development for bioethanol production in C. Wyman, ed., *Handbook on Bioethanol — Production and Utilization*, chapter 16, pp. 351–379. Routledge, 2018.
39. C.E. Wyman, B.E. Dale, R.T. Elander, M. Holtzapple, M.R. Ladisch, and Y. Lee, *Bioresource Technology*, Vol. 96, p. 1959, December 2005.

40. S. Kim and M.T. Holtzapple, *Bioresource Technology*, Vol. 96, p. 1994, December 2005.
41. T.A. Lloyd and C.E. Wyman, *Bioresource Technology*, Vol. 96, p. 1967, December 2005.
42. J. Larsen, Non-sterile fermentation of bioethanol, US Patent 8 187 849, assigned to Inbicon A/S (Fredericia, DK), May 29, 2012.
43. D.J. Flannigan and K.S. Suslick, *Nature*, Vol. 434, p. 52, March 2005.
44. International Union of Biochemistry, *Enzyme Nomenclature*, Elsevier, Amsterdam, New York, 1965.
45. A. Day, F. Goedegebuur, P. Gualfetti, C. Mitchinson, P. Neefe, M. Sandgren, A. Shaw, and J. Stahlberg, Variant Hyprocrea jecorina CBH1 cellulases, US Patent 8 232 080, assigned to Danisco US Inc. (Palo Alto, CA), July 31, 2012.
46. M.P. Coughlan and L.G. Ljungdahl, Comparative biochemistry of fungal and bacterial cellulolytic enzyme systems, in J. Millet, ed., *Biochemistry and Genetics of Cellulose Degradation*, number 43 in FEMS Symposia, pp. 11–30, Delft, the Netherlands, 1988. Federation of European Microbiological Societies, Academic Press, Inc.
47. K.M.H. Nevalainen and M.E. Penttilä, Molecular biology of cellulolytic fungi in U. Kück, ed., *Genetics and Biotechnology*, Vol. 2 of *The Mycota*, pp. 303–319. Springer, Berlin, Heidelberg, 2004.
48. D. Shallom and Y. Shoham, *Current Opinion in Microbiology*, Vol. 6, p. 219, June 2003.
49. P.K. Foreman, *Journal of Biological Chemistry*, Vol. 278, p. 31988, June 2003.
50. J. Wang, R. Fu, and T.C. Cheung, Method of making bioethanol by using glucose isomerase mutants, US Patent 8 012 726, assigned to BioRight Worldwide Company Limited (Tortola, VG), September 6, 2011.
51. J. Larsen and H. Jorgensen, Methods for reducing enzyme consumption in second generation bioethanol fermentation in the prescence of lignin, US Patent 7 972 826, assigned to Inbicon A/S (Fredericia, DK), July 5, 2011.
52. A. Berlin, N. Gilkes, A. Kurabi, R. Bura, M. Tu, D. Kilburn, and J. Saddler, *Applied Biochemistry and Biotechnology*, Vol. 121-124, p. 163, 2005.
53. H. Palonen, *Role of lignin in the enzymatic hydrolysis of lignocellulose*, VTT, Espoo, 2004.
54. D. Lindsay, *Br. Antarct. Surv. Bull.*, Vol. 37, p. 105, 1973.
55. H. Skult, *Ann. Botanici Fennici*, Vol. 22, p. 201, 1985.
56. A. Lazaridou, C. Biliaderis, M. Micha-Screttas, and B. Steele, *Food Hydrocolloids*, Vol. 18, p. 837, September 2004.
57. D. Kushwaha, N. Srivastava, I. Mishra, S.N. Upadhyay, and P.K. Mishra, *Reviews in Chemical Engineering*, Vol. 35, p. 475, 2019.

58. A.P. Burgard, S.J. Van Dien, and M.J. Burk, Methods and organisms for the growth-coupled production of 1,4-butanediol, US Patent 8 470 582, assigned to Genomatica, Inc. (San Diego, CA), June 25, 2013.
59. C. Weizmann, Production of acetone and alcohol by bacterial processes, US Patent 1 315 585, September 9, 1919.
60. H. Amiri and K. Karimi, Biobutanol production in M. Hosseini, ed., *Advanced Bioprocessing for Alternative Fuels, Biobased Chemicals, and Bioproducts*, Woodhead Publishing Series in Energy, chapter 6, pp. 109–133. Woodhead Publishing, 2019.
61. S. Behera, S. Koushalya, S. Kumar, and A.B.M. Jafar, Potential and prospects of biobutanol production from agricultural residues in L. Singh and G. Chaudhary, eds., *Liquid Biofuel Production*, chapter 9, pp. 285–318. John Wiley & Sons, Ltd, 2019.
62. V. García, J. Päkkilä, H. Ojamo, E. Muurinen, and R.L. Keiski, *Renewable and Sustainable Energy Reviews*, Vol. 15, p. 964, February 2011.
63. T.C. Ezeji, N. Qureshi, and H.P. Blaschek, *Current Opinion in Biotechnology*, Vol. 18, p. 220, June 2007.
64. A.F. Abomohra and M. Elshobary, Biodiesel, bioethanol, and biobutanol production from microalgae in M. Alam and Z. Wang, eds., *Microalgae Biotechnology for Development of Biofuel and Wastewater Treatment*. Springer, Singapore, 2019.
65. M. Kumar and K. Gayen, *Applied Energy*, Vol. 88, p. 1999, June 2011.
66. T.S. Jeong, W. Choi, J. Lee, and K.K. Oh, Production of biobutanol from hydrolysate of red algae (Gelidium amansii) treated with acid saccharification, in *32nd Symposium on Biotechnology for Fuels and Chemicals*, Clearwater Beach, Florida, US, 2010.
67. T. Ezeji, N. Qureshi, and H.P. Blaschek, *Process Biochemistry*, Vol. 42, p. 34, January 2007.
68. O.I. Shapovalov and L.A. Ashkinazi, *Russian Journal of Applied Chemistry*, Vol. 81, p. 2232, December 2008.
69. T. Wallner, S.A. Miers, and S. McConnell, A comparison of ethanol and butanol as oxygenates using a direct-injection, spark-ignition (DISI) engine, in *ASME 2008 Internal Combustion Engine Division Spring Technical Conference*, pp. 129–139, Chicago, 2008. Tech. Conf. ASME Internal Combustion Engine Division.
70. P. Dürre, *Biotechnology Journal*, Vol. 2, p. 1525, December 2007.
71. J.C.S. Remi, T. Rémy, V.V. Hunskerken, S. van de Perre, T. Duerinck, M. Maes, D.D. Vos, E. Gobechiya, C.E.A. Kirschhock, G.V. Baron, and J.F.M. Denayer, *ChemSusChem*, Vol. 4, p. 1074, August 2011.
72. P.H. Pfromm, V. Amanor-Boadu, R. Nelson, P. Vadlani, and R. Madl, *Biomass and Bioenergy*, Vol. 34, p. 515, April 2010.
73. A. Ranjan and V.S. Moholkar, *International Journal of Energy Research*, Vol. 36, p. 277, March 2012.

74. A.P. Mariano and R.M. Filho, *BioEnergy Research*, Vol. 5, p. 504, June 2012.
75. C.-L. Cheng, P.-Y. Che, B.-Y. Chen, W.-J. Lee, C.-Y. Lin, and J.-S. Chang, *Applied Energy*, Vol. 100, p. 3, December 2012.
76. G. Black, H.J. Curran, S. Pichon, J.M. Simmie, and V. Zhukov, *Combustion and Flame*, Vol. 157, p. 363, February 2010.
77. P. Oßwald, H. Güldenberg, K. Kohse-Höinghaus, B. Yang, T. Yuan, and F. Qi, *Combustion and Flame*, Vol. 158, p. 2, January 2011.
78. M. Karabektas and M. Hosoz, *Renewable Energy*, Vol. 34, p. 1554, June 2009.
79. M. Mascal, *Biofuels, Bioproducts and Biorefining*, Vol. 6, p. 483, July 2012.
80. W. Cai, N. Homs, and P. Ramirez de la Piscina, *Green Chemistry*, Vol. 14, p. 1035, 2012.
81. W. Cai, P.R.d.l. Piscina, and N. Homs, *Bioresource Technology*, Vol. 107, p. 482, March 2012.
82. M. Fernández-Delgado, P.E. Plaza, M. Coca, M.T. García-Cubero, G. González-Benito, and S. Lucas, *Industrial Crops and Products*, Vol. 130, p. 409, 2019.
83. J.C. López-Linares, M. García-Cubero, S. Lucas, G. González-Benito, and M. Coca, *Chemical Engineering Journal*, Vol. 368, p. 1045, 2019.
84. P. Wang, Y.M. Chen, Y. Wang, Y.Y. Lee, W. Zong, S. Taylor, T. McDonald, and Y. Wang, *Applied Energy*, Vol. 236, p. 551, 2019.
85. M.K. Alavijeh and K. Karimi, *Industrial Crops and Products*, Vol. 129, p. 641, 2019.
86. G. Qi, L. Xiong, H. Li, Q. Huang, M. Luo, L. Tian, X. Chen, C. Huang, and X. Chen, *Biomass and Bioenergy*, Vol. 122, p. 76, 2019.
87. M.S.M. Salleh, M.F. Ibrahim, A.M. Roslan, and S. Abd-Aziz, *Scientific Reports*, Vol. 9, p. 7443, 2019.
88. W.D. Morgan, Production of ester-based fuels such as biodiesel from renewable starting materials, US Patent 8 105 398, assigned to Endicott Biofuels II, LLC (Houston, TX), January 31, 2012.
89. R. Selvaraj, R. Praveenkumar, and I.G. Moorthy, *Biofuels*, Vol. 10, p. 325, 2019.
90. S. Ishak and A. Kamari, *International Journal of Environmental Science and Technology*, Vol. 16, p. 2481, May 2019.
91. B. Kim, H.Y. Heo, J. Son, J. Yang, Y.-K. Chang, J.H. Lee, and J.W. Lee, *Algal Research*, Vol. 41, p. 101557, 2019.
92. F. Moazeni, Y.-C. Chen, and G. Zhang, *Journal of Cleaner Production*, Vol. 216, p. 117, 2019.
93. D.-T. Tran, H.R. Lee, S. Jung, M.S. Park, and J.-W. Yang, *Algal Research*, Vol. 31, p. 525, 2018.
94. D.N. Thoai, C. Tongurai, K. Prasertsit, and A. Kumar, *Biocatalysis and Agricultural Biotechnology*, Vol. 18, p. 101023, 2019.

95. C.G. Tsanaktsidis, E.P. Favvas, A.A. Scaltsoyiannes, S.G. Christidis, E.X. Katsidi, and A.V. Scaltsoyiannes, *Fuel Processing Technology*, Vol. 114, p. 135, October 2013.

Index

Trade names

Acronyms

Chemicals

Boldface numbers refer to Figures

General Index

MPM 160620
Printed in Singapore